Intelligent Microservices

An Agentic-AI Approach

Anil Kumar Sharma
distinguished fellow & engineering leader

Shafi Khan
CEO & founder at autonomOps

Paperback: 978-1-967828-90-6

Published by:

Pine Book Writing

www.PineBookWriting.com

R-10225 Yonge St Suite #250, Richmond Hill, ON L4C 3B2, Canada.

Printed in the United States of America

To,

My daughters,

Anukriti Sharma and Aaradhya Sharma

Thank you for inspiring this book with your futuristic curiosity, kindness, and perseverance.

Anukriti you were also first reviewer of this book; your insightful questions and empathy kept me grounded.

Aaradhya, your laughter, art, and fearless imagination reminded me that invention should always make room for joy.

Building this manuscript was possible because of our meaningful moments together. Any wisdom here is thanks to you.

Stay curious, kind, and courageous as you shape your future. Everything I create is a tribute to you both.

Yours truly, Dad (Anil Sharma)

Shafi Khan

For my children, Iqra and Ishaq. Your playful energy and giggles kept me moving when the road felt long. If this book helps anyone, it's because of you. Stay curious, be kind, and build things that help others.

Yours truly, Dad (Shafi Khan)

Disclaimer

- Purpose: This book is for educational and informational purposes only and is not legal, security, compliance, financial, medical, or other professional advice. Apply any concept with qualified professionals who understand your environment and obligations.

- No Warranties: Content (including patterns, diagrams, metrics, and checklists) is provided "as is." Software, AI models, and regulations change. The authors and publisher make no warranties and accept no liability for loss or damages arising from use or reliance.

- Safety & Human Oversight: Examples involving autonomy are illustrative. Deploy only with human-in-the-loop controls, testing, monitoring, and rollback. Do not use in safety-critical or high-risk contexts without independent validation.

- Security & Compliance: You are responsible for threat modeling, testing, privacy impact assessments, and regulatory review. Process only data you are permitted to use. Follow least-privilege access, auditing, and redaction practices.

- Responsible AI: AI systems can produce inaccurate, biased, or unsafe outputs. Perform evaluation, guardrails, red-teaming, bias/fairness assessments, and continuous monitoring aligned with your ethics policies and applicable law.

- Code & Artifacts: Scripts, configurations, prompts, and code snippets are examples for learning. Test in isolated, non-production environments first. They are provided without support or guarantees and may be incomplete or simplified.

- Case Studies: Some scenarios are composites; details may be changed to protect confidentiality. Any resemblance to real systems or incidents is coincidental unless stated.

- Third-Party References: Product names and trademarks belong to their owners; mentions do not imply endorsement or affiliation. Links may change without notice.

- Views Expressed: Opinions are the authors' and do not represent any past or present employer or organization.

- Forward-Looking Statements: Roadmaps and projections are uncertain and may change. No obligation is assumed to update such statements.

- Acceptance: By using this book and related materials, you accept responsibility for decisions and outcomes and agree to hold the authors and publisher harmless to the extent permitted by law.

"This book perfectly describes that modernization is not just a tech upgrade but it's a strategic necessity. It ensures agility, scalability, and competitiveness, while avoiding the compounding risks of technical debt, inefficiency, and slower innovation."

– **Abanish Mishra @ Director- Data Science & Analytics**

"I found it both ambitious and inspiring. It does an excellent job connecting current enterprise challenges with emerging research and paints a clear picture of where intelligent microservices could be heading." - **Avitesh Kesharwani @Senior Principal Consultant, Genpact**

"Industry expects that, introducing AI in multiple facets of an organization should have governance

embedded deeply into the root of those systems, this book highlights the importance of it in most of the case studies by mentioning the importance of Humans in the Loop

" - Saurabh Kulkarni – Full Stack Engineer @ ETQ

"The case study presented is an example of how AI and microservices can work together in an enterprise. It reflects modern tech patterns and includes thoughtful architecture, cloud-agnostic tools, and a strong understanding of automation workflows." – **Jyothsna Devi @** E Tech Group

""Microservices are not a silver bullet…." is a great use of the grounding fact that no technology, principle or framework in software is perfect. This treats readers as professional and avoids selling an idea that seems too good to be true." - Ramesh Sinha @ at Zscaler

"The story presented gives it a strong narrative thread, making complex ideas easier to follow. Readers can see the shift from manual processes to automation, intelligence, and eventually autonomy in a way that feels concrete" - **Sweta Mulji , Co-founder@SalesPlay**

"The book has comprehensive coverage of scaling topics, including hierarchical structures, complexity management, observability, AI performance, and real-world AWS examples, providing a well-rounded guide" – **Ajay Borra** Lead Engineer @Salesforce

Preface

Enterprises have developed complex systems over time through migrations, integrations, and temporary solutions. Incidents often reveal underlying issues: runbooks may be inconsistent, manual intervention is frequently needed, and automation is attempted without addressing architectural causes.

This book presents an approach that focuses on developing systems capable of perception, decision-making, action, and learning within business constraints, referred to as Intelligent Microservices. These services are designed with embedded decision-making and defined responsibilities, coordinated for ongoing improvement and accountability.

Current Context

Microservices support incremental change, cloud infrastructure offers scalability, and MLOps enables model deployment. Despite these advancements, many production environments rely on manual decisions across various platforms. Recent developments in agentic AI allow for contextual reasoning and tool coordination. Combined with observability, identity management, and security practices, intelligence can now be incorporated at critical architectural points rather than through reactive scripts.

Purpose of This Book

This work provides guidelines for transforming microservice architectures into intelligent service ecosystems. Topics include:

- Placement of intelligence: centralized decision services, embedded agents, and AI-first orchestrators, with analysis of appropriate contexts.

- Safe coordination: negotiation protocols, compensation mechanisms, and human-in-the-loop (HITL) checkpoints for important outcomes.

- State and dataflow management: event sourcing with semantic evaluation, online/stream learning, and adaptive contractual agreements.

- Feedback integration: converting telemetry into explanations and learning signals for continuous improvement.

- Trustworthiness: agent identities, restricted tool access, redaction and provenance, and auditable decision records.

• Resilience strategies: early detection, graceful degradation, self-healing mechanisms, and relevant metrics (SLOs, MTTR, adverse outcome rates).

The material includes patterns, checklists, and examples applicable to various domains.

Intended Audience

• Enterprise architects and platform teams updating legacy environments.

• Engineering managers establishing effective boundaries for autonomy.

• SRE/DevOps professionals designing rollout procedures, incident protocols, and feedback loops.

• Security, risk, and compliance personnel seeking traceability and defensible audits.

• Data/ML specialists implementing accountable, goal-oriented models.

This book Avoids vendor or product specifics, highlights practical, decision-enabling services over exhaustive ML techniques, Stresses the need for human oversight, with advice on HITL integration.

Approach to Safety

Design recommendations align system autonomy with business objectives and operational safeguards. Explanations, reversibility, and progressive delivery (shadow → canary → widen) are emphasized. Incremental improvements and clarity for technical stakeholders are prioritized.

Starting Point

Begin with an existing workflow, instrumenting for decision tracking. Introduce an agent proposing actions, log decision rationales, and maintain human oversight. Measure reduction in manual effort, mean time to repair, and rate of negative outcomes. Expand autonomy gradually when evidence supports reliability.

Acknowledgments

Content draws from experiences of teams managing operational systems under varying constraints. Appreciation is extended to practitioners who contributed insights and to those highlighting the role of systems in supporting people.

The aim is to assist readers in building systems that are efficient, reliable, and responsive to stakeholder needs.

— Anil Kumar Sharma & Shafi Khan

October 2025

Table of Contents

Chapter 1: Broken Enterprise IT Systems and Workarounds

The 3 A.M. Outage Nobody Wants

It's 3:00 AM on a Tuesday, and your phone is ringing incessantly on the nightstand. You groggily fumble for it, squinting at the caller ID. It's the on-call pager system. Not good. Moments later, you're logging into a VPN with one hand while rubbing sleep out of your eyes with the other. A mission-critical batch process failed, again.

Picture a scene all too familiar in large IT departments: a frazzled engineer named Sam is hunched over his laptop in a dimly lit home office, trying to restart a data transfer script that should have run hours ago. The script is one of dozens of fragile nightly jobs chained together to shuffle data between siloed systems. Tonight, **Order Sync Script #42** decided to crash, and now sales orders from the website haven't made it into the fulfillment system. If this isn't fixed by morning, warehouse trucks will be empty and **everyone from operations to the CEO will be livid.**

Sam mutters under his breath, "Not again…" as he scans through a 500-line mash of code hastily written years ago by someone who left the company. Comments? Hardly any. Documentation? Nonexistent. His heart is pounding as lines of error logs scroll across the screen, a jumble of stack traces and "out of memory" messages. This isn't the first time, either – the same job hiccupped just last week. Each time, the team applied a quick patch and promised to "address it properly soon." But as usual, 'soon' never came, and now the temporary fix has unraveled. He finally pinpoints the issue – an out-of-memory error caused by an unexpectedly large file. A quick fix — splitting the file — gets the pipeline moving again. Crisis averted… for now. Sam knows he'll be back to bandage this same wound again another night, because **nothing fundamentally changed**.

Sound familiar? Many enterprise IT professionals have lived a version of this 3 A.M.,nightmare. It's a symptom of deeper issues: brittle scripts, fragile integrations, data scattered across multiple systems, and ancient legacy software all tangled together. By dawn, systems are back on their feet, but the uneasy truth remains: **the enterprise is held together by digital duct tape.**

In this opening chapter, we'll explore why these kinds of predicaments are so common. From data silos and hard-coded scripts to legacy systems and mounting technical debt, we'll shine a light on the systemic issues that create such 3 A.M. emergencies. Along the way, we'll see how well-intentioned workarounds – those quick fixes like manual file transfers, spreadsheets, and cron jobs – often become permanent crutches. Through illustrative stories and analogies, we'll understand how enterprise IT sometimes resembles a house of cards, ready to topple at the worst time.

Let's dive into the broken state of many large IT environments and set the stage for why a new approach (one involving intelligent automation and modern architectures) is urgently needed.

Data Silos: Islands of Isolated Information

One of the most pervasive issues in enterprise IT is the *data silo*. Just as silos on a farm store grain separately, data silos refer to disparate stores of information that remain isolated, each accessible only to one team or system . In a large organization, it's common to find customer data in a CRM, financial data in an ERP, analytics data in a data warehouse, and perhaps additional information locked away in various SaaS tools. These systems often don't talk to each other well, leading to islands of information that are tough to bridge.

When data is siloed, employees spend inordinate amounts of time wrangling information. Have you ever heard someone say, "I'll email you the spreadsheet"? That's a tell-tale sign of a silo. For example, the sales team might export customer orders from the CRM to CSV files and email them to Finance at month-end because Finance's system can't directly pull the data. Meanwhile, the support team manually copies customer issue logs from their helpdesk software into a SharePoint list so engineering can see them. Each department is like a fiefdom with its own data kingdom, and collaboration becomes an exercise in *manual data ferrying.*

The consequences of silos are serious. They cause delays, errors, and frustration. Studies have shown that employees can waste *5+ hours a week* just waiting on or searching for data stuck in another team's system . Important decisions get made with incomplete information because no one has the "full picture" handy at the right time. In fact, lost productivity from poor data accessibility and quality is estimated to cost U.S. businesses trillions annually . Data silos don't just inconvenience staff; they hurt the bottom line.

And then there's the issue of *consistency*. When each system is a silo, you inevitably get multiple versions of the truth. The classic joke is that no one at an executive meeting can agree on something as simple as last quarter's revenue because Sales, Finance, and Analytics each came with different figures. One database wasn't updated, another used a different formula, a spreadsheet had a typo – and now an hour is wasted reconciling whose data is "right".

How do organizations typically deal with silos? Unfortunately, with *workarounds*. Instead of a unified solution, you get a patchwork of nightly jobs and manual processes to sync data.

It's common to see a contrived pipeline like this:

```
CRM System (Sales data)
     ↓ nightly export (CSV)
Data Warehouse (Consolidated data)
     ↓ ETL script (transforms data)
Reporting Database (for dashboards)
```

In the above ASCII sketch, the CRM doesn't directly integrate with the reporting tools., Instead, a nightly CSV export acts as the courier – a batch file that dumps data out of the CRM, which is then picked up by an ETL (Extract-Transform-Load) script to shove into a reporting database. This indirect approach eventually delivers a unified view to management, but it's slow (data is a day old) and brittle (if the CSV format changes or the script has a bug, the whole chain breaks). Yet, setups like this are astonishingly common.

Case Study: Marketing vs. Sales – The Silo Effect

At **Acme Corp**, the marketing department kept its leads and campaign data in a cloud SaaS tool, while the sales team tracked opportunities in a separate CRM. For years, managers complained that marketing's numbers never matched sales' figures. Why? Each quarter, a marketing analyst would dump data from the SaaS tool into Excel and massage it to produce a leads report, while a sales operations rep did the same from the CRM for pipeline reports. The two would then huddle in a meeting to manually crosscheck which leads turned into deals. Inevitably, something was missed – a row hidden here, a filter applied there – and their reports painted different pictures. This led to tense executive meetings and finger-pointing. The workaround to fix it was to create yet another spreadsheet that merged the two reports, maintained by a weary analyst who spent days reconciling data every quarter. Everyone depended on that analyst's custom spreadsheet, which lived on her desktop. It was a classic data silo symptom: no single source of truth, and a "heroic" manual effort to stitch together information after the fact.

In the end, Acme Corp's leadership realized the cost of these silos was not just inconvenience – it was causing lost sales insights. Marketing was targeting leads based on outdated info, and Sales was missing context on lead sources. This mini-crisis became a catalyst for them to consider an integrated data strategy. The pain of the silo had finally outweighed the pain of fixing it.

Data silos also breed **shadow IT** – when impatient teams take tech matters into their own hands. If the official IT solutions don't provide a needed integrated view, business users might build their own Access database or start using a cloud app without IT's blessing. While this can temporarily relieve the pain, it often creates new problems: yet another silo, potential security risks, and unsupported tools. It's like squeezing a balloon – push the problem in on one side, it bulges out on the other.

In short, silos turn what should be seamless data flows into relay races with baton handoffs (often analog, like emails or CSV files). As organizations grow, these islands of data become harder to connect, and the maze of workarounds becomes more complex. The more silos, the more brittle scripts and manual steps appear (as we'll explore next), further entrenching the fragility of the enterprise architecture.

Transitioning from siloed data, let's examine the next layer of the problem: the hard-coded scripts and fragile integrations that attempt to tie these silos together.

Hard-Coded Scripts and Fragile Integrations: A House of Cards

To connect all those siloed systems, enterprises often resort to hard-coded scripts and adhoc integrations. Think of these as the duct tape and glue holding the digital world together. Individually, a script might seem like a neat solution to automate a task. But in aggregate, an enterprise ends up with hundreds (or thousands) of little scripts and point-to-point connectors forming a shaky lattice connecting systems. It's a house of cards – one small change can bring down a whole section of the structure.

In many organizations, if you peel back the layers of how data flows, you'll find things like custom Python scripts, batch files, or even old ETL jobs that move data around every hour or every night. These scripts are often written by whoever needed a quick fix at the time – maybe an ops engineer, a savvy analyst, or that one programmer who knew just enough to get it working before moving on. Rarely are these "integrations" built with long-term maintainability in mind. Hard-coded file paths, IP addresses, credentials, magic numbers – you name it, it's probably lurking in the code.

Over time, this leads to what architects grimly nickname the "spaghetti architecture." Imagine a bowl of spaghetti where each noodle is a connection between two systems. It's tangled and hard to follow. For instance, System A needed data from System B, so someone wrote a script for that. Then System B needed to feed System C, another script. Later, System A also needed to send data to System C, so rather than redesigning everything, yet another point-to-point integration was bolted on. Multiply this by dozens of systems and you get a chaotic mesh. One bank famously found they had over 800 distinct point-to-point data integrations for their core applications – a complexity no one fully understood.

Let's illustrate a simplified example. Imagine four systems in a company: CRM, ERP, a Data Warehouse, and a Reporting tool. In an ideal world, they might all connect through a central hub or share a common interface. But in a quick-fix world, you could end up with something like this:

```
[CRM] <--> [ERP]
  \           |
   \__→ [Data Warehouse] ←__/
            |
        [Reporting]
```

Every arrow here is a custom integration (a script or direct database link). The CRM directly updates the ERP, the ERP dumps data to the Data Warehouse, the CRM also feeds the Data Warehouse via another path, and the Reporting system queries both the warehouse and ERP. It's not fully symmetrical, but it's definitely messy. If the person who coded the CRM-ERP sync left, and something changes in the CRM's API, that sync fails. If the Data Warehouse structure changes, several of those arrows break. This web of interdependencies becomes *brittle*.

We've already seen how Sam's 3 AM fiasco was triggered by a script failure. Such failures are not the exception – they are almost inevitable in a big tangle of hard-coded integrations. *Why do they break so often? Because they are fragile by design:*

Assumptions change: A script might assume a file will never exceed a certain size or a field name will remain the same. Eventually, reality deviates (bigger data, renamed fields) and the script chokes.

Lack of error handling: Many DIY scripts lack robust error checking. They fail silently or just stop, and only hours later someone notices something's wrong.

No monitoring or alerting: Unless someone added monitoring, a nightly job might fail unseen. In some companies, the "monitoring" is just an email from a user saying "Hey, yesterday's data looks off."

Single points of failure: Often, only the author truly knows how the script works. If they're on vacation (or have left the company), troubleshooting falls to whoever draws the short straw. The bus factor (how many people need to get "hit by a bus" before a system has no knowledgeable caretaker) is uncomfortably low for many such scripts.

Chain reactions: These scripts frequently form dependencies. If one fails at 1 AM, the 3 AM downstream job has no data to process and also fails. By morning, you have a cascade of failures – a domino effect that takes hours to untangle.

Pitfall: "It Worked Yesterday!"

A common scenario illustrating fragility is the infamous date format bug. Consider a script that generates a daily report and names the output file with yesterday's date. It concatenates strings like Report_20250421.csv. This worked fine until the end of the year when 20241231 rolled over to 20250101. The script hadn't accounted for the year change and used a fixed length for the date string. On January 1, it produced

Report_202501.csv (missing the last two digits). Downstream, another job looking for the exact file name didn't find it and crashed. No one noticed until a manager wondered why the New Year's Day report was missing. The moral: these quick scripts often lack resilience for edge cases. They happily run for 364 days and then faceplant on the 365th, leaving everyone scrambling.

Case Study: The Payroll Interface Fiasco

At a retail chain, the timekeeping system (where employees' hours and overtime are recorded) was connected to the payroll system by a custom nightly script. One week, HR added a new category for "special overtime" in the timekeeping system to handle a holiday rush. The change seemed minor, but the integration script wasn't updated to recognize the new category. As a result, the hours tagged as "special overtime" were simply dropped from the data transfer.

Come payday, hundreds of employees found their paychecks short. Overtime hours hadn't been paid out. The mistake sparked outrage – employees flooded their managers and HR with complaints. The company had to issue emergency corrections and mollify angry staff. The root cause? A fragile integration that failed silently. A one-line change in the timekeeping system (adding a new code) cascaded into a major payroll blunder because the script on the other side was brittle.

This fiasco became a cautionary tale within the company. It highlighted how a tiny technical oversight (not updating the script) led to a huge human impact. The loss of trust and morale, plus potential legal penalties for delayed wages, drove home the point that these systems needed more robust, resilient integration methods. It was a painful lesson on the real-world cost of a brittle workaround.

Ironically, many of these scripts started as temporary fixes – meant to be replaced by a "real solution" later. But in the pressure cooker of enterprise IT, "later" can be elusive. What was a stopgap becomes a permanent fixture, labeled "DO NOT TOUCH" in fear that any change will break it. In one organization, a comment in a script read: "This is a hack. FIXME in Q4 2019."

The comment survived into 2025, a grim reminder of how temporary fixes become permanent legacy.

The integration fragility issue doesn't stop at scripts. Some enterprises invest in heavier integration middleware or enterprise service buses but then misuse them as dumping grounds for complex logic, essentially creating monolithic integration hubs that are just as hard to maintain. The result? When that central piece has a hiccup, it can paralyze multiple systems at once.

The house of cards can collapse in other ways too. Consider cron jobs (scheduled tasks) that overlap or conflict. If one job takes longer than usual and a second job starts before the first finishes, they might contend for the same file or database record, corrupting data. These timing-related bugs are nightmare-inducing because they might only happen on rare occasions (like end of quarter when data volumes peak).

All told, fragile integrations create an environment where firefighting becomes the norm. IT staff find themselves spending more time reacting to breakages than building new capabilities. It's hard to plan ahead when you're busy putting out yesterday's fires. And every new system added to the mix threatens to introduce more spaghetti into the bowl.

So far, we've focused on data and integration pain points. But underpinning a lot of this is the reality of aging systems and accumulating technical debt. Let's turn to the challenges posed by legacy systems and the weight of past technology choices.

Legacy Systems: Living on Borrowed Time

Lurking beneath many of these integration woes are the legacy systems – the old, trusty (or rusty) software and hardware that form the backbone of the enterprise. These are the aging ERP platforms, mainframe applications, or homegrown systems that have been running since the 90s (or even the 70s!). They are sometimes joked about as the "living fossils" of IT. Everyone knows they're ancient, yet they continue to chug along, day after day, executing core business processes.

Legacy systems are often mission-critical. Banks, for example, still run a significant portion of their operations on COBOL programs written decades ago. In fact, the COBOL language (dating back to 1959) still powers **43% of banking systems and 95% of ATM swipes worldwide** . Government agencies and airlines likewise rely on mainframes and vintage software. These systems are *stable workhorses* in that they reliably handle high volumes of transactions. The philosophy has long been **"if it ain't broke, don't fix it."**

The trouble is, over time, *it does break* – or rather, it becomes **very hard to fix or improve**.

Legacy environments present a few big challenges:

- **Integration difficulty:** Many legacy systems weren't built with modern integration in mind. They predate APIs and assume they are the center of the universe. Getting data in or out might require reading obscure flat files, or even screen-scraping through an emulator. This is why some teams resort to RPA bots clicking through green-screen interfaces – it's literally the only way to interface with a 40-year-old program .

- **Lack of agility:** Adapting a legacy system to new requirements is often slow and expensive. A simple change might require a rare specialist developer, testing in a fragile environment, and coordination with multiple dependent modules. It's like trying to renovate a house built in 1900 – you worry that **every change might knock loose something else**.

- **Knowledge drain:** The people who built and maintained these systems are retiring or have left. Younger IT staff may have never worked with something like IBM iSeries or Oracle Forms. A recent report highlighted that over a quarter of COBOL programmers are nearing retirement age . With fewer experts, the risk of maintaining legacy code grows. Who will debug a production issue in a language no one on the current team fully knows?

- **Operational cost:** Legacy infrastructure can be costly to run and maintain. Often it sits on old hardware in private data centers. Organizations spend a fortune just "keeping the lights on." Studies have found that, on average, **60–80% of IT budgets** end up dedicated to maintaining existing systems (mostly legacy), leaving only a sliver for innovation . In other words, **technical debt interest payments** consume the budget, with little left to build new capabilities.

- **Security and compliance:** Older systems might not meet modern security standards. They may run on out-of-support operating systems or use weak encryption. Upgrading them is non-trivial, so companies take on additional risk or costly compensating controls (like isolated networks or additional firewalls) to guard legacy apps. Compliance regulations (GDPR, HIPAA, etc.) can also pose challenges if the system can't provide needed auditing or data controls.

In practice, legacy systems often become surrounded by layers of **wrapper applications and patches**. It's common to see a legacy core that has a web front-end bolted on, an export tool tacked on the side, and numerous scripts (here they are again!) funneling data in and out. Picture an old building that has had many renovations: new plumbing strapped to the outside walls because the internal pipes are too hard to replace. Each add-on solves a short-term need but also adds complexity.

Case Study: Robotic Patch Automation

A mid-sized insurance company had a legacy claims management system running on a mainframe. It had no APIs and could only be operated via a terminal interface. Rather than undertake an expensive rewrite, the IT team implemented **Robotic Process Automation (RPA)** bots to act as digital users. These bots would literally log in each night, navigate the screens, and copy data from the mainframe system into a modern cloud database where data could be more easily reported and analyzed. This workaround did the job for a while, but it was fragile. When the mainframe screens were slightly updated in an upgrade, the RPA bots started mis-clicking – they were essentially "typing" into the wrong fields. The team spent weeks retraining the bots. This RPA solution was, as some called it, a band-aid on a bullet wound – it kept things going but didn't address the underlying issue that the legacy system needed a better integration approach. The story had a semi-happy ending: the pain of maintaining these bots finally convinced leadership to invest in a modernization project for the claims system.

This case illustrates how far teams will go to avoid touching a legacy system's code – even automating at the screen level. It's ingenious, yet it underscores the **unsustainable nature of relying on brittle workarounds** long-term.

While legacy systems carry the baggage of the past, they also hold critical business logic that still has value. That makes ripping them out risky. Many enterprises face a dilemma:

modernize or maintain? The result, too often, is a stalemate where nothing big is done; just small patches continue indefinitely. This leads to accumulating pain and increasing risk.

Knowledge Preservation

One way to mitigate legacy risk is to invest in **knowledge transfer**. Pair up veteran engineers who understand the old system with newer team members. Encourage documentation of tribal knowledge before it walks out the door. Even if you can't replace a legacy system yet, making sure more than one person knows how it works is crucial. Also, consider training a few staff in the legacy technology (yes, even if it's COBOL or some old database) to buy breathing room. While not a permanent solution, it can prevent being one retirement away from a crisis.

The takeaway is that legacy systems, despite their reliability so far, **are living on borrowed time.** They grow harder to interface with and maintain each year. Every new band-aid applied (be it an RPA bot, a wrapper API, or a nightly extract) is a reminder that the core system isn't keeping up with modern needs. And this accrues what we call **technical debt**, which we'll explore next.

Accumulated Technical Debt: Yesterday's Quick Fix, Today's Crisis

The concept of **technical debt** is often used to describe the cumulative effect of all these shortcuts and Band-Aid solutions over time. The term was coined by software pioneer Ward Cunningham, who famously said that shipping code without refactoring it later is like borrowing money: **you gain speed now, but incur a debt that must be repaid with interest later** . In simpler terms, every time we choose an expedient fix over a robust solution, we are **taking on debt**. If we don't "pay it back" by improving the code or system later, the interest shows up as slower development, more bugs, and nasty surprises down the road.

For instance, the **Year 2000 (Y2K)** crisis was essentially technical debt writ large. For decades, many programs stored years as two digits to save memory – a small shortcut that acted like a loan taken against the future. As the millennium approached, that debt came due: companies worldwide had to spend enormous effort rewriting and testing code to avert catastrophic date bugs when '99' turned to '00'. In the late 1990s, entire teams were mobilized to pay off this massive accumulated debt from a design shortcut made in the 1970s and 80s.

In an enterprise context, technical debt isn't just in the code – it's in the architecture, processes, and even documentation. All the data silos, fragile scripts, and ancient systems we've discussed contribute to technical debt. They represent **deferred work** – things that really should be cleaned up or improved, but haven't been. Like a credit card balance, the debt grows and compounds. Maintaining a heavily indebted IT system costs more and more effort each year.

How do you know if your organization is swimming in technical debt? Here are a few signs:

- **Change is hard and slow:** Even a small feature request or update requires touching many systems or navigating convoluted code. Engineers joke that there's a "tax" on every change.

- **Frequent hotfixes:** The team is frequently in fire-fighting mode, issuing hotfixes to patch broken things that ideally should be redesigned entirely. (Recall Sam's 3 AM patch – that's interest being paid on an old design decision.)

- **Inconsistent architecture:** You have a mix of old and new technologies that don't mesh well. Perhaps half the app is microservices, but it still depends on a monolithic core that can't scale – a sign that past modernization efforts were only partial.

- **Reliance on heroes:** Certain staff are constantly relied upon to "save the day" because only they know how to navigate the mess. This hero culture often signals underlying debt; without heroic efforts, things would fall apart.

- **Outdated documentation (or none at all):** When systems are built in a rush, documentation lags. Years later, nobody has the full picture written down. This makes any onboarding or transfer of work painful.

Accumulated technical debt tends to **bite at the worst times**. It might be during a highstakes project when you discover integration issues that stem from years of neglect. Or an outage happens because an old module finally reached a breaking point under new load. Many enterprises have had that "wake-up call" incident – maybe not quite an apocalyptic failure, but a serious enough scare – that forces leadership to acknowledge the debt problem.

On Architecture and Cost

"If you think good architecture is expensive, try bad architecture." – Brian Foote & Joseph Yoder, *Big Ball of Mud*

This quote poignantly reminds us that skimping on proper design and cleanup (i.e., letting technical debt fester) may save budget now, but **will cost exponentially more later**. Enterprises often discover this truth the hard way.

Sometimes teams even maintain a **technical debt register** – a list of known issues to fix when time permits. It's a sober inventory of cracks in the foundation. But if that to-do list keeps growing instead of shrinking, it's a red flag that the team is sinking further into debt.

It's worth noting that not all technical debt is intentional or due to laziness. Business realities often force trade-offs. Maybe a feature had to go live in Q4 to meet a compliance deadline, so corners were cut. Or a startup acquired by the enterprise came with a shaky codebase that now must be integrated. Technical debt is often an inevitable byproduct of moving fast. The key is **acknowledging it and having a plan** to address it before it becomes unmanageable.

Left unchecked, technical debt can reach a tipping point where the system becomes almost unchangeable – what some call **analysis paralysis** or gridlock. Imagine a Jenga tower of blocks where each block is a quick fix; as the tower grows, it becomes perilously unstable. Eventually, only a radical teardown (a complete rewrite or re-architecture) will solve the issues, akin to declaring bankruptcy and starting fresh.

All these systemic pain points – silos, fragile integrations, legacy drag, and accumulated debt – create a perfect storm that holds companies back. They spend most of their IT effort just keeping systems running, rather than innovating or improving. It's a frustrating situation: the **status quo is unsustainable** but changing it seems daunting.

This is precisely why modernization and intelligent automation have become rallying cries in recent years. In the next section, we'll conclude with a recap of these challenges and discuss how organizations can gear up to break free from this cycle.

Conclusion: From Patchwork to Progress

> *"The most dangerous phrase in the language is, 'We've always done it this way.'"* — Grace Hopper

We began this chapter with a 3 A.M. wake-up call – a visceral reminder of how fragile enterprise IT systems can become after years of patchwork fixes. The journey through data silos, brittle scripts, aging legacy platforms, and compounding technical debt has painted a clear (if daunting) picture of the status quo in many organizations. Let's summarize the key pain points we've uncovered:

- **Data Silos:** Information trapped in isolated systems, leading to duplicate work, inconsistencies, and a reliance on manual data wrangling.

- **Fragile Integrations:** Hard-coded scripts and point-to-point connectors that break under slight changes, turning IT into a constant firefight.

- **Legacy Systems:** Outdated but critical systems that are costly to maintain, hard to adapt, and increasingly unsupported – acting as anchors slowing down progress.

- **Technical Debt:** The accumulation of quick fixes and neglected improvements that make each subsequent change more difficult, akin to an ever-growing maintenance burden.

Individually, each of these is a headache. Together, they form a **systemic malaise** that stifles innovation and increases operational risk. An organization stuck in this quagmire finds itself spending most of its energy just to stand still, rather than moving forward.

Why Modernize Now?

Continuing with the current state is like driving a car with the check engine light on, hoping it doesn't break down. The **cost of inaction** can be immense. Small inefficiencies add up to huge losses over time – remember the billions in lost productivity due to inaccessible data . There's also the looming threat of a major outage or security breach. It's not melodramatic to say that without modernization, some businesses could face existential crises when a critical system fails or when a nimble competitor overtakes them by leveraging more agile technology.

The Cost of Doing Nothing

Ignoring these problems doesn't make them go away; it makes them worse. Technical debt interest compounds. Legacy systems inch closer to failure with each passing year of postponed upgrades. And every manual workaround is a potential error waiting to happen. In one company, a series of minor data entry mistakes (from manually consolidating spreadsheets) accumulated until they resulted in a multi-million dollar reporting error – a fiasco that made headlines. The lesson? **Choosing not to modernize is often the most expensive choice of all, just in slow motion.**

On the positive side, the upside to modernization is huge. Firms that successfully tackle these issues become more responsive, efficient, and innovative. Imagine saying goodbye to nightly panic fixes and instead deploying new features or insights weekly, or having accurate, real-time data at your fingertips whenever you need it. Modern architectures and intelligent automation can turn those dreams into reality – but first, you need to be ready to embark on the journey.

Modernization Readiness Checklist

Before diving headlong into modernizing and intelligent microservices (which we'll explore in the next chapters), it's wise to assess where you stand. Below is a **Modernization Readiness Checklist** to gauge how prepared your organization is to tackle its IT woes:

- **Pain Point Inventory:** Have you documented the major pain points (frequent outages, long lead times for changes, security risks, etc.) in your current systems? Knowing *what* hurts is the first step to healing.

- **Stakeholder Buy-In:** Are business and IT leaders aligned on the need for change? Modernization often stalls without executive sponsorship and a shared vision between IT and business units.

- **User Feedback:** Have you spoken with the employees and end-users who work with these systems day-to-day? Their firsthand insight can highlight pain points and practical needs that might not be apparent from a purely technical analysis.

- **Critical Systems and Risks:** Do you know which legacy systems pose the highest risk or cost? Identify systems that are unsupported, have known failure points, or consume disproportionate resources.

- **Skilled Team & Knowledge:** Do you have people who understand both the old and new technology stacks? If not, can you train staff or bring in expertise? (For instance, if you're retiring a COBOL system, you'll need someone who can interpret its logic for the replacement solution.)

- **Data Assessment:** Is your data locked in silos? Audit where key data resides and who needs access. Plan for how data will be migrated or integrated in a new world (e.g., cleaned, consolidated, and made accessible).

- **Defined Goals:** Have you clearly defined what "success" looks like? Whether it's faster time-to-market, reduced downtime, cost savings, or enabling AI capabilities, clear goals will guide the modernization effort.

- **Phased Game Plan:** Modernization is not a one-shot project, it's a multi-phase journey. Do you have a roadmap that prioritizes quick wins (like modularizing a small service or automating a tedious task) while laying groundwork for bigger changes? This plan should include a timeline, responsibilities, and checkpoints to measure progress.

- **Safety Nets:** Have you prepared contingency plans? For example, backups and rollback strategies during system migrations, or parallel runs of old and new systems to validate outcomes. Being ready for hiccups ensures a bump in the road doesn't derail the whole initiative.

If you find you can check off many of the above items, you're in a strong position to start modernizing aggressively. If not, these checklist items highlight areas to shore up before (and as) you proceed. Modernization is as much about mindset and preparation as it is about technology.

Looking Ahead: A New Hope with Intelligent Microservices

The problems outlined in this chapter set the stage for why a new approach is needed. Fortunately, the industry has been converging on solutions. **Microservice architectures** break down those monolithic systems and spaghetti integrations into cleaner, more manageable services. **Intelligent automation and agentic AI** promise to handle routine tasks and even make autonomous decisions, reducing reliance on manual workarounds. In essence, there's a path forward that addresses data integration, system resilience, and agility all at once.

Throughout the coming chapters, we'll explore how designing systems as collections of independent, communicating services can eliminate data silos and allow for more flexible integrations. We'll also introduce the idea of AI agents that can orchestrate processes and adapt to changes in real-time – bringing a level of smartness to your IT operations that manual scripts simply can't match.

Modernization isn't easy – but as we've detailed, clinging to the old patchwork is a recipe for 3 A.M. emergencies and stagnant progress. By understanding the broken state of things (as we have in this chapter), you've taken the first step. From here on, this book will guide you through the art of transforming that fragile house of cards into a robust, intelligent system of microservices.

Chapter 2 awaits, where we'll begin that transformation journey and explore the foundations of building an enterprise architecture powered by microservices and agentic AI. The vision is a system that doesn't break at 3 AM – in fact, with intelligent automation, that dreaded outage could become a non-event, resolved by an AI agent before anyone even wakes up.

Chapter 2: Introduction to Microservices

Introduction

In the previous chapter, we witnessed the struggles that **WebkinsTech** – our fictional enterprise – faced with its monolithic systems and brittle integrations. The company's all-inone platform had become a tangled web of interdependent parts, where a small change or integration could send shockwaves through the entire system. WebkinsTech's IT team felt the pain of slow deployments, frequent outages from minor bugs, and an inability to move quickly in a fast-paced market. Their monolithic application was like a heavy **freight train**: powerful and steady in its early days, but increasingly hard to steer or stop as it grew. What WebkinsTech needed was the agility of a fleet of race cars – nimble **microservices** – each handling a specific job with speed and independence.

This chapter introduces microservices as an antidote to the monolithic woes highlighted earlier. We will explore how large applications evolve from monoliths to microservices, using WebkinsTech's journey as an anchoring example. Along the way, we'll dive into the principles of microservices design (like bounded contexts and loose coupling), examine how business growth drives complexity, and outline the architecture of microservice-based systems (covering API gateways, service registries, load balancers, and more). We'll discuss the **benefits** microservices can bring – from agility to team autonomy – as well as the **tradeoffs and challenges** they introduce (such as operational complexity and data consistency issues). To separate hype from reality, we'll debunk common **myths** about microservices.

Finally, we'll provide a practical **readiness checklist** and guide for teams considering a move to microservices, ensuring you know how to plan and pilot such a transformation. Throughout, we'll use WebkinsTech's story and other real-world analogies to keep things engaging and concrete.

By the end of this chapter, you'll have a solid understanding of what microservices are and why they matter. More importantly, you'll see how adopting a microservices architecture can alleviate the pains of monolithic systems – setting the stage for **intelligent automation and AI-driven orchestration**, which we'll explore in Chapter 3. Let's begin our journey by looking at how monolithic applications evolve and why enterprises like WebkinsTech reach a breaking point that compels them to consider microservices.

Evolution from Monoliths

Most enterprise software journeys start as a monolith. A **monolithic application** is a single, unified unit of software – all features and components packaged together, typically sharing one codebase and database. In the early days of an application, a monolith can be **simple and convenient**. There's just one thing to deploy, one place to look for debugging, and all the functionality lives in one cohesive code repository. WebkinsTech followed this common path in its startup phase: the company built a single core application, the "WebkinsTech Portal," which handled everything – from user authentication and product catalog, to order processing and reporting – all in one **big codebase**. In the beginning, this monolith served WebkinsTech well. It was easy for developers to add features by touching the one project, and deployment was straightforward: package the whole application and release it as one unit. Early on, a monolith even had performance advantages, since calls between components were just in-process function calls. In short, **monoliths shine at the start** because of their simplicity.

However, as WebkinsTech's business expanded and the codebase grew, cracks began to show in the monolithic approach. A monolith that was once a nimble sports car became a lumbering freight train. When applications grow large, the **downsides of monoliths** start to outweigh their initial convenience. WebkinsTech's single codebase became large and complex, making it **slower to develop and test** changes. Multiple development teams had to work in the same code repository, often stepping on each other's toes. A minor tweak in one module (say, updating the billing logic) required rebuilding and redeploying the *entire* application – a time-consuming and risky process. This tightly coupled design meant **slower development speed**; as the code grew, so did the cognitive load on the engineers. Every new feature risked breaking something else, since all modules were intertwined. The once agile WebkinsTech Portal turned into a **"big ball of mud"** – a term architects use for a haphazardly structured monolithic system where everything is interconnected in messy ways.

One major pain was **scalability**. Monoliths can only scale as a whole; you can't easily scale individual pieces of functionality. WebkinsTech noticed that at peak times, the product search and ordering parts of their app were heavily used, whereas other parts (like the admin reporting console) sat mostly idle. With a monolith, however, they had to **scale the entire application** to handle the peak load of one component. That meant deploying multiple copies of the huge application on new servers just to support the busy part, which was inefficient. They couldn't just add resources to the search function alone – it was inseparable from the rest of the system. This "all or nothing" scaling was like having to add more locomotives to pull the entire freight train, when really only one wagon needed extra horsepower.

Another growing issue was **reliability and fault isolation**. In a monolith, all components run in one process (or a tightly coupled set of processes). If any part crashes or runs into a performance bottleneck (e.g. a slow database query in the reviews module), it can **bring down the whole application**. WebkinsTech experienced this the hard way. A tiny bug in the inventory module (a null pointer exception) once caused the entire production site to go down, since the error bubbled up and crashed the single monolithic process. Customers couldn't browse products or place orders – all because of an unrelated glitch in inventory tracking.

WebkinsTech's Monolith Meltdown

Late one Black Friday, the WebkinsTech Portal went down for two hours. The culprit? A single misconfigured integration to a third-party shipping API in the monolithic application caused memory leaks. Because everything was deployed as one unit, the memory issue in the shipping component eventually crashed the **entire** app server. Customers experienced a full outage – no part of the site was available. This incident was a wake-up call: in a monolith, a small failure isn't isolated; it can cascade into a system-wide crash. WebkinsTech's CTO remarked that it felt like "one faulty car derailed our whole train." This brittle all-or-nothing reliability was no longer acceptable as the business grew.

The monolithic architecture also became a **barrier to technological change**. WebkinsTech's platform was initially built with a specific programming stack. As new technologies emerged (say, a new machine learning library for better product recommendations), it was hard to incorporate them into the monolith without affecting everything else. Any change in frameworks or languages would require rewriting or adapting the entire application – a daunting task. Thus, the tech stack of the monolith tended to **stagnate**, locked into early decisions. For WebkinsTech, this meant sticking with an old version of a web framework longer than they wanted, because upgrading it in the huge monolith was risky and timeconsuming.

Integration with other systems was another point of pain. As WebkinsTech started partnering with external services (for payment processing, CRM, analytics, etc.), plugging these into the monolith led to **brittle integrations**. Each integration meant adding more code into the big codebase and often tight coupling with external APIs. If an external service changed or went down, it could directly break the monolith's functionality. The lack of modular boundaries meant there was no elegant way to contain integration-related failures or evolve those connections independently.

Over time, WebkinsTech's monolith had grown into a **bottleneck to the company's agility and growth**. Deployments became so painful that the team moved from daily deploys to weekly or even monthly deploys – they'd bundle many changes together to "reduce the overhead," which only increased the risk when a deployment happened. The integration points in the monolith were so fragile that developers joked about a "butterfly effect" – a small change in one corner of the system could break something completely unrelated elsewhere. The friction was evident: new product ideas took too long to release, scaling the business required heavy upfront planning, and the engineering team was firefighting production issues rather than innovating.

Faced with these realities, WebkinsTech's leadership knew something had to change. **Microservices architecture** began to emerge as a promising solution to this monolithic quagmire. Before diving into how microservices work, it's important to understand that WebkinsTech's story is not unique – many companies (Netflix, Amazon, eBay, to name famous examples) have gone through a similar evolution. Monoliths that served them well initially became impediments at scale, prompting an architectural shift. In the next sections, we will break down what microservices are and the design principles behind them. First, let's define microservices in contrast to monoliths: microservices involve splitting an application into a **suite of small, autonomous services**, each running independently and focusing on a specific business capability. How do we effectively split a big system? That's where core design principles come in.

Principles of Microservices Design

Moving from a monolithic architecture to microservices is not as simple as chopping a codebase into pieces. It requires a thoughtful design approach. The goal is to create services that are **independent, cohesive, and loosely coupled**, so that each can evolve and scale on its own. In this section, we'll explore the key principles that guide a successful microservices design. We'll also see how WebkinsTech's engineering team applied these principles to plan their migration. Keep in mind that microservices are as much about **organizing your software by business domain** as they are about technology. Let's go through the major principles one by one.

Bounded Contexts and Domain-Driven Design

A fundamental concept in microservices design is the idea of a **bounded context**, drawn from Domain-Driven Design (DDD). In essence, a bounded context defines a clear boundary around a particular subset of the business domain, within which a certain set of concepts and rules apply. In a monolith, all business domains (users, orders, inventory, billing, etc.) are mushed together. In a microservices architecture, each bounded context often corresponds to a microservice (or a set of closely related microservices). The service is responsible for everything within that domain boundary.

For example, WebkinsTech identified distinct business domains in their platform: **User**

Management, **Product Catalog**, **Ordering**, **Inventory & Shipping**, and **Billing/Payments**. Each of these became a candidate for a separate microservice, because each domain had its own logic and data. In the monolith, these domains were just modules in one codebase, but their boundaries were leaky – e.g., the Order module could directly query the User module's database tables, creating tight coupling. By establishing bounded contexts, WebkinsTech aimed to untangle these domains. The **Product Catalog service** would own all things related to products (item descriptions, pricing, etc.) and nothing else. The **Order service** would handle order placement, status, history, and not attempt to manage product details or user profiles. This clear separation enforces an **ubiquitous language** within each context – terms like "order status" mean something only to the Order service, whereas "inventory level" is a concept only the Inventory service deals with internally.

Bounded contexts help ensure each microservice has a **single responsibility** in terms of business capability. This not only minimizes ambiguity and overlap between services, but also makes it easier for different teams to own different services without stepping on each other's toes. In microservices, it is often said that services should be organized around **business capabilities** rather than technical layers. This is exactly what bounded contexts push us to do: model services after real business domains (like "Inventory" or "Billing") instead of generic technical components (like a "database service" or "UI service").

Design for Bounded Contexts

When splitting a monolith, start by mapping out your business domains. Aim to give each important domain its own microservice. For example, separate "Customer Accounts", "Orders", and "Payments" if those are distinct areas of your business. This aligns your architecture with the business structure and avoids creating microservices that mix unrelated responsibilities. A good rule of thumb is that each microservice should correspond to a **noun** in your business domain (e.g., Order, Inventory, Shipping) and own that concept end-to-end. This bounded context approach prevents the new system from turning into a distributed big ball of mud.

Service Autonomy and Single Responsibility

Each microservice should be **autonomous** – able to function independently of other services as much as possible. Autonomy means a service can be developed, tested, deployed, and scaled on its own, without requiring coordinated actions with other services. To achieve this, services must have a **single, well-defined responsibility (high cohesion)** and **minimal dependencies (loose coupling)** on other services.

High cohesion within a service means that the functionality a service provides is strongly related. For instance, everything the Product Catalog service does is related to products: creating a product, updating its info, fetching product details, etc. It doesn't also randomly handle user logins or order payments, because those belong elsewhere. By keeping each service focused (single responsibility principle), the internal complexity of each service is manageable, and the team working on it can specialize in that domain.

Loose coupling between services means that changes in one service should not ripple into others. If the Order service changes how it calculates discounts, that change should ideally *not* break the Product or Inventory services. How do we ensure loose coupling? By designing **clear interfaces or contracts** between services. Services communicate with each other through well-defined APIs (often REST endpoints or message queues), and as long as those APIs remain stable, the internal workings of each service can change freely. WebkinsTech, for example, defined API endpoints like /orders for the Order service and /products for the Product service. The Order service might call the Product service's API to get product details when needed, but it treats it like a black box – only relying on the contract (e.g., "provide a product ID, get back name and price"). The internals of how the Product service retrieves that data (database queries, cache, etc.) are encapsulated within that service.

Service autonomy is also reinforced by **decentralizing data management** (more on that soon) and by each service running in its own process or container. In a properly designed microservices system, you could deploy the Inventory service on one server (or container) and the Order service on another, and they would communicate over the network. Each can be restarted or scaled independently. In fact, if one service is temporarily down, other services should ideally continue operating (perhaps with degraded functionality) rather than crashing altogether. That isolation is a direct result of autonomy and loose coupling.

Another aspect of autonomy is that each service team can choose the **technology stack** that best suits that service (this is sometimes called *polyglot* architecture). Because services communicate only via language-agnostic protocols (like HTTP or messaging), one service could be written in Java, another in Python, another in Go, if there's a benefit to doing so. This allows teams to use the right tool for each job. WebkinsTech initially built all services in the same stack for simplicity, but appreciated that, for example, a future **Recommendation Service** (to suggest products) could use Python with a machine learning library, without forcing the entire company onto that stack. This polyglot freedom is only viable because services are independent processes with well-defined interfaces.

Decentralized Data Management

One of the most pivotal principles in microservices is that each service **owns its data**. In a monolith, there is often a single, huge database shared by all parts of the application (e.g., one relational database with dozens of tables, accessible by all modules). This shared database becomes a coupling point – changes to the data schema for one feature can impact others, and it's a source of contention between teams. Microservices push towards **decentralized data management**, meaning each microservice keeps and manages the data it needs within its boundary, and other services are not allowed to directly access that data.

In practice, this often means each microservice has its **own database or data store**. For example, WebkinsTech's Order service has its own Orders database (with tables for Orders, Line Items, etc.), and the Product service has its own Product database (with tables for

Products, Categories, etc.). If the Order service needs product information, it must ask the Product service via an API call – it cannot query the Product database directly because it doesn't have access to it. This isolation of data ensures that the Product service is free to change its internal schema or even migrate to a different type of database if needed, without breaking other services (as long as it continues to fulfill its API contracts). It also means that each service can choose the type of data store best suited for its needs; for instance, Product service might use a document database if it needs flexible schemas for product info, while Order service might use a relational database for transactional consistency on orders.

Decentralized data does introduce challenges – particularly around maintaining data **consistency** across services – which we will discuss in the Tradeoffs section. However, it's a critical principle to achieve true independence of services. It forces architects to carefully design **service boundaries** such that each piece of data has a single home. If multiple services heavily need the same data, maybe they shouldn't be separate services or maybe that data should be split (like Product info vs. Inventory levels might be split between Product service and Inventory service). WebkinsTech's architects used DDD concepts to decide data ownership: customer profile data lives in the User service's database, order records live in Order DB, and so forth. Where cross-service queries were needed in the monolith (say, generating a report that joined customers and orders), the new approach would be to have one service call the other's API or to consolidate data in a reporting service via events. The rule of thumb became: **no direct database sharing** between microservices.

Loose Coupling and High Cohesion

We touched on this under autonomy, but it's worth emphasizing as its own design principle because it's so vital. **Loose coupling** means services interact with each other in minimal, standardized ways, and are not tightly bound to each other's internals. **High cohesion** means all the logic within a service is closely related to that service's purpose. These principles guide how you split the system and how services communicate.

Loose coupling is achieved by designing services to communicate through narrow, welldefined interfaces (like a REST API endpoint or a message queue topic) rather than through sharing complex data structures or through frequent chatty interactions. In a loosely coupled system, Service A might send a request to Service B and not assume too much about how B will process it – just that it will get a response in a certain format. Or Service A might emit an *event* ("Order Placed") that other services listen for, without knowing who exactly will consume that event. This way, services can be added or removed without requiring changes in others, as long as the contract is maintained.

High cohesion, on the other hand, affects how we scope a service. It argues against splitting a service purely by technical concerns (for example, having one service handle *all* database interactions for all domains would be low cohesion, because that service's responsibilities span multiple unrelated domains). Instead, each service should fully handle one area. WebkinsTech ensured high cohesion by bundling related functions together: the Inventory service manages stock levels, restock events, and warehouse info – all inventory-related concerns in one place. It doesn't, say, try to send marketing emails (that would be a Marketing or Notification service's job). High cohesion often comes for free if you follow bounded contexts properly.

One practical approach WebkinsTech took to maintain loose coupling was to avoid creating **tightly coupled synchronous dependencies** whenever possible. For instance, if the Order service needed to confirm payment via the Payment service, doing this synchronously (Order calls Payment and waits) can couple their runtime behaviors – if Payment is slow or down, Order might get stuck. To mitigate this, they designed an **asynchronous pattern** for some interactions: the Order service would place an order and emit an "OrderPlaced" event, the Payment service would consume that event and process payment independently, then emit a "PaymentConfirmed" or "PaymentFailed" event that Order service could listen to. This event-driven approach decouples the timing of the two services (we'll talk more about communication patterns in the Architecture section). The key point is, by carefully choosing how services depend on each other, WebkinsTech aimed to keep them as loosely coupled as possible. If tomorrow they decided to use a third-party payment gateway instead of their own

Payment service, only the Payment service would change (to call that gateway) and the Order service wouldn't need any modification – that's the power of loose coupling.

Independent Deployability and DevOps Alignment

A true microservice should be **independently deployable**. This means you can change and deploy one service without needing to deploy any other services. If your "microservices" are such that deploying service A always requires deploying service B (perhaps because of some hidden dependency or a version lock-step), then you haven't really achieved microservices benefits – you've just broken a monolith into pieces that still operate as a single unit, often called a *distributed monolith*. Independent deployability is a litmus test for proper microservice boundaries and loose coupling.

For WebkinsTech, embracing independent deployments was a cultural shift as much as a technical one. It required adopting DevOps practices like continuous integration/continuous deployment (CI/CD) pipelines for each service, comprehensive automated testing to ensure a service update doesn't break its contract, and robust monitoring to deploy services frequently with confidence. They reorganized their engineering teams so that each microservice (or a small group of related microservices) was owned by a particular **twopizza team** – a term popularized by Amazon meaning a team small enough to be fed with two pizzas, typically 6-8 people. Each team had all the skills needed (developers, QA, ops) to build and run their services. This team autonomy aligns with the microservices principle of independent deployability: if Team A owns Service A, they can decide when to release a new version of it without coordinating with Team B (as long as they don't break the public API contract that others depend on).

Because each service can be deployed on its own schedule, microservices enable **continuous delivery**. WebkinsTech's Order service team could push out a fix or a new feature as soon as it was ready and tested, without waiting for a "big bang" release of the entire platform. This dramatically increased agility (one of the benefits we'll detail later). However, to make this safe, they had to implement strong **observability** (logging, monitoring, tracing) per service, and practices like *canary deployments* (releasing a new version to a subset of users first) to ensure that if something went wrong in a deployment, it wouldn't affect the whole system.

Independent deployability also implies that microservices architecture often goes hand-inhand with technologies like **containers** (Docker) and container orchestration (Kubernetes, etc.), which make it easier to package and deploy services in isolation. While the technical stack is not mandated by microservices (you could, in theory, deploy microservices as separate processes on VMs or even physical servers), modern microservices deployments leverage cloud-native tools to automate scaling and deployment. WebkinsTech containerized each service, which meant, for example, the Inventory service could be running version 2.3 while the Order service is on version 2.8, and they communicate over the network seamlessly.

Lastly, independent deployability enforces **backward compatibility** of service interfaces. If the User service provides an API that the Order service calls, the User service team must ensure that deploying a new version doesn't suddenly change the API in a way that breaks Order (unless they coordinated and updated Order beforehand). This often means supporting multiple versions of an API during transitions or providing clear versioning in APIs. It's an extra responsibility on service teams but a necessary discipline to keep the ecosystem of services working smoothly without lock-step deployments.

With these design principles – bounded contexts, autonomy, decentralized data, loose coupling, high cohesion, and independent deployment – WebkinsTech mapped out a blueprint for their microservices architecture. They imagined a constellation of small services, each true to a business purpose, communicating through stable interfaces, and able to be changed on the fly as business needs evolved. Before we see the end results and benefits, let's consider why WebkinsTech (and many enterprises) felt the *need* to break apart their monolith in the first place. It often comes down to **business growth and integration complexity**, which we'll discuss next.

Business Growth and Integration Complexity

Growing businesses introduce new demands on software systems. What works for a small startup or a single product can buckle under the strain of a larger enterprise with multiple product lines, global users, and numerous integrations. In WebkinsTech's case, initial success led to expansion – and expansion led to **complexity** that their monolithic architecture struggled to handle.

Imagine WebkinsTech's trajectory: They started with a niche product (say, smart lighting devices) sold in one country. Their monolithic system was built around this single line of business. As the company found success, they **launched new products** – perhaps branching into smart thermostats and security cameras, each with its own features and data. The engineering team kept bolting these onto the existing monolith. The database grew to accommodate new product types, new order workflows, new analytics. Soon, parts of the codebase were littered with if statements or modules specialized for one product vs. another. The once-unified platform started to resemble Frankenstein's monster, with pieces stitched together.

Then came **geographical expansion**. WebkinsTech entered new markets in Europe and

Asia, which meant handling multiple currencies, languages, and region-specific regulations. The monolithic application became peppered with locale-specific conditions ("if region == EU apply VAT calculation"), further complicating the code. Deployment schedules had to consider time zones and regional peaks. A change made for Asia could inadvertently affect the US operations due to shared code. The complexity was growing non-linearly.

The company also engaged in a **merger and acquisition (M&A)**: WebkinsTech acquired a smaller competitor to broaden its portfolio. That competitor had its own software system (with its own data formats and processes). Integrating this into WebkinsTech's platform was a nightmare. One approach would be to throw away the acquired system and fold its functionality into the WebkinsTech monolith – but that would take a long time and risk losing important capabilities during the rebuild. The other approach was to maintain it separately and create integration points between the two systems. WebkinsTech opted for interim integration – which meant writing a bunch of glue code: scripts that synced data between the monolith's database and the acquired system's database, and batch jobs that tried to keep orders and inventory aligned across systems. This resulted in **brittle integrations** galore. The systems were never designed to talk to each other, so the integration code was complex and error-prone. Data inconsistencies began to appear (e.g., a product that existed in one system but not properly synced to the other), causing customer and staff frustration.

Every new business initiative – whether adding a feature for a partner, customizing the product for a key client, or adopting a new SaaS tool for marketing – resulted in *increased coupling* to the monolithic core. Over time, WebkinsTech's IT architecture diagram started to look like a **spaghetti** of interconnected components, with the monolith in the middle entangled with many subsystems. It became evident that scaling the business further would exponentially increase the risk of major system issues. The **integration complexity** was throttling the company's ability to innovate and respond to market changes.

WebkinsTech Expands and Faces Complexity

Let's consider a specific scenario: WebkinsTech expanded into offering a subscriptionbased service for ongoing device firmware updates and support. This new service required integrating the monolithic ecommerce platform with a cloud-based subscription management system (from a third-party) and a separate billing service for recurring payments. Initially, the team added subscription features into the monolith and wrote custom connectors to the third-party systems. Soon, things went wrong – subscription data in the monolith would sometimes fail to sync with the billing service, causing some customers to lose access erroneously. Why? The brittle, ad-hoc integration layer had so many moving parts that keeping it reliable was a constant battle. Whenever the thirdparty subscription API changed, the team had to scramble to update the monolith's integration code, delaying product launches. This scenario underscored that the **monolith + patchwork integrations** approach could not sustain WebkinsTech's growing business. It was time to rethink the architecture.

Business growth not only adds complexity but also highlights organizational issues. With more employees and teams, WebkinsTech found that a single monolithic codebase became a coordination choke point. Multiple teams working in the same code meant lots of merge conflicts, divergent coding styles, and difficulties in ownership (who "owns" a piece of code that is used by many features?). Teams were stepping on each other's toes, and planning releases required a **coordination meeting across all teams** to bundle changes. This slowed down the company's ability to deliver new value. Essentially, the software architecture was misaligned with the organizational structure – a concept known as **Conway's Law**, which states that systems tend to mirror the communication structures of the organizations that build them. WebkinsTech's organization had grown into semiindependent units (product lines, regional teams, etc.), but the monolith forced them to work as one large unit, creating friction.

Microservices architecture offered a way to tame this complexity by **breaking the problem space into smaller, manageable pieces**. Instead of one monster system handling everything, WebkinsTech could have a network of services, each dedicated to a specific business capability (which nicely mirrors how their business units were forming). Integrations with external systems could also become more modular – for example, a **Subscription Service** microservice could be created specifically to handle subscriptions, bridging between the ecommerce core and third-party billing, encapsulating that complexity in one place. If the third-party API changes, only the Subscription Service needs updating, not the entire platform.

To paint a clearer picture of how microservices architecture looks, let's move on to an overview of a typical microservices system. We'll see how different components (like API gateways, discovery services, etc.) come into play to support this modular architecture. We'll also illustrate examples in different domains (e-commerce and fintech) to show how microservices can address complexity in various enterprise scenarios.

Microservices Architecture Overview

Designing a microservices system involves more than just chopping a monolith into services. You also need an ecosystem of supporting components and a strategy for communication between services. In a monolithic app, all function calls are in-process and you don't worry about network issues, discovery, or load balancing – it's one big program. In a microservices architecture, by contrast, you have many separate programs that must work together, often over a network. This requires a thoughtful architecture to make sure services can find each other, communicate efficiently, and that the system as a whole can be observed and managed. Let's outline the key components and concepts of a microservices architecture:

- **API Gateway**: Rather than exposing dozens of services directly to client applications, most microservices setups use an API Gateway as a single entry point. The API Gateway is a layer that receives external requests (from a web or mobile client, for instance) and routes them to the appropriate microservice on the backend. It can also handle cross-cutting concerns like authentication, rate limiting, and response aggregation. For WebkinsTech, instead of the client calling Product Service and Order Service separately, the client talks to the WebkinsTech API Gateway, which then proxies or fowards the request to the correct internal service. This simplifies the client and allows internal changes without affecting external consumers.

- **Service Registry and Discovery**: In a dynamic microservices environment (where services may scale up and down, or move between servers/containers), we need a mechanism for services to find each other. A Service Registry is like a phonebook where each service instance registers itself (with its network address). Other services or the API Gateway can query the registry to discover where, say, "Order Service" is currently running. Tools like Netflix Eureka, Consul, or Kubernetes' built-in DNS serve this purpose. Discovery can be client-side (the client or gateway looks up and calls the instance) or server-side (a load balancer sits in front and uses the registry to direct traffic).

- **Load Balancers**: Since each microservice can have multiple instances for scalability and reliability, load balancing is crucial. A load balancer distributes incoming requests for a service across its instances. This can be done by the API Gateway or a dedicated component. For example, if WebkinsTech's Product Service is running 3 instances, the gateway or a load balancer will spread user requests across those instances so none is overwhelmed and if one fails, traffic is routed to the others.

- **Inter-Service Communication**: Services need to talk to each other. There are two primary communication patterns:

- *Synchronous calls (often HTTP/REST or gRPC)*: One service calls another and waits for a response, similar to how function calls worked in the monolith but now over a network. RESTful APIs over HTTP are common – e.g., Order Service calling GET /products/{id} on the Product Service to fetch details. gRPC (a highperformance binary protocol) is another option, especially for internal service-toservice calls.

- *Asynchronous messaging (events or message queues)*: Services emit events or messages to a message broker (like Apache Kafka, RabbitMQ, AWS SNS/SQS, etc.) without waiting for a response. Other services subscribe to relevant events and react. For instance, when an order is placed, Order Service could emit an "OrderPlaced" event. Inventory Service, Shipping Service, and Notification Service might all subscribe to that event to update stock, schedule shipment, and inform the customer, respectively. Asynchronous communication enables **loose coupling** and can improve resilience (services continue working even if others are temporarily down, processing the event when available).

- **Data storage per Service**: As mentioned, each service has its own database or data store. In the architecture overview, this means your system will likely have a variety of databases. One might be a MySQL instance for orders, another a MongoDB for products, maybe a Neo4j graph DB for a recommendation service, etc., depending on what fits each service. The architecture needs to account for managing these multiple data sources (backups, consistency where needed, etc.). There's no single "database layer" – it's decentralized.

- **Observability and Monitoring**: With many moving parts, it's vital to have a way to see what's happening across the system. Observability in microservices typically includes: - *Centralized Logging*: Each service writes logs, which are aggregated (using tools like ELK stack – Elasticsearch/Logstash/Kibana – or cloud logging services). This way, you can search across all service logs when debugging an issue.

- *Metrics and Monitoring*: Each service emits metrics (e.g., request count, error rate, memory usage) which are collected by a monitoring system (like Prometheus, Datadog, etc.). Dashboards can show the health of each service, and alerting systems can notify engineers if a metric goes out of expected range (e.g., the error rate on Payment Service spikes).

- *Distributed Tracing*: To diagnose performance or failures in a call that involves multiple services, distributed tracing tools (like Zipkin or Jaeger) tag requests with IDs and track them as they flow through multiple services. For example, a user request that goes through API Gateway -> Order Service -> Payment Service can be traced end-to-end to see where time was spent or where an error occurred.

- **Deployment and Container Orchestration**: While not strictly part of "architecture" in a diagram sense, the platform that runs microservices is key. Typically, microservices are deployed in containers and orchestrated by systems like Kubernetes, Docker Swarm, or cloud orchestration services. These platforms handle scheduling service instances on machines, restarting them on failure, scaling them up/down based on load, and so on. In WebkinsTech's case, they might use Kubernetes to ensure if an instance of User Service crashes, another is automatically brought up, or to scale out Order Service during Black Friday traffic and scale it back in afterward.

Let's visualize a **generic microservices architecture** with some of these components:

```
                    [ Client Application ]
                              |
                [ API Gateway / Load Balancer ]
                              |
       -------------------------------------------
       |                      |                      |
  [ Service A ]          [ Service B ]          [ Service C ]
  (e.g. Users)           (e.g. Orders)          (e.g. Products)
       |                      |                      |
   [DB for A]             [DB for B]             [DB for C]
```

Figure: A simplistic view of a generic microservices architecture. In this diagram, the client (could be a web app or mobile app) talks to an **API Gateway**, which then routes requests to various microservices: Service A, Service B, Service C (each representing a different business capability). Each service has its own database. The gateway might also handle load balancing across multiple instances of each service (not shown). In a real system, there would also be a **Service Registry** (so that the gateway or services know where each service is located) and possibly a **Message Broker** for event-based communications between services (not depicted above). But this basic figure illustrates the core idea of microservices: independent services behind a unified entry point, each with its own data.

Now, to ground these concepts, let's look at two concrete examples of microservices architecture in specific domains: e-commerce and fintech. This will help demonstrate how an enterprise can lay out microservices for different business functions.

Microservices in an E-Commerce Platform

Consider an e-commerce platform (not unlike WebkinsTech's online store). In a microservices world, such a platform would be composed of specialized services. A possible breakdown is:

- **User Service** – manages user accounts, profiles, authentication.
- **Product Catalog Service** – manages product listings, details, categories, search functionality.
- **Cart Service** – handles users' shopping carts (items picked but not yet ordered).
- **Order Service** – processes orders, order history.
- **Inventory Service** – keeps track of stock levels in warehouses.
- **Payment Service** – processes payments (or integrates with payment gateways).
- **Shipping Service** – manages shipment of orders, tracking, logistics.
- **Review/Rating Service** (optional) – handles customer reviews for products.
- **Notification Service** – sends out emails/SMS for order confirmations, etc.

Let's illustrate a simplified version with some of these services:

```
                        [ Customer Web/Mobile App ]
                                     |
                              [ API Gateway ]
                                     |
        --------------------------------------------------------
        |            |             |              |             |

    [User Service] [Product Service] [Cart Service] [Order Service]
 [Inventory Service]
        |            |             |              |             |
  [User DB]     [Product DB]   [Cart DB]      [Order DB]    [Inventory DB]
```

Figure: Microservices in an E-commerce domain. Here, the **API Gateway** routes requests to the appropriate service. For example, a request to "/products/123" would go to the Product Service, while a login request might go to the User Service. The **User Service** might call an internal authentication component (or a library) to verify credentials and manage user profiles in its User DB. The **Product Service** handles queries for product info, and keeps that data in a Product DB. A **Cart Service** could store the contents of each user's shopping cart in a fast database (perhaps an in-memory store or NoSQL, since cart data is ephemeral and needs quick access). When a customer proceeds to checkout, the **Order Service** would be invoked to create an order. The Order Service would likely call the **Payment Service** (not shown in the diagram for brevity) to charge the customer's credit card. This could be a synchronous call – e.g., Order sends payment details to Payment and waits for confirmation.

Or it could be async: Order submits an order and an event goes to Payment for processing.

Upon successful payment, the Order Service finalizes the order and emits an "OrderPlaced" event. That event could be picked up by the **Inventory Service**, which then decrements stock counts in the Inventory DB for the items sold, and by the **Shipping Service** (not shown) to initiate fulfillment. The Inventory Service might also expose an API that the Product Service calls to get current stock when displaying product availability – or they might choose to keep it decoupled by having Product Service subscribe to inventory updates and maintain an in-memory cache for display. Different integration approaches can be used depending on consistency needs.

This microservices e-commerce architecture would allow WebkinsTech (or any retailer) to scale and evolve each piece separately. If suddenly the marketing team wants to overhaul how search works, they can focus on the Product Service (perhaps swap out the database for a search-optimized engine) without touching Order or User services. If the volume of orders spikes during a sale, they can scale out the Order Service and maybe the Cart Service more, without unnecessarily scaling the entire system. If a new feature is needed, like a **Wishlist Service**, it can be added alongside these services without modifying the existing ones – just integrate via the gateway and events as needed.

WebkinsTech's initial move towards microservices followed this kind of breakdown. They carved out the **Product Catalog** and **Inventory** as separate services first, since those were relatively self-contained domains and sources of a lot of load. This relieved the monolith from having to serve all product lookup requests. Next, they tackled the **Order and Payment** flows. Over time, their architecture started to resemble the diagram above, rather than one giant application.

Microservices in a FinTech Platform

Microservices are equally popular in fintech (banks, payment processors, etc.), where there are clearly delineated domains like account management, transactions, fraud detection, and so forth. Let's sketch a microservices architecture for a hypothetical fintech platform, for example, a digital banking system:

Possible services:

- **Account Service** – manages customer bank accounts, balances, personal info.
- **Transfer Service** – handles fund transfers between accounts (could also be part of Account, but let's assume separate for complexity).
- **Payment Service** – maybe handles bill payments or credit card transactions.
- **Fraud Detection Service** – monitors transactions for suspicious activity.
- **KYC Service** – handles "Know Your Customer" verification processes.
- **Reporting Service** – generates statements, regulatory reports.
- **Notification Service** – sends alerts for transactions, etc.
- **Audit Service** – logs all transactions for compliance.

For simplicity, consider Account, Payment, and Fraud services and how they interact:

[Banking Mobile App]

```
                         |
                 [ API Gateway ]
                         |
  ---------------------------------------------
  |                      |                      |
[Account Service]   [Payment Service]    [Fraud Service]
  |                      |                      |
 [Accounts DB]       [Payments DB]          [Fraud Rules
                         DB]
                          \
     --> (publishes events) --> [Fraud Service]
```

Figure: Microservices in a FinTech domain. In this example, a mobile banking app might hit an **Account Service** to get account details and balance. When the user initiates a payment or fund transfer, that request goes to a **Payment Service**. The Payment Service will execute the transaction (update balances in Accounts DB or its own Payments DB, communicate with external networks if needed). Simultaneously, it could emit a **Transaction Event** that the **Fraud Service** listens to. For instance, if a very large transfer occurs or one that fits certain fraud patterns, the Fraud Service (upon receiving the event) can flag it or even call back to Payment Service or Account Service to freeze funds. The Fraud Service might use its own database of rules or a machine learning model to decide what's suspicious. By decoupling fraud monitoring into its own service, the Payment Service remains simpler and focused on the mechanics of moving money, while the Fraud Service can evolve sophisticated algorithms without impacting the rest of the system's operation.

If we include an **Account Service**, note that it might maintain overall account state. In some designs, the Payment Service might directly update account balances as well (if each service can touch account data). However, a cleaner design is often to let Account Service be the source of truth for balances and have Payment Service call Account Service to adjust balances. Alternatively, Payment Service could handle balances for the accounts it's transferring between and then emit events that Account Service uses to reconcile. There are multiple patterns (saga patterns for transactions, etc.), but they all revolve around ensuring consistency while keeping services separate.

In a fintech microservices architecture, **data consistency** and **transaction management** become a challenge – ensuring that if Payment Service says “money moved from A to B”, the Account Service reflects that in both A and B’s balances. This often requires careful orchestration or eventual consistency approaches (like two-phase commits or sagas). We will discuss these challenges later (under Tradeoffs), but the key is that microservices can still work for such scenarios, they just need patterns to coordinate state.

The benefits for fintech are similar: each service can be scaled and secured differently. Fraud detection might require a lot of compute (maybe analyzing patterns with AI), so you scale it independently. If one part – say Fraud Service – goes down, you might configure the system to continue processing payments but mark them for later review when Fraud comes back (graceful degradation). Each service could even be developed by different specialized teams (perhaps in different departments of a bank). This modularity improves resilience and allows the platform to expand functionality by adding new services (for new financial products) without messing up existing ones.

Having explored these domain-specific examples, you can see that microservices architecture encourages thinking about the **components of your enterprise as separate products or services** in themselves. Each has a clear purpose and interface. This architecture does introduce complexity in infrastructure, but it **simplifies** the mental model within each service. Instead of 100 developers all grappling with one huge monolith, you might have 10 teams of 10, each fully owning a service – and each service is small enough to be understood and maintained more easily.

Now that we’ve covered what microservices architecture looks like and how it addresses evolution and complexity, let’s summarize the **benefits** that microservices can bring to an enterprise like WebkinsTech. After that, we’ll candidly address the **tradeoffs and challenges** because, as with anything, microservices are not a silver bullet and come with their own difficulties.

Benefits of Microservices

Microservices have been embraced by many organizations because they offer several compelling benefits, especially for enterprises needing to innovate and scale quickly. By breaking a large system into smaller services, companies can achieve a level of agility and resilience that is hard to obtain with a monolith. Let's go through some key benefits of microservices architecture and illustrate how WebkinsTech (and real-world companies) reap these advantages:

- **Agility and Faster Time-to-Market:** Microservices enable development teams to work in parallel on different components without waiting on each other. Each team can build, test, and deploy their service independently. This often translates into faster release cycles. WebkinsTech, after migrating to microservices, noticed that teams could introduce new features or improvements in one service (say, the Product recommendation algorithm in the Product Service) and deploy it immediately, rather than waiting for a monolithic release train. This agility means the business can respond quickly to customer needs or market changes. In practice, companies like Amazon and Netflix have reported deploying code **hundreds or thousands of times a day** across their microservices – an impossible feat in a monolithic world. The decoupling of services reduces the coordination overhead for releases, so features go from development to production faster, giving the company a competitive edge.

- **Independent Scaling:** With microservices, you can scale the parts of the application that need more resources without scaling the entire app. This **fine-grained scalability** is more efficient. For example, if WebkinsTech sees heavy read traffic on its Product Catalog (many users browsing products) but relatively lighter usage on the Checkout, they can allocate more servers/instances to the Product Service. In a monolith, they would have had to scale up everything, which wastes resources on the less busy parts. Microservices also allow using different scalability strategies per service – one service might benefit from in-memory caching, another might scale by partitioning data, etc., tailored to their needs. This leads to better performance under load and cost optimization in infrastructure.

- **Fault Isolation and Resilience:** In a microservices architecture, if one service fails, it doesn't necessarily bring down the whole system (unlike a monolith where one bug can crash everything). This **fault isolation** improves overall system uptime. For instance, if the **Review Service** (customer reviews) in WebkinsTech's platform has a bug and crashes, the rest of the site (browsing and ordering) can continue to function; maybe the reviews just temporarily don't show up. Microservices, especially when designed with network timeouts and fallback logic, can **degrade gracefully**. Netflix famously engineered their system so that if the recommendation service is down, the UI just skips showing recommendations rather than failing the whole page. Additionally, microservices encourage the use of resilience patterns (like circuit breakers that stop calling a service if it's not responding, bulkheads that isolate resources, etc.). This benefit was huge for WebkinsTech – after breaking apart their system, they saw a reduction in full outages. Problems became more localized and easier to fix without user-visible downtime in other areas.

- **Polyglot and Right Tool for the Job:** Microservices allow teams to choose the most suitable **technology stack** for each service. This means you aren't locked into one database or one programming language for the entire system. If one part of your system would perform better with a graph database or a high-performance compiled language, you can do that just for that part. At WebkinsTech, their core services remained in the original language, but for a new **Analytics Service** they needed to process large data sets, they chose a different stack that was more data-crunching efficient. They also adopted specialized databases: for example, the Search team introduced an Elasticsearch engine for the Product Service to enable more advanced search queries, without affecting any other service. This **polyglot flexibility** leads to technical **innovation** within the organization – teams can experiment with new tech on a small scale (one service) without risking the entire platform. It also helps with hiring and using existing skills: if you have a great Node.js team, they can work on one service while a Java team works on another.

- **Team Autonomy and Alignment with Business:** Microservices, when done right, align with small, cross-functional teams (as mentioned with the two-pizza team concept). Each team has **ownership** of a service and can work more autonomously. This tends to increase accountability and pride in work – the team fully owns the success of "their" microservice. It also reduces the coordination headache of large teams all working in the same codebase. In WebkinsTech's case, after moving to microservices, the front-end developers, back-end developers, and QA for a given domain (like "Orders") all sat together (physically or virtually) and focused on the Order Service end-to-end. They didn't have to attend meetings about unrelated features in other parts of the system. Team autonomy drives **productivity** because each team can make decisions faster (choosing their release schedule, tools, etc., within boundaries).

It also fosters **innovation**; teams can try new approaches in their service more freely. From an organizational perspective, microservices allowed WebkinsTech to **scale the engineering organization** – you can add more teams working on different services without them colliding in one codebase. This maps well to growing a business: you can form a new team to build a new microservice for a new initiative, rather than putting more cooks in the same kitchen of a monolith.

- **Better Maintainability through Modularity:** Each microservice is a smaller codebase than the monolith, which can be easier to understand and maintain. New team members can get up to speed with one service's code faster than reading a gigantic monolithic codebase. Bug fixing is often simpler – you know exactly which service (and thus which code repository) to look at for a given issue. Over time, microservices can reduce the accumulation of **technical debt** because each service can be refactored or even rewritten on its own if needed (it's feasible to rewrite a small service without boiling the ocean). This modularity ensures that the whole system's quality can be improved piece by piece.

- **Agility in Deployment and Experimentation:** Because services can be deployed independently, you can do experiments (A/B testing, trying a new feature with a small user group) by deploying a new version of a microservice just for a subset of traffic.

This is harder with a monolith where any change goes to all users. Microservices facilitate advanced deployment strategies like **blue-green deployments** (switching between two environments for zero-downtime release) or **canary releases** (gradually rolling out a change). WebkinsTech used this to test a new checkout process: they deployed a new version of the Cart and Order services and directed 10% of users to it via the API Gateway's routing rules. When metrics showed it improved conversion, they scaled it up to everyone. Such controlled experimentation is a huge business benefit – you can validate ideas faster and more safely.

Amazon's Two-Pizza Teams and Microservices

Amazon's e-commerce platform is a famous real-world example of microservices benefits. In the early 2000s, Amazon moved away from a large monolithic application towards a microservices architecture. They organized small "two-pizza teams" – each responsible for a specific service, such as the

Catalog Service, the Payment Service, or the Review Service. This reorganization allowed each team to iterate quickly and deploy their service independently. The result? Amazon could innovate at a much faster pace. Reportedly, Amazon deployments grew to occur every 11.7 seconds on average across their many services. Features like "Customers who bought this also bought that" could be developed and released by a team without waiting on the entire platform's next release. The microservices approach enabled Amazon to scale both its technology (handling millions of users seamlessly by scaling individual services) and its organization

(hundreds of teams working in parallel). WebkinsTech's move to microservices mirrors this logic – albeit on a smaller scale – seeking the same kind of agility and scalability that Amazon achieved.

All these benefits sound great – and they are – but it's important to acknowledge that they come with a cost. Microservices solve many problems of monoliths, but they introduce new complexities and challenges of their own. In the spirit of being thorough, the next section will act as a cautionary discussion on the **tradeoffs and challenges** that come with microservices. Understanding these will help in planning mitigations and knowing whether microservices are right for your situation.

Tradeoffs and Challenges

Microservices architecture is not a silver bullet. While it provides many benefits, it also introduces new complexity and potential pitfalls. It's crucial to go into a microservices journey with eyes wide open about these challenges, so you can address them proactively. Let's examine some of the main tradeoffs and challenges that WebkinsTech encountered (and any team is likely to face) when moving to microservices:

- **Increased System Complexity:** You are replacing a single application with a distributed system composed of many moving parts. The overall complexity of the *system* increases. Things that were simple function calls in a monolith are now remote calls over a network. This means dealing with network latency, possible network failures, serialization of data, and so on. The **mental model** of the system becomes more complex – developers and architects must consider inter-service interactions, network protocols, and data consistency across services. In a monolith, if you call a function, you can almost always assume the function runs and returns (or throws an error you can catch). In a microservices world, a call to another service might **timeout** or fail due to network issues, and you have to handle that gracefully. Essentially, some complexity is moved out of the code and into the space between services.

- **Operational Overhead:** Operating dozens or hundreds of services is undeniably more work than operating one monolith. Each service needs to be deployed, monitored, logged, and managed. You'll likely need an advanced **DevOps setup**: automated deployment pipelines for each service, container orchestration, service discovery mechanisms, etc. The tooling and infrastructure investment is significant. WebkinsTech had to set up a whole CI/CD pipeline per service, configure Docker and Kubernetes, and implement a service mesh for observability. Organizations might find that they need a platform engineering team or to adopt cloud platforms to handle this complexity. Also, debugging issues can be harder – when something goes wrong, you have to figure out which service (or which combination of services) is the culprit. This often requires sophisticated tracing and logging. The **cost** (in both infrastructure and people time) of running microservices can be higher than a monolith, especially if not managed well.

- **Network Latency and Communication Overhead:** Splitting a process into multiple services means what used to be in-memory function calls are now network calls, which are orders of magnitude slower. If your microservices are too fine-grained or chatty (i.e., they need to call each other a lot to complete a single user request), the user-facing performance can suffer. For example, if rendering a page requires calls to 5 different services, the latency of each call adds up, and it could be slower than the monolith was. You have to design APIs and service boundaries carefully to minimize synchronous call chains. Techniques like **caching**, aggregating data (maybe via the API Gateway or a composite service), or using async patterns can mitigate this, but it's a challenge to solve. There's also overhead in data serialization/deserialization in each call (e.g., converting objects to JSON and back). In high-throughput systems, this overhead can be non-trivial. Engineering effort is needed to ensure microservices interactions remain efficient.

- **Data Consistency and Distributed Transactions:** In a monolith with a single database, it was relatively easy to have ACID transactions that updated multiple parts of the data together – ensuring consistency. In microservices, with each service owning its data, maintaining **consistency across services** is difficult. For instance, when an Order is placed, you might need to decrement inventory and charge payment. In a monolith, you might do that in one database transaction ensuring either all or nothing happens. In microservices, Order Service, Inventory Service, and Payment Service each have their own DB – how do you ensure they either all succeed or roll back? Distributed transactions (two-phase commit across databases) exist but are complex and can hurt performance; they are usually avoided in microservices because they introduce tight coupling and failure modes that are hard to manage. Instead, systems use eventual consistency and **saga patterns** (a saga is a sequence of local transactions with compensating actions to undo them if a later step fails). Implementing a saga is not trivial – you need to write logic to handle partial failures. For example, if Payment fails after Order is created, perhaps Order Service must cancel the order (or mark it failed) and Inventory Service must put back the stock. Designing and testing these failure compensation paths is challenging. WebkinsTech had to create a robust retry and compensation mechanism for scenarios like payment failures or inventory mismatches. This area – managing distributed data consistency – is often one of the hardest in microservices adoption.

- **Testing Complexity:** Testing a microservices system can be more complex than testing a monolith. In addition to unit tests for each service, you need to test the interactions between services (**integration tests** and **end-to-end tests**). Setting up test environments that have multiple services running (or using test doubles/mocks for services) is a new challenge. You may need contract testing to ensure that if Service A expects Service B's API in a certain format, Service B's changes don't break that contract. Ensuring that a new version of a service is compatible with others can require diligent testing and sometimes backward compatibility code. Automation in testing is key; many companies invest in elaborate test harnesses or use Docker

Compose/Kubernetes test clusters to deploy a full stack for testing. All of this is doable, but it's more effort than testing one monolith in isolation.

- **Deployment and Versioning Challenges:** While independent deployment is a benefit, it can also be a headache if not managed well. If Service A and Service B have dependencies (for example, A calls B's API), you must handle versioning of those APIs.

What if Service B wants to change an API in a way that Service A must also change? You either need to **coordinate deployments** (which is what we wanted to avoid) or ensure backward compatibility until Service A updates. This means you might have to maintain multiple versions of an API in a service until all consumers migrate – adding complexity inside the service. There's also the scenario of **configuration management** – each service might have its own configurations, feature toggles, etc. Managing these consistently (especially across different environments like dev, staging, prod) requires discipline or tooling (like a central config server). Deployment automation must be nearly flawless to handle many services – the chance of human error is higher if deploying 50 services manually versus 1, so scripting and infrastructure-as-code become important. WebkinsTech learned that investing in an internal developer platform (with templates and tools for launching new services and managing them) was essential to avoid a chaotic devops situation.

- **Observability and Debugging Distributed Systems:** As mentioned, finding the root cause of issues can be tricky. Take an example: a user reports an error at checkout. In a microservices world, that single action might involve half a dozen services. The error could have originated in any one of them (or in the network between them). Without good observability, it's like finding a needle in a haystack. Teams must implement **distributed tracing**, correlated logs (use IDs that flow through calls to tie together log entries), and monitors on each service's health metrics. It's a challenge to set up but not optional – flying blind in a microservices system is a recipe for disaster. Moreover, the sheer volume of logs and metrics can be much larger, since each service emits its own – so scalable monitoring systems are needed. WebkinsTech experienced this: when they first split into 10 services, their log volume multiplied roughly by that factor. They had to upgrade their logging infrastructure and define log retention policies to keep costs reasonable. Debugging also sometimes requires running a subset of the system in isolation – developers might need to run 5 services on their machine to simulate a scenario, which can be cumbersome (though technologies like service virtualization or using test doubles can help).

- **Organizational Challenges:** Adopting microservices isn't just a technical shift; it's a cultural and organizational one. Teams need to embrace **DevOps practices**, meaning developers are often more involved in deployment and monitoring than in traditional siloed setups. Not every developer may have experience with distributed systems, so training is needed. Communication between teams (which now each own services) becomes vital – they must agree on APIs and data contracts. There is the risk of siloing knowledge – each team is deep in their service but might lose sight of the overall system behavior. To mitigate that, WebkinsTech rotated some engineers across teams for cross-pollination and held architecture guild meetings to ensure everyone understood the big picture. Also, with autonomy comes the need for **discipline**: if one team doesn't follow best practices (say, they skip adding proper monitoring), their service can become the weak link that hurts the whole system. Governance without stifling teams is a tricky balance. Some organizations create *platform teams* that provide common tools and guidelines to all service teams (for logging, security, etc.) – effectively building a paved road so individual teams don't have to reinvent wheels or go off the rails.

Microservices Pitfalls to Watch Out For

Distributed Monolith: One common anti-pattern is breaking a monolith into services, but still letting them be tightly coupled – for example, sharing databases or requiring all services to be deployed together because of synchronous dependencies. This "distributed monolith" has all the downsides of microservices (complexity) with none of the upsides. Avoid it by enforcing clear boundaries and independence (no sneaky backdoor database access between services!).

Over-Granularity: Making services too fine-grained (e.g., a separate service for every single function) can lead to excessive inter-service calls and complexity. It's better to have a slightly larger service that handles a cohesive domain than dozens of tiny ones that constantly talk to each other. In practice, many successful architectures have "microservices" that are more like **mini-services or macroservices** – not enormous, but not absurdly small either. Find the right slice size by trial and observation.

Lack of Automation: Operating microservices manually is a non-starter. If you don't invest in automation (CI/CD, auto-scaling, failover, etc.), you'll drown in operational toil. Ensure you automate build, test, and deployment for each service, and use infrastructure tools to manage them. If deploying a fix to 10 services is a 10-step manual process, things will break.

Inadequate Monitoring: A microservices system without centralized monitoring and alerting is like a city without a traffic control or emergency services – accidents will happen and you won't know where. Set up your logging, metrics, and tracing from day one. Also, implement health checks for services so that your orchestrator (or load balancer) can detect and replace unhealthy instances.

Ignoring Network Failures: Eventually, something will go wrong in the network. If your code doesn't handle timeouts, retries, and fallbacks, users will face errors. Use timeouts on calls, implement retry logic with backoff for transient issues, and consider circuit breakers (which stop repeated attempts to call a service that is down). Design your client-service interactions with the assumption that "anything can fail at any time." This mindset helps maintain a robust system.

Despite these challenges, organizations often find that the **benefits outweigh the costs** once they mature in handling microservices. The key is to be aware of the tradeoffs and address them with good design, tooling, and practices. Not every company is ready for microservices; sometimes a well-structured monolith (or a modular monolith) might serve better until the team and product grow larger. This leads us to the next part of our discussion:

understanding if *you* are ready for microservices and what it takes to get there. Before leaping, it's wise to check some readiness factors.

Microservices Myths vs. Realities

Microservices have been a buzzword for years, and with that buzz come a lot of misconceptions. In planning a move to microservices, it's important to clear the air of some common myths and set realistic expectations. Let's debunk a few of these myths by contrasting them with the realities:

- **Myth: "Microservices will automatically make my application faster." Reality:** Microservices can improve *development speed* and scalability, but they don't inherently make the application's response time faster. In fact, the introduction of network calls and inter-service communication can add overhead. If not designed carefully, microservices might even slow down request processing due to serialization and network latency. The performance benefit of microservices comes from the ability to scale out bottleneck components and isolate heavy workloads, not from some magical speed boost. A well-optimized monolith can often be very fast. Microservices help you keep performance good as the system grows in complexity, by scaling and isolating concerns, but you still need to design for performance (using caching, efficient protocols like gRPC, etc. where appropriate).

- **Myth: "Microservices = SOA (Service-Oriented Architecture), just with a new name." Reality:** Microservices did evolve from some SOA principles, but there are differences in scale and philosophy. Traditional SOA in the 2000s often meant a few large services, often with an Enterprise Service Bus (ESB) handling communication, and was sometimes heavyweight with XML/SOAP messages. Microservices usually refer to much smaller, fine-grained services, often using simpler protocols (JSON/REST, gRPC) and with decentralized governance (no big ESB – rather, smart endpoints and dumb pipes). The goals are similar – modularity, reuse – but microservices emphasize *independent deployment* and *decentralized data*, which not all SOA implementations achieved. One way to put it: all microservices architectures are a form of SOA, but not all SOA implementations would qualify as microservices. Microservices generally push the concepts to a more granular level with modern DevOps practices.

- **Myth: "Every part of our application should be a microservice." Reality:** Not necessarily. It's perfectly valid to start or even remain with some parts of the system as a monolith or grouped in a service. Some systems use a **hybrid approach**: a core monolith for certain tightly coupled functionality and microservices for extending around it (sometimes called the "modular monolith" approach). The key is to split where it makes sense – typically at **bounded context** boundaries – and not split if it's going to create artificial fragmentation. Also, some utility or support functions might not need their own service (for example, you might not create a separate microservice just to do a small calculation that only one other service uses; that could just be a library). Use microservices where they solve a problem (team scaling, independent scaling, complexity management). It's about **serving the architecture's needs**, not following a dogma. WebkinsTech, for instance, kept their admin back-office functions in a small monolith because only one team used it and it wasn't performance critical – it didn't justify becoming a microservice.

- **Myth: "Microservices guarantee scalability and uptime." Reality:** Microservices provide the *architecture* to improve scalability and uptime, but they must be accompanied by the right practices to realize those promises. Simply chopping into services won't help unless you also implement auto-scaling, load balancing, failover, etc. A poorly built microservice can still be a single point of failure if all requests funnel through it and it's not replicated or if it has a critical bug. The system is only as strong as its weakest link. Yes, microservices let you scale out pieces independently – but you have to *do* that scaling and capacity planning. Similarly, microservices can enhance uptime by isolation, but you need robust monitoring and perhaps self-healing in place (like automatic restarts) to truly achieve high availability. In short, microservices are an enabler for scalability and reliability, not a guarantee – you must design and operate the system well to achieve those outcomes.

- **Myth: "We should move to microservices because did it." Reality:** The fact that companies like Netflix, Amazon, and Uber use microservices doesn't mean it's the right choice for everyone. Those companies had very specific needs – hyper-scale, rapid team growth, global distribution – that pushed them to adopt microservices. If you're a smaller organization or your system isn't that complex yet, jumping into microservices could be overkill and actually slow you down. Martin Fowler, a respected authority on software architecture, famously said: *"You shouldn't start a new project with microservices, even if you're sure your application will be big enough to make it worthwhile."* The point is to **start with a simple architecture** (which might be a modular monolith) and only break into microservices as pain points of scale or complexity emerge. Adopting microservices because it's trendy can lead to unnecessary complexity. It's better to have a well-factored monolith than a poorly executed microservices architecture. The decision should be driven by concrete problems that microservices help solve (like independent team deployment or scaling a rapidly growing product) – not just because others are doing it.

"You shouldn't start a new project with microservices, even if you're sure your application will be big enough to make it worthwhile." – Martin Fowler (Microservices pioneer)

Insight: Fowler's advice underscores that microservices come with a cost. If you can manage with a simpler architecture initially, do so – you can always evolve to microservices when needed. In fact, having a well-structured monolith can be a stepping stone to microservices: you can split it along its modular boundaries when the time is right. This evolutionary approach often works better than trying to build a complex microservices system from scratch without having a clear understanding of your domains and needs.

Myth: "Microservices allow any team to use any technology with no restrictions." Reality: While microservices do support polyglot development, completely unfettered choice can lead to a fragmented ecosystem that's hard to maintain. In practice, many companies set some **guardrails**: for example, they might allow a few approved languages or frameworks (maybe a primary one and a couple of secondary ones). If every microservice used a different tech stack, the operations burden (different build tools, libraries, deployment processes) could become enormous. Also, people transfer between teams, and having totally distinct tech in each service could silo knowledge. So, yes, you aren't forced into one stack for everything, but it's usually wise to standardize parts of the stack (logging, monitoring, RPC frameworks, etc.) across services. WebkinsTech gave teams freedom in certain areas (the choice of database or internal libraries) but kept consistency in others (all services expose RESTful APIs in a similar style, all containerized in the same way, using the same CI pipeline tooling). Microservices are not a free-for-all; they require discipline in conventions to avoid chaos.

- **Myth: "If one microservice is slow or broken, the system will automatically isolate it and never affect the user." Reality:** This is partly true – the architecture isolates failures, but it's not automatic without thoughtful design. If a service is slow and other services are calling it synchronously, those calls will be slow too unless you implement timeouts or fallbacks. A failure in one service can still cascade if, for

example, other services hang waiting for it. Proper use of circuit breakers, bulkheads, and asynchronous messaging where appropriate is needed to truly contain faults. Basically, microservices give you the *opportunity* to isolate faults, but you must design for resilience. Otherwise, you might just create a situation where a failure in Service A triggers failures in all services that depend on it – causing a **domino effect** (often called a cascading failure). So, the reality is you have to be very deliberate to get the fault tolerance benefits.

In summary, microservices are powerful but not panacea. They come with *tradeoffs* and require engineering maturity to implement successfully. The decision to use microservices should be justified by specific context and needs. With those myths dispelled, you can evaluate microservices more objectively. If you still feel it's the right direction (as WebkinsTech did to overcome their monolithic pains), the next step is to assess **readiness** and plan the journey carefully.

Are You Ready? Microservices Readiness Checklist

Not every organization or application is ready for microservices. It's important to assess both technical and organizational readiness before diving in. Below is a practical **checklist** of factors to consider. If you can tick most of these boxes (or are consciously addressing them), you're more likely to succeed with microservices. If not, you might decide to hold off or prepare more before breaking things apart.

- ✓ **Monolith Pain Points Identified:** Do you have clear pain points with your current monolithic (or tightly-coupled) system? (e.g., very slow deployment cycles, inability to scale certain functions, teams tripping over each other in the codebase). Microservices should address specific problems; be sure those problems are real and significant.

- ✓ **Clear Domain Boundaries:** Have you identified logical boundaries in your application that could become service boundaries? It's crucial to have a decent understanding of your domain model (perhaps via Domain-Driven Design) so you can split along natural seams (for example, "Orders" separate from "Catalog" separate from "Users"). If your domain model is a jumble, splitting may just create tightly-coupled services.

- ✓ **Sufficient Team Size and Skills:** Microservices make most sense when you have (or plan to have) multiple teams that can own different services. If you have a very small dev team, managing many services can be overwhelming. Also, does your team have (or can learn) the skills in distributed systems, DevOps, containers, etc.?

Upskilling and possibly hiring may be necessary.

- ✓ **DevOps Culture & Automation:** Are you practicing or willing to adopt DevOps principles? Automation of testing and deployment is a must. If currently deployments are manual or infrequent, you'll need to invest in CI/CD pipelines. Also, the team needs a mindset of taking ownership from code to operations (you build it, you run it). Without this culture, microservices could flounder.

- ✓ **Infrastructure/Tooling in Place or Planned:** Do you have the foundational infrastructure for microservices, or a plan to get it? This includes containerization (Docker), orchestration (Kubernetes or similar, or using cloud services), service discovery, monitoring systems, etc. You might use cloud platforms that provide some of these out-of-the-box, or you might build up your own. But going in, you should have a blueprint of the tooling stack for your microservices environment.

- ✓ **Monitoring & Logging Strategy:** Are you prepared with a strategy for centralized logging, monitoring, and tracing? You should have or choose tools (like ELK/EFK stack for logs, Prometheus/Grafana for metrics, OpenTracing/Jaeger for traces) and ensure you can deploy them. Defining SLAs/SLOs for services might also be part of readiness (e.g., "this service should respond within 200ms 99% of the time").

- ✓ **Organizational Buy-In:** Do the stakeholders and leadership understand what microservices entail and support the move? This is important because microservices might require investment in infrastructure and temporarily slower feature development as you refactor. Everyone should be on board with the long-term benefits and shortterm costs. Additionally, product management should understand that deployments will be more frequent and perhaps more granular.

- ✓ **Security and Governance Considerations:** Splitting into services means calls that were in-memory are now over networks – possibly needing encryption (TLS), authentication, and authorization for inter-service communication. Are you ready to manage service identities, API keys or tokens, and secure communication channels? Also, governance in terms of ensuring consistency (like how to do error handling, or common libraries) should be thought through. You don't want completely divergent practices that cause chaos later.

- ✓ **Testing Strategy:** Ensure you have a robust automated testing approach. Unit tests for each service, plus integration testing harnesses. Consider how you will test services in isolation (e.g., using mocks or stubs for their dependencies) and how you will test the system as a whole (maybe end-to-end tests in a staging environment). If your current testing is ad-hoc or mostly manual, that's a gap to fill as part of readiness.

- ✓ **Plan for Data Management:** Think about how you will split the data. If you have one big database now, consider how to migrate to each service having its own store. You might plan to give each service its own schema in the same database initially and later split databases, or do it in one go. Also plan for handling **migrations** and **duplication** of data if needed (some data might be duplicated across services for efficiency, which is fine if managed). Essentially, have a data architecture plan alongside your service architecture.

- ✓ **Tolerance for Initial Complexity Introduction:** Understand that things might get a bit more complex before they get simpler. The team should be mentally prepared for a learning curve and some bumps as you implement microservices. If you're in a phase where you absolutely cannot afford any slowdown or risk (say you have looming deadlines or unstable finances), it might be better to stabilize those before venturing a major architectural change. The organization should have the patience and commitment to see the transition through.

If you go through this checklist and find many unchecked boxes, you might decide to postpone microservices or address the gaps first (e.g., implement CI/CD and improve modularity within the monolith as stepping stones). If most boxes are checked, you're likely ready to embark on a microservices journey.

For WebkinsTech, the checklist looked promising: they had well-defined domains, their team was growing and already adopting DevOps culture, and leadership was fully behind the architectural modernization. Thus, they green-lit the microservices migration. The next step is planning that journey carefully to mitigate risks. We'll now provide a **readiness guide** for how to actually approach implementing microservices, which is useful once you've decided you're ready.

Readiness Guide: Planning the Microservices Journey

Breaking apart a monolith into microservices is a significant project that requires strategy. You don't want to end up in "microservices hell" (a state where you have a broken system more painful than the monolith was). Below is a guide that outlines how to plan, execute, and roll out microservices in a prudent way, gleaned from WebkinsTech's approach and industry best practices.

Start Small – Pick a Pilot Service: Rather than splitting everything at once (the "big bang" approach, which is risky), identify one part of the system to extract first as a proof of concept. A good candidate is something that has clear boundaries and interacts with the rest of the system through a fairly narrow interface. For WebkinsTech, the **Product Catalog** was chosen as the pilot microservice. It had a well-defined function (serve product data) and could be separated with minimal impact on other parts (just replacing direct DB calls with API calls). By carving out one service, you can build out the pipeline and tooling for microservices and solve unforeseen issues on a small scale. It's like dipping your toes in the water before jumping. Ensure the first service isn't your most critical (not payments or orders if you can avoid it), so that any hiccups have limited blast radius.

Use the Strangler Pattern for Gradual Extraction: The "Strangler Fig" pattern (named after a type of vine) is a strategy to slowly refactor a monolith by extracting pieces out to services over time, until the monolith is "strangled" (either much smaller or completely gone). In practice, this might mean: you build a new microservice, then reroute appropriate calls from the monolith to that service. For example, deploy the new Product Service and have the monolith call it (or have the gateway call it) for any product-related requests, rather than using the old in-process module. Over time, more and more functionality moves out. You might leave some parts of the monolith running until the very end (possibly indefinitely for low-risk pieces). This incremental approach means you're never doing a full system replacement in one shot. WebkinsTech did exactly this – they gradually moved features out and decommissioned those parts in the monolith as each service took over.

Parallel Runs and Feature Flags: When extracting a new service, it can be wise to run it in parallel with the old monolith functionality for a bit, to verify it produces the same results.

Using feature flags or config switches, you can route a small percentage of traffic to the new service and compare outcomes. For instance, initially WebkinsTech's Product Service ran in shadow mode – it received read requests alongside the monolith's product module, and responses were compared behind the scenes to ensure the new service was correct. Once confident, they switched 100% of read requests to the Product Service. This approach mitigates risk by validating in production-like conditions.

Invest in Automation and Tooling Early: As you plan the migration, set up the infrastructure you'll need. Containerize the monolith itself early on (if it isn't already) – that will get the team used to Docker and orchestrators, and it provides consistency (the monolith and microservices then deploy in similar ways). Establish your CI/CD pipeline for microservices with the very first service. It's worth taking a little extra time to create templates or scripts for new services (for example, a standardized way to create a new service repository with all boilerplate ready – CI config, Dockerfile, logging setup, etc.). This way, when you proceed to service #2, #3, and #10, you're not reinventing the wheel each time. WebkinsTech created an internal "Microservice Starter Kit" – a repository template that had a basic service structure and common libraries for things like auth and metrics already included. This saved time and enforced consistency.

Plan Data Migration and Synchronization: One of the toughest parts is splitting the database. You'll need to decide how to migrate data to each service's database without downtime or data loss. Strategies include:

- **Table-by-table migration:** e.g., the Product Service gets its own DB and you migrate the product tables over to it. The monolith then either stops using those tables (using the service instead) or, during transition, the service might update both its DB and the monolith DB to keep them in sync until cutover.

- **Change Data Capture (CDC):** using CDC tools to feed updates from the monolith's DB to the new service's DB (or vice versa) in real-time until you can fully switch.

- **Dual-write or Event sourcing:** temporarily have both monolith and service write changes to each other's domain. Dual-writes can be tricky (risk of inconsistency if not done carefully), but events can help – e.g., Order Service emits "OrderCreated" event which Product Service listens to update inventory, and Product Service emits "ProductUpdated" events which Order Service might listen to for price changes, etc.

- **Database view or Gateway**: Sometimes, as an intermediate step, the monolith might still hold the primary DB but the service is created and the calls go through a gateway that either queries the service or falls back to monolith, gradually shifting where truth lives.

Each case is different. The key is to not lose data or let data drift. Often, you start with a read-only service (for reads it's fine to have a copy of data) and then move write responsibility once stable. For WebkinsTech, product data was mostly readonly for the app (the product team managed it offline), so copying product info to the new service's DB was straightforward and the monolith was then changed to fetch from the service. For more transactional data like orders, they saved that for later when they had more experience, and planned a careful migration (pausing order creation briefly during a low-traffic window to migrate order records to the Order Service DB, etc.).

Implement Robust Inter-Service Communication: Early in the design, choose how services will communicate. Likely a mix of synchronous REST/gRPC for request-response flows and asynchronous messaging for events. Set up a message broker if you plan to use one. Standardize on how you structure API calls and events (e.g., JSON payloads and an event schema registry). WebkinsTech decided on REST+JSON for core APIs and used a lightweight event bus (cloud messaging service) for publish/subscribe of events like “OrderPlaced”. By having this in place from the first service, they created a nucleus of an event-driven architecture that others could plug into. Decide on timeouts and retries globally to avoid services waiting indefinitely on each other. Basically, establish a **contract** for service interactions.

Security and DevSecOps: When your services start talking over a network, enforce security from the get-go. Possibly use mTLS (mutual TLS) between services or at least ensure connections happen within a secure network environment. If using an API gateway, that’s a good choke point for authentication/authorization for external calls. But also consider service-to-service auth (like using JWT tokens or an internal identity system) so that, for example, only the Order Service can post to the Inventory Service’s stock update endpoint. This prevents a compromised service from easily affecting others. Also, build security scanning into the pipeline (container image scans, dependency vulnerability checks) because with many services, it’s easy for one to lag behind on patches.

Monitor and Adjust: As you migrate one service, monitor how it affects the system. Did latency improve or worsen? Is the team able to deploy more frequently now? Are there any new pain points? Use these lessons for the next service. Also monitor team workload – operating microservices might initially seem more burdensome; ensure no one team is overwhelmed. Possibly introduce the role of a Site Reliability Engineer (SRE) or a “platform team” to assist with common concerns so service dev teams can focus on features.

Incremental Rollout to Users: When transitioning functionalities to microservices, do it gradually for users too. This relates to the strangler pattern – you might route a small percentage of live traffic through the new service initially. If you have a canary release system or A/B test framework, leverage it. WebkinsTech did a beta program where some power-users were intentionally served by the new microservices stack and their feedback and behavior were closely watched, while others were still on the legacy path. This helped catch issues early.

Keep the Monolith Stable (if it's still partly in use): During the migration period, you'll have a hybrid architecture – part monolith, part microservices. It's important to keep the monolith stable and not churn it too much with new features (unless absolutely necessary). Ideally, treat it as "legacy" – only vital fixes – while new features go into services. This freeze prevents thrash in something you're actively pulling apart. However, sometimes you need to update the monolith to call the new services, or to remove code that's been offloaded. Do those carefully and remove dead code from the monolith to reduce confusion. Over time, the monolith shrinks or at least becomes less critical, and you might eventually retire it (or leave a stub for some rarely touched functionality).

Best Practices for a Smooth Microservices Migration

One Domain at a Time: Resist the urge to split everything at once. Migrate domain by domain, and declare victory incrementally. This also provides morale boosts as you see parts of the system running on microservices successfully.

Automate Testing in CI: As you add each service, add its tests to your CI pipeline and also have an integration test that spins up that service with a stub of the monolith or necessary collaborators. Ensuring each piece works in isolation and together will save headaches.

Documentation and Catalog: Maintain a living document or diagram of services, what their responsibilities are, and how they interact. After a few services, it's easy for knowledge to fragment. WebkinsTech created an internal wiki page listing each microservice, its API (with link to API docs), who owns it, and its current version.

This helped everyone keep track of the new ecosystem.

Refine and Refactor: After a couple of services are live, revisit your initial service boundaries if needed. You might find that two services are chatting too much – perhaps they should be merged, or a new service should be introduced to reduce the chatter. Microservices architecture itself can evolve. It's okay to consolidate services if you split too finely – the end goal isn't to have the maximum number of services, but the right decomposition.

Consider Team Structure: Align service ownership clearly. If one team owns multiple services, that's fine, but ensure they have capacity. Avoid one service being worked on by multiple teams frequently, as that reintroduces coordination issues. If that happens often, maybe that service should be split along team lines. Basically, try to follow Conway's Law in a positive way: structure teams such that their communication patterns result in the desired service boundaries.

Communication is Key: Hold periodic check-ins (architecture reviews, post-mortems of incidents) to share lessons learned across teams. In microservices, something one team learns (like how to improve database connection pooling under load) could benefit all. Fostering a culture of sharing will prevent each team from solving the same problems in parallel silos.

By following a careful plan, WebkinsTech transitioned to a microservices architecture over several quarters without major outages or missed business commitments. They started seeing benefits after the first few services: deployments of those services were decoupled from the rest, giving teams a taste of the new agility, and system performance improved in the areas that were split (for instance, product pages loaded faster under load due to the dedicated, optimized Product Service).

Finally, after completing the migration of core services, WebkinsTech's platform was composed of microservices and a much smaller core (what remained of the monolith was eventually just an admin interface and some cron jobs, which they left as-is). They could then fully focus on **orchestrating these services and adding intelligent automation** – which is the perfect segue into our next chapter.

Conclusion

Microservices are not just an architecture choice; they represent a shift in how we think about building software systems. In this chapter, we introduced microservices by examining why architectures evolve away from monoliths and how intelligent, well-designed services can address the pain points of growth and change. Through WebkinsTech's story, we saw a practical example of an enterprise grappling with a monolithic "freight train" and gradually transforming it into a fleet of nimble services – each like a specialized vehicle in a wellcoordinated convoy.

We covered the **principles of microservices design** – emphasizing domain separation, autonomy, loose coupling, and independent deployability. These principles guided us in structuring a microservices system that mirrors business capabilities and enables teams to work efficiently. We explored how **business expansion** increases complexity and how microservices, when done right, help manage that complexity by breaking it into digestible pieces. The **architecture overview** gave us a glimpse of the moving parts in a microservices ecosystem, from API Gateways to service registries and from synchronous REST calls to asynchronous events. By delving into e-commerce and fintech examples, we made the abstract concrete, showing real-world slices of microservice architectures in action.

The **benefits** of microservices – agility, scalability, fault isolation, polyglot flexibility, and team autonomy – are indeed compelling. WebkinsTech began to unlock these benefits as they progressed: deployments that used to take weeks were happening in days or hours, and system outages became far less frequent due to isolated failures. Teams felt a greater sense of ownership and were able to innovate more freely within their service boundaries. These improvements translate directly into business value: faster feature releases, more uptime for customers, and the ability to tackle new product ideas without being shackled by a fragile core.

Yet, we also confronted the **challenges** and realities head-on. Microservices bring complexity in distribution, require strong DevOps maturity, and demand new solutions for data consistency and monitoring. By understanding these tradeoffs, WebkinsTech was not blindsided during their transition – they anticipated the need for better tooling and made necessary investments (like setting up comprehensive monitoring early, and training the team on distributed system concepts). We dispelled myths to ensure microservices are viewed not as a cure-all, but as a tactical approach that solves certain problems very well while introducing others that must be managed.

The readiness checklist and guide provided a roadmap for organizations considering this path. It's clear that microservices adoption is as much a journey as a destination – it needs careful planning, incremental steps, and a willingness to adapt and learn. WebkinsTech's journey reaffirmed that one of the keys to success is **evolution, not revolution**: by strangling the monolith gradually and continuously delivering value (even during the transformation), they kept the business running and stakeholders supportive.

At the end of Chapter 2, you should have a solid understanding of what microservices entail and how to approach adopting them. You've seen why enterprises like WebkinsTech choose this route and how they mitigate the risks. The value of microservices lies in enabling **intelligent, agile, and resilient systems** – qualities that modern enterprises need in order to innovate and respond in real time to market changes. Microservices architecture lays the foundation for something even more powerful: orchestrating complex workflows and injecting intelligence into the system via automation and AI.

In **Chapter 3**, we will build upon this foundation. With microservices as independent building blocks, we can explore how to orchestrate them into end-to-end business processes, how to coordinate their interactions (or let them choreograph themselves via events), and how to introduce **intelligent automation** and **agentic AI** capabilities. We'll see how an "Intelligent Microservices" approach leverages AI agents and workflow engines to manage the flow between services, handle failures gracefully, and even make autonomous decisions. WebkinsTech, having modernized its architecture, is now poised to inject advanced automation – such as an AI-driven recommendation engine service, or an orchestration agent that optimizes order fulfillment across inventory locations.

The journey doesn't end with splitting the monolith; it opens new horizons. With a microservices architecture in place, WebkinsTech can now plug in machine learning services, implement smart orchestration of business processes, and achieve levels of automation that were impossible in their old system. As we conclude this chapter, remember that microservices are a means to an end: they enable scalability in both software and organization. They set the stage for continuous evolution. And in the spirit of continuous evolution, we move forward to the next chapter, where those microservices will not just work – they'll *work together* in orchestrated harmony, augmented by AI and intelligent workflows to drive the business forward.

End of Chapter 2. In Chapter 3, **"Orchestration and Intelligent Automation,"** we will dive into how to coordinate microservices and empower them with AI-driven agents and workflows, turning our flexible architecture into a truly intelligent, autonomous system. Get ready to see how orchestration ties our services together and how an agentic-AI approach can elevate WebkinsTech's microservices to the next level of smart automation.

Chapter 3 : Think Workflows, Automation, and Need for Intelligence

Introduction: From Automation to Intelligence

WebkinsTech's vendor onboarding process was once the envy of its competitors. A new vendor would fill out a web form, and behind the scenes a sequence of microservices would spring into action: creating records, sending emails, and provisioning accounts. It was a workflow anyone in IT could be proud of – automated from end to end. But as WebkinsTech grew, cracks started to show. One afternoon, a single unexpected data format from a new vendor brought the entire onboarding workflow to a halt. No new vendors could be added until an engineer updated a validation rule. This kind of brittleness was becoming all too common in their rule-based system. The workflow was automated, **yes**, but not intelligent.

In conversations around the office, the WebkinsTech team realized a hard truth: microservices had enabled them to automate and orchestrate complex workflows, but these workflows couldn't *adapt* or learn. Each microservice did exactly what it was told – nothing more, nothing less. If a situation fell outside the predefined rules, the system couldn't handle it. One team member joked, "Our workflow is like a robot that malfunctions when you ask it a question it wasn't programmed for." They didn't just need automation; they needed **intelligence** within their workflows.

This chapter explores that journey – from workflows that are merely automated to those that are truly intelligent. We'll see how microservices laid the groundwork by making complex workflows possible, and then we'll delve into the limitations of these traditional, static automations. Through WebkinsTech's story and other real-world examples, we'll illustrate why the next leap in software architecture is to infuse workflows with context-awareness, adaptability, and a spark of autonomy. By the end, you'll understand how intelligent workflows can transform brittle automation into resilient, **agentic** systems that not only execute tasks, but also make wise decisions. This will set the stage for Chapter 4, where we dive into the world of agentic AI and how it builds upon these intelligent workflows.

Evolution of Enterprise Workflows

Not so long ago, enterprise workflows were often nothing more than nightly batch jobs in a monolithic system. Imagine a giant all-in-one application processing the day's orders at midnight, or a mainframe program running an end-of-day payroll. These early workflows were **sequential and rigid**, but at the time they solved the need for automation within a single system. However, as businesses grew more complex, the limitations of monoliths became apparent. Scaling or changing one part meant touching the entire system. This paved the way for an architectural evolution: breaking big systems into **microservices**.

Microservices enabled enterprises to break down business processes into smaller, focused services that could be developed and scaled independently. Instead of one colossal program handling an entire workflow, you might have a dozen microservices each responsible for a slice of the process – one for orders, one for payments, one for inventory, one for shipping, and so on. These services communicate with each other to complete the workflow. The result? Workflows became more modular and easier to maintain. For example, in a modern e-commerce pipeline, placing an order triggers a series of microservice interactions:

- The **Order Service** validates and creates the order.
- A **Payment Service** charges the customer's card.
- An **Inventory Service** updates stock levels.
- A **Shipping Service** arranges delivery logistics.
- An **Email Service** sends the order confirmation.

All these steps form a workflow pipeline that spans multiple services. One way to coordinate such a pipeline is through **orchestration**: a central service (or an orchestration engine) tells each microservice what to do and when. Alternatively, some architectures use **choreography**, where each service emits events and reacts to events from others, without a central controller. Let's unpack these two modes because they are fundamental to how workflows are implemented in microservices:

- **Orchestration:** A central orchestrator service coordinates the workflow, invoking each microservice in turn and aggregating results. It's a bit like a conductor in a symphony – the orchestrator signals when each part should play. For instance, an orchestrator might call the Payment Service, wait for a response, then call Inventory, and so on. This approach gives a clear, centrally managed sequence of steps.

- **Choreography:** There is no central controller. Instead, services communicate by emitting and listening to events. This is analogous to dancers in a ballet who each know their steps and respond to the music and movements around them. For example, the

Order Service could publish an "Order Placed" event. The Payment, Inventory, and

Email services, upon hearing that event, would each independently perform their tasks. There's no single service telling them "do this now"; they autonomously react based on predefined rules.

To visualize the difference, consider the following illustration. In the **orchestration** diagram

(top), a central "Orchestrator" service (in blue) invokes Service A, B, C, and D in a controlled sequence, waiting for replies. In the **choreography** diagram (bottom), there is no orchestrator; Service A, B, C, and D all emit and respond to events (arrows indicate the flow of messages) without a single point of control.

Orchestration vs Choreography of Microservices. In orchestration (top), a central orchestrator service coordinates calls to each microservice (A, B, C, D) and awaits their replies, much like a conductor cues each musician. In choreography (bottom), services communicate through events (e.g., A publishes an event that B, C, D listen to), akin to dancers following a common rhythm without a single director. Orchestration offers centralized control and simpler error handling, while choreography yields more autonomy and flexibility at the cost of more complex debugging.

Both approaches have their strengths. Orchestration can be easier to understand since one component has the whole workflow logic, and it can enforce order and handle errors centrally. In fact, companies like Netflix have built orchestration engines (e.g., Netflix Conductor) to manage their complex microservice workflows. Choreography, on the other hand, leads to loosely coupled systems; services only know about the events, not the implementation of other services. This makes adding a new service simpler – just subscribe it to the relevant events, no other service needs to change. It's like adding another dancer who already knows the dance, rather than updating the conductor's script for a new musician. As one engineer described it, in orchestration the central service acts as the conductor telling all parts what to do, whereas in choreography each service is a dancer listening to the music (the event stream) and performing its part independently.

Real-world enterprise workflows often use a mix of both patterns. For example,

WebkinsTech's vendor onboarding initially used orchestration: a Vendor Onboarding service called each step (document verification, compliance check, account setup) in sequence. As the system grew, they introduced event-driven choreography for parts of the process – for instance, whenever a new vendor was created (event), other services like reporting and analytics would automatically update without the orchestrator's involvement. The key takeaway is that microservices gave enterprises the building blocks (small services) and the patterns (orchestration/choreography) to automate complex workflows more robustly than ever before.

However, even with this evolution from monoliths to microservices, something was still missing. The workflows could be **automated** and **distributed**, but they weren't yet **smart**. In the next section, we'll examine traditional automation's pros and cons, which will highlight why something more intelligent is needed on top of these microservice workflows.

Traditional Automation: Pros and Limits

When we talk about "traditional automation," we mean systems that execute predefined steps based on static rules. These are the workflows that do exactly what we tell them, nothing more, nothing less. Traditional automation has been a cornerstone of IT and business process management for decades, and it comes in many forms:

- **Scripting and Batch Jobs:** Simple automation can be a script that, say, runs every night to generate reports or transfer files. Cron jobs, PowerShell scripts, or Python scripts fall here. They follow a set sequence of commands.

- **Business Process Management (BPM) and Workflow Engines:** Tools like those following BPMN (Business Process Model and Notation) allow modeling workflows as diagrams – e.g., "First do X, then if Y go to Z, otherwise go to W." Workflow engines (like Camunda, Pega, or IBM BPM) can execute these models, orchestrating multiple systems according to a flowchart of rules.

- **Robotic Process Automation (RPA):** RPA tools (e.g., UiPath, Automation Anywhere, Blue Prism) mimic human actions on user interfaces. They are often used to automate repetitive tasks like data entry between systems, using predefined rules to, for example, click here, copy data there, etc. They are rule-based in the sense that an RPA bot will only do what it's been scripted to do on the screen.

The **pros** of traditional automation are evident: it can execute tasks faster than humans, 24/7, with zero fatigue. It enforces consistency (every transaction is processed the same way), and it frees humans from drudgery to focus on more complex tasks. Traditional automation shines in well-understood, repeatable processes. If you have a high-volume task that rarely changes, like converting file formats or checking form entries for completeness, a script or workflow engine can handle it reliably. In fact, WebkinsTech's initial vendor onboarding automation saved their operations team countless hours by automatically creating vendor accounts and sending welcome emails instead of doing those manually.

However, the **limits** of these systems become clear when the environment is not perfectly predictable. Traditional automated workflows are typically **brittle** in the face of change or exceptions. Why? Because they are only as smart as the rules we gave them upfront. If something even slightly outside those expectations happens, the automation often doesn't know how to cope.

When Static Rules Break: A large telecom company deployed an automated workflow to process phone number porting requests. The process was scripted step-by-step: validate the request, check the number format, update databases, notify the customer. It worked fine for thousands of requests – until one day an unusually formatted number (from an international SIM) slipped in. The script hadn't seen that format before, so it threw an error and halted. Because the workflow was linear, that single failure blocked subsequent requests. Customers experienced hours of delay until an engineer modified the rule to accept the new format. This incident became an internal case study of how a minor unexpected input crashed a "foolproof" automated system.

Such failures illustrate how **rigid** traditional automation can be. They handle the "known knowns" well, but struggle with the "unknowns." The telecom example shows an explicit rule mismatch, but sometimes the brittleness is more insidious. Automated workflows often don't realize they're producing bad outcomes because they weren't programmed to monitor the **outcomes**, only to follow the steps. As one engineer quipped, "it will happily do the wrong thing very consistently if you tell it to."

Rigidity and Fragility: Traditional workflows and RPA bots can't improvise. They follow their script even if the situation quietly changed. This rigidity means if a step in the process fails or an input is out of the ordinary, the automation may crash or, worse, silently produce incorrect results. Small changes – a field added in a form, a slight process variation – can require reprogramming the workflow. Static workflows can't learn or adapt, so any scenario not anticipated by their creators becomes a potential point of failure. The result is fragile automation that requires continuous maintenance to keep up with business changes.

A vivid concept that has emerged in industry talks is the idea of "**watermelon metrics**" for legacy automation: green on the outside, red on the inside. Everything looks okay in the dashboard (tasks are executed, boxes checked – green status everywhere), but deep down the process is failing in ways the metrics aren't catching (problems fester – red). For example, an automated customer support workflow might mark all tickets as "processed" (green), but customers could be unhappy because the responses didn't actually solve their issues (a red reality). The static workflow wouldn't catch that because it's not built to understand customer satisfaction; it only knows it followed the script. This highlights a core limitation of rule-based automation: **lack of context and insight**. It can execute predefined steps flawlessly without understanding the broader goal or outcome.

In summary, traditional automation using microservices or scripts made things faster and more consistent, but at the cost of flexibility. As long as conditions are ideal, the machine hums along. When something unexpected happens – a new type of input, a subtle change in business policy, an external event – these systems often can't cope. The process either stops, or someone has to manually intervene and then update the code or rules to handle the new scenario going forward.

This is why organizations like WebkinsTech started yearning for something more. They had tasted the efficiency of automation; now they craved the adaptability of intelligence. In the next section, we'll explore why adding intelligence is the natural next step, and how it can address these very limitations.

Why Intelligence Is the Next Leap

Why can't we just add more rules to cover the exceptions? That's a fair question and indeed the first approach many teams try. If the workflow breaks on a new vendor type, add a rule for that type. If an RPA bot doesn't recognize a new screen, update the script for that scenario. But this approach has diminishing returns. The real world can throw infinite variations at you – eventually you can't anticipate them all. Enter intelligent workflows.

Intelligent workflows are processes that incorporate AI or machine learning to make context-aware decisions in real time. Instead of following a static if-this-then-that map, an intelligent workflow can evaluate what's happening and choose a path or action based on **data**, not just predefined logic. It's the difference between an assembly line robot and a skilled human operator:

- A static assembly line robot will tighten exactly five bolts on every product that comes by, at the exact same spots, every time. If a product comes by that doesn't even have bolts (say the design changed), the robot will still go through the motions, likely screwing into thin air or halting in confusion.

- A skilled human operator, on the other hand, can adapt on the fly. If they see a product missing a bolt hole, they might flag it or drill a new hole, or fetch a different tool. They can handle variation because they understand the **goal** (ensure the product is assembled correctly) and can adjust their actions to achieve that goal under new conditions.

An intelligent workflow is like that skilled human operator. It maintains the speed and efficiency of automation (no human needed for routine work), but it can adapt to new information and conditions in a way traditional automation cannot. Let's break down some characteristics that make a workflow "intelligent":

- **Context-Awareness:** The system considers context before acting. For example, an intelligent order fulfillment workflow might route high-value orders to a priority shipping service, but only if it's before 5 PM and the customer is VIP. It's aware of time, customer status, and perhaps warehouse load – and adapts accordingly.

- **Data-Driven Decision Making:** Instead of only human-defined rules, it leverages patterns learned from data. Imagine an inventory replenishment workflow that doesn't have a fixed reorder point for all products. Instead, it analyzes sales trends and market data to decide when to reorder each product. If the data shows a spike in demand for item X, the workflow orders more of X earlier than usual. If demand for item Y is dropping, it postpones or reduces the reorder. This is a dynamic decision that traditional rule systems struggle with, but an AI model can handle by forecasting.

- **Real-Time Decisions:** Intelligent workflows often operate in or near real-time, thanks to event streams and fast analytics. They can detect anomalies or changes as they happen. For instance, an intelligent fraud detection workflow in a banking system might flag a transaction as suspicious not by a hard-coded rule, but by an AI model that considers dozens of factors (location, amount, user's spending history) in milliseconds and decides it doesn't fit the normal pattern. The workflow could then automatically halt that transaction and ask for additional verification, preventing fraud in the moment.

- **Adaptability and Learning:** Perhaps the most exciting aspect – these workflows can learn from new cases. If our intelligent workflow encounters a scenario it's not confident about, it could ask for human input (e.g., a human approves a weird insurance claim this one time). But importantly, the system can take that outcome and incorporate it into its knowledge. Next time, it might handle a similar claim automatically. Over time, the need for human exceptions goes down as the system learns. The workflow essentially **improves itself** through feedback.

To put it simply, adding intelligence means your workflow isn't just a fixed assembly line; it becomes a living process that can make judgements. It transforms automation from doing things right (efficiency) to also doing the right things (effectiveness in achieving the goal under various conditions).

Consider anomaly detection, dynamic routing, and self-healing – three capabilities often cited as low-hanging fruits for intelligent workflows:

- **Anomaly Detection:** An intelligent monitoring workflow in IT operations might watch metrics and log streams, using machine learning to detect when something is off (maybe CPU usage pattern doesn't match any known normal scenario). Instead of waiting for a rule threshold (CPU > 90%), the AI might catch a subtle combination of signals that indicate a coming issue. It can then alert or even trigger a mitigating action. Traditional automation would never notice until a hard limit is crossed (if it was set at all).

- **Dynamic Routing:** In customer support or logistics, routing is key. A static system might always assign tickets in a round-robin manner, or always send delivery trucks on pre-planned routes. An intelligent system can route based on current conditions. For support tickets, that might mean analyzing the email content to detect an angry tone or urgent issue and skipping it ahead in the queue to a specialist agent. In delivery, as we'll detail later, AI can reroute vehicles on the fly if it learns of a traffic jam or a storm (no human dispatcher needed for routine changes).

- **Self-Healing:** Intelligent workflows can achieve a form of "self-healing" in systems. If a service fails, an intelligent workflow might automatically spin up a new instance or switch to a backup. For example, imagine a microservice in a payment pipeline crashes. A traditional orchestrator will mark the workflow as failed and alert someone. An intelligent orchestrator could detect the failure, realize that historically simply retrying on a new server fixes it 90% of the time, and do just that. It might also log the incident for developers, but it keeps the workflow moving with minimal disruption.

Start Small: Add Intelligence Gradually – You don't have to rebuild your entire workflow with AI from day one. A practical first step is to add intelligent components to handle exceptions or edge cases. For example, keep your normal rule-based process for 95% of cases, but integrate an AI module to handle the "weird" 5%. That AI could be an anomaly detector that watches for unusual inputs or outcomes. When it spots something odd, instead of the process failing or producing bad output, the AI flags it for human review or routes it through a different path. By doing this, you create a safety net that *learns* from exceptions. Over time, the AI might learn enough to start handling some of those exceptions automatically. This approach improves resilience without risking the reliable core of your existing automation.

To illustrate why this leap to intelligence is so powerful, let's revisit WebkinsTech's vendor onboarding. With the old rule-based workflow, any deviation (like that unexpected data format) caused trouble. If WebkinsTech adds an intelligent layer, say an AI that scans new vendor data for anomalies or missing info, the workflow could adapt. If a vendor's data is incomplete, the AI might automatically reach out to the vendor for clarification or fill in the blanks from trusted data sources. If a vendor is from a region that usually requires additional compliance checks, the AI could flag it and trigger an extra step dynamically, rather than relying on a static list of regions.

In short, intelligence in workflows is the bridge between **automation** (fast and consistent execution) and **autonomy** (making sound decisions in varying contexts). It addresses the brittleness of static automation by introducing flexibility and learning. Instead of breaking on the unexpected, an intelligent workflow can often handle the unexpected – or at least fail gracefully and learn from it.

Next, we'll delve deeper into this idea of autonomy: what differentiates mere automation from a truly autonomous system, and how that sets the stage for agentic AI.

From Automation to Autonomy

It's one thing for a system to execute tasks automatically; it's another for it to decide **which** tasks to execute and **when** – on its own. This is the difference between **automation** and **autonomy**. We've been inching toward autonomy by adding intelligence to workflows, but let's clarify the concepts:

- **Automation** means tasks are done without human intervention, but the steps are predefined by humans. The system doesn't initiate anything by itself; it simply follows a script or responds to triggers in a predetermined way. It's like a player piano: it plays music on its own, but only the song that's encoded in the punched paper roll. You "set it and forget it," until you need to change the roll.

- **Autonomy** means the system has an agency – it can make decisions and take actions towards a goal without being explicitly told each step. An autonomous system can figure out a new sequence of actions to respond to a new situation, within its domain, aiming to achieve the goal it's been given. It's more like a jazz musician who, given a theme, can improvise a melody rather than following a fixed score.

In technical terms, autonomy in software often implies **AI-driven decision-making**. The system might use machine learning models, optimization algorithms, or even cognitive-like reasoning to choose actions. It monitors the environment, evaluates possible responses, and then acts in a way that it predicts will best meet its objectives.

Let's consider a couple of examples to make this concrete:

Loan Approval Systems: Many banks use automated workflows to process loan applications. A traditional automated system might automatically approve or reject loans up to a certain amount based on fixed criteria (credit score above X, income above Y -> approve, else -> escalate to human). That's automation. An **autonomous** loan system, however, could handle a broader range of decisions. It might analyze a customer's credit history with a machine learning model, cross-reference with macroeconomic data, and decide not just binary approve/reject, but also decide on custom loan terms (interest rate, amount, payment plan) that best fit the customer's risk profile. It might even decide to request additional documents if something looks unusual – decisions a static system wasn't explicitly programmed to do. It's operating within bounds set by the bank (it won't do something crazy like approve a loan the bank can't afford), but it's making nuanced decisions on its own.

Dynamic Pricing Engines: Think about ride-sharing apps or airlines ticket pricing. An automated pricing system might follow a rule like "if demand > supply, increase price by 20%." That's simple automation responding to a trigger. A more autonomous pricing engine, possibly powered by reinforcement learning, might explore different pricing strategies at different times, learn from purchase patterns, and come up with dynamic price adjustments that weren't directly coded. It has a goal (maximize revenue or balance load), and it takes actions (changing prices) based on real-time data (current demand, competitor prices, even weather or local events) and learned experience. The system might decide at 5 PM to lower prices slightly to attract more early riders and avoid a bigger surge at 6 PM – a subtle strategy a human didn't explicitly predefine.

The leap to autonomy brings great power, but also challenges (which we will explore more in Chapter 4). An autonomous system is, by design, more unpredictable than a fixed one, because it can change its behavior as it learns or as conditions change. This is where concepts like **agentic AI** come into play – AI agents that act in an environment toward goals, sometimes collaborating or negotiating with other agents or humans. Agentic AI represents the cutting edge of autonomy, where systems can initiate actions on their own to fulfill highlevel objectives given by humans.

To ground this, it's worth noting what experts describe as agentic AI. One definition is: *Agentic AI refers to AI systems that can act autonomously – making decisions or taking action – to achieve a certain goal, marking a departure from static rule-based automation. These systems are goal-driven, self-adaptive, and capable of independent decision-making based on context and real-time data*. In essence, agentic AI is the realization of autonomy using AI techniques.

Let's compare traditional automation and intelligent autonomy side by side, to summarize:

Aspect	Traditional Automation	Intelligent Autonomy (Agentic Systems)
Control	Predefined sequence of steps (the system has no say in altering the sequence)	Goal-oriented; the system can choose different paths to reach the goal
Decision-Making	Based on fixed rules and logic coded by humans	Based on AI/ML models, can evaluate options and select actions
Aspect	**Traditional Automation**	**Intelligent Autonomy (Agentic Systems)**
Adaptability	Low – struggles with scenarios not explicitly handled	High – can adapt to new scenarios by generalizing from data or experience
Learning	None – performance remains the same unless humans update the code	Yes – continuously learns from new data/outcomes to improve decisions over time
Data Handling	Primarily structured, expected inputs; fails on unstructured or unexpected data	Can handle unstructured data (text, images) and find patterns invisibile to static rules

Example Use Case	Invoice processing with set approval rules (if amount < $1000, auto-approve)	Autonomous supply chain agent that reroutes shipments and adjusts inventory targets on the fly based on realtime conditions
Human Role	Define all rules upfront; intervene on failures	Set high-level goals/policies; monitor or guide when AI encounters novel extremes

As the table suggests, intelligent autonomy doesn't replace automation so much as extend it into areas of uncertainty. The autonomous system still *automates* the execution (no humans turning the cranks), but it also has a kind of **judgement**. It answers the question: "What should I do now?" whereas traditional automation assumes that question has been answered by a human in the form of a flowchart or code.

To avoid a misconception: Autonomous systems aren't necessarily unbounded or magical. They operate within **constraints and domains** set by their designers. For instance, our autonomous loan approval system might have a constraint that it cannot approve a loan above a certain limit without human sign-off. Autonomy doesn't mean an AI can do *anything*; it means within its allowed scope, it can figure things out itself.

A useful way to think of it is like an autopilot in a plane. Traditional automation is like cruise control in a car – it keeps speed constant but a human must still steer and brake. Autonomy is the full autopilot – it can maintain course, adjust altitude, avoid turbulence, maybe even land the plane, though a human pilot sets the destination and can take over if needed. Agentic AI is pushing toward the scenario where the plane could potentially plan its own route to the destination given general guidance.

To tie this back to microservices and workflows: enabling autonomy might mean having certain microservices or orchestrators that are powered by AI agents. For example, a microservice in charge of "promotion decisions" might not follow a static policy but instead use an AI model to decide which promotion to offer to a customer, and it keeps evolving that strategy. Or an orchestrator might not always call services A -> B -> C in order, but, if autonomous, could choose to skip B or do C->A in a particular case because it has learned that sometimes yields better outcomes (admittedly a complex scenario, but conceivable with agent guidance).

Before we move on, it's important to note that with great power (autonomy) comes great responsibility. Trust, ethics, and oversight become crucial. A fully autonomous workflow that approves loans or routes medical ambulances needs a lot of testing and likely human oversight at first. Many systems start as decision support (the AI suggests, human disposes) and only gradually move to full autonomy as confidence grows.

Now that we understand the distinction between simply automating tasks and creating autonomous intelligent agents, let's look at how different industries are already applying these ideas in intelligent workflows, and what benefits they're seeing.

Intelligent Workflow Use Cases Across Industries

Intelligent workflows aren't just theoretical; they're being applied across various industries to great effect. Let's tour a few sectors – healthcare, finance, retail, logistics, and IT operations – to see concrete examples of what "intelligence" in workflows looks like and the value it provides. Along the way, we'll also drop in on WebkinsTech's continuing journey in a different area of their business.

Healthcare: Smarter Patient Triage

In a busy hospital emergency department (ED), triage is a critical workflow – deciding which patients need the most urgent care. Traditionally, triage is done by nurses following protocols. It's a kind of workflow: take vitals, ask questions, assign a priority level. But it can be subjective and variable. Enter AI-driven triage assistance. Hospitals are now experimenting with intelligent triage systems that help nurses make more consistent, datadriven decisions. For example, Johns Hopkins researchers developed an AI tool that integrates with the patient's digital health record. When a nurse enters the patient's vitals and answers to questions, the AI instantly predicts the risk of severe outcomes and recommends a triage level, along with an explanation. This all happens in seconds, and the nurse can then factor that into their decision. The AI is essentially an extra brain in the workflow, having learned from thousands of past patients. It catches subtle patterns (maybe a combination of slightly elevated heart rate and certain answers that historically led to complications) that a busy human might miss. The result: more accurate triage and better patient flow. Those low-risk patients can be confidently sent to fast-track treatment and discharged sooner, while high-risk ones get flagged for immediate attention. In short, the workflow becomes adaptive to the individual patient's data, rather than a one-size-fits-all questionnaire.

Another emerging approach in healthcare is using AI chatbots for initial triage in telehealth. A patient starts by chatting with an AI, which asks questions, assesses urgency (perhaps detecting if symptoms sound like a heart attack vs a minor cold), and then routes the patient: "You should go to the ER" vs "We scheduled you a tele-appointment tomorrow." These are intelligent front-doors to care. They don't replace clinicians, but they ensure the workflow of patient intake and routing is smarter and more efficient.

Finance: Real-Time Fraud Detection and Prevention

Financial transactions – whether credit card swipes, online purchases, or bank transfers – are now often guarded by intelligent workflows that detect fraud in real time. In the old days, fraud detection might have been a batch process (flag transactions overnight) or based on very basic rules (block if amount > $10,000 from a new location). Today, machine learning models analyze transactions on the fly, scoring them for fraud risk in a fraction of a second, and the workflow adapts accordingly. If you swipe your credit card and something seems off to the AI (maybe the location, time of day, purchase pattern don't match your usual behavior), the system might pause the transaction or route it for verification. Perhaps you get a text: "Did you attempt this purchase? Reply YES or NO." That entire process is an intelligent workflow – it took an action (challenge the transaction) based on an AI prediction.

The impact is huge: AI-based workflows catch far more fraudulent activities with fewer false alarms than rule systems. For example, one e-commerce fraud detection case study by DataVisor reported their AI solution helped businesses detect over **30%** more fraudulent attempts with about **90%** accuracy and only 1.3% false positives. That means more stolen cards or scams caught (saving money and hassle) while legitimate customers are less likely to be wrongly flagged (saving frustration). The workflow here involves real-time data streaming to an AI, which is a microservice in itself (a fraud detection service), and then dynamic routing: transactions scored as risky might trigger additional steps like the text message verification, or sending an alert to a fraud analyst, whereas safe transactions go through normally.

Another example in finance is loan or insurance underwriting. Companies use AI to intelligently workflow the application process. A simple application might get automatically approved by an AI model's judgment, a risky one might get automatically declined, and borderline ones get routed to human underwriters with AI-provided insights ("This application is unusual because X, Y, Z"). This triaging and decision-making workflow, augmented by AI, makes processing faster and often fairer (if the AI is well-trained, it can reduce human bias by focusing on data).

Retail: Smart Inventory Replenishment and Personalization

Retail businesses live and die by their supply chain efficiency and customer satisfaction. Intelligent workflows have made big inroads here. One great example is inventory management. Large retailers like Walmart use AI-driven systems to manage inventory and restocking in an incredibly dynamic way. Instead of a store manager just reordering stock when shelves look low (the old manual workflow), modern systems predict demand for each product at each location and orchestrate the entire supply chain to meet that demand. Walmart, for instance, has publicly discussed how their AI-powered inventory management places items in distribution centers and stores strategically by leveraging historical sales data combined with predictive analytics. During the holiday season, their system might foresee a spike in toy demand in Region A and start routing extra shipments there weeks in advance, while throttling back shipments of say, winter coats to a region experiencing a warm spell unexpectedly. These decisions happen through a workflow that involves data collection from stores (sales, weather, local events), running predictions, and then automatically creating or adjusting orders to suppliers, scheduling trucks, etc. The entire chain from forecasting to replenishment can be an intelligent, largely autonomous workflow.

On the customer-facing side, retail uses intelligence in workflows for personalization. Think of an online shopping experience: the moment you place an order, there's an opportunity for an intelligent cross-sell workflow. A traditional system might have a static rule: "if buying phone, recommend case". An intelligent workflow, however, might analyze your browsing history and purchase patterns in real time and personalize an offer: "I see you bought a

DSLR camera, often customers also get an extra battery – here's a one-click add-on with 10% off." If you accept, it modifies the order workflow on the fly to bundle that item. If not, no harm done. This decision is made by an AI recommendation engine service plugged into the order workflow. The key is that the process isn't linear and fixed; it has branches and options that only trigger when data suggests they should, aiming to maximize either the customer's satisfaction or the sales outcome (ideally both).

Another retail example: dynamic pricing (which we touched on). Online retailers adjust prices dynamically with AI models; their pricing workflow might run every hour, pulling in competitor pricing data, inventory levels, and conversion rates, then updating prices automatically. Traditional workflows set prices manually or on a fixed schedule. The AI-driven one reacts to market conditions. If a competitor runs out of stock, the system might raise price a bit to maximize margin; if everyone else has a sale, it might algorithmically decide whether to match or beat it based on predicted elasticity.

Logistics: Adaptive Delivery Routing

In the world of logistics – shipping, delivery, supply chain – even small efficiencies can save millions. Intelligent workflows have found a home in routing and scheduling problems. Shipping companies and ride-sharing services alike use algorithms to optimize routes (that's already automation), but now many incorporate real-time AI to adjust those routes on the fly (that's the intelligence).

For instance, consider last-mile delivery (getting packages from a local depot to customers). A traditional route planning system might give a driver a route in the morning and that's it for the day. A modern AI-backed workflow will continue to adjust as conditions change. If suddenly a road closure or heavy traffic is reported, the system can re-calculate and push an updated route to the driver's navigation. If a severe weather alert comes up (thunderstorm, blizzard), an intelligent system can even reroute deliveries preemptively. One real-world example: FedEx has been noted to employ AI for real-time rerouting, where their systems monitor conditions and recalc routes instantly to avoid delays . Similarly, an AI agent in charge of shipments could detect a snowstorm forecast along a long-haul truck route and advise (or automatically arrange) an alternate route or delay to prevent a pile-up of issues.

WebkinsTech Gets Smart in IT Ops: WebkinsTech isn't only dealing with vendors; like any tech-enabled enterprise, it has an internal IT operations center handling incidents (server downtimes, network blips, etc.). They decided to introduce an AI Ops agent into their incident response workflow. This AI agent monitors all the server logs and metrics across their data centers. One weekend, a database server started slowing down due to a memory leak. In a traditional setup, maybe a human on call would get an alert once a threshold was breached and then scramble to restart the service. But WebkinsTech's AI agent recognized the pattern of the memory usage creeping up abnormally (it matched a known issue signature) and automatically triggered a failover workflow. Within seconds, it shifted the load to a standby database and gracefully restarted the troubled instance. By the time the oncall engineer saw the alert, the issue was already resolved. Over a few months, this intelligent incident workflow cut the average incident resolution time by 60%. It would detect failures and orchestrate remediation steps (like restarting services, clearing caches, reallocating workloads) autonomously. Humans were still in the loop for complex or unclear situations, but many routine outages became non-events visible only in retrospect. This not only saved downtime but also let the IT team sleep a bit easier at night, knowing a tireless agent was watching the fort.

The logistics industry also benefits from **predictive maintenance workflows**. Trucks, planes, ships all have sensors (IoT). An intelligent maintenance workflow might analyze sensor data daily to predict if a vehicle needs service before a breakdown happens. For example, an airline might use AI to predict when a jet engine component will likely fail, and automatically schedule maintenance (rerouting that plane's assignments and lining up a replacement plane) before that failure ever causes a cancellation. This turns maintenance scheduling into an intelligent workflow that balances safety, costs, and timing optimally.

Summary of Use Cases

Across these industries, a pattern emerges: intelligent workflows use data and AI to **sense, analyze, and act** far more dynamically than before. They break the rigid "if X then Y" mold and instead do "if X then maybe Y or Z or W, depending on what's best given the circumstances." The ROI can be seen in faster processing (e.g., instant loan decisions), higher accuracy (fewer false fraud flags, better triage), cost savings (optimized inventory, less downtime), and improved customer experience (personalization, quicker responses).

Each domain might use different AI techniques (NLP for text, computer vision for images, predictive models for time-series, etc.), but the principle is the same: the workflow is **augmented with an intelligent decision-making step** (or multiple steps). A microservice architecture makes this easier because you can slot in, say, a Fraud AI service into your payment pipeline or a Recommender service into your shopping cart pipeline. The next section will delve into what you need in place (data, technology, process) to enable these intelligent pieces – the foundations for intelligent workflows.

Before that, it's worth noting: we've hinted at autonomous, agent-driven actions (like the AI Ops example). That's edging into agentic AI territory, which will be our focus in Chapter 4. WebkinsTech's journey will continue with an even more agentic approach, but for now, let's equip ourselves with the foundations that make any of this possible.

Foundations of Intelligent Workflows

So, you're convinced of the potential of intelligent workflows. But how do you get from a bunch of microservices and databases to that AI-enhanced harmony? There are a few foundational ingredients and infrastructure considerations to get right:

1. Data Readiness and Quality: It's often said that *"data is the bloodstream of intelligent workflows."* If your workflows are going to make smart decisions, they need a steady flow of high-quality data. This means you should be capturing the relevant data from your processes and making it available for analysis. For WebkinsTech's vendor onboarding, relevant data included things like the time it takes for each onboarding step, the types of vendors (by region, size, etc.), and where bottlenecks or failures occur. Many organizations start by improving their **observability** – centralizing logs, metrics, and event traces from all microservices. Why? Because an AI can't learn what it can't see. If you deploy an anomaly detection service, it needs rich data: e.g., API response times, error rates, payload contents perhaps. Data quality is also crucial – if your data is full of gaps or errors, the AI will learn the wrong lessons or be unable to learn at all. Thus, having pipelines to clean and preprocess data, and maybe a **feature store** (a repository for the key input variables that models use, kept up-to-date in real time), is a best practice. For instance, a feature store might serve an up-to-date count of how many orders a customer placed in the last month, so any service or model needing that info gets a consistent answer.

"Data is the bloodstream of intelligent

> *(In other words, without a healthy flow of data, your intelligent withe*

2. Real-Time Event Processing: Traditional BI (Business Intelligence) operates on historical data (yesterday's sales, last month's trends). Intelligent workflows often need to act on data the moment it's produced. This is where technologies like event streaming (Apache Kafka, Amazon Kinesis, etc.) and real-time analytics frameworks (Spark Streaming, Flink, etc.) come into play. Instead of storing data and querying it later, you stream events through processors that can detect patterns or trigger actions on the fly. For example, in a fraud detection workflow, each transaction can be an event that gets scored in milliseconds as it streams through an AI model service. In a manufacturing workflow, sensor readings might continuously stream and an anomaly detector flags readings that deviate from the norm by 5 sigma, triggering a control system adjustment. Setting up a robust event-driven architecture is a common step. Microservices should emit events for important things that happen (order placed, payment failed, vendor onboarding started, etc.), which then can be consumed by intelligent components or even simple monitoring dashboards. The motto here is **streaming over batch** when aiming for responsive intelligence.

3. ML Model Integration: At the heart of many intelligent workflows are machine learning models – the algorithms that make predictions or classifications. Integrating these into your architecture requires some thought. You might have a model that predicts something (e.g., risk score, recommended action). To use it in a workflow, you need to operationalize it, a process often called MLOps (Machine Learning Operations). This involves:

- **Model Deployment:** packaging the model (which a data scientist trained in, say, Python) into a service or microservice that other parts of the system can call. This could be a REST API that returns predictions. Tools and frameworks (TensorFlow Serving, PyTorch Serve, or cloud AI services) help with this.

- **Scalability:** If a model is going to score every user click or every financial transaction, it must handle high throughput. This might mean deploying multiple instances, using GPUs for heavy tasks, etc.

- **Latency considerations:** Some workflows are okay with a model that takes 2 minutes to run (maybe a nightly scheduling optimization). Others need 20 milliseconds (fraud checks on a payment). The infrastructure should be designed accordingly – more hardware, simpler models, or approximate methods for low-latency needs.

- **Versioning and Updating:** Models aren't static. You'll retrain them as you get more data or as behavior changes. The workflow platform should allow updating the model without disrupting the whole system (canary release new model versions, for example). Also, sometimes multiple models might run in parallel (one in production, one learning, etc.).

- **Inference vs Training separation:** Typically, training (learning the model from data) is separate from inference (using the model to make decisions). Training might happen offline on big data sets and the result is deployed for inference in the live workflow. However, some advanced systems do online learning (continuously updating models as new data comes). It's a spectrum – you might retrain nightly, weekly, or live.

4. Knowledge Representation & Business Rules Integration: Not every intelligent workflow component is a black-box ML model. Sometimes it's about codifying knowledge in a more symbolic way or mixing rules with AI. For example, you might have an expert system component or a decision engine that uses a rules engine (like Drools or Camunda's DMN engine) in tandem with AI. A healthcare workflow might have a rule "if patient allergic to X, never recommend medication Y" – that's a strict business rule, whereas another part of the workflow uses AI to rank possible diagnoses. **Hybrid approaches** (rules + AI) are common in intelligent workflows to ensure critical domain knowledge is respected by the AI decisions.

5. Human-in-the-Loop and Feedback Loops: Intelligent does not mean 100% automated with no human oversight. In fact, many successful intelligent workflows are designed with **human-in-the-loop** steps where appropriate. For instance, if an AI in a loan processing workflow is not very confident about a case, the workflow might automatically route that case to a human officer for review. The system thus knows its limits. Human-in-the-loop can serve two purposes:

- **Safeguard / Oversight:** Ensure quality, compliance, or ethics where AI isn't trusted fully.

- **Training data generation:** Each time a human handles an exception or makes a decision the AI couldn't, that data can be fed back as a new training example so the AI gets better. Over time, the goal is the AI asks for human help less often, and when it does, those are truly novel cases.

Feedback loops are crucial. You want your workflow to learn from both success and failure. Did the AI recommend a product to a customer and the customer ignored it? That's feedback to possibly not recommend it again or refine the model. Did the AI flag 100 transactions as fraud and 5 were later confirmed by humans to be false alarms? Feed that back in to reduce false positives. Setting up mechanisms to capture outcomes and feed them into model updates closes the learning loop.

One of the challenges organizations face is how to test and iterate on intelligent workflows. Traditional software you can unit test with predetermined inputs and outputs. But an AIbased component might be probabilistic. This is why many recommend deploying AI in shadow mode first (it makes decisions in parallel to humans or existing system, to compare outcomes, without actually affecting the real process), or gradually ramping up autonomy.

Brandon Dey, an engineering leader, aptly noted the need for a "reproducible process of testand-learn" within automated workflows – basically, treat the workflow as an experiment that you continuously tweak and improve. Rather than expecting to get it perfect on day one, the successful teams set up the infrastructure to measure how the intelligent workflow is performing (KPIs, monitoring), and the agility to adjust it (retrain models, add new data features, tweak thresholds) in response to that feedback. This iterative improvement process is arguably more valuable than the initial automation itself, because it ensures the workflow stays effective as conditions change.

6. Technical Architecture Considerations: Enabling all the above often means some added pieces in your architecture:

- **Feature Store:** mentioned earlier, to serve up-to-date features for models.
- **Model Registry:** a system that stores different versions of models, along with metadata about their training (data used, accuracy, etc.), and possibly an automated way to roll back if a new model underperforms.
- **Workflow Orchestration Engine:** even if you have microservices, you might need an orchestration layer (like Netflix Conductor, Apache Airflow, or Temporal) especially if you're doing complex multi-step flows that involve waiting for events, invoking AI services, and handling timeouts or retries. Modern orchestration engines can include human tasks (pause workflow until human approves) which is handy for human-in-loop integration.
- **A/B Testing Framework:** for intelligent decisions, you might sometimes want to compare two strategies (maybe two different models or a model vs. rule). Being able to route a percentage of workflow instances to a "variant" and measure outcomes is valuable to improve the system.
- **Monitoring and Observability:** goes beyond just data collection. You want specific monitoring of the AI's decisions too. For instance, track the percentage of predictions above a certain confidence, drift in input data distributions (are we seeing very different data now than the model was trained on?), etc. This crosses into ML monitoring or AI observability – a growing practice to ensure your intelligent components are healthy and not silently degrading.

This might sound a bit overwhelming, but not every intelligent workflow needs all of these from the start. The key is to be aware of them and build up the capabilities as you scale. Many cloud providers now offer integrated platforms (for example, Azure's Cognitive Services, AWS SageMaker, GCP's AI Platform) that handle a lot of the MLOps and even data pipelines, so you can focus on the business logic.

A simple blueprint to begin is: **identify a decision point in a workflow that could be improved with data, implement a model or smart rule for it, integrate it as a microservice, and ensure you can measure its impact.** Then iterate. Over time, that might grow into multiple decision points and a more elaborate architecture around it.

In the next section, we will discuss specifically how microservices architecture helps enable all this intelligence (hint: the modularity and clear interfaces of microservices are a big advantage for plugging in AI), and some best practices in designing intelligent microservices.

The Role of Microservices in Enabling Intelligence

Microservices aren't just a way to chop up an application; they can be key enablers for injecting intelligence into systems. Let's examine why the microservice architecture style pairs so well with creating intelligent workflows:

Modularity for Specialized Intelligence: In a microservices architecture, each service has a well-defined responsibility. This modularity means you can introduce a new microservice whose sole job is to provide an intelligent capability, without rewriting the entire application. Want to add image recognition to a retail app for detecting products on a shelf? You can create an Image Analysis microservice and call it from relevant workflows. Because microservices communicate via APIs or events, the intelligent component can be treated like a black box that you plug in. This encourages building libraries of intelligent services – like a Fraud Detection Service, a Recommendation Service, a Language Translation Service – which can be reused across workflows. If these were all part of one monolith, integrating, deploying, and updating them would be much harder.

Separation of Concerns: A best practice in system design is to decouple the decisionmaking from the process flow. Microservices let you do exactly that. For instance, consider a payment processing pipeline: the flow might be "Process Payment -> Update Ledger ->

Send Receipt". Now, say you want an intelligent step to decide if the payment might be fraud and require verification. Rather than bloating the Payment Processing service with that logic, you can have the Payment service call an **FraudScoring** microservice. The FraudScoring service encapsulates the AI model. The Payment service just gets a response "risk = low" or "risk = high" and acts accordingly (perhaps calling a Verification service). By decoupling, you can evolve or replace the FraudScoring logic (maybe switch to a better ML model later) without touching the core payment code. This decoupling also means teams can work independently – data scientists can update the FraudScoring service, while payment engineers focus on reliability of payment transactions.

Design Best Practice: Decouple decision logic from control flow. Structure your workflows so that the "brainy" decisions are made by dedicated components (or services), separate from the "muscle" that executes the process. This way, improving or changing the decision-making (e.g., training a better model) doesn't require rewriting the workflow orchestration, and vice versa. For example, use a rules engine or AI service that your main workflow calls, rather than hard-coding lots of decision rules directly into the orchestration code.

Interfaces for AI Services: Microservices communicate through well-defined interfaces (API endpoints, message formats). This provides a natural point to integrate AI. If you define an interface like GET /recommendations?user=123, the service behind it could be simple rules or a complex AI – the workflow calling it doesn't need to know. This abstraction is powerful. It means you can start perhaps with a simple solution (like a rule-based recommendation) and later upgrade the microservice to a machine learning-based recommender without changing the consumers of that service. Microservice interfaces also force clarity about what data is given to the AI and what it returns, which helps in maintaining control (the AI can't just do anything; it serves a specific contract).

Scalability and Independent Deployment: Some intelligent components might be heavy on computation (training models, doing big computations for optimization). With microservices, you can scale those parts independently. If your "Route Optimization Service" that uses AI for logistics needs a lot of CPU at 2 PM daily when routes are planned, you can scale it up separately, without scaling the entire system. Moreover, if an AI service crashes or has a bug, it ideally doesn't bring down the whole workflow – the orchestrator can handle that gracefully or default to a simpler path. Independent deployment is also key for rapid iteration: the team managing an AI service can deploy updates frequently (perhaps to improve accuracy) without having to redeploy the whole platform.

Workflow Orchestration & Choreography Revisited: In the context of intelligent workflows:

- If you use **orchestration**, say with a tool like Conductor or Temporal, you can model your workflow with "decision tasks" that call AI services and branch based on results. The orchestrator can wait for responses and then choose a path. This explicit orchestration is useful when you need to ensure certain order and have fallback logic (e.g., “if AI service times out or says ‘unsure’, then assign to human task andContinuing from the discussion of microservices and intelligence:

If you use **orchestration**, say with a tool like Netflix Conductor or Temporal, you can model your workflow with “decision tasks” that call AI services and branch based on results. The orchestrator acts as a smart coordinator: it invokes a microservice for a prediction, waits for the response, and then chooses the next step. This explicit orchestration is useful when you need to ensure certain steps happen in order and have clear fallbacks. For example, in WebkinsTech’s case, the orchestrator of the vendor onboarding could call an AI risk scoring service and, if the risk comes back high or the AI is unsure, route the workflow through a manual review step; if low risk, skip straight to approval. Modern orchestration frameworks even support mixing automated steps with human tasks easily. (In fact, Orkes Conductor, an orchestration platform, has features to incorporate AI-driven tasks and human review checkpoints in one flow.)

In a **choreographed** (event-driven) intelligent system, you might embed intelligence within the services themselves or as event subscribers. For instance, in an event-driven architecture, the moment a “New Vendor Registered” event is emitted, an AI compliance service could listen and automatically pull external data to enrich the vendor profile (e.g., perform a quick background check) then emit a “Vendor Risk Assessed” event. Other services react accordingly. There’s no central brain coordinating, but each service has some smarts. Alternatively, you could have an AI service subscribing to the event bus and making decisions that it then publishes as new events (like “Order Fraud Alert” event emitted after it sees an “Order Placed” event and determines something is off). Choreography can make intelligent workflows very flexible and scalable: you can add more AI services listening to events without disrupting existing flows. The challenge is ensuring the overall system’s behavior is observable and understandable – when lots of smart services dance together, you need good monitoring to see what’s going on.

Many systems actually use a **blend**: orchestrating high-level processes while choreographing the low-level event handling. For example, you might orchestrate a longrunning process with explicit steps (some of which involve waiting for events or AI decisions), and within each step, services might use event-driven mechanisms. The key is to choose what fits each scenario best. The good news is microservices give you the freedom to choose per component – some microservices might be orchestrated, others pure event subscribers.

Another benefit of microservices for intelligence is **technology heterogeneity**. One service might be written in Python using scikit-learn or TensorFlow for AI, another in Java for transaction handling. Microservices communicate via language-agnostic APIs, so you can use the best tool for each job. The AI teams can iterate with their ML frameworks independently. This is huge in practice: it means your data scientists can productionize a model as a microservice without having to convert it to the primary language of the monolith or platform. They just need to meet the API contract.

Finally, microservices make it easier to **sandbox and experiment**. Want to try a new recommendation algorithm? Deploy it as RecommenderV2 service, route 10% of traffic to it (via API gateway or orchestrator logic), and compare results. In a monolithic setup, that kind of A/B test could be risky or difficult. With microservices, you spin up a new instance, and the rest of the system doesn't mind as long as you direct some calls to it. This agility is important for developing intelligent workflows because there's inherently more trial-and-error and tuning involved in AI features compared to deterministic code.

In summary, microservices architecture provides the **scaffolding** to insert intelligence at will: clear interfaces to plug into, isolation to iterate on AI without breaking everything, and scalability to support heavy compute when needed. Designing intelligent microservices does require discipline: keep them focused, handle failures gracefully (e.g., what if the AI service is down or returns "I don't know"?), and monitor their outputs. But the payoff is a system where smart capabilities can evolve rapidly.

Next, let's step back and consider how ready your organization or team is to embark on this intelligent workflow journey. The following checklist will help assess readiness and identify areas to strengthen before (or while) diving in.

Checklist: Readiness for Intelligent Workflows

Implementing intelligent workflows isn't just a technical endeavor; it's an organizational journey. Here's a checklist of questions and factors to consider to gauge your readiness:

- 🤔 **Clarity of Purpose:** Do you have a clear understanding of why you want intelligence in your workflows? Identify the pain points in your current processes (e.g., high exception rates, slow manual decision steps, frequent rule updates) and the value intelligent automation could bring (faster service, cost savings, better accuracy, etc.). Having a concrete use case or KPI target is important to rally effort and measure success.

- 📊 **Data Availability and Quality:** Are you collecting the data needed to make intelligent decisions? Review the data inputs an AI model or smart system would require. For example, for an AI to optimize a supply chain, you need historical sales, lead times, maybe weather or social trends data. Is that data accessible, integrated, and of good quality? If not, invest in data integration and cleanup first. Also, do you have a data infrastructure (data lake/warehouse, streaming pipeline, etc.) in place to feed AI models in real time?

- 🛠 **Technology Stack:** Does your current tech stack support adding AI/ML components? Consider if you have (or can introduce) components like an event streaming platform, model serving infrastructure (or cloud AI services), and workflow/orchestration tools. You don't need everything from day one, but you should have a plan. For instance, if your architecture is already microservices, do you have an easy way to deploy new services? If you're using serverless functions, can they call out to AI services or host models? Ensure your platforms are extensible for these needs.

- **Skill Sets:** Does your team have experience with data science or machine learning, or do you have access to those skills? An honest assessment here is crucial. If not, you may need to hire or partner or start with simpler machine-learning-as-a-service approaches. Similarly, do your developers know how to integrate and maintain ML in production (MLOps skills)? If not, plan for training or bringing in new talent. Intelligent workflows live at the intersection of software engineering and data science.

- 🗐 **Governance and Ethics:** Are you prepared for the governance aspect of AI decisions? In regulated industries like finance or healthcare, you may need explainability for decisions (why did the AI reject this loan?). Even outside of regulation, consider the ethical implications – e.g., ensuring the AI isn't biased, and having a process to handle appeals or corrections. Put in place governance: which team is responsible if the AI makes a bad call? How will you monitor outcomes for fairness and accuracy?

- **Monitoring and Feedback Loops:** Do you have (or can you build) a strong monitoring culture? As discussed, you'll want to monitor not just uptime of services, but also the quality of decisions. Think about metrics like accuracy, false positive/negative rates, and business KPIs (e.g., reduction in processing time). Set up dashboards and alerts for these. Additionally, plan how feedback will be gathered – e.g., capturing when humans override AI decisions or when outcomes indicate the AI was wrong – and fed back into improving the system.

- ⚙ **Existing Automation Maturity:** How mature are your current processes? It's generally easier to add intelligence to a process that is already well-understood and stable. If you're still firefighting basic automation issues or your processes are constantly changing in fundamental ways, it might be wise to stabilize those first. On the other hand, a very well-oiled automation might be ripe for an intelligent upgrade.

- 💡 **Culture and Buy-In:** Is there buy-in from stakeholders and end-users for introducing AI? Intelligent workflows might change how people work. For example, if an AI starts triaging support tickets, support staff need to trust it (and not feel threatened by it). Managing change is key – involve the end-users early, perhaps present the AI as assisting them, not replacing them. Ensure management supports the initiative not just for one prototype, but for the long haul (AI projects can start slow as they learn).

- 🔄 **Iterative Mindset:** Are you ready to iterate and experiment? Unlike deploying a new static app feature, deploying an AI-driven feature often requires tuning and iteration. Set expectations that the first version might only be somewhat better than the old way, but will improve. You might need to try different models or approaches. Having a sandbox or A/B testing mechanism in place is a plus. The organization should be ready to learn

  and adapt (treating it as a continuous improvement process, not a one-and-done project).

If you go through this checklist and find many "no" or "not sure" answers, don't be discouraged. Use them as a roadmap. Perhaps start with improving data logging, or providing some AI training for staff, or implementing a pilot on a smaller scale to build confidence.

Getting Started – Crawl, Walk, Run: If you're just starting out, pick a small, contained workflow that would clearly benefit from some intelligence but is low-risk. Maybe something internal or a segment of a process. Implement an AI augmentation there and measure the impact. This pilot will teach your team a lot. Also, don't rip out all the old rules on day one. Often the best approach is a hybrid: keep the reliable parts of your existing automation and augment with AI where it adds value. In practice, the most successful organizations combine the speed and reliability of traditional automation with the adaptability of AI. Use AI to handle the cases or decisions where static logic struggles, and use traditional automation for the rest. This blended strategy ensures you get improvements quickly without sacrificing what already works. As your team gains experience and trust in the intelligent components, you can gradually expand their scope. And always maintain a fallback (even if it's manual) for when the AI is unsure or something goes wrong – this keeps risk low.

Armed with this readiness checklist and tips, you can embark on the journey in a pragmatic way. Now, let's bring it all together with a fictional case study. We'll return to WebkinsTech and follow the evolution of their vendor onboarding process from manual to automated to intelligent, and even a glimpse of an agentic future, to solidify the concepts we've discussed.

Fictional Case Study: WebkinsTech's Journey to Intelligence

Meet **Elena**, the Director of Operations at WebkinsTech. A few years ago, Elena oversaw a *mostly manual* vendor onboarding process. It was paperwork-heavy: new vendors emailed PDFs of tax forms, a coordinator manually entered data into the system, compliance officers cross-checked watchlists by hand, and approvals were done via email chains. This process was slow – taking weeks sometimes – and error-prone (typos, missed emails). Elena knew there had to be a better way.

Stage 1: Basic Automation (Rule-Based Microservices)

Elena first pushed for an **automation overhaul**. The company's IT team adopted a microservices approach to redesign vendor onboarding. They created services for each part of the workflow: a Vendor Portal for input, an Onboarding Orchestrator to drive the process, a Compliance Check service, an Approval service, and integrations with the internal ERP system. The new workflow looked like this:

1. **Vendor Portal** – The vendor fills out an online form and uploads required documents. Upon submission, the system kicks off an "Onboarding Request" event to the orchestrator.

2. **Onboarding Orchestrator (Automated)** – This central service coordinates the next steps. It first calls the **Compliance Check Service**.

3. **Compliance Check Service** – This service runs a series of rule-based checks (e.g., verify the tax ID format, ensure required documents are provided, run the vendor name against a sanctions list using an API). It returns a pass/fail result.

4. The Orchestrator then calls the **Approval Service** – essentially a workflow that notifies a WebkinsTech manager to review the vendor info if needed. Since they wanted to automate, they made a simple rule: if the vendor's contract value is below $10,000, auto-approve; otherwise, require a manager's approval via an inbox task.

5. Once approved (or auto-approved), the Orchestrator creates the vendor record in the ERP and finalizes the process (maybe sending a welcome email via an Email Service, etc.).

This automated workflow was a huge improvement. What used to take weeks shrank to days or hours. The team even visualized it with a flow diagram to celebrate their success:

WebkinsTech Vendor Onboarding Workflow – Before vs. After Intelligence. The **left side** shows the original rule-based automation: the Onboarding Orchestrator calls the Compliance Check and Approval services in sequence, strictly following predefined rules (e.g., always do step A then B then C). The **right side** shows the later intelligent workflow, where an AI Risk Scoring Service (yellow) is introduced. In the intelligent flow, the orchestrator still coordinates, but it first consults the AI for a risk analysis. Depending on the AI's report, it can adapt: e.g., auto-approve low-risk vendors or require extra steps for high-risk ones (dashed lines represent those dynamic paths). This illustrates how adding an AI component made the once linear process more flexible and context-aware.

In practice, the automated process worked well for straightforward cases. However, Elena's team discovered some pain points remained. For one, whenever a new type of compliance requirement came up (say a new regulation to check environmental certifications), the development team had to modify the Compliance Service code or add a new step, which took time. The system also couldn't prioritize or differentiate vendors except by the crude contract value rule. A vendor dealing in a sensitive country versus a local small business both went through the same steps. And remember that incident from the introduction: an unexpected data format (a vendor from abroad with a different tax structure) caused the workflow to fail. The automation was efficient, but **inflexible**.

Stage 2: Introducing Intelligence

Having seen the benefits of automation, Elena was keen to tackle its limitations. She collaborated with WebkinsTech's newly formed "AI Task Force" (a small team of a data scientist, a ML engineer, and a savvy developer) to inject intelligence into the onboarding workflow.

They identified two key areas to improve:

- **Risk and Prioritization:** Not all vendors are equal. Some pose higher risk (compliance, security, financial) and should get more scrutiny; others are low risk and could be fasttracked. Instead of a single rule (like contract value), they wanted a more nuanced risk assessment.

- **Document Processing:** Vendors uploaded documents (like compliance certificates, IDs, tax forms). Previously, humans checked these off-system. They aimed to automate that using AI (extracting text, verifying authenticity) to speed things up and reduce manual labor.

The AI Risk Scoring Service: The team trained a machine learning model on past vendor onboarding data combined with external data. The model considered factors like the vendor's country, type of product/service, industry risk databases, and any anomalies in their application (e.g., missing fields or certain keywords in descriptions). It produced a risk score 0 to 100 and a simple recommendation: "Green – low risk", "Yellow – moderate", or "Red – high risk". They deployed this model as a new microservice.

The **workflow** changed as follows:

- After a vendor submits the form, the **Orchestrator** now first calls the **AI Risk Scoring Service** before doing other checks.

- If the risk comes back "Green (low)", the orchestrator actually bypasses the manager Approval step for more cases (for example, now auto-approving vendors up to $50k contract if low risk, expanding the auto-approve scope safely). It also possibly skips or simplifies some compliance checks (maybe for low-risk vendors it doesn't require a heavy background check).

- If "Yellow (moderate)", it proceeds as normal through compliance and
- approval.

If "Red (high risk)", it might trigger additional steps: for instance, ask the Compliance service to run extra checks or require not just manager approval but also security team review. High-risk vendors could even be routed into a sub-workflow involving an interview or additional documentation. These branches were new and defined in the orchestrator's logic based on the AI output.

Intelligent Document Processing: They integrated an OCR and NLP capability (using a third-party AI API) into the Compliance Check service. Now, when a vendor uploads, say, a tax certificate, the system automatically extracts key fields and verifies authenticity (crosschecking numbers with government databases where available). If the AI is unsure about a document (maybe it's a type of document it hasn't seen), it flags it for manual review. Over time, as the AI model was trained on more docs, it handled more of them automatically. This saved the compliance team a lot of tedious work.

These intelligent additions made the workflow more **adaptive**. A low-risk vendor with perfectly filled data sailed through in minutes, delighting them and the WebkinsTech business owners waiting for that vendor. A potentially risky vendor got the scrutiny needed (and the AI's risk reasoning gave the team heads-up why, e.g., "vendor registered in country X with history of sanctions – flagged"). The system also started learning from outcomes: whenever a high-risk vendor was reviewed by a human, they fed that result (approved or rejected and why) back into the model training, so the risk predictions would improve.

There were some hiccups. Initially, the AI risk model was too cautious (flagging too many yellows), so it had to be tuned. Also, the team had to build trust in the AI. Elena organized weekly reviews of the AI's decisions where the team would randomly sample cases to see if they agreed with the AI – this helped them calibrate it and also build confidence as they saw it was mostly reasonable. They also kept **audit logs**: for any automated approval or rejection, the system logged the factors and scores, so nothing was a black box. This proved vital the first time a vendor contested a rejection – WebkinsTech could explain it was due to a failed compliance check that the AI spotted.

The benefits were tangible. The average onboarding time dropped further, especially for lowrisk vendors. The compliance team reported they spent 40% less time on manual checks, focusing only on the tricky cases. And importantly, the number of "surprises" (workflow breaking or a bad vendor slipping through) reduced – the AI was catching weird cases early.

Elena's favorite example was when the AI flagged a vendor as high risk because the business address in the form was in a location known for shell companies. Upon manual review, that vendor did turn out to have some red flags in background checks. The system had effectively given an early warning that might have been missed in a purely rules-based approach.

By the end of this stage, WebkinsTech's onboarding workflow was not just automated, but **intelligently automated**. It could make distinctions and adjust its path in ways the old system never could. Still, Elena wasn't done. She had a futuristic vision: could the system become even more autonomous, almost like a virtual project manager for onboarding? This leads to the final stage.

Stage 3: Towards Autonomy (Agentic Workflow)

In the intelligent workflow, the AI components made decisions, but the overall process was still following a defined orchestrator script. What if the system could truly self-organize parts of the workflow? This is where **agentic AI** concepts came in.

WebkinsTech started a pilot project with an "Onboarding Agent" – an AI agent whose goal was to get a vendor from application to active as quickly as possible **while** ensuring all policies are met. They gave it certain powers: it could communicate with people (e.g., email the vendor for missing info or ping a WebkinsTech employee if an approval was delayed), and it had access to trigger various services in the workflow. Essentially, they wrapped the orchestrator logic into a more flexible agent that could re-order steps or take extra steps as needed.

One scenario they tested: A vendor's insurance certificate (required document) was missing. In the old process, the compliance check would fail and stop there, requiring someone to reach out and then restart the process. The new Onboarding Agent, upon noticing the missing document, itself emailed the vendor's contact with a polite message requesting the certificate and instructions to upload it. It then paused the onboarding process waiting for the event that the document was uploaded. Once received, it resumed the process. If too much time passed, it would send a reminder, and if still no response, it would alert a WebkinsTech coordinator to follow up by phone. This kind of **autonomous handling of exceptions** was a big step up: the system didn't just throw an error and wait; it took initiative to solve the issue.

Another example: The agent learned to dynamically rearrange tasks. Suppose a manager was taking too long to approve a low-risk vendor who only needed a rubber stamp. The agent would ping a backup approver after a certain timeout, or escalate by notifying Elena that an approval is pending for too long. Earlier, the orchestrator had fixed wait times and escalation rules; the agent made these more context-aware (e.g., it might wait less time if it knows the manager is out-of-office based on calendar integration).

One of the most interesting aspects was the agent's ability to **learn optimization**. WebkinsTech defined a metric called "onboarding cycle time" and another for "compliance confidence". They trained the agent (using reinforcement learning techniques) in simulation to try different policies (like varying which checks to do first, or how to parallelize tasks) to see what minimizes time while keeping confidence high. The result was the agent discovered that by initiating a credit check and background check in parallel (which no human had originally set in the workflow, as they were sequential), it could save hours. In the new agent-driven flow, as soon as a vendor comes in, multiple checks happen concurrently when possible. The orchestrator script was rewritten by the AI, in a sense, for greater efficiency.

Internally, this Onboarding Agent was implemented with a combination of rules, ML models, and event-driven triggers – essentially an advanced orchestrator with a bit of decisionmaking autonomy. It's not sentient or anything, but it acts more proactively than the old system.

The outcome? WebkinsTech's vendor onboarding became **self-driving** to a large extent. The agent handled routine cases end-to-end with no human intervention at all, and it handled many exceptions by itself (like chasing down missing info). Humans mainly oversaw and handled truly novel or sensitive cases. Onboarding time became highly predictable and short. WebkinsTech's partners noticed – some vendors even commented, "Wow, your onboarding process is smooth and even reminded me to do stuff promptly; it kept us on track."

Elena proudly presented the journey in a company town hall:

- Initially, onboarding was manual and chaotic.
- Then automated and orchestrated (faster, consistent).
- Then intelligent with AI (adaptive and risk-aware).
- Finally agentic (autonomous and self-optimizing).

Not every workflow at WebkinsTech has reached the final stage, but vendor onboarding became a showcase of what's possible. It also laid the groundwork (data, trust, experience) for applying similar ideas to other processes (like customer onboarding, or internal employee onboarding).

Through WebkinsTech's story, we see that achieving intelligent and autonomous workflows is a gradual process. Each stage built on the previous one, delivering value at each step:

- Automation gave efficiency.
- Intelligence gave adaptability.
- Autonomy (agents) gave proactivity.

This mirrors how many real organizations will evolve – step by step, proving value and expanding.

Conclusion: The Road to Agentic AI

We began this chapter with a simple premise: doing things fast is good, but doing the right things is even better. Microservices and automation gave us speed and scale. By thinking in terms of workflows, we learned to orchestrate complex processes across systems. But traditional automation alone was like a fast vehicle with a rigid steering wheel – it could accelerate, but not turn easily when the road changed.

Through exploration of intelligent workflows, we saw how adding an "AI copilot" or sometimes letting the AI drive, makes our systems far more effective. They can handle the unexpected detours and roadblocks, not by stopping, but by intelligently rerouting or adjusting course. We explored how enterprises are already benefiting: hospitals triaging patients with AI assistance, banks catching fraud in the blink of an eye, retailers stocking shelves perfectly in time for demand surges, and logistics routes redrawing themselves around storms. These are not sci-fi; they're happening today, powered by the synergy of microservices architectures and AI technologies.

Some key takeaways from this chapter:

- **Workflows are the backbone**: By viewing processes as orchestrated/choreographed workflows, we can systematically improve them. Microservices make these workflows modular. The next step is not to just automate the steps, but to make the flow of steps itself adaptive.

- **Traditional automation has limits**: It's excellent for the predictable, but brittle for the unpredictable. Static rules can't cover every scenario, and maintaining them becomes a burden as complexity grows.

- **Intelligent workflows fill the gap**: By leveraging data and AI, we make workflows context-aware. They can ask, "What's different about this case and how should I handle it?" This leads to better outcomes (fewer errors, faster exception handling, personalized service).

- **Autonomy vs Automation**: We distinguished autonomous decision-making from mere automated execution. Autonomy (agentic AI) is the ultimate goal for many – systems that can handle high-level goals with minimal micromanagement. It's like graduating from a basic autopilot to a fully self-driving process.

- **Microservices + AI = ♥** : Microservices provide the structure that AI can fit into. They allow us to insert intelligence precisely where it's needed and keep everything else decoupled. Think of microservices as the body and AI as the brain – together you get a living system.

- **Iterative Journey**: The WebkinsTech case study reinforces that it's a journey. Start with what you have, incrementally add intelligence, and maybe one day you have autonomous agents running around improving your business.

As we look ahead to **Chapter 4 on Agentic AI**, we'll dive deeper into those autonomous agents – the next frontier. Agentic AI takes the concepts here (autonomy, context, learning) and scales it up: multiple agents, collaboration, negotiation, and goal-driven systems that could fundamentally change how we design software.

One thing to remember is that technology is only half the equation. People, processes, and mindset complete the picture. Intelligent automation works best when humans trust and understand it, and when it's used to elevate human work, not just cut costs. Elena from WebkinsTech didn't aim to eliminate the operations team; she aimed to free them to focus on strategic vendor relationships while the system handles the grunt work. That's the ideal partnership of human and AI in workflows.

In closing, consider this thought:

> *"Automation made things faster. Intelligence will make them wiser."*

As you modernize your architectures, don't just ask, "How can I automate this task?" – also ask "How can this workflow learn and improve over time?" That mindset shift is what "Think Workflows, Automation, and Need for Intelligence" is all about. We stand at an inflection point where our software can become not just tools that do what we tell them, but collaborators that figure out what to do with us. And that opens up a world of possibilities in every domain of enterprise software.

The road to agentic AI is a journey of increasing capability and autonomy. With the foundations we've laid out – and perhaps a bit of inspiration from WebkinsTech's story – you're now prepared to explore that road. In the next chapter, we'll step fully into the world of agentic AI, where systems act with agency, and we'll learn how to design and harness these powerful new building blocks in our microservices ecosystems. The future is calling, and it sounds intelligent. Let's answer.

Chapter 4: Core Concepts of Agentic AI

Imagine you had an assistant who not only answers your questions, but anticipates your needs, makes decisions on your behalf, and learns from experience. Such is the promise of **agentic AI** – a new paradigm where AI systems exhibit **agency** and autonomy in pursuing goals. In this chapter, we'll explore what makes an AI system "agentic", how autonomous agents differ from traditional AI models, and the frameworks that drive their goal-oriented decision-making. We'll also dive into multi-agent systems and the ingenious ways agents coordinate with each other. Throughout, we'll follow **WebkinsTech**, a fictional tech company, and its lead architect **Elena** as they navigate the practical realities of building intelligent microservices with agentic AI. Let's embark on this journey into the core concepts of agentic AI in a style that's part friendly conversation, part authoritative guide.

Autonomous Agents vs. Traditional AI Models

Before we get into the nuts and bolts, let's clarify what **agentic AI** actually means and how it contrasts with more traditional AI approaches. **Traditional AI models**, such as a trained machine learning classifier or a rule-based expert system, are often like *script followers* – they do exactly and only what they are designed or instructed to do, when they are prompted to do it. In contrast, an **autonomous agent** is more like a *skilled assistant* – capable of taking initiative, operating on its own in a dynamic environment, and continuously working towards a goal without needing step-by-step instructions.

At its core, an **AI agent** is a software program that **perceives** its environment, **decides** on actions, and **acts** upon the environment to achieve certain goals . When we say an agent is *autonomous*, we mean it can perform these tasks **independently**, without constant human oversight or input. For example, a simple AI model might classify images when asked, but an autonomous agent could proactively monitor a camera feed and decide to sound an alarm if it detects an intruder – all on its own.

To put it formally, *agentic AI* refers to AI systems *"*designed to autonomously make decisions and act, with the ability to pursue complex goals with limited supervision."* These agents have what we call **agency** – the capacity to act in the world and make choices in pursuit of objectives. Unlike a typical program that reacts only when prompted, an agentic system can keep running, observing, and acting according to its built-in drives or goals.

Key Characteristics of Intelligent Agents

What specific traits distinguish an intelligent agent from other software? Researchers and practitioners generally highlight a few **key characteristics** that agentic AI systems possess:

- **Autonomy:** Agents operate without the need for frequent human guidance . They have control over their actions and internal state. For instance, once you deploy an agent to manage network traffic at WebkinsTech, it should handle routine decisions on its own (like rerouting traffic or balancing load) without asking an engineer each time.

- **Reactivity:** Agents are **responsive** to changes in their environment . They don't just follow a fixed script blindly; they observe what's happening and adjust their behavior. If our network agent sees a server is down, it notices that change and reacts (perhaps by shifting traffic to healthy servers) in real-time.

- **Proactiveness:** Beyond reacting, agents also take initiative **in pursuit of goals** . They can *anticipate* problems or opportunities and act on them. Think of a smart personal assistant that doesn't just wait for you to ask "Do I have any meetings today?" but instead proactively warns you: "You have back-to-back meetings, shall I order lunch for you at noon?" An agent in a cloud system might predict high load in advance and spin up extra servers *before* performance degrades.

- **Goal-Oriented Behavior:** An agent is driven by objectives or *motivations*. It isn't random or purely reactive; it has a **goal** (or set of goals) it's working toward. A goal could be as simple as "keep temperature at 22°C" for a thermostat agent, or as complex as "maximize user satisfaction in our e-commerce site" for a recommendation agent. Everything the agent does is (or should be) in service of its goal.

- **Learning and Adaptability:** Many agents can **learn from experience** and improve over time . Through techniques like machine learning or reinforcement learning, an agent refines its behavior based on feedback. For example, WebkinsTech's customer support agent might learn from past interactions that certain solutions resolve issues faster, and thus get better at helping users as it gains experience.

- **Continuous Operation:** Unlike a one-off function call, an agent often runs continuously, **persistently** sensing and acting. It's "always on duty," much like a guard on watch. This continuity is what lets it be proactive and responsive over time.

- **Communication (Social Ability):** In multi-agent scenarios, an intelligent agent can often **communicate** with other agents or systems. This isn't a must for a single-agent system, but it's a common trait in more complex environments – the agent might send messages, negotiate, or collaborate with peers to achieve its goals . (We'll explore multi-agent interactions later in the chapter.)

These characteristics set agents apart from simpler software. A traditional program might have some of these traits in isolation (for instance, a thermostat is reactive but not proactive or learning). An agentic AI system tends to combine all of them to some degree.

To make this more tangible, let's use an **analogy**: Consider the difference between a basic **scripted chatbot** and an **AI assistant**. A basic chatbot follows a predefined script – if you say X, it responds with Y (and if it's something unexpected, it might get confused). It's like a customer support representative who can only read from a script. In contrast, a true AI assistant (an agent) understands your intent, keeps context from previous interactions, can handle various tasks proactively, and might even remind you of things you forgot. It's more like a savvy personal secretary who not only responds to requests but also plans and organizes on your behalf. As one industry expert put it, *"While chatbots follow scripted conversations, AI agents evaluate data, recognize patterns, and decide the best course of action when executing complex tasks based on real-time inputs."* The agent is not limited to canned responses; it has the autonomy and intelligence to figure out what to do.

We can also frame this difference in terms of **traditional AI vs. agentic AI** using a quick comparison:

Aspect	Traditional AI Model	Agentic AI (Intelligent Agent)
Initiative	Acts only when prompted (passive)	Acts on its own, can initiate actions (proactive)

Scope of Operation	Single-step tasks (e.g., one-shot prediction)	Continuous process (persistent goal pursuit)
Adaptability	Typically fixed once trained or programmed	Learns and adapts from experience and feedback
Goal Orientation	Solves a specific, narrow task as instructed	Operates towards higher-level goals (may involve many sub-tasks)
Reaction to Environment	Pre-defined responses to inputs; limited awareness	Senses environment actively and reacts to changes (reactive to context)
Examples	A classifier that labels an image; a rule-based system for loan approval.	A virtual travel agent that plans a trip (finds flights, books hotels proactively); a robot vacuum that navigates and learns the room layout over time.

Elena often describes this to new team members at WebkinsTech as the difference between a **tool** and a **colleague**. A traditional AI model is like a very advanced tool – you pick it up and use it for a task. An agent is more like a junior colleague or assistant – you give it an objective, and it figures out a lot of the *how* on its own, coming back to you with results or when it needs guidance.

From Scripts to Agents at WebkinsTech

In WebkinsTech's early days, their cloud management system was a collection of scripts. If a server's CPU usage went above 90%, a rule-based script would trigger an alert email. Another script could be manually run to add more servers. These were useful, but they were *static*. They only did exactly what they were coded to do, and only in those specific conditions.

Fast forward to today: Elena's team has replaced many of those scripts with an **autonomous cloud manager agent**. This agent runs 24/7, monitoring dozens of metrics (CPU, memory, request latency, etc.). If it notices unusual spikes, it doesn't just send an email – it proactively spins up new server instances, balances the load, and later spins them down if they're no longer needed. The agent learned over time the typical daily usage patterns and can even **anticipate** traffic surges (like the Monday morning login storm) and prepare resources in advance. Unlike the old scripts that only reacted after a threshold was breached, the new agent embodies *proactiveness* and *goal-oriented autonomy*: its goal is to keep the system healthy and responsive, and it uses whatever actions (within pre-set safety limits) necessary to achieve that.

In the above example, we see clearly how the **agentic approach** offers a more powerful, flexible solution. The traditional scripts were like obedient workers doing exactly one task, while the agent is like a smart manager handling situations as they evolve. This doesn't mean traditional AI models are obsolete – far from it. In fact, agents often *incorporate* traditional AI models internally (for example, an agent might use a machine learning model to predict future load). The difference lies in **how they operate** and **how much initiative they have**. Agentic AI is about wrapping those capabilities in a layer of decision-making and autonomy so that the system can operate intelligently **without micromanagement**.

To cement this understanding, let's highlight an insightful quote:

On Agentic AI's Autonomy

"As its name suggests, agentic AI has 'agency': the ability to act, and to choose which actions to take. Agency implies autonomy, which is the power to act and make decisions independently. When we extend these concepts to agentic AI, we can say it can act on its own to plan, execute, and achieve a goal — it becomes 'agentic.' The goals are set by humans, but the agents determine how to fulfill those goals." – *Deloitte Insights, 2025*

This quote nicely summarizes that agentic AI is all about independent decision-making in pursuit of human-defined objectives. In the next sections, we'll delve into how agents actually do this: how they represent and pursue goals, how they make decisions (immediately or via planning), and how they're structured internally to perceive and act.

Agency, Goals, and Decision-Making Frameworks

So we know agents have **goals** and can act autonomously. But how exactly does an agent *decide* what to do at any given moment? What does it mean for an agent to have "agency" in practical terms? In this section, we'll peel back the curtain and look at the internal workings of an intelligent agent. We'll discuss the components that give an agent its senses, its reasoning ability, and its means to affect the world. We'll explore different decision-making approaches (from reflexive reactions to long-term planning) and how agents balance reacting *now* vs. planning for the *future*. We'll also share best practices in designing goals and decision frameworks, with examples from Elena's projects at WebkinsTech.

The Anatomy of an Agent: Sensors, Logic, and Actuators

At a high level, an agent interacts with its environment through a **sense-think-act** cycle. This is sometimes called the *perception-decision-action* loop. The agent continuously perceives the state of its environment, decides on an action based on its goals and observations, and then executes that action through some effect on the environment. This loop then repeats, allowing the agent to respond to changes and progress toward its goals over time.

Let's break down the essential components involved:

- **Sensors (Perception)** – These are the agent's means of observing the environment. They could be literal hardware sensors (like a camera, microphone, temperature sensor for a robot) or software "sensors" (like API calls, event streams, database queries that give the agent information about its world). The sensor inputs inform the agent of **"what the world is like now."** For example, WebkinsTech's cloud manager agent has sensors in the form of monitoring data: it "perceives" CPU usage, memory usage, network traffic, error rates, etc. as its inputs.

- **Decision-Making Logic (Brain)** – This is the core of the agent: the algorithms, rules, or models that take the perceived information and decide **"what should I do next?"**. This logic can range from simple condition-action rules (if X, do Y), to complex planning systems, to machine learning policy networks, or a combination of all. Often, this decision module is where the agent's **goals** are represented (e.g., a reward function in reinforcement learning, or objectives in a planner). We might also call this component a **policy engine**, a **planner**, or simply the agent's "AI". This is effectively the agent's mind.

- **Actuators (Actions)** – These are the mechanisms by which an agent **affects the environment**. Just as a robot might have motors or grippers as actuators, a software agent might have actuators in the form of the ability to send commands, trigger API calls, or modify data. The actuators allow the agent to execute the decisions it made. For the cloud manager agent, the actuators are things like the cloud provider's API – the agent can call an API to launch a new server instance, or call another to shut one down, or invoke a script that clears a cache. Those are its "hands" in the digital environment.

We can visualize this as follows:

Figure: A conceptual diagram of an agent interacting with its environment. The agent perceives inputs via sensors (e.g., monitoring data), decides on an action using its decisionmaking logic (condition-action rules, or more complex reasoning), and affects the environment through actuators (e.g., API calls). This loop repeats continuously as the agent pursues its goals.

In the figure above, you see how the **environment** provides percepts (inputs) to the agent, the agent's logic determines an action, and then through actuators it sends actions back to the environment. This loop is the heartbeat of an autonomous agent.

Elena often refers to this as the **sense-plan-act** loop (with "plan" encompassing any decision process). When designing any agent at WebkinsTech, her team clearly defines:

1. **What inputs the agent can observe** (and how it gets them),
2. **What goals and decision rules/algorithms the agent uses** to map those inputs to actions,
3. **What actions the agent is allowed to take** (and how it executes them).

Defining these three aspects becomes a design checklist to ensure the agent will function correctly and safely.

Checklist: Designing an Autonomous Agent

Define the Environment: What external context will the agent operate in? Identify the boundaries of the environment (e.g., the cloud infrastructure, a game world, a user's smartphone).

Sensors (Inputs): List what data the agent will perceive. For each piece of information the agent needs (e.g., "current CPU load", "number of active users", "position of nearby obstacles"), establish how the agent gets it (monitoring APIs, event feeds, sensor devices, etc.). Ensure the agent has sufficient visibility into the environment to make informed decisions – blind spots can be dangerous.

Goals: Clearly specify the agent's objectives or performance metrics. Is it trying to maximize something (uptime, throughput)? Minimize something (response time, cost)? Achieve a specific outcome (win a game, keep temperature stable)? A welldefined goal helps align the agent's decisions.

Decision-Making Framework: Choose an approach for how the agent will make choices. Will it use simple rules, a state machine, a planning algorithm, or a learned policy (or a hybrid)? We'll discuss options shortly (reactive vs deliberative, etc.).

Select what fits the complexity of the task and the predictability needed.

Actuators (Actions): Enumerate the actions the agent can take to affect the environment. For each action (e.g., "add server", "send alert", "move forward"), define how the agent invokes it (API endpoint, function call, device control signal) and any limitations or guardrails (perhaps the agent can add at most 5 servers at a time, or it can't delete data without confirmation).

Learning and Adaptation: Decide if the agent will learn from experience. Will it update its strategy over time (via machine learning or feedback rules)? If yes, plan how it gathers feedback (reward signals, labeled data) and how to ensure it learns safely (avoid wild experimentation in production!).

Safety and Constraints: Identify anything the agent must *not* do. Implement checks in the decision logic or actuators to prevent harmful actions. For example, even a very autonomous cloud agent at WebkinsTech has a rule "never shut down more than 30% of servers at once" as a safety constraint, no matter what its ML brain might suggest.

Testing and Simulation: Before unleashing the agent in the real world, test it in a simulated or controlled environment. Verify it behaves as expected, achieves its goals, and handles edge cases gracefully.

With these design considerations in mind, let's examine different **decision-making frameworks** an agent might use internally. Not all agents decide things the same way – some are reflexive and fast-reacting, others are more thoughtful and planning-oriented. Understanding this spectrum helps in designing the right type of agent for a given problem.

Reactive vs. Deliberative Decision-Making

One of the classic distinctions in agent design is between **reactive** and **deliberative** decision-making. It's essentially a spectrum from quick reflexes to careful planning.

- **Reactive Agents:** A purely reactive agent decides what to do *almost instantaneously* based on the current situation, with little or no internal modeling of the world. It's the "see something, do something" approach. Reactive agents often rely on simple rules or conditioned responses: *if condition X is true, perform action Y.* They don't plan ahead or consider the long-term consequences explicitly – they just respond to the here-and-now stimuli. For example, an obstacle-avoiding robot might just have a rule "if an obstacle is detected to your front, turn right." Classic AI examples include simple thermostats (if too cold, turn heater on; if too hot, turn it off) or enemy NPCs in a video game that attack when the player is in range. These agents are typically **stateless** beyond the immediate moment (no memory of past events) and **don't have an explicit internal world model** . The upside is that they are **fast** and **simple** – there's no heavy computation, which is great for real-time reactions. The downside is they can't handle complex tasks that require planning or considering future states; they might also get stuck in loops or exhibit unintelligent behavior in tricky scenarios (imagine a reactive robot that keeps turning left forever because it doesn't "realize" it could try backing up).

- **Deliberative Agents:** A deliberative agent, on the other hand, has an internal representation of the world and can **reason** about the future. It's like the agent pauses

to think, "Given what I know and what I want to achieve, what sequence of actions should I take?" Deliberative agents engage in processes like **planning** and **search**. They might simulate possible outcomes, evaluate which action will lead closer to the goal, and so forth . These agents maintain an **internal state** (memory of the past, beliefs about the current world) , and they update this state as they gather new observations. Because they consider objectives and have an internal model, they can handle more **complex** and long-horizon tasks that reactive agents cannot . For instance, an autonomous Mars rover that needs to plan a path to a target 500 meters away will deliberate: it will map the terrain, perhaps plot a route around obstacles, and only then execute the plan. This is far beyond a simple "reflex" reaction. The downside is deliberation takes **time and computational resources** – it can be slower to act. There's also the risk of overthinking (or getting stuck in planning). These agents are "the philosophers of the AI realm" who might sit and contemplate their next move if not carefully designed.

Most real-world agents combine **both approaches** in a balanced way, known as **hybrid architectures**. It's often useful for an agent to have some reactive behaviors (for urgent, time-critical responses) and some deliberative capabilities (for strategic planning). For example, a self-driving car agent will react immediately to an obstacle appearing suddenly (hard brake, that's reactive), but it will also deliberatively plan its route and even plan an overtake maneuver considering what other cars might do in the next few seconds (that's deliberative planning).

To illustrate, Elena's team at WebkinsTech implemented a **data center cooling agent**. This agent's goal is to keep server temperatures in range while minimizing energy usage. It uses a deliberative layer to plan optimal cooling strategies (like pre-cooling before an expected heavy load based on forecast), but it also has reactive rules for safety: e.g., *if any server temperature exceeds 85°C, immediately ramp up all fans to max.* The reactive rule is a simple safeguard (no time to deliberate when a server is overheating badly – just act!). The deliberative part handles efficiency over time (figuring out how to cool proactively to avoid reaching that emergency). This hybrid ensures **robustness** (safety through quick reaction) and **optimality** (lower energy through planning).

From an implementation perspective, deliberative decision-making often involves algorithms from AI planning or search. This could be:

- **Rule-Based Systems & Expert Systems:** Early AI agents, especially in defined domains, used a set of logical rules. These can be seen as either reactive (if rules) or part of a larger reasoning system (like chaining rules to deduce a conclusion).

- **State-Space Search:** If an agent can model states of the world and actions that lead to new states, it can perform a search (like BFS, DFS, or A* algorithm) to find a sequence of actions leading to a goal state. For instance, a puzzle-solving agent might search for moves to solve a Rubik's Cube.

- **Planning Algorithms:** These include specialized approaches like **STRIPS planning** or modern PDDL-based planners in AI that take a description of actions and goals and compute a plan. There's also Goal-Oriented Action Planning (GOAP) used in game AI which is similar in spirit.

- **Optimization & Scheduling:** Some agents formulate their decision problem as an optimization problem (like linear programming or constraint solving) – e.g., allocate resources to minimize cost while meeting demands.

- **Reinforcement Learning (RL):** An agent learns a **policy** (a mapping from observed states to actions) through trial and error and feedback (rewards). A fully trained RL agent might then act in a way that appears reactive (just quickly evaluating its learned policy), but the process of arriving at that policy was deliberative during training. RL sits somewhat between reactive and deliberative – at runtime the agent is typically fast (just a neural network forward pass, for example), but its behavior can be very sophisticated because it has internalized a lot of knowledge. One challenge: pure RL agents sometimes have trouble with long-term planning unless designed carefully, but techniques like hierarchical RL or Monte Carlo Tree Search with learned policies are bridging that gap.

In summary:

- Reactive approach = **fast response**, no planning, good for simple or
- time-critical tasks.

Deliberative approach = **planned response**, considers future, good for complex tasks, but needs time/compute.

Elena finds it useful to explain this with another analogy for her team: A reactive agent is like a **reflex** (touch a hot stove, you jerk your hand away immediately), while a deliberative agent is like **problem-solving** (planning a weekend trip, considering various factors). Human intelligence combines both – we have reflexes and we have thoughtful planning – and so do the best AI agents.

WebkinsTech's Cloud Optimizer: A Hybrid Agent

Earlier, we mentioned WebkinsTech's cloud manager agent that autoscaling servers. Let's look at how it mixes reactive and deliberative decision-making:

Reactive layer: For immediate incidents, the agent has simple triggers. If a server's load goes beyond 95% for more than 2 minutes, the agent *immediately* adds a new server instance. If an instance goes unresponsive, the agent takes it out of rotation at once. These are hard-coded rules – quick reflexes to ensure reliability. Elena added these after an incident where an overloaded server caused a cascade because the system waited too long. Now, these reflexes act as a safety net.

Deliberative layer: Every 5 minutes, the agent runs a planning routine. It forecasts the expected load in the next hour (using a predictive model) and calculates an optimal scaling plan: how many servers to have running in each region, to balance performance and cost. It considers things like time of day, a marketing event schedule, etc. It uses this plan to gradually scale up or down, *ahead* of demand changes. This planning avoids oscillations and keeps the system efficient.

The result: The reactive rules handle surprises and emergencies, while the planning logic handles routine scaling smoothly. This agent effectively demonstrates a *layered architecture*, where a reactive sub-agent handles low-level decisions and a deliberative sub-agent handles high-level strategy. Working together, they achieve robust and efficient performance that neither approach could accomplish alone.

Goal Setting and Policy Design

An agent's behavior is ultimately driven by its **goals**. Defining the right goal (or set of goals) is perhaps the most crucial part of designing an agent, because the agent will do *whatever it takes* (within its capabilities) to achieve what you specify as its goal. Mis-specified goals can lead to problematic behavior (there's the classic cautionary tale in AI about a hypothetical agent told to "make as many paperclips as possible" that starts transforming everything, including humans, into paperclips – a scary extreme of goal mis-specification!).

For practical systems, goal setting usually comes down to one or a combination of:

- **Objective Functions:** If you can quantify the goal, you might define a reward or objective function that the agent tries to maximize (or minimize). In reinforcement learning terms, this is the reward signal. In control systems, it could be a cost function. For example, a ride-sharing company's dispatch agent might have an objective function that rewards serving rides quickly and penalizes long idle times or customer wait times.

- **Explicit Goals/Tasks:** Sometimes the goal is a specific state or task to complete (e.g., "find a solution to this puzzle" or "deliver this package to location X"). Planners work with such goals, trying to find a sequence of actions to reach a goal state. These can also be dynamic – an agent might get new tasks over time.

- **Heuristics or Priorities:** In more heuristic systems, the "goal" might be encoded as a priority system. For instance, a domestic robot could have a set of tasks with priorities: (1) if someone is at the door, answering it is top priority (goal = greet visitor); (2) keep the floor clean (goal = vacuum any dirty area); (3) recharge when battery < 20%. This is more of a rule/priority-based goal system rather than a single numeric objective.

Designing a **policy** or decision mechanism for the agent means telling it *how to pursue those goals*. There are a few frameworks to note:

- **Finite State Machines (FSMs) & Behavior Trees:** Common in robotics and game AI, these allow an agent to switch between different modes or behaviors based on conditions. For example, an agent can be in "Patrol" state or "Chase Enemy" state; transitions between states happen based on triggers (like seeing an enemy). Behavior Trees are a hierarchy of conditions and actions often used in game NPCs to create complex behavior from simpler chunks. These approaches are partly reactive

(transitions triggered by environment) but can encode some structured behavior (like a sequence of actions when in a certain state).

- **Utility-Based Agents:** These agents have a utility function that evaluates how desirable a certain state is, and they choose actions that increase utility. This is akin to constantly asking "what's the best thing I can do (highest utility) right now?" It's more flexible than simple goal attainment – it captures trade-offs. For instance, an AI trading agent might have a utility that balances profit vs. risk; it will choose trades that maximize that utility (some profit, acceptable risk).

- **Belief-Desire-Intention (BDI) Model:** This is a popular architecture from academic agent research for modeling rational agents. In BDI, agents have **Beliefs** (information about the world), **Desires** (outcomes they would like to bring about – essentially goals), and **Intentions** (the plans or actions they are currently committed to). A BDI agent will continually update its beliefs from perceptions, use its desires to filter what's relevant, and form intentions (plans) that it then executes. If something changes, it can reconsider and adapt its intentions. BDI is a way to structure deliberation and is useful when designing agents that might have multiple competing goals and need to select among them. Many agent programming frameworks (like JADE or JACK in the Java world) use BDI concepts. For example, Elena considered a BDI approach for WebkinsTech's incident response agents: **Beliefs** = current system status and alerts, **Desire** = resolve any incident affecting uptime, **Intention** = currently working on mitigating a specific outage with a chosen procedure. This can model human-like decision cycles (recognize problem, decide to act, then act).

- **Planning and Scheduling Systems:** We touched on this earlier – if the agent's task can be expressed in a planning domain, then integrating a planner is a powerful way for it to find sequences of actions. These systems typically take a goal state and figure out the steps to get there (like a GPS finding a route on a map, but more general). The agent's policy in that case might be simply "execute the current plan step; if plan is finished or fails, compute a new plan."

- **Reactive Rules & Overrides:** Even in complex agents, it's common to include some specific reactive rules (often as **overrides for safety** or domain-specific hacks for efficiency). We saw this in the cloud optimizer example. Another example: In a digital assistant agent, you might hard-code a rule that if the user says "STOP", the agent immediately stops whatever it's doing – a reactive override for user safety/comfort, regardless of what the agent's current high-level goal was.

Best Practices in Goal Design: One of the best practices in designing the goal or reward for an agent is to ensure it aligns with what you *truly* want to happen – a concept known as **alignment** in AI. In practical terms: define success criteria clearly, and consider edge cases. If WebkinsTech creates an agent to manage disk storage with the goal "keep disk usage below 80%", you don't want it to achieve that by simply deleting logs or data indiscriminately! That would meet the letter of the goal but violate the spirit. Instead, you refine the goal or add constraints: "keep disk below 80% without losing critical data; if cleaning, only remove files older than X days and always keep backups." A well-specified goal often comes as a **combination of metrics and rules**.

Another practice is to include some level of **human oversight or feedback loop**, at least in early deployments. Elena's team often runs new agents in a **shadow mode** where the agent makes decisions but doesn't execute them without human approval, just to see what it would do. This can reveal if the goal design is off. They had a case where an energy-saving data center agent tried to consolidate all load to just a few servers to let others sleep – great for energy, but it risked performance if those few servers spiked. In shadow mode, they caught this and modified the reward function to also include a penalty for having any server above 70% load, achieving a more balanced outcome.

The Perception-Action Pipeline in Practice

To give a concrete example of an agent's internal decision pipeline, let's walk through what happens within WebkinsTech's **Incident Response Agent** (a simplified scenario):

1. **Perception:** The agent receives inputs: an alert that Service A is down, CPU on Server X is 0% (perhaps crashed), and users are complaining on Twitter (sentiment feed input). These come in through various sensors: monitoring APIs and social media feed analysis.

2. **Belief Update:** The agent updates its internal state: it now *believes* Service A is likely offline unexpectedly.

3. **Goal/Desire Check:** The agent's overarching goal is "restore service availability." It recognizes this goal is now triggered (because a service is down).

4. **Option Generation (Deliberation):** The agent considers possible actions. It queries its knowledge base for Service A's recovery procedures. Options might include: reboot Server X, switch to backup instance, notify on-call human, etc.

5. **Evaluation:** The agent evaluates which action best fits the goal. Perhaps it knows from past learning that 80% of the time, simply rebooting the service fixes the issue. It also knows notifying a human

takes longer (and it desires to minimize downtime). So it chooses "attempt automated reboot of Service A on Server X" as the first intention.

6. **Action (Actuation):** The agent executes the reboot via an API call (actuator).

7. **Observation:** It then closely watches the service status (sensor input) to see if it comes back up. If yes, great – goal achieved, maybe send a summary report. If not, it updates its beliefs (reboot didn't solve it) and possibly escalates to the next plan: perhaps deploy a fresh instance of Service A or notify an engineer, etc.

8. **Learning (If applicable):** After the incident, the agent logs the outcome: reboot failed, a redeploy was needed, and that solved it. It can use this to update its knowledge (maybe next time try redeploy first if similar symptoms).

This cycle shows how an agent can combine reactive steps (immediate recognition of an outage triggers a straightforward action) with deliberative ones (considering multiple options to resolve it, planning steps if first attempt fails).

Throughout all this, what gives the agent "agency" is that no one told it at 3 am "Service A is down, go reboot it." The agent autonomously *detected* the problem and *decided* on a course of action within the parameters and goals it was given. It's like a diligent team member working the night shift, handling issues as they arise.

Pitfall: Overly Rigid vs. Overly Unconstrained Agents

Designing an agent's decision-making requires finding a sweet spot between rigidity and freedom. If the agent's rules are **too rigid** (like a pure reactive system with no ability to adapt), it may fail when conditions deviate slightly from what you anticipated. For example, if you coded an agent with a specific threshold (CPU > 90% triggers action) and one day 89% usage causes a failure, the agent would blindly do nothing because the exact rule wasn't met. Rigid agents can also be brittle to noise or slight changes in the environment.

Conversely, if an agent has **too unconstrained a goal or policy space**, it might find unintended solutions. This is sometimes called the *"alignment problem"*. An infamous anecdote: an experiment with a cleaning robot agent that was rewarded for picking up trash started taking the trash and hiding it under a couch – technically the floor was clean (goal achieved) but not in the intended way! Always think about corner cases: *What weird shortcut might the agent take to achieve its goal?* Add constraints or tweak the objective to prevent this. In the cleaning robot case, adding a penalty for moving trash outside of designated bins would align the goal with the designer's intent (i.e., *trash properly disposed of* vs just *out of sight*).

Lesson: Test your agent's decision-making logic with scenarios and adversarial cases. Iterate on the rules and objectives to ensure the agent's "common sense" aligns with real-world expectations.

Example: Agentic Decision Frameworks at WebkinsTech

To bring these abstract ideas to life, let's examine a scenario from WebkinsTech where Elena had to choose a decision-making framework for a new agent. WebkinsTech was deploying a **Smart Deployment Agent** – an agent to decide when to roll out new versions of microservices and to which servers, in order to minimize user impact (a sort of release orchestrator).

They had two prototypes:

1. A **Rule-based reactive prototype**: It had rules like "If traffic low and error rate low, then deploy new version to one server." "If that server's error rate stays low for 5 minutes, deploy to rest." This was straightforward and worked in most simple cases.

2. An **AI planning prototype**: This one would look at the dependency graph of microservices, the current load, etc., and plan an optimal sequence of rolling upgrades (maybe choosing an order that minimizes cross-service impact). It used a deliberative approach considering many factors.

During testing, the reactive prototype did fine when conditions were normal, but it failed to deploy during an unusual scenario (it was waiting for "traffic low" which never happened because traffic was 24/7 high due to a campaign, so it never deployed the needed fix!). The planning agent handled that case by finding a plan to deploy gradually even under high traffic (it planned to temporarily redirect some traffic elsewhere, then deploy, etc.). However, the planning one was a bit slow and occasionally would come up with overly complex plans for simple scenarios.

Elena's solution: **combine them**. The final agent used a rule-based approach for known safe scenarios (nighttime low traffic – just do the simple rolling deploy) and switched to a planning mode when conditions were outside the norm. It could even do a quick evaluation: if no simple rule applies, trigger the planner. This way they got the best of both worlds.

This example shows that understanding various decision frameworks (from simple to complex) allows you to tailor the agent's "brain" to the problem at hand. Always ask: does this task need reasoning about future steps? Or is it mostly a real-time response? Can we enumerate all conditions or should the agent learn/adapt?

By carefully designing the agent's internal decision logic and goals, WebkinsTech ensures their agents are both **effective** (achieve what's needed) and **safe** (don't do something crazy). In the next part, we'll expand our view from a single agent to **multiple agents working together**. Many of WebkinsTech's systems involve not one, but many agents – each handling a piece of the puzzle. How do these agents coordinate? What patterns of interaction emerge? Let's find out.

Multi-Agent Systems and Coordination Patterns

So far, we've considered single agents in isolation. But many real-world problems are too large or complex for a lone agent. This is where **Multi-Agent Systems (MAS)** come in – ensembles of agents that interact, collaborate, or even compete within an environment. In a multi-agent system, each agent might have its own specialty or role, and together they can tackle problems beyond the reach of any individual agent .

WebkinsTech's architecture is a great example: they didn't just deploy one monolithic AI agent to run the entire company's IT. Instead, they have a **team of agents**: some monitor infrastructure, others manage deployments, some handle security scans, others optimize costs. These agents must work in concert, much like different departments in an organization.

However, coordinating multiple autonomous agents is no small feat. If you think getting humans to work together on a project is hard, imagine dozens of AIs each doing their own thing – we need structures and protocols to avoid chaos. This section explores how agents can coordinate, the patterns that enable effective teamwork, and some of the fascinating mechanisms (like marketplaces and pheromone-like signals) that allow distributed agents to self-organize. We'll look at both **cooperative** scenarios (agents share goals) and **competitive** ones (agents have individual goals, possibly at odds), though our focus in microservices will mostly be cooperative or at least *coordinated* behavior.

Cooperative vs. Competitive Agents

First, let's clarify the distinction between **cooperative** and **competitive** multi-agent setups:

- **Cooperative Agents:** These agents share a common goal or are at least working towards a collective objective. They will coordinate their actions to jointly solve tasks or maximize a global utility . In cooperative settings, communication tends to be about sharing information and helping each other. Examples: a team of robot rescuers in a disaster scenario, working together to find survivors and allocate areas to search; or multiple server-managing agents at WebkinsTech that collectively try to keep the whole system stable (one might manage database, another manage web servers, but they both want overall uptime). In cooperative MAS, the challenge is often how to **decompose the problem** and ensure agents help rather than hinder each other (avoiding duplication or interference).

- **Competitive Agents:** These agents have different (sometimes opposing) goals. This is the domain of game theory and economics. Competitive agents might be trying to maximize their own utility even at the expense of others. They may still interact (e.g., negotiate, bid, fight), but not necessarily to help each other – rather to achieve their own ends. Examples: automated trading bots in a market (each wants to profit, maybe at others' expense), or opposing AI players in a strategy game. In such settings, you might have concepts like adversarial planning, auctions, or contracts. Coordination here can mean forming temporary alliances or just adhering to shared protocols (like rules of an auction or game).

There are also **mixed settings** (also called *cooperative-competitive* or *collaborative with self-interest*) – for instance, two departments (agents) in a company each want to maximize their performance, but they also need to collaborate for the company's overall success.

Many real systems fall in between pure cooperation and pure competition.

For the purpose of intelligent microservices, we often design agents under a unified goal (the system's reliability and efficiency). So we lean heavily into **cooperative multi-agent patterns**. Even if agents have specialized sub-goals (one focusing on cost saving, another on performance), we design them under an overarching objective that aligns them (like a weighted balance between cost and performance). This avoids agents working at crosspurposes.

However, even cooperative agents can conflict accidentally without coordination. For example, a cost-optimizer agent might want to shut down servers at night to save money, while a performance agent might want to keep them all running "just in case". If each blindly did its thing, they'd constantly undo each other's actions! Coordination mechanisms are needed to reconcile such differences – perhaps a higher-level orchestrator sets a policy: maintain minimum 3 servers at night, more if needed. Or the agents negotiate a compromise (cost agent says "I propose shutting 5 servers", perf agent counters "that will hurt latency, how about shutting 3", etc.). This leads us to coordination patterns.

Patterns of Coordination

Over the years, researchers have identified common **interaction patterns** for multi-agent systems . These are like design patterns in software, but for how agents communicate and collaborate. Let's explore a few key patterns (some of which were touched on earlier by the Confluent team in an event-driven context):

1. **Orchestrator-Worker (Master-Slave) Pattern:**

One agent is the boss (orchestrator) and the others are workers that carry out tasks assigned by the orchestrator . This is analogous to a manager giving orders to a team. The orchestrator has the overview of the global goal and breaks it into pieces, delegating to worker agents. The workers execute and report back results. This pattern simplifies coordination because one central brain makes decisions, but it can be a single point of failure or bottleneck. In microservices, this could be like a central scheduling agent distributing jobs to multiple processing agents. WebkinsTech uses this in their batch processing system: a **Batch Orchestrator Agent** looks at pending jobs every hour and assigns tasks to several **Worker Agents** that handle the jobs in parallel. The orchestrator ensures no two workers duplicate the same job and that the workload is balanced. This pattern is effective but not fully decentralized.

Orchestrator-Worker at WebkinsTech

WebkinsTech's AIOps includes a **Monitoring Orchestrator Agent** that acts as an overseer. It receives all alerts and issues detected in the system (like high CPU, low memory, security flags). It doesn't fix anything itself; instead, it delegates to specialized worker agents: e.g., a **Scaling Agent** for performance issues, a **Healing Agent** for crashed services, a **Security Agent** for security alerts. When multiple issues come in, the orchestrator prioritizes them and tells the appropriate agent to take charge. This way, the worker agents don't step on each other's toes – they only act when instructed. The orchestrator makes sure two agents aren't trying to fix the same thing in different ways. This clear division of labor (with one "brain" coordinating) has worked well, though Elena is mindful that the orchestrator agent must be highly reliable. They even have a backup orchestrator agent in hot standby, just like you'd have for any critical service.

2. **Hierarchical Agents Pattern:**

This is essentially a multi-level orchestrator-worker pattern . Instead of one master and many flat workers, you have layers of agents. A top-level agent might delegate to midlevel agents, which in turn delegate to lower-level agents. Think of an army command structure, or a corporate org chart. This can mirror complex problem breakdowns. For instance, in a smart city simulation, a top agent handles city-wide policy, mid-level agents handle each district, and low-level agents manage individual intersections or resources. The hierarchy can help scale coordination because each agent only manages a subset at its level. However, it introduces multiple points of coordination and potentially latency in decisions. At WebkinsTech, they don't use a very deep hierarchy because their domain doesn't demand it, but one could imagine a hierarchy like: Global Orchestrator for all data centers -> Data Center Manager agents -> individual Server agents. Each level abstracts the complexity below.

3. **Blackboard (Shared Memory) Pattern:**

In the blackboard pattern, agents communicate indirectly by reading and writing to a common **shared data structure** called the "blackboard" . You can think of it literally like a blackboard in a meeting room where anyone can jot down notes or read what others have written. Agents post information, hypotheses, partial solutions, etc., to the blackboard, and other agents pick up tasks or contribute based on that information . This pattern is great for problems where you can accumulate pieces of a solution from different experts. It was inspired by early AI systems where multiple knowledge sources incrementally solved a problem (like different experts solving parts of a puzzle and writing their findings on a board for others to see). Blackboard systems are **asynchronous and decoupled** – agents don't talk to each other directly, they talk via the board. This can reduce the complexity of pairwise communication and avoid the need for one central orchestrator. The drawback is you need a well-defined schema for what's on the blackboard and how agents know when to act on it. Plus, if too many agents write at once, it could get messy without some control or locking mechanism.

In practice: WebkinsTech implemented a blackboard-style system for their complex incident analysis. They have an **Incident Blackboard** where various diagnostic agents write their findings. For example, one agent monitors network logs and if it sees anomalies (say, a spike in errors between Service A and B), it writes an entry: "ServiceA->ServiceB communication errors high." Another agent monitoring server metrics might post: "CPU on ServiceB maxed out." A separate **Analysis Agent** reads these and infers a bigger picture ("Possibility: Service B is overloaded, causing downstream errors"). That agent writes a hypothesis: "Hypothesis: Service B bottleneck". Yet another agent, seeing that hypothesis, decides to test it by temporarily increasing Service B's resources, and then writes the result ("After adding capacity to B, errors went down"). All agents effectively converse through this shared knowledge base. The issue gets resolved through collaborative posts rather than direct messaging. This asynchronous collaboration is powerful for investigative or incremental problem solving.

To cite a formal definition: *"The blackboard pattern provides a shared knowledge base – a 'blackboard' – that agents use to post and retrieve information. This pattern enables agents to collaborate asynchronously without direct communication."* It's especially useful for solving complex problems requiring incremental, multi-agent contributions .

4. **Market-Based (Auction) Pattern:**

This pattern borrows concepts from economics – treat tasks or resources as goods in a marketplace, and agents as buyers/sellers or bidders. Agents **negotiate or bid** for tasks, and some mechanism (an auctioneer or a market logic) allocates tasks or resources based on bids . For example, imagine a fleet of delivery drones (agents) and packages that need delivery (tasks). A market-based system might have each package "announce" itself for delivery and drones bid (perhaps bidding lower time or cost to deliver it). The package is assigned to the drone with the best bid. Or in resource allocation, multiple service agents might bid for extra CPU based on how badly they need it, and a central auction allocates CPU to those who value it most (this ensures resources go where they bring the most benefit). Market-based coordination can be very effective for decentralized decision-making, and it inherently handles trade-offs and competition in a fair way (if designed well). It does require that you can quantify the "bids" or utility in comparable terms.

WebkinsTech experimented with a market approach for scheduling background jobs. Each **Worker Agent** would bid for a new job based on its current load (like, "I can finish it in estimated 5 minutes"). The job is assigned to the agent that promises the quickest completion. This turned scheduling into a self-organizing auction rather than a central scheduler. They found it worked but also ended up similar in outcome to a centralized solution; still, it was a fun experiment in robustness (if one worker mis-estimates, others will win bids next time, etc.).

In literature, it's described: *"This pattern models a decentralized marketplace where agents negotiate and compete to allocate tasks or resources."* For example, solver agents might exchange proposals and an aggregator picks the best solution . A realworld analogy is different contractors bidding for a project, and the best bid wins the contract. In multi-agent terms, it's a way to achieve global allocation optimizations through local decisions and bids.

One subset of this is the **Contract Net Protocol (CNP)**, a well-known coordination protocol where a manager agent announces a task, contractor agents bid, and manager awards the contract to one (similar to an auction).

Market-based systems are quite useful when you have a **competitive element** or at least a need to distribute work or resources efficiently without a heavy central authority.

5. **Stigmergy (Environmental Coordination):**

Stigmergy is a less direct, but very intriguing coordination mechanism inspired by social insects like ants and termites. The idea is that agents coordinate by **changing their environment** in a way that influences others, rather than by direct communication. In an ant colony, for instance, ants leave pheromone trails on the ground when they find food; other ants sense this chemical and it guides them to the food source. Over time, paths with stronger pheromone (because more ants traveled there) attract more ants, effectively coordinating the colony's foraging. The term *stigmergy* describes *"a mechanism of indirect coordination, through the environment, between agents or actions. The principle is that the trace left in the environment by an action stimulates subsequent actions by the same or other agents."*. In simpler terms, agents leave **signals in a shared environment** that others respond to, enabling group behavior to emerge.

In computational systems, we mimic this by using things like digital pheromones or shared data points that have properties like evaporation (fading over time). For example, suppose we have a swarm of robotic vacuum cleaners. Instead of them communicating directly about which areas are cleaned, they could drop a virtual marker in the map when they clean an area. That marker can slowly fade (evaporate). Vacuums then tend to go where markers are weaker (not recently cleaned) and avoid strong markers (just cleaned), achieving a natural distribution of work. No central planner, no direct messages – they coordinate by the "traces" they leave.

Another example: network routing can use stigmergy. Packets (or agents representing them) lay down "pheromones" on paths. Good paths get reinforced, bad ones fade. Eventually, routes emerge. This is akin to **ant colony optimization algorithms** in computer science.

In WebkinsTech's context, stigmergy isn't an obvious choice since their agents have more direct ways to communicate. But Elena once applied a stigmergic approach to load balancing: each server agent would write an "availability score" to a shared table that decays over time. When a server was very busy, its score dropped; idle servers had high scores. Other agents (or a distributor) would route new tasks to servers with higher scores. The continuous update and decay of scores acted like a pheromone system guiding tasks to free servers. It was robust in that it didn't require precise realtime communication – even if a server's status update was slightly delayed, the residual "pheromone" system smoothed it out. The system self-adjusted as busy servers "repelled" new tasks (low score) and free servers "attracted" them (high score). This is a form of stigmergic coordination through an environmental variable (the score).

Stigmergy often results in **emergent behavior** – complex global patterns arising from simple local rules. It's powerful but can be tricky to design since you're relying on indirect effects. If done right, it offers a very scalable and fault-tolerant coordination (ants don't need a leader to know what to do; likewise, agents following stigmergy just do their local thing and the group behavior emerges).

On Stigmergy

"In these insect societies, individuals influence each other's behavior by leaving traces in the environment, which others then respond to. For instance, ants leave pheromone trails to food sources, guiding other ants to the same location. This form of communication is indirect but highly effective, allowing complex behaviors to emerge from simple interactions." – *Hank Holiday (explaining stigmergy in organizational design)*

This quote, while discussing ants, beautifully captures why stigmergy can be useful even in AI systems: it enables simple agents to coordinate on complex tasks by using the environment as a shared medium of communication.

6. **Direct Communication and Teaming:**

Beyond these structural patterns, agents can also directly message each other in a peer-to-peer fashion. For instance, agent A can send a message "I need help" to agent B, and B can respond or change behavior. This is flexible but requires careful protocol design if many agents are involved (to avoid a chaotic web of messages). Often, multiagent frameworks have communication languages (like FIPA-ACL, an agent communication language) and protocols (request, inform, propose, etc.). In microservices, we might implement direct comms via APIs or messaging queues. While not a fancy pattern, direct messaging is important for explicit coordination when needed (e.g., a cache agent telling a database agent "I've updated my cache with key X, you might invalidate your old copy").

Coordination in Practice: WebkinsTech's Multi-Agent Ecosystem

Now let's tie these patterns to a practical scenario. WebkinsTech's internal AIOps architecture involves multiple agents as mentioned. How do they actually coordinate in daily operations? We already saw bits of this with the orchestrator-worker and blackboard examples. Let's expand it into a cohesive narrative:

WebkinsTech AIOps Agents:

- **Monitoring Orchestrator Agent** – collects signals and delegates to others (orchestrator pattern).
- **Scaling Agent** – adds/removes servers to maintain performance.
- **Healing Agent** – restarts or replaces failed components.
- **Security Agent** – monitors for security issues (intrusions, vulnerabilities) and can take containment actions.
- **Analysis Agent** – does deeper analysis on complex issues (this one often uses the blackboard to compile evidence).
- **Cost Optimizer Agent** – looks at resource usage and costs, identifies waste (e.g., idle resources to turn off).

Coordination scenario: A sudden traffic spike hits, causing some microservices to struggle.

- The **Monitoring Orchestrator** gets alerts of high latency and error rates. It immediately engages the **Scaling Agent**: "Service X needs more instances." (Direct delegation as orchestrator-worker).
- The **Scaling Agent** uses its logic (which might include a bit of planning to not overshoot) and adds two instances of Service X. It posts an update on the Incident

Blackboard: "Added 2 instances to Service X, monitoring results."

- Meanwhile, the **Monitoring Orchestrator** also notifies the **Analysis Agent**: "We have a traffic spike scenario, keep an eye on system-wide metrics." The Analysis Agent starts gathering data (it might query other agents or just monitor the blackboard entries).
- The **Cost Optimizer Agent** notices a sudden increase in resource usage (more servers = higher cost). Normally, it might try to shut down something else to compensate, but it also sees from the context (could be through a shared state or blackboard) that this is intentional due to high load. Perhaps there's a rule that during incidents, cost optimization is secondary. So the Cost

Agent holds off any action – an example of implicit coordination by prioritization.

- Now, suppose the traffic spike also caused some cache misses or DB stress. The **Healing Agent** might see an alert that a database read latency is very high. It decides to flush a cache or restart a cache server that's stalled (it has authority in that area). Before doing so, it could put a note on the blackboard: "Cache server Y unresponsive, restarting it." This prevents the Analysis Agent from double-counting that as a new problem – they know it's being handled.

- The **Security Agent** in this case might not be involved (no security issue), but let's say in parallel it found a possible malware file on a server. It would ideally handle that without interfering with the traffic spike mitigation (different domain). If it needs to quarantine a server, it might check with the Orchestrator or mark it in the blackboard, so the Scaling Agent knows that server is off-limits for new tasks (coordinating via environment marking).

- Throughout this, agents use a **shared "language" of events** to stay in sync. In fact, WebkinsTech uses an event bus (like Kafka) and all agents produce and consume relevant events (similar to a blackboard but streaming). This is akin to the **common operating model** described in modern designs, where *"events serve as the system's group chat: keeping agents synchronized and integrating new ones smoothly."* Each agent emits events like "Scaling performed" or "Service X healthy now", and other agents listen and react accordingly, maintaining a shared context.

The outcome: within a few minutes, the system stabilizes under the load. The Scaling Agent added capacity, the Healing Agent fixed the cache, the orchestrator managed priorities, and the Analysis Agent compiled a report that this was due to an unexpected marketing campaign mention causing a traffic surge. The agents all did their parts without a human in the loop, and importantly, without working at cross purposes. This is coordination success.

To facilitate this dance of agents, Elena's team put significant effort into designing the **interfaces and communication** between agents:

- They defined the events and blackboard entries format clearly (so all agents interpret them consistently).

- They set up rules for precedence (like scaling actions take priority over cost-cutting during high load).

- They test multi-agent scenarios frequently, much like how one would do integration testing in microservices, to see that the agents together produce the desired system behavior.

One more pattern worth noting that emerges in multi-agent systems is **emergent leadership or dynamic roles**: sometimes one agent might take lead in a situation not because it's preassigned but because it has the right information. For example, if an Analysis Agent deduces the root cause of an issue, it might temporarily instruct other agents ("stop adding more servers, it's a database issue, focus on DB"). In a way, the one with the insight becomes an orchestrator for that incident. Designing for such flexibility – where any agent can emit suggestions or constraints that others heed – is an advanced approach and can be implemented via something like the blackboard or a publish/subscribe priority system (e.g., an agent publishes "flag X means do not scale further", and scaling agent subscribes and respects that flag).

Visualizing Coordination

It often helps to visualize how agents coordinate. If we were to draw diagrams for some patterns:

- **Orchestrator-Worker Diagram:** A central node (orchestrator) connected to multiple worker nodes. Arrows from orchestrator to workers saying "assign task", and arrows back from workers saying "result done". (Imagine a star topology.)

- **Blackboard Diagram:** A central blackboard entity (like a chalkboard icon), and multiple agents around it. Each agent has an arrow going to the blackboard (to read/write). One agent might write a fact, another reads that fact. It shows a hub but the hub is just data, not a controller.

- **Market Diagram:** Perhaps tasks as one set of nodes, agents as another. Lines representing bids from each agent to each task, then one agent wins each task. Or show an auctioneer agent in the middle if using an explicit auctioneer.

- **Stigmergy Diagram:** Agents moving in an environment (like dots moving), leaving behind traces (like colored pheromone trails on a grid). Other agents later follow those trails. (Picture ants and pheromone lines.)

- **Event-Driven (Pub/Sub) Coordination:** Could be drawn like a bus or a topic that all agents connect to. They put events on it and react to events from it. Similar to blackboard but more dynamic.

While we may not embed all such diagrams here, mentally picturing them helps to solidify how these patterns function.

One figure Elena often shows in presentations is a **coordination timeline**: a timeline with multiple agents and messages between them (like a sequence diagram in UML). It highlights how, for a given scenario, messages/events flow between agents. It helps find any deadlocks or unnecessary chatter between agents.

Advantages and Challenges of Multi-Agent Coordination

Advantages:

- **Scalability:** Multi-agent solutions can scale out with more agents handling subdivided tasks. If one agent can't handle a load, multiple cooperating agents might.
- **Robustness:** If designed well, failure of one agent doesn't cripple the system – others can fill in or the system degrades gracefully. Decentralized patterns (like stigmergy or market) especially have no single point of failure.
- **Specialization:** Agents can become experts at sub-tasks, potentially using different methods (one might use ML, another rule-based, etc.). Together, they cover a breadth of functionalities.
- **Parallelism:** Tasks can be done in parallel by different agents – speeding up overall operation.
- **Emergent Intelligence:** Sometimes a solution emerges from agent interactions that wasn't obvious in a monolithic approach. Multi-agent reinforcement learning has shown agents can collaboratively solve tasks like playing football or handling traffic lights in a city by emerging strategies.

Challenges:

- **Communication Overhead:** Agents need to exchange info; too much communication can bog down the system or complicate design.

- **Coordination Complexity:** Avoiding conflicts and ensuring alignment can be tricky. You often need to implement conflict resolution (who wins if two agents want opposite actions?) and consensus mechanisms.

- **Design and Debugging:** Distributed systems are harder to debug. A bug might not be in one agent but in an interaction between two. It's like debugging a team: individual members might be fine, but their combo causes an issue.

- **Performance Tuning:** If not careful, multiple agents could unintentionally thrash the system (e.g., two agents oscillating state by undoing each other). Achieving a stable equilibrium is an art – sometimes requiring adding slight delays, damping responses, or arbitration logic.

- **Emergent Unpredictability:** Emergent behaviors are a double-edged sword: you might get clever solutions, or you might get weird unintended behaviors that are hard to predict from the get-go. Simulating a wide range of scenarios is important.

WebkinsTech confronted some of these challenges. Early on, they had an issue where the Cost Optimizer Agent kept shutting down servers that the Scaling Agent would then immediately restart (since performance degraded). It was a tug-of-war that oscillated every 10 minutes. The fix was to introduce a **shared policy**: during business hours, performance wins over cost; after hours, weight cost a bit more but with a smoothing factor (don't remove more than one server per hour). Essentially, they made the agents "agree" on a schedule and moderation. This stopped the oscillation.

Another challenge was ensuring the **Security Agent** didn't quarantine too aggressively without telling others. At one point it isolated a machine for suspected malware, but the Scaling Agent saw a server drop and provisioned a new one – which copied over the same malware because the root cause wasn't fixed! Now the Security Agent, when it isolates a server, broadcasts a "quarantine event" that instructs the Scaling Agent not to replace that server until cleared. Coordination and communication again solved the conflict.

Multi-Agent Architectures and Microservices

The phrase "Intelligent Microservices" in the book title hints that each microservice could be thought of as an intelligent agent. Indeed, in a modern distributed system, we can imagine each service is instrumented with an agent that manages it and communicates with other service-agents. This starts to look like a multi-agent system on top of a microservice architecture. Each service might have goals (like ensuring its own reliability, meeting its SLA). They might coordinate via patterns we discussed, likely through event-driven messages (which is akin to a blackboard or pub/sub).

Many of the patterns in multi-agent coordination have analogs in microservices communication:

- Orchestrator-worker is akin to an API gateway or orchestrator service calling other services (like a central coordinator service).

- Event-driven pub/sub is common in microservices for loose coupling (very similar to a blackboard or stigmergy if you think of events as digital traces).

- Market-based could be compared to dynamic service discovery and load balancing where instances sort of compete for traffic (though typically not implemented as auctions, but conceptually).

- Hierarchical is like layered services (maybe a root service calling mid-tier services which call lower services).

- Direct comms is just synchronous RPC calls between services.

So, the knowledge of multi-agent systems can inform how we design microservice interactions, especially as we inject more autonomy into them. The next chapter will likely delve deeper into how to architect a system of intelligent microservices (agents) at a system level, ensuring they form a coherent whole.

To wrap up this section, let's consider a final *big picture* analogy: a multi-agent system is like an **organization or team**. You have individuals (agents) with certain roles and skills. They communicate and coordinate via meetings (blackboards), emails (direct messages), market decisions (bidding for resources, like budgeting internally), and managers (orchestrators). The organization works well if roles are clear, communication is effective, and everyone is aligned to a common mission. If those break down, you get inefficiencies or conflicts. Designing MAS is somewhat like organizational design for software entities. Elena, acting somewhat like the **chief AI-architect**, ensures that WebkinsTech's agent team is structured and "managed" in a way that the overall system goal (keep services running optimally) is achieved.

Having explored agents alone and in teams, we have a solid grasp of the core concepts of agentic AI. We understand how autonomous agents differ from traditional software, how they internally operate with goals and decisions, and how multiple agents can coordinate through various patterns.

To ground these ideas, let's turn to a concrete case study that encapsulates everything we've discussed. In the following case study, we'll walk through a real (fictional) incident at WebkinsTech where agentic AI played the hero, highlighting the transition from a traditional system to an agent-driven solution, and how the coordinated agents handled the situation. This will not only reinforce the concepts but also set the stage for looking at **system-level thinking** in the next chapter.

Case Study: Night of the Agents – WebkinsTech's Transformation

It was 2:00 AM on a Saturday when an unexpected chain of events put WebkinsTech's new agentic AI approach to the test…

Elena is jolted awake by a notification on her phone. It's an automated message from the system: "Multiple services degraded. Agent team responding." Normally, Elena would dread a late-night incident, but this time she's more curious than scared – it's the first major incident since deploying the full fleet of intelligent microservice agents. Will her "digital colleagues" manage it?

Background

A few months ago, WebkinsTech's infrastructure was more manual and reactive. One notorious incident – the "Black Friday Meltdown" – had exposed the limitations of their old setup. During a peak usage period, a database became a performance bottleneck. The monitoring scripts noticed too late, engineers were woken up, and by the time a manual fix (adding read replicas) was applied, users had experienced slowdowns and some even timeout errors. It was a dent in WebkinsTech's reputation, and Elena vowed to never let it happen again. That crisis spurred the adoption of an **agentic AI approach**.

Elena and her team introduced multiple agents (as described earlier) for proactive monitoring, scaling, healing, etc. They also set up an **incident command agent** (the Monitoring Orchestrator) to coordinate responses.

Now, on this unassuming Saturday night, a perfect storm is hitting: a routine deployment during low traffic hours coincided with an unexpected surge of traffic from a partner integration, plus a latent bug in a new release causing memory leaks. A trifecta of trouble.

The Incident Unfolds (Traditional vs. Agentic response)

Old World (What would have happened before):

In the pre-agent era, a surge + memory leak would first trigger some CPU and memory alarms. A human on-call might get paged (if the alarms were loud enough). They'd have to wake up, diagnose which services are leaking or struggling, perhaps restart services or roll back the deployment. Meanwhile, users might already be facing slow responses or errors. Coordination depended on the on-call person pulling in others if needed ("This looks like a DB issue, I need a DB specialist to help"). Each step was manual and slow relative to computer speeds.

Agentic Response (What happened now):

2:05 AM – The **Scaling Agent** has already noticed the traffic surge (coming from the partner's API calls that were mistakenly hitting WebkinsTech's service en masse). It starts adding a couple of application servers preemptively, seeing the trend of rising load. It also emits an event: "Scaling up Service A – high incoming request rate."

2:06 AM – The new deployment had a memory leak in Service B (a downstream service). Memory usage is climbing steadily. The **Healing Agent** sees that one instance of Service B is nearing its memory limit and will likely crash. It doesn't wait – it gracefully restarts that instance to clear the leak, and notifies the Deployment Agent that this instance was recycled. It then schedules a staggered restart of the other instances of Service B, one at a time, to preemptively clear their memory before they fail ungracefully.

2:08 AM – The **Monitoring Orchestrator** agent correlates events: multiple services are now involved (Service A scaling, Service B being restarted by Healing Agent). It declares an "incident context" and ensures all relevant agents know a coordinated incident response is in effect. It pings the **Analysis Agent**: "Investigate root cause – multiple services abnormal."

2:09 AM – The **Analysis Agent** pulls logs and finds that right before this incident, a new version of Service A and B were deployed (the routine deployment). It also sees the surge of requests from Partner X's IP range. It forms a hypothesis: "New deployment plus traffic surge causing strain. Possibly a bug (memory leaks observed in Service B)." It posts this on the Incident Blackboard.

2:10 AM – The **Deployment Agent** (part of CI/CD automation) is an observer here; it sees that the new version might be problematic. Upon the Analysis Agent's hypothesis, the Deployment Agent decides to **pause any further rollouts** and prepares a rollback for Service B's new version as a contingency.

2:12 AM – The system is still under heavy load, but users haven't noticed much disruption.

Thanks to the Scaling Agent, there were enough instances of Service A to handle the surge (just some slightly increased latency until scale-out caught up). The Healing Agent's proactive restarts of Service B kept that service from crashing outright, though some user requests might have been slightly delayed during each restart. So far, minor impact.

2:15 AM – The **Cost Optimizer Agent**, which normally might fret about so many new instances (the cloud bill is spiking), knows an incident is ongoing (it subscribed to incident events). It stands down entirely – no actions to cut costs now; it even helps by identifying any non-critical batch jobs or dev/test instances that can be terminated to free up capacity for the main workload.

2:20 AM – The surge subsides (turns out it was a one-time batch from the partner, finished now). The memory leak, however, remains a concern. The Healing Agent rotated all Service

B instances, but the leak will build up again. The Analysis Agent recommends a rollback on Service B to the previous version (which didn't have the leak). The Deployment Agent executes the rollback automatically (since it was pre-authorized in such conditions). Service B is returned to last known good state.

2:30 AM – All services are stable. The agents have scaled Service A back down to normal levels (gradually, to ensure the load was truly gone), and confirm no further leaks in Service B after rollback.

2:45 AM – Incident is resolved. The agents even handle the **post-incident reporting**: The Analysis Agent compiles a summary (timeline of events, actions taken, likely root cause being the bug in new Service B plus unexpected partner load). It stores this in the knowledge base and sends a summary to the engineering team (so when humans wake up, they're informed). The system even creates a ticket to follow up with the partner about rate limiting, and another ticket for developers to fix the memory leak before re-deploying that update.

Reflection: Human Involvement

Elena, watching much of this from home on her laptop (since she woke up from the alert), is amazed. The incident was essentially handled by the agents. She intervened minimally – mostly to double-check that the decisions made were reasonable (they were). She gave a quick acknowledgment in the system dashboard: "Agents handled incident – no human intervention needed, good job." (She jokes by sending a thumbs-up emoji in the incident channel, even though only bots are there to see it.)

This case study showcases:

- The autonomy of agents (scaling, healing, analysis all doing their jobs without being told).
- The coordination (through orchestrator oversight, shared incident context, blackboard hypothesis posting, etc.) ensuring all agents were aligned to solve the problem, not working at cross purposes.
- Proactiveness (fixing things *before* massive user impact) as opposed to the old reactive script-based approach.
- Learning/Improvement potential: The knowledge base now has a scenario logged that can train future responses (maybe next time, the system will automatically not deploy when a partner surge is detected, or will autoscale even faster).
- The reduced need for human firefighting – freeing engineers to sleep at night or focus on planned work.

One can imagine if something went outside the agents' scope, Elena or another on-call could step in. But importantly, the agents handled the bulk, and the system stayed resilient.

The Transformation Journey

It's worth noting how WebkinsTech got here:

- They started with identifying key pain points (like slow incident response, missed signals).
- They gradually introduced agents in parallel with existing systems, tested them in shadow mode (initially the agents would observe and recommend, while humans still made final calls).
- Over time, confidence grew and agents were allowed to act autonomously, first in lowrisk areas, then progressively in critical paths.
- They set up fail-safes: any agent action could be overridden by a human, and agents would defer if a human command conflicted (just as a co-pilot AI defers to a pilot's direct command in a plane).
- They built trust as the agents proved themselves in minor incidents.
- Finally, on a night like this, the true payoff is seen.

Elena presents this internally as **"Incident 47: The Night Our Agents Stood Watch"**. It has become a shining example within WebkinsTech of why an agentic AI approach is powerful. It wasn't about removing humans entirely; it was about leveraging the speed and tirelessness of AI to handle routine or fast-moving tasks, while humans design and supervise the overall logic.

She also notes areas to improve: the system did send her a 2 AM alert initially, which perhaps it didn't need to until if things had gotten worse. They might tweak the alerting so that if agents handle it and things recover in, say, <30 minutes, it can just notify in the morning with a summary instead of waking someone. However, it was the first big trial, so a cautious alert was understandable.

Looking Ahead

By the end of the case study, we see a microservice architecture not as a static set of programs but as a living, **agent-based ecosystem**. The agents, each with specific roles and expertise, cooperated to navigate a challenge. This is the essence of the agentic approach: building systems that are **adaptive, autonomous, and aligned with our goals**.

WebkinsTech's next steps involve scaling this approach further – more predictive agents, maybe using advanced learning to anticipate issues days in advance, and extending coordination to not just IT ops but business processes as well. Elena is already thinking about an "Agent Operations Center" where agents might even coordinate with external agents (like the partner's systems) for smoother collaboration across organizational boundaries.

As we conclude this chapter, we have not only explored agentic AI in theory but also seen it in action. We saw how an agent differs from traditional systems, how it internally makes decisions, and how multiple agents can form a robust network that self-manages a complex system.

Segue to Chapter 5: In the next chapter, we will take a broader architectural view. Now that we grasp the core concepts of agentic AI, we'll look at how to **design entire systems** with these agents – essentially **agentic architectures** for intelligent microservices. We'll discuss best practices in layering agents with traditional components, how to handle data and knowledge in an agent-driven system, and dive deeper into case studies of system-level design. We'll also address questions of governance: how do we oversee a system of many agents? Chapter 5 will bridge the gap from individual agent patterns (which we now have under our belt) to holistic system thinking – ensuring all the pieces come together in a cohesive, intelligent architecture. The journey from reactive scripts to proactive agents that we saw at WebkinsTech is just the beginning of reimagining software systems as intelligent, autonomous ecosystems. Let's carry these core concepts forward as we architect the future in the upcoming chapter.

Chapter 5 : Microservices Architecture Revisited

Having explored the core concepts of microservices and agentic AI in previous chapters, we now revisit microservices architecture with fresh eyes. In this chapter, we'll reinforce the fundamentals of microservices design and examine the challenges practitioners face today. More importantly, we'll explore how adding an **intelligence layer** – making services more *agentic* – can address those challenges. By the end, you'll see how traditional microservices can evolve into **intelligent microservices** that are more adaptive, context-aware, and autonomous.

We'll balance theory with practice: grounding concepts in the story of **WebkinsTech**, a fictional enterprise on its journey from a basic microservices implementation to an AIpowered ecosystem. Along the way, watch for callouts with tips, warnings, examples, and quotes to highlight key insights. Let's dive in!

1. Review of Microservices Principles and Best Practices

Before leaping into the future, it's important to firm up our foundation. Microservices architecture is built on well-established principles that ensure systems remain modular, scalable, and manageable. We'll briefly review these core principles and best practices:

Single Responsibility and Bounded Contexts

One of the hallmarks of microservices is aligning each service with a **single responsibility** or business capability. In practice, this often means defining clear **bounded contexts** (from Domain-Driven Design) for each service's scope. Each microservice owns a specific piece of the business domain and concentrates on doing that one thing well. This high cohesion makes services easier to understand, develop, and maintain .

In a monolith, it's easy for different concerns to intermingle; in microservices, we intentionally separate them. If a service starts accumulating multiple unrelated responsibilities, it risks becoming a mini-monolith. Conversely, splitting functionalities too finely can lead to excessive fragmentation. Striking the right granularity is key. Research suggests that services should be **not too coarse-grained** (to avoid the dreaded "*distributed monolith*" where you simply have a tightly-coupled monolith spread over the network) but also not **too fine-grained** (which can introduce needless complexity and overhead) . In essence, each service should have **one logical reason to change**, rooted in a well-defined business context.

WebkinsTech's Bounded Contexts: After breaking up their legacy system, WebkinsTech defined services around distinct business capabilities. For example, an Order Service handles order processing logic, an Inventory Service manages stock levels, and a Billing Service takes care of payments and invoices. Each service corresponds to a bounded context – "Order Management", "Inventory Management", "Billing" – with its own data and rules. This alignment ensured that when the WebkinsTech team updates how inventory is tracked, only the Inventory Service is affected (single responsibility), and other services remain untouched. The clear separation paid off: developers could work on different services in parallel without stepping on each other's toes.

To identify these boundaries, teams often start with **domain analysis**: looking at the business domain and finding natural divisions. Techniques like event storming and domain modeling help in discovering the bounded contexts. A good litmus test for your service boundaries is the **High Cohesion, Loose Coupling** principle: everything within a microservice should relate to a unified purpose, and the service should minimize tight dependencies on other services . When done right, each microservice can be understood without reading the whole system – a big boost for agility.

Service Communication: Synchronous vs. Asynchronous

Microservices must work together to fulfill business workflows, and how they communicate is a fundamental design choice. There are two primary communication styles in microservices architectures:

- **Synchronous communication:** Services call each other directly, often using requestresponse protocols like HTTP/REST or gRPC. This is straightforward – akin to a function call over the network – and is a good fit when you need an immediate reply. For example, an Order Service might synchronously call a Payment Service to process a payment and wait for the result before proceeding. Synchronous calls are simple to understand but can introduce tight runtime coupling: if the downstream service is slow or unavailable, the caller has to wait or handle the failure. Techniques like client-side load balancing and circuit breakers (e.g., Netflix OSS's Hystrix) are often used to mitigate risks of cascading failures in synchronous chains.

- **Asynchronous communication:** Services communicate via messaging or event streams, which decouples the sender and receiver in time. A service can publish an **event** (e.g., "OrderPlaced") to a message broker or event bus without caring which service will consume it, or when. Consumers process the event whenever they receive it. This pattern enables **event-driven architecture**: it's great for workflows where subsequent steps don't need to block the initial request. For instance, after an order is placed, the Order Service can emit an "OrderPlaced" event. Inventory and Shipping services can listen for this event and react (reserve stock, schedule shipment) on their own timeline. The original Order Service doesn't wait for these to complete; it can immediately respond to the user that the order is received. Asynchronous messaging increases decoupling and resiliency – services can even handle bursts of events at their own pace – but it introduces complexity in ensuring eventual consistency and debugging (since the flow is no longer a single call stack).

Both styles have their place, and often a hybrid approach is used. **Critical request flows** that need immediate results (like charging a credit card) may use synchronous calls, whereas **follow-on processes** (like sending a confirmation email or updating analytics) happen asynchronously. The guiding principle is to minimize direct dependencies where possible to keep services independent, but not at the expense of business correctness or simplicity.

Illustration – WebkinsTech's Order Workflow: The diagram below shows a simplified communication flow in WebkinsTech's microservices. The user's request comes through an

API Gateway to the Order Service. The Order Service uses a synchronous call to the Payment Service (it needs an immediate confirmation of payment). After successfully placing the order, it publishes an asynchronous event to an Event Bus ("OrderPlaced"), which the Inventory Service and Shipping Service consume in order to update stock and arrange delivery, respectively. Each service has its own database. Solid arrows denote synchronous HTTP calls, and dashed arrows denote asynchronous event flows.

```
flowchart LR
    U[User] -->|REST API| G[API Gateway]
    G -->|HTTP| O[Order Service]
    O -->|Charge Credit| P[Payment Service]
    P -->|Payment OK| O
    O -.->|OrderPlaced event| EB((Event Bus))
    EB -.-> I[Inventory Service]
    EB -.-> S[Shipping Service]
    I -->|Update stock| IDB[(Inventory DB)]
    O -->|Write order| ODB[(Order DB)]
```

In the above workflow, even if Inventory or Shipping services are momentarily down, the Order Service and Payment still function – the events will be processed later when those services are up, thanks to the asynchronous decoupling. WebkinsTech's architects chose this mix to ensure a smooth user experience (fast order confirmation) while allowing downstream processes to scale and recover independently.

Infrastructure: Containers, Orchestration, and Automation

Microservices rose to prominence alongside advances in deployment automation and containerization. With dozens of small services, manually managing servers for each would be impossible. **Containers** (like Docker containers) became the de facto unit of deployment for microservices, because they package a service and its environment consistently. Each microservice can be built and tested in isolation, then shipped as a container image. This fosters technology heterogeneity too – one service might be in Go, another in Python, and each runs in its own container without conflicts.

To manage all these containers in production, organizations rely on **container orchestration platforms** such as Kubernetes or Docker Swarm. Orchestrators handle the heavy lifting of scheduling containers on machines, scaling them up or down, restarting them on failure, and providing service discovery (so services can find each other). Kubernetes, for example, can ensure that if you require 5 instances of the Inventory Service, it keeps 5 running – replacing any that crash – and uses a service registry and load balancers so that other services or the API Gateway can call InventoryService without knowing the actual instance addresses. Orchestrators also facilitate rolling updates (deploying new versions with zero downtime) which is crucial for independent service deployments.

Another essential piece of infrastructure in microservices is an **API Gateway**. Rather than exposing dozens of services directly to clients, WebkinsTech uses an API Gateway as a single entry point. The gateway routes incoming requests to the appropriate backend service, handles concerns like authentication, rate limiting, or protocol translation, and simplifies the external interface. This is a common pattern – API Gateways or service meshes (like Istio, Linkerd) help manage cross-cutting concerns for microservices, so each service can remain focused on its core logic.

Microservices also encourage a culture of **automation** – often summarized as "**You build it, you run it**." Each service team should be able to build, test, and deploy their service independently. Continuous integration/continuous deployment (CI/CD) pipelines automate building and deploying container images for each service. Infrastructure-as-code automates provisioning of environments. In fact, microservices architecture is only feasible when backed by a high degree of operational automation; otherwise the overhead of managing many moving parts would overwhelm the team.

Did You Know? Pioneers like Amazon popularized the idea of "two-pizza teams" – small, autonomous teams that own a service end-to-end (so small that two pizzas would feed the team). This organizational principle goes handin-hand with microservices architecture. Each team has the freedom to choose appropriate technologies for their service (within guardrails) and is responsible for its deployments and runtime health. This decentralization of governance and technology choices is another microservice best practice . It allows services to evolve with minimal centralized bottlenecks.

In summary, a solid microservices architecture stands on a foundation of **containers, orchestrators, API gateways, and automation tools** that together provide a platform for independent services to thrive. WebkinsTech, for example, containerized all their services and deployed them on a Kubernetes cluster. They implemented automated health checks and used Kubernetes' self-healing to restart crashed service instances. By doing so, they ensured that no single service failure would bring the whole system down

– a stark contrast to their previous monolith where one bug could crash the entire application.

Observability: Monitoring and Tracing

With great power (of distributed systems) comes great responsibility: you must be able to **see** and **understand** what's happening across dozens or hundreds of services.

Observability is the umbrella term for the tools and practices that allow teams to monitor, trace, and debug microservices in a production environment. This includes logging, metrics, and distributed tracing.

- **Centralized Logging:** Each microservice writes logs, but in a microservices setup, aggregating these logs is critical. Platforms like ELK (Elasticsearch-Logstash-Kibana) or cloud logging services collect logs from all services so that operators can search and correlate events. Adding **correlation IDs** to requests (a unique ID that follows a request through all the microservices it touches) is a best practice to trace a single transaction across service boundaries . WebkinsTech learned to tag all log entries with an OrderID when processing an order, which made it much easier to diagnose issues that spanned multiple services.

- **Metrics and Monitoring:** Each service should emit metrics (like request rates, error counts, memory/CPU usage, etc.). A monitoring system (Prometheus, Datadog, CloudWatch, etc.) scrapes or collects these metrics for analysis and alerting. **Servicelevel indicators** (SLIs) such as latency, throughput, and error percentage help assess if a service is healthy and meeting its Service Level Objectives (SLOs). In microservices, monitoring becomes more challenging – instead of one process to monitor, you have many. It's important to set up dashboards that can display the health of each service and the system as a whole. WebkinsTech's ops team, for instance, set up an "Orders Dashboard" that shows the end-to-end order pipeline health: if an order is failing, they can quickly see if it's the Payment Service throwing errors or maybe the Shipping Service lagging.

- **Distributed Tracing:** When a single user request flows through multiple services, understanding the end-to-end latency and where bottlenecks occur requires tracing. Tools like Jaeger or Zipkin allow capturing a trace ID and timing information at each hop of a request. This way, you can see that "Order #1234" took 500ms total: 100ms in Order Service, 200ms in Payment, 150ms in Shipping, etc. Tracing is invaluable when debugging complex interactions. Without it, a microservices app can feel like a black box of network calls. With tracing, it becomes clearer how a request traverses the system.

All these aspects of observability are not just "nice-to-haves" – they are essential best practices. In fact, one lesson learned in the industry is that **operational complexity** is the tax you pay for microservices' benefits. Teams must invest in observability from day one. WebkinsTech initially underestimated this; after moving to microservices, they experienced a few incidents where they struggled to pinpoint the failure among 20 services. They quickly realized that good dashboards, aggregated logs, and tracing would save countless hours during outages. Over time, they adopted a service mesh which gave them built-in metrics (like request success rates) and even more fine-grained control over traffic for debugging purposes .

Pitfall: Ignoring Monitoring – A common mistake when adopting microservices is focusing solely on development and ignoring monitoring and debugging until problems occur. Remember that in surveys, 75% of teams reported significant debugging and tracing issues in microservices deployments, with over half citing the distributed nature of microservices as the primary challenge . The takeaway? Plan and budget for a robust observability stack; it's non-negotiable for running microservices in production.

In review, the principles of single responsibility, well-defined communication patterns, containerized deployment, and thorough observability form the backbone of successful microservices architectures. WebkinsTech's early microservices implementation, following these best practices, resulted in faster deployment cycles and more resilient services compared to their old monolith. However, it also brought to light new **challenges** that weren't as pronounced before – which we'll explore next.

2. Current Challenges in Microservices Architecture

Microservices architecture has undoubtedly delivered benefits to organizations like WebkinsTech: agility in development, independent scaling, resilience to some failures, and alignment of services with business domains. Yet, it's not a silver bullet. By breaking systems into many moving parts, we introduce new complexity. In this section, we'll examine some of the current challenges and pain points teams face with microservices. These include **fragmentation and cognitive overload** for developers, **siloed logic with no**

learning or adaptation, difficulties in **monitoring and debugging** distributed systems, and **hard-coded workflows that lack flexibility**. Understanding these gaps is the first step to addressing them – perhaps with new approaches like agentic AI.

Fragmentation and Cognitive Overload

Ironically, the very thing that makes microservices attractive – splitting a large system into smaller parts – can become a double-edged sword. As the number of services grows, developers and operators can struggle with **cognitive overload**: there's simply so much more to keep track of. Instead of one codebase, there might be fifty repositories. Deployments move from a single pipeline to dozens of pipelines. When something goes wrong, one has to consider "Is it in Service A? Or Service B? Or the interaction between C and D?"

WebkinsTech experienced this first-hand. After they decomposed their application into many microservices, new team members were initially overwhelmed. One developer joked, "The monolith was a tangled spaghetti, but at least I could follow one noodle. Now I have a whole bowl of tiny noodles to trace!" In more formal terms, the team lost some **system-wide visibility**. Keeping a mental model of the entire system became tough, and documentation had to work overtime to map out service interactions.

Studies of microservice adopters note this challenge: as systems grow, maintaining clear oversight and traceability becomes difficult . There is also a risk of teams specializing too narrowly in their service and losing sight of the big picture (leading to suboptimal integration). This fragmentation can slow down troubleshooting and even development, because understanding the impact of a change requires knowledge of many services. In extreme cases, the system can devolve into a so-called **"distributed monolith"** – an architecture that has all the drawbacks of a monolith (tight coupling, synchronous dependencies, need for coordinated deployments) but with the added network complexity on top. This happens when services are not truly independent – e.g., if Service A cannot function without always calling Service B synchronously, and B calls C, etc., you effectively have a monolithic chain split by network calls. The result is high complexity with few benefits.

Pitfall: The Distributed Monolith – Be wary of splitting a system into microservices without proper decoupling. If every service is directly dependent on several others to function, you haven't gained true independence. For example, if WebkinsTech's Order, Inventory, and Payment services all had to be deployed together because of implicit dependencies, that would indicate a distributed monolith. This scenario combines the worst of both worlds: you incur the overhead of multiple services (network latency, ops complexity) but still lack modularity. To avoid this, enforce loose coupling: services should communicate through well-defined APIs or async events and handle unavailability gracefully (timeouts, retries). Each service should be able to start, stop, and deploy on its own as much as possible.

Managing configuration and versioning across many services is another aspect of fragmentation. Without disciplined practices, you might find that Service A is using v1.2 of a library and Service B uses v1.3, leading to inconsistencies. Or one service might process data slightly differently than another, causing data mismatches. Organizations mitigate this by establishing some standards (for logging formats, error handling, etc.) and using automation to propagate changes. Still, the governance overhead is non-trivial.

From a **team perspective**, microservices can also increase the communication overhead. Instead of one big team working on a monolith, you might have 5 teams owning 5 services, which is fine – until an initiative spans multiple services. Teams then must coordinate their changes and releases, sometimes reintroducing synchronous planning cycles that microservices were supposed to minimize. This is why the *organizational structure* must evolve with the architecture (we'll come back to this in Section 5's quote about organizational evolution).

In summary, fragmentation is a real challenge. It demands excellent engineering discipline to manage effectively. Tools like service catalogs (for registry of all services), automated dependency mapping, and robust documentation of service contracts can help alleviate cognitive load. But the industry has started to ask: can some of this complexity be *managed in smarter ways*? Could there be an approach where the system helps reduce our cognitive burden? We'll hint at that when we discuss intelligent orchestration.

Siloed Logic and Lack of Learning

Each microservice encapsulates some business logic and rules. But out of the box, microservices are inherently **static** in their behavior. They will only do what they were programmed to do, and nothing more. If conditions change in the environment or if there are opportunities for optimization, a traditional microservice will not adjust itself – someone has to notice the issue, change the code or config, and redeploy the service.

In other words, microservices **lack any built-in learning or adaptation**. The logic inside them is siloed and often hard-coded. For example, WebkinsTech's initial Order Service had fixed rules for how to allocate an order to a warehouse: East coast orders went to the New York warehouse if stock was available, otherwise to Chicago. This was coded as an if/else in the service. It worked, but it was rigid. If stock levels or shipping times changed, the code didn't automatically adjust; someone would have to update those rules. Essentially, the microservice had no memory of past outcomes or any foresight – it was blind to whether its decisions were optimal over time.

This is not a "bug" in microservices per se – they were never designed to learn. But in an era of rapidly changing conditions and abundant data, this starts to look like a limitation. Modern systems accumulate loads of data (on user behavior, system performance, failures, etc.). Traditional microservices architectures don't inherently capitalize on that data to improve themselves. They are not **context-aware** beyond what you explicitly program. Each service is a bit like a diligent worker that will do the same thing day in, day out, according to the training (code) it was given, with no ambition to improve or adapt on its own.

The result is that opportunities for optimization or personalization can be missed. Imagine if instead of fixed rules, the Order Service could learn which warehouse fulfillment leads to fastest delivery for certain regions by analyzing historical data – and adapt its strategy dynamically. In a vanilla microservice world, that requires adding a whole new component or logic (essentially an ML model and the code to use it). It's doable, but it's an **add-on** to the architecture, not intrinsic to it.

Another aspect of this "siloed logic" challenge is that microservices often encode business processes in a somewhat fragmented way. The "workflow" (e.g., order fulfillment steps) either lives in an orchestrator service or is implied by a chain of calls. These workflows are typically *explicitly* defined by developers. If a process needs to change (say we want to add a fraud check after payment), developers must modify the orchestrator or insert a new service call. There is little flexibility for the system to alter its workflow on the fly in response to conditions. In essence, **traditional microservices are great at doing tasks, but not at deciding how to do them differently under novel circumstances.**

To be clear, this predictability is partly by design – we generally want deterministic, testable services that do known things. However, it becomes a disadvantage when the environment is unpredictable or if we want the system to handle scenarios we didn't explicitly code for. For instance, if a surge of traffic comes in, microservices will scale up if you set auto-scaling rules – but those rules might be simplistic (e.g., CPU > 80% triggers adding instances). They won't, on their own, realize "hey, this traffic pattern looks like a DDoS attack, maybe I should behave differently" or "these requests seem to be all failing, maybe it's due to a downstream outage – perhaps pause and retry later". Without some intelligence, the microservices just keep doing their thing or failing in place.

This lack of learning can also impact user experience. Consider a simple recommendation feature. A traditional approach might have a Recommendation microservice with a hardcoded algorithm or maybe a periodically updated model that developers deploy. It's static until the next update. If users suddenly all start liking a new product, the service might not adapt recommendations until a developer retrains and redeploys a model. There's latency in adaptation.

In summary, **microservices by themselves don't learn from outcomes**. Each is an island of compute that does what it's told. The overall system can only get smarter if the humans intervene or if we augment the services with something more dynamic. As we will see, this is exactly where injecting AI or agent-like capabilities can help – by giving services the ability to observe, learn, and make smarter decisions inside or around their fixed logic.

Monitoring and Debugging Distributed Systems

We touched on observability best practices earlier; now let's frame them as a **challenge**. When you move from a monolith to a microservices architecture, debugging issues becomes *orders of magnitude harder*. Instead of checking one log file or attaching a debugger to one process, you might need to comb through a dozen services to find where a breakdown happened.

One concrete scenario: A user reports that "my order didn't go through". In a monolithic app, one could trace the order function and perhaps find an exception in the log. In a microservices app, that order might have passed through the Order Service, Payment Service, maybe an external tax calculation service, etc. You might find "order failed" in the

Order Service log, which says "Payment declined" because it got an error from Payment Service. Then in Payment Service logs you discover it timed out waiting for a response from the third-party credit card API. And perhaps the root cause was a network glitch or a bug in how the timeout was handled. The point is, **the failure path is distributed**, and so is the information needed to diagnose it. This is typical – each service only knows its part, so debugging means reassembling the story from many pieces.

Another challenge is **latency and performance issues** in distributed calls. If a page is slow, you have to figure out which service in the chain is the bottleneck. Without good tracing, this can be guesswork. And if multiple services are slow, it compounds.

Moreover, microservices can produce an explosion of **monitoring data** – metrics, logs, traces – which can overwhelm Ops teams. It's not uncommon that teams set up dozens of dashboards but still have trouble pinpointing anomalies because there's just so much data (the "needle in a haystack" problem). One report noted that adding more dashboards without a clear strategy can actually increase cognitive overload for Ops teams . Teams can drown in alerts if they aren't carefully tuned, leading to alert fatigue.

WebkinsTech encountered debugging pain during a holiday sale event. An order processing slowdown occurred, and they saw elevated error rates. They had to coordinate engineers from four microservice teams on a Zoom call to hunt it down. Eventually, they found that the root cause was a subtle memory leak in the Notification Service that caused a cascade of slowdowns (because that service was hogging resources and caused the others to time out waiting for it). That incident prompted them to invest in better tooling. They introduced distributed tracing across all services, which later allowed them to quickly visualize call flows and pinpoint where time was spent. They also adopted a **service mesh** which provided outof-the-box metrics like request latency for each service and even features like circuit breaking to contain failures.

It's worth highlighting that new complexity requires new approaches: for example, **service meshes** (like Istio) have emerged as a way to abstract some of the networking and observability concerns by adding a layer that can handle service discovery, metrics, and fault injection uniformly. Tools and platforms continue to evolve to ease microservice operations.

Despite all available tools, the fact remains that debugging microservices is **hard**. It requires a mindset shift from debugging a program to debugging a *distributed system*. Each request is a mini-travelogue through various components, and you, as the sleuth, need to gather clues from each stop. Many organizations mitigate this by rigorous testing (so issues are caught earlier) and by building fallback mechanisms (so if one service fails, perhaps the system degrades gracefully rather than completely breaking). But when a novel issue surfaces in production, it's all hands on deck.

This challenge sets the stage for one of the promises of intelligent systems: what if the system could **assist in its own debugging or healing**? For example, could an AI agent notice anomalies across many services and automatically flag the likely culprit? Or even restart a misbehaving service (self-healing) before humans notice? These ideas belong to the realm of **AIOps (Artificial Intelligence for IT Operations)**, which we'll touch on later. The bottom line is, with microservices, you trade simpler debugging for other gains – and you need strategies to handle that trade-off.

Hard-Coded Workflows and Limited Adaptivity

Microservices architectures often involve orchestrating multiple services to accomplish a business process. Typically, the orchestration is done either by a dedicated **orchestrator service** or implicitly through a chain of calls. In both cases, the sequence of steps is usually **hard-coded** or configured in a fixed manner. For instance, consider a purchase workflow: maybe the Order Service calls Payment, then if payment ok calls Shipping, then sends a notification. This sequence is coded in the orchestrator (or as a fixed saga). It works well for normal operations, but if something unusual happens, the system might not handle it gracefully without explicit programming.

Let's say in the above workflow, the item is out of stock after payment succeeded. A typical coded workflow might then mark the order as failed and trigger a refund, or escalate to manual intervention. It's a static outcome. However, one can imagine a smarter approach: perhaps check an alternative warehouse for stock, or delay shipment and inform the customer with an apology and a coupon. Traditional microservices won't do that out of the box because they only know the path we gave them.

This rigidity means **limited adaptivity**. The system can't easily change its behavior or reconfigure the workflow on the fly. Why does this matter? In today's environments, we often need systems to react to changing conditions: spikes in demand, downstream service outages, supply chain issues, etc. With microservices, handling each such scenario requires coding more logic, more "if this, then that" paths, or creating new services. It's doable but can become complex to manage all those conditional flows.

Another angle is **policy changes** or **experimentation**. Businesses might want to experiment with different flows (A/B test a new process) or update rules (say, skip a step for premium customers). Implementing these in a microservices landscape can mean significant coordination: every service or orchestrator involved might need an update. There's no central brain that can adjust behavior dynamically based on high-level goals or learnings.

WebkinsTech noticed this when trying to optimize their vendor onboarding (another process in their system). They had automated it with microservices and an orchestrator, but when they wanted to make it smarter (e.g., skip certain approval steps for low-risk vendors), they had to introduce an AI Risk Scoring Service and then modify the orchestrator to use it. This was essentially a mini-project, modifying the code to add a branch: *if risk < threshold then skip step X*. The improvement was great (faster onboarding for low-risk cases), but the point is it required a human-driven change. Until they did that, the workflow was static.

Adaptivity implies the system can adjust to meet goals under varying conditions. Traditional architectures rely on pre-defined rules for adaptation (like auto-scaling rules, or fallback rules). There's no learning from experience unless explicitly built in.

To put it succinctly: microservices gave us *modularity* and *deployability*, but not *adaptability*. They are not inherently *reflexive* or *intelligent*. When faced with conditions not anticipated in their code, they either fail or behave suboptimally. From an architectural viewpoint, there is a **gap between what our systems do and what we wish they could do in complex scenarios**.

This gap is exactly what has led architects to start looking at agentic and AI-driven approaches to layer on top of microservices. We want systems that can **sense, analyze, and react** – possibly even reconfigure themselves – without every decision being preprogrammed. Could a service decide “I will route traffic differently because I sense an upstream issue” or “I will change the workflow because the usual path is clogged”? These ideas sound futuristic, but they’re increasingly within reach using AI techniques.

Before we leap into solutions, let’s recap the challenges:

- We love microservices’ modularity, but **managing many services taxes our mental bandwidth** (and operational tooling).

- Each service is a static expert at one thing, but **doesn’t learn or adapt** without explicit reprogramming.

- Observability helps, yet **debugging and operating** a distributed system is still complex and error-prone.

- Workflows are executed exactly as coded, with **little flexibility or intelligence** in-themoment.

These pain points set the stage for a new approach. In the next section, we’ll look at how making microservices more *agentic* – infusing them with some AI-driven intelligence – can address these gaps.

3. The Gaps Agentic Approaches Can Fill

How can we make our microservices smarter and more autonomous? Enter **intelligent microservices**, which leverage AI and agentic principles to extend the capabilities of a traditional microservice architecture. The idea is not to throw away what we have, but to **augment** it: keep the modular, scalable structure of microservices, but imbue some services (or new components) with the ability to sense, reason, and learn. In this section, we’ll introduce the concept of agentic (intelligent) microservices and discuss how they can fill the gaps we identified.

From Static Services to Intelligent Agents

First, what do we mean by *agentic* or *intelligent* microservices? An **intelligent microservice** is essentially a microservice that has been enhanced with AI capabilities – it could be via an internal machine learning model, or by integrating with an AI service, or by actually being an **autonomous agent** that makes decisions. Agentic implies that the component can act with a degree of autonomy, making context-aware choices rather than just following a script.

Consider a normal microservice as a diligent clerk that follows a checklist. Now imagine upgrading that clerk with some intelligence: it can observe its environment, recognize patterns, and make some decisions on its own to better achieve its goals. In AI terms, it can **perceive, decide, and act**.

For example, a traditional Recommendation Service might return product recommendations based on static rules or a model updated once a week. An intelligent version of this service could continuously learn from user interactions (perceive patterns), update its recommendations in real-time (decide using a learning model), and perhaps even personalize content for each user session (act differently depending on context). It's a microservice from an architecture standpoint – it has an API, runs in a container, etc. – but under the hood, it has a "brain" that evolves.

So, where exactly can AI help in a microservices architecture? Let's map the gaps to potential AI solutions:

- **Predicting failures and self-healing:** One of the challenges was catching issues in distributed systems. AI can be used to analyze logs and metrics to predict failures or performance issues before they become critical. For instance, an intelligent monitoring agent could notice that Service X's error rate is climbing in a pattern that historically preceded an outage, and then preemptively alert or even trigger a mitigation (like spinning up new instances, or toggling a feature flag). AI-driven anomaly detection systems do exactly this – they learn what "normal" looks like and flag deviations automatically . Some advanced systems will even attempt to **self-heal** by restarting services or rerouting traffic when an anomaly is detected. Imagine if WebkinsTech's system had an agent that saw the Notification Service's memory leak early and automatically restarted that service or removed it from rotation, preventing the cascade of failures – that's the promise of intelligent self-healing microservices.

- **Optimizing routing and resource usage:** Microservices often involve routing (load balancing between instances, or choosing a path for a request). Traditionally this is random or round-robin or based on static weights. AI can optimize this by learning from data. For example, Netflix OSS's Ribbon provides client-side load balancing, and an AI extension could learn which instance of a service is likely to give the fastest response (taking into account current load, past performance, etc.) . In fact, researchers have used reinforcement learning to choose service instances or schedule requests for optimal throughput. Similarly, for network routing, AI can predict which route might be congested and pick an alternate. These optimizations are difficult to maintain manually, but an AI can continuously adjust decisions to current conditions.

- **Personalizing behavior:** One size doesn't fit all, and hard-coded rules treat every situation or user the same. AI allows microservices to personalize outcomes. In a retail scenario, a simple example is an intelligent Pricing Service that can adjust discounts or offers based on customer segment or behavior (within preset guardrails). Or an intelligent UI microservice that rearranges content based on what the AI thinks the user will find most engaging. Under a microservices hood, this could mean each request triggers a model inference to decide something instead of using a fixed rule. WebkinsTech considered adding a personalization engine to their marketing emails – rather than every user getting the same "weekly specials", an AI-infused Email Service could select products each user is likely to buy. The microservice calls an AI model trained on past purchases to do this. The key is, the service is not just executing static logic; it's making a decision using learned knowledge in real-time.

- **Dynamic workflows and decision-making:** Perhaps the most transformative aspect is letting the system decide **how** to achieve a goal under varying conditions. Consider an **orchestrator agent** that oversees a workflow (like order fulfillment). Instead of a fixed sequence, this agent could have a goal ("fulfill the order as efficiently as possible") and some smarts to adjust the process. If one path is blocked (warehouse A out of stock), it finds another (warehouse B or drop-ship from supplier) without needing an explicit if/else coded for that scenario. Essentially, the workflow becomes **adaptive**. This is a leap from current systems – it requires encoding goals and using AI planning or decision policies. But it's not science fiction; even simple rule-based expert systems in the past tried to do dynamic decision-making. Modern AI can take it further with planning algorithms or reinforcement learning that figure out optimal paths. We will see an example of this in the next section with an agentic fulfillment.

- **Context-awareness and goal orientation:** Agentic AI components shine in being aware of the broader context and pursuing objectives. A microservice normally handles a request in isolation – it doesn't "remember" that perhaps 100 similar requests failed earlier, or that it's night time and maybe a partner service is offline for maintenance. An AI-enabled service could be context-aware (e.g., aware of time, location, past interactions). It could also operate towards a **goal** rather than just a single-step task. For example, a Delivery Scheduling Agent might have a goal to minimize delivery time. It could reschedule routes dynamically if traffic data indicates delays, calling various services (maps, traffic API, notification service to update customer) to achieve its goal. Traditional microservices wouldn't do that unless we explicitly programmed all those steps. The agentic approach is about giving the system a higher-level directive and the ability to figure out the steps or adjustments itself.

In summary, the **mental model shift** here is significant: from thinking of our services as static building blocks to thinking of (some of) them as **adaptive, goal-driven entities** within the system. We are essentially moving up the ladder of intelligence:

- **Reactive** (basic microservice): Do something when asked, rigidly.
- **Context-responsive** (with simple AI): Adjust behavior based on current context or simple learned patterns (e.g., change a threshold if load is high).
- **Goal-driven** (agentic service): Choose different actions or sequences to meet an objective (e.g., ensure orders complete successfully, even if it has to try alternate steps).
- **Learning** (continuously improving): Improve performance or decisions over time by learning from data (e.g., predict failures better as it sees more examples, recommend better as it learns preferences).

Lightbulb: Think of your microservices not just as workers following a script, but as potential problem-solvers. A traditional service asks: "What do I need to do, step by step?" An intelligent service asks: "What am I trying to achieve, and how might I achieve it given what I know?" This shift from imperative to goal-oriented can be powerful. It means designing some components to have a bit of "mind" of their own – bounded by rules you set, of course. Embracing this mental shift is key to architecting truly intelligent microservices.

It's also important to highlight that injecting AI doesn't mean every service becomes a complex agent. We can target specific pain points. For example, you might introduce an AI component purely for predictive maintenance of the system (monitoring agent), or an AI for dynamic pricing, etc. These become **intelligent services** in the ecosystem. Over time, more of the architecture could become adaptive as confidence grows.

A concrete case was referenced by a 2025 report: companies combining microservices with AI found improvements in scalability and fault tolerance by using AI to dynamically adjust configurations . In Netflix's case, they moved toward AI-driven circuit breakers and load balancers – so rather than fixed rules (trip circuit after 5 failures), the AI learns the optimal time to trip based on patterns . This prevented outages by catching issues earlier than a static threshold would. Similarly, reinforcement learning in load balancing led to better latency by routing traffic smartly .

Design Differences: Not Just Automation, but Cognition

Let's highlight the design philosophy differences between traditional microservices architecture and an agentic, intelligent microservices architecture:

- **Predefined vs. Emergent Behavior:** Traditional design is all predefined. Every decision path is drawn by architects/developers. In an intelligent design, we allow some behavior to emerge from the system's learning. For instance, we might not explicitly code "use server A at 2pm and server B at 3pm", but an AI load balancer might develop a pattern to prefer A in the afternoon because it learned B gets slower then. We didn't program that schedule; the system discovered it. Architecting for this means we focus more on setting up the learning mechanisms and constraints rather than enumerating all behaviors.

- **One-size-fits-all vs. Adaptive responses:** Traditional services treat all inputs the same way (unless broken into separate services). Intelligent services might differentiate – e.g., an AI-powered Customer Service bot microservice might handle a VIP customer vs a new customer differently (perhaps prioritize, give special offers, etc.), because it has learned or been trained to. This brings more **dynamic behavior** into what was a uniform service.

- **Central Orchestration vs. Distributed Decision Making:** Currently, if we need complex decision logic, we often centralize it (like in an orchestrator or a rules engine). With agentic components, decision-making can be more distributed. Each intelligent service might make local optimizations (like an autonomous team). They cooperate to achieve system-wide goals but without a single central brain dictating every step. This is analogous to how in an ant colony, each ant follows simple rules and the colony as a whole adapts to conditions (finding food, etc.). We could have multiple agents (services with AI) coordinating via events. This is pushing boundaries, but it's being explored in **multi-agent systems** for enterprise workflows .

- **Static rules vs. Learning models:** Instead of a config file of rules, you might have a machine learning model embedded in a service. Designing the system then includes designing the **feedback loop** to retrain or update that model. You begin to consider data pipelines and model serving as part of your architecture. For example,

WebkinsTech's Risk Scoring Service uses a machine learning model to predict vendor risk. The architecture now had to include how that model is trained (using historical vendor data), how it's versioned, and how the service falls back if the model is unsure.

These are new design elements that pure microservices didn't concern themselves with.

It's worth noting that agentic or intelligent components introduce new challenges too (which Chapter 4 likely covers regarding core concepts of agentic AI, like ensuring they behave as intended, dealing with unpredictability, etc.). We have to design with **guardrails**. Just because an AI suggests something doesn't mean we blindly execute it if it could be harmful (e.g., an AI that optimizes throughput might end up starving some low-priority tasks indefinitely – we'd need to prevent bad side-effects). In architecture terms, it means **policy** and **ethics** considerations might get baked into the design of intelligent microservices. We might have "approval" steps for certain AI decisions or use explainable AI models for transparency.

To wrap up this section, let's envision the **benefits** if we successfully add agentic capabilities to microservices:

- The system becomes more **resilient**: It can anticipate failures and adapt, reducing downtime without human intervention.
- It becomes more **efficient**: Learning-based optimizations could use resources better (scale down when not needed, avoid over-provisioning, route traffic optimally, etc.).
- It becomes more **user-centric**: Personalization and adaptive workflows lead to better user experiences (faster responses, relevant content, etc.).
- It handles **complexity** better: By offloading some decision-making to the system itself, the cognitive load on humans might decrease. Instead of manually tuning dozens of parameters, the AI tunes itself to some extent. We move from micromanaging the system to supervising it at a higher level.

These are the gaps agentic approaches aim to fill in the microservices world. Of course, achieving this requires careful integration of AI into our architecture. In the next section, we will explore how to practically layer intelligence into microservices workflows and infrastructure, with concrete examples from WebkinsTech's journey and system diagrams to illustrate the changes.

4. Workflows, Automation, and the Intelligence Layer

Now that we understand *what* intelligent microservices can do in theory, let's explore *how* to incorporate them into an architecture. In this section, we'll map out how microservices, workflows, and an intelligence layer can work together. We'll use an example of a **customer order fulfillment** process at WebkinsTech to illustrate traditional vs. agentic approaches. We'll also discuss the different layers at which you can infuse AI – inside individual services, as separate agent services, or in the infrastructure itself. To tie it together, we'll compare a traditional microservices workflow with an intelligent one, and present a table summarizing the differences.

Intelligent Orchestration in Action: WebkinsTech's Order Fulfillment

Let's walk through a scenario with and without intelligence. WebkinsTech has an order fulfillment workflow as part of their e-commerce operations. The simplified process is: when a customer places an order, the system needs to charge payment, check inventory, arrange shipping, and confirm back to the user. In the current setup, an orchestrator service (or the Order Service itself) coordinates these steps in a fixed sequence.

Traditional Workflow:

1. **Payment Step:** Charge the customer's credit card via the Payment Service.

2. **Inventory Step:** Deduct the items from inventory via the Inventory Service. If item is out of stock, mark order as failed (or backorder, depending on business rules).

3. **Shipping Step:** If payment and stock are fine, trigger the Shipping Service to schedule shipment.

4. **Confirmation Step:** Send confirmation to the customer (e.g., via Email/SMS Notification Service).

This is a straightforward saga. But notice, some logic is rigid: if out-of-stock, the chosen action is to fail/backorder; there's no alternative path. The orchestrator doesn't try to get stock elsewhere, for example, unless that was explicitly coded.

WebkinsTech's goal was to make this process more **adaptive and efficient**. They identified a few improvement opportunities:

- If the primary warehouse is out of stock, maybe an alternate warehouse or drop-ship from a vendor could fulfill the order.

- If the Shipping Service predicts a delay (maybe due to weather or overload), perhaps upgrade the shipping method proactively or ship from a different location.

- For very high-value customers, maybe route the order differently (e.g., manually flag to customer service or prioritize it somehow).

- If the Payment Service is slow or down, perhaps try an alternate payment processor or at least not hold the order indefinitely.

Implementing all of the above with static rules would explode the complexity of the orchestrator. Instead, they decided to introduce an **Order Fulfillment Agent** – an AI-driven orchestrator that has a goal of fulfilling the order under various constraints. This agent uses a combination of business rules and AI models to decide on the fly how to handle exceptions or optimizations.

Intelligent Workflow with an Agentic Orchestrator:

1. The **Order Agent** receives the order request from the API Gateway. It understands the goal: fulfill the order successfully while maximizing customer satisfaction (a somewhat abstract goal that can be translated into rules like "prefer faster delivery if cost allows", etc.).

2. The agent first consults an **Inventory AI** component to check stock availability across all warehouses (and even external suppliers). Suppose the main warehouse is out of stock but a secondary warehouse has it. The agent decides to source from the secondary warehouse rather than backorder.

3. The agent then calls the Payment Service to charge the card. If Payment Service is not responding quickly, the agent has a policy (learned or configured) to retry up to 3 times and then, if still failing and if an alternate payment method is on file, try that. (In a static system, you might not bother with alternate methods unless coded explicitly.)

4. Once payment is confirmed, the agent asks a **Shipping AI** for the best shipping option.

The Shipping AI might take into account current delivery network status. It might say, "Warehouse B can deliver in 2 days via ground, but there's a storm coming – recommend using air freight to meet the 2-day delivery promise." The agent then invokes the Shipping Service with the chosen method (perhaps upgrading the shipping priority automatically because the AI predicted a delay with the standard method).

5. The order is fulfilled and a confirmation is sent, as usual, but now maybe the

Notification Service personalizes the message (since we have an AI theme: maybe the confirmation email is tailored by an Email AI that picks related product recommendations to include).

Let's visualize a portion of this intelligent flow focusing on the decision points for inventory and fulfillment:

```
flowchart TD
    Start([Order Placed])
    Start --> ICheck{Item in stock?}
    ICheck -->|Yes| Pay[Process Payment]
    ICheck -->|No| AltInv[Find alternate source]
    AltInv -->|Found stock| Pay
    AltInv -->|None| Abort[Abort Order]
    Pay --> PCheck{Payment Successful?}
    PCheck -->|Yes| ShipOpt[Select Shipping Option]
    PCheck -->|No| RetryPay[Try alternate payment]
    RetryPay --> PCheck
    ShipOpt --> Ship[Arrange Shipping]
    Ship --> Confirm[Send Confirmation]
    Abort --> Confirm
```

In this flowchart, the diamonds represent decision points where an intelligent agent or model might intervene:

- If not in stock, AltInv finds an alternate source (another warehouse or supplier) – something a smart agent does via inventory AI.
- If payment fails initially, the agent tries an alternate payment (maybe a stored PayPal account or simply retries after a short delay).
- ShipOpt is where the agent chooses the best shipping option, possibly using a predictive model.

The result is a more flexible workflow. Notice that the *structure* of the flow is similar to a flowchart we could hard-code, but the difference is **who is making the decisions**. In a static scenario, those diamonds are predetermined branches. In the agentic scenario, those decisions are made by AI components at runtime, which consider the context (stock levels across warehouses, payment retry logic, shipping forecasts, etc.).

WebkinsTech's Agentic Fulfillment in Practice: Initially, if an item was out of stock, WebkinsTech's system automatically marked the order as backordered and emailed the customer about a delay. After introducing the Order

Fulfillment Agent, they noticed a change: the agent would often find inventory in a different warehouse or even trigger a just-in-time shipment from a supplier to fulfill the order on time. In one case, a sudden spike in demand emptied the main warehouse of a popular gadget. The agent, having been fed inventory data from all warehouses, immediately routed subsequent orders to the West Coast warehouse which still had stock, avoiding backorders. Customers in the East got their orders shipped from the West with expedited shipping – a decision the agent made by weighing the options (delay vs. cost). This kind of agility was not possible before; it would have required manual intervention or a lot of conditional code. Now it happened automatically, guided by the objectives given to the agent (minimize delays, keep costs in check, etc.). The development team simply updated the agent's decision policies and provided it access to company-wide data; the heavy lifting of deciding was done at runtime by the agent.

This example shows how **workflows can be enhanced with an intelligence layer**. The orchestration is no longer just a static sequence, but a sequence with dynamic branches chosen by AI. From a design perspective, one might implement the Order Agent as a microservice that orchestrates others (like a traditional orchestrator) but internally it contains logic that uses AI models and heuristic rules to make decisions. It could use, say, a rule engine for policy and a couple of ML model calls for predictions (like an inventory prediction model, a shipping time predictor, etc.).

Layered AI Infusion Points

There are multiple ways to infuse intelligence into a microservices architecture. They are not mutually exclusive; you can mix and match according to needs:

1. **Intelligence inside a microservice (internal AI):** This is when you embed an AI model or logic within the service itself. The service's external interface might not change, but internally it uses machine learning to perform its function better. For example, an Authentication Service might use ML to detect anomalous login attempts (potential fraud) instead of just rule-based checks. Or a Database caching service might use an AI to predict which data will be frequently accessed and pre-cache it. The service remains a single deployable unit, just smarter internally. This approach is often the easiest entry point – you improve one service's behavior by adding AI to it. It doesn't disrupt the overall architecture much. However, you need to manage model training/deployment for that service.

2. **Agent as a Service (external AI service):** Here you introduce new microservices that are themselves AI-driven or agentic, which other services can call. A clear example is adding a **Recommendation Service** that uses collaborative filtering ML to return personalized recommendations. None of your existing services might have had ML, but you add this new service which others (like the frontend or product service) call to get AI-powered results. Another example: a **Risk Scoring Service** for fraud detection, which Payment or Order services call. Or the Order Fulfillment Agent we described – that's essentially a new service in the ecosystem that provides an intelligent orchestration capability. This approach treats intelligence as a separate layer of services that augment the system. The advantage is you isolate the AI complexity in dedicated services (which can be maintained by an AI/ML team perhaps). It's also a way to retrofit intelligence without heavily modifying every service. The downside is an increase in the number of services and potentially new inter-service call patterns. But this is often the approach taken: e.g., add an AI-based monitoring service, add an AI decision service, etc.

3. **Intelligence in the infrastructure (AIOps and platform intelligence):** This involves using AI at the platform level – not in the business logic, but in the management of the system. Examples: an AI that manages auto-scaling (predictive scaling). Instead of scaling only on CPU, an AI learns usage patterns and pre-scales or consolidates proactively . Or an AI that schedules services to servers in a way that optimizes resource usage (beyond what a simple bin-packing algorithm might do). Another example is AIOps for incident management: logs from all services are analyzed by an AI which then creates alerts or even resolves known issues (like auto-killing a stuck container). These kinds of intelligence don't change what the microservices do functionally, but they improve reliability, performance, or cost of the overall system. Big cloud providers are adding such features – e.g., AWS has predictive scaling, and tools can do anomaly detection on metrics to trigger alerts. In WebkinsTech's case, they integrated a machine learning based scheduler that predicted traffic surges for their ecommerce site and would warm up instances in advance – thus avoiding slow response during sudden spikes that used to happen mid-day during promotions.

In practice, an intelligent microservices architecture will likely include all three to some extent. For instance, WebkinsTech's evolution: they started embedding ML in a couple of services (Risk scoring inside Approval Service for vendor onboarding), then added separate AI services (the risk service became standalone to be reused, a recommendations service for marketing), and they started using AI-driven monitoring (in their DevOps tooling) to catch anomalies in their microservices cluster. So the intelligence layer is both **horizontal** (affecting cross-cutting concerns like scaling, monitoring) and **vertical** (specific services doing smarter things).

One must design carefully to ensure these AI additions don't become new single points of failure or opaque black boxes. It's crucial to maintain **observability of the AI as well** – i.e., monitor your intelligent components for correctness. For example, if an AI model starts giving bad recommendations due to a data drift, you want to catch that.

Observability and Feedback Loops in an Intelligent System

When microservices gain the ability to change their behavior (based on learning or decisions), having feedback loops becomes important. A feedback loop means the system learns from its own outputs or from the outcomes of actions. In architecture, this often translates to collecting outcome data and feeding it back into model training or decision policy updates.

For instance, if our Order Fulfillment Agent tries alternate sourcing for out-of-stock items, we should track the results: Did that improve delivery times? How often? If an alternate source frequently ends up delayed, the agent should learn that perhaps it's not a good option unless necessary. Implementing this might involve logging those decisions and building a dataset, then periodically retraining the agent's decision model to incorporate success/failure of alternate sourcing.

Another example: an AI-based scaling system should be evaluated – did its predictions match reality? If it over-provisioned and wasted resources, that error can be used to adjust the prediction model.

In short, the architecture should enable **data flows** from runtime back to training environments. This might be as simple as storing events in a database or data lake for offline analysis, or as complex as an online learning system that updates models continuously. Many organizations do this offline: e.g., they log all recommendations shown and whether users clicked them, then retrain the recommendation model nightly with that data.

WebkinsTech set up feedback loops for their AI components. The Risk model was retrained monthly with new fraud cases to keep it sharp. The Order Agent's performance (successful fulfillment vs. backorders, etc.) was reviewed by architects quarterly to tweak its rules and training data – effectively a human-mediated feedback loop. As they mature, they plan to automate more of that.

One more point on observability: When an AI is making decisions, it's important to have **explainability**, especially in a business context. If the Order Agent decides to incur extra cost (like using air freight), the system should log why (e.g., "predicted ground shipment would miss ETA by 2 days, so upgraded"). This is vital for trust. As an architect, you might design your agent to output a reason code or probabilities that can be logged or even presented to an admin. This way, if something goes awry, you can debug not just the technical calls, but the decision process of the AI.

Traditional vs Intelligent Microservices: A Comparison

To crystallize the differences, here's a comparative table highlighting how a traditional microservices architecture contrasts with an intelligent microservices architecture across various dimensions:

Aspect	Traditional Microservices	Intelligent Microservices (Agentic AI)
Orchestration & Workflow	Predefined static workflows (sequence of calls or events is hard-coded). Changes require code updates. **Example:** Order process follows a fixed A -> B -> C sequence.	Dynamic, goal-driven workflows. Agents can choose alternate paths in real time based on context. **Example:** Order agent can skip or reorder steps (e.g., source from alternate warehouse) to meet the fulfillment goal.

Decision-Making	Rule-based logic and simple conditionals. All decisions are explicitly programmed by developers. **Example:** "If stock X > 0, do Y, else do Z."	Autonomous, context-aware decisions using AI models and policies. Some decisions emerge from learned patterns. **Example:** Agent uses a trained model to decide the best shipping method given current weather and load, not just an if/else.
Adaptability	Reactive only to conditions foreseen in code. Limited ability to handle novel scenarios without human intervention or new code. **Example:** Out-of-stock scenario always triggers backorder because that's the only coded path.	Proactive and adaptive. Learns from data and past events to handle scenarios, even novel ones, more gracefully. **Example:** After encountering several backorders, the system learns to proactively check alternate suppliers when stock is low, reducing future backorders.
State & Context	Typically stateless per request (beyond stored data in a database). Services don't remember past requests unless explicitly built to (e.g., session state stored externally). Context is limited to the current request info.	Rich context awareness. Agents maintain internal state or memory of relevant info (recent events, knowledge base) and consider broader context (time of day, user profile, global system status) in decisions. They can aggregate knowledge over time.

Failure Handling	Predefined retries, timeouts, and fallbacks. E.g., use circuit	Self-healing and intelligent fault management. Agents predict and detect failures early, adjust
Aspect	**Traditional Microservices**	**Intelligent Microservices (Agentic AI)**
	breakers, fixed retry counts. If those fail, escalate to humans.	behavior (e.g., reroute requests, spawn new instances) and only escalate if needed. The system can autonomously recover from certain failures .
Personalization	One-size-fits-all or segmentation handled by separate logic paths. Any personalization is explicitly coded (like "if premium user, then X").	Fine-grained personalization via AI. Services tailor responses per user or situation using models (without needing separate code branches for each variation). E.g., recommendation or content delivered is unique to each user based on learned preferences.
Evolution of Behavior	Improvements via code changes and deployments. The system behavior is static between deployments (it doesn't get "better" on its own).	Continuous learning loops. The system can improve over time as models retrain or agents learn from new data. Behavior can evolve without new code deploys, as long as the learning infrastructure is in place .

Team Structure Impact	Teams focus on code and predefined test cases. Clear separation of concerns, but also a gap between dev and data science if any ML is needed (often handed off).	Requires collaboration between software engineers and data scientists. Teams may include "AI engineers" and need monitoring of AI performance. Organizationally, there's a shift towards crossfunctional teams that handle both the service and its learning components.

※ **Table:** Traditional vs. Intelligent Microservices. Note how the intelligent approach introduces adaptability and learning at the cost of added complexity in managing AI models and ensuring reliable autonomous decisions.

As we see, intelligent microservices bring a host of benefits in adaptability and potential performance gains, but they also introduce new considerations in design, testing, and governance. For example, testing an intelligent service is trickier – you have to not only test fixed functionality but also verify that the AI decisions are reasonable across scenarios (often requiring simulation or shadow testing). Deploying an update to a model is not the same as deploying code; it may need its own validation pipeline.

WebkinsTech addressed this by gradually rolling out their intelligent features. They ran the Order Fulfillment Agent in a shadow mode first – it would make decisions in parallel to the existing orchestrator, but not actually execute them, just log what it *would* have done. This was compared against the real outcomes to see if the agent's decisions would have been beneficial or not. Only after tuning the agent with these insights did they let it control a subset of orders (like 10% of traffic as a trial), and eventually full deployment. This cautious approach is advisable when introducing AI into critical workflows.

Recap of Section

We examined how to weave an intelligence layer into microservices architecture:

- By using an example, we saw the tangible differences in an order workflow when an agentic approach is applied (more branches handled automatically).

- We identified different levels at which AI can be injected: within services, as separate intelligent services, and at the platform level.

- We stressed the importance of observability and feedback in these systems to keep the AI-driven parts accountable and improving.

- A comparative table highlighted the shift in capabilities and mindset.

With this understanding, the final piece of the puzzle is preparing your organization and system for this evolution. The next section will provide a readiness guide and design checklist to help you evaluate if you're ready for intelligent microservices, and how to approach the design in practice. We'll also discuss a maturity model to help you gauge where you are and what the next steps might be in your journey towards an AI-augmented microservices architecture.

5. Readiness Guide and Design Checklist

Adopting intelligent, agentic microservices is not just a technical refactor – it's a transformation in how we build and operate systems. Before diving head-first, it's wise to assess readiness and proceed methodically. In this final section, we'll provide a **practical design checklist** to ensure you have the necessary pieces in place. We'll also outline stages of maturity, so you can identify where you currently stand and how to progress. Remember, introducing AI into your architecture is a journey that involves technology, people, and processes.

Checklist: Are You Ready to Intelligent-ize Your Microservices?

Use the following checklist as a guide when planning to infuse AI into your microservices or to build new intelligent services. These points serve as design considerations and organizational prerequisites for success:

- ✓ **Solid Microservices Foundation:** Ensure your existing microservices implementation is stable and well-managed. You should have good observability (logs, metrics, tracing) already (as discussed) and a handle on deploying and updating services easily (CI/CD). If your current microservices are a house of cards, fix that foundation first. Adding AI to a shaky system will only amplify problems.

- ✓ **Clear Use Cases for Intelligence:** Identify specific pain points or opportunities where AI could add value. Is it reducing manual work (like automatic anomaly detection)? Improving user experience (like personalization)? Optimizing operations (like predictive scaling)? Having clear targets (e.g., "reduce order fulfillment time by 20%" or "catch 50% more fraud") will focus your efforts and also help get buy-in from stakeholders.

- ✓ **Data Availability and Quality:** AI runs on data. Do you have the data needed to train models or inform decisions? This could be historical logs, user behavior data, product data, etc. Ensure you have a way to collect and store relevant data from your microservices. Also, consider data quality and preparation – you might need to invest in cleaning or labeling data. If a needed dataset is missing (say, you want to predict failures but haven't been monitoring certain metrics), set up those data pipelines first.

- ✓ **Talent and Team Readiness:** Do you have people who understand AI/ML in addition to your services? This might mean upskilling some developers in machine learning or hiring ML engineers. Also, your teams should be prepared to collaborate more closely (Dev with Data Science, Ops with ML engineers). Fostering a culture of cross-functional teams will help – perhaps embedding an ML engineer into the microservice team that's adding an AI component. An organizational design where "the AI team" is totally separate from "the microservice team" can lead to integration hiccups. Aim for shared ownership of intelligent features.

- ✓ **Governance and Ethical Considerations:** Introducing AI decisions (especially those affecting customers or critical operations) requires thinking about governance. For example, decide on **fallback mechanisms**: if the AI model is uncertain or unavailable, what does the system do? Also, consider ethics and bias – if your AI is making customer-impacting choices (like credit approval, content display, etc.), ensure you have checked for biases in the model and have transparency. At WebkinsTech, any AI that affects external users goes through an additional review. They implemented simple safeguards, like if the Order Agent isn't at least, say, 80% confident in a decision, it defaults to a safe rules-based path or flags human oversight.

- ✓ **Testing Strategy for AI:** Traditional testing (unit, integration, end-to-end) still applies, but you need additional testing for AI components. Plan how you will validate models – perhaps using historical replay ("would our model have made the right call on past incidents?"). Also consider a phased rollout for risky features (shadow mode, A/B testing, etc. as described earlier). Make sure you can monitor the outcomes of the AI decisions specifically. Having a way to turn off the AI decisions (feature flag) and fall back to manual or old approach is prudent in case things go wrong. Essentially, **don't deploy and pray** – deploy, observe, and be ready to revert or adjust.

- ✓ **Infrastructure and Tools for AI:** Beyond your normal DevOps tools, you might need ML tools: e.g., environments for training models (could be as simple as some scripts using scikit-learn/TensorFlow, or as elaborate as an ML pipeline on a platform like SageMaker or Azure ML). You'll also need a way to deploy models (maybe as part of a service or using a model server). If using external AI APIs (like cloud AI services), ensure your architecture accounts for those calls (latency, cost, reliability of external APIs). Essentially, extend your DevOps to MLOps (Machine Learning Ops) – versioning models, monitoring model performance drift, and retraining workflows should be considered.

- ✓ **Security and Compliance:** Check how adding AI might affect security. For instance, if an agent can execute actions, ensure it cannot do anything outside its scope (avoid creating a "rogue AI" even accidentally – e.g., an agent that triggers scaling should not exceed certain limits). Also consider data security: training data might include sensitive info, and AI services may introduce new attack surfaces (like someone spoofing inputs to manipulate the model). Compliance is key if in regulated industries – e.g., explanations for decisions might be legally required. So design for traceability of AI decisions.

- ✓ **Gradual Introduction Plan:** It's usually wise to introduce intelligence in increments. Maybe start with one service or one use-case as a pilot. For example, WebkinsTech first added a simple ML-based recommendation service for their marketing emails – a noncritical path – to get their feet wet. Then they tackled the more critical order fulfillment agent. By starting small, you can develop best practices and confidence. Have a roadmap: which components will you make intelligent first, and how will you phase others in?

- ✓ **Measure Impact:** Define metrics to measure the impact of intelligent components. Otherwise, how do you know it's "smarter" or just more complex?

If you add a failure prediction agent, measure MTTR (Mean Time to Recovery) or number of incidents mitigated. If you add personalization, measure engagement or conversion lift. This ties back to having clear use-cases: for each, decide how you'll quantify success. This not only justifies the effort but also helps tune the system (if the metric isn't moving, maybe the model needs improvement or the approach needs tweaking).

This checklist is not exhaustive, but covers the key areas. Essentially, **get your house in order, align the people, secure the data, and have a plan**. If you check most of these boxes, you are in a good position to proceed with designing intelligent microservices.

Crawl, Walk, Run: It's tempting to envision a fully autonomous system from day one, but the practical approach is incremental. Start with a "crawl" – a small internal AI feature or a pilot in a low-risk area. Learn from it, then "walk" – integrate a more impactful AI service or agent in a core workflow, still keeping a close eye and perhaps a safety net. Finally, "run" – gradually let the AI have more autonomy as trust and experience grow. This phased approach lets your organization absorb the changes and avoids jeopardizing system stability. It mirrors how WebkinsTech rolled out their intelligent features: carefully and with a lot of monitoring at first, more freely later.

Maturity Stages of Intelligent Microservice Architecture

To understand progress, it's useful to think in terms of maturity levels. Below is a simplified maturity model for microservice architectures evolving towards intelligence:

Maturity Stage	Characteristics & Focus
0. Traditional Automation (Pre-AI)	Microservices in place but no AI integration. Automation is rule-based. Scaling is reactive (e.g., threshold-based). Workflows are static. The focus here is on getting the microservices architecture and DevOps practices right. *(Many organizations are here or just starting to consider AI.)*

1. Augmented Insights (Analytics)	The system gains visibility with analytics but not yet automated decisions. For example, dashboards with predictive insights or anomaly alerts via AI, but humans still make decisions. Focus on data collection and monitoring improvements. *(This stage builds the data foundation – e.g., using AI offline to identify patterns, but not yet in real-time operations.)*
2. Intelligent Assist (AI inloop)	AI components assist humans or services but are not given full control. E.g., a recommendation service suggests actions, an AI flags a transaction as suspicious, but final decision might be manual or via simple rules. Some workflows have AI decision points, but with oversight or easy override. Focus on initial AI integration and proving value. *(WebkinsTech's early risk scoring and recommendation systems fit here – they provided guidance that could be overridden.)*
3. Autonomous Decisions (Agentic)	The architecture has multiple agentic components making autonomous decisions within bounded domains. The Order Fulfillment Agent actively manages orders, the monitoring agent auto-resolves certain incidents, etc. Workflows can reconfigure dynamically. The organization trusts AI for many decisions, with proper guardrails. Focus on optimization and expansion of intelligent capabilities across the system. *(WebkinsTech reached this stage when their Order Agent and other AI*

	services were fully deployed and handling cases end-toend.)
4. Continual Learning & Adaptation	The system not only makes autonomous decisions, it also continuously learns and adapts its strategies. Feedback loops are fully automated – models retrain regularly from new data, policies evolve. The system might simulate and self-improve. At this stage, the architecture is highly adaptive: performance improves over time with minimal human tuning. Focus on self-optimization and possibly multiagent coordination for complex goals. *(Few organizations are here yet – it's the cutting edge. This would be an AI-driven enterprise nervous system.)*

※ **Table:** Stages of Maturity for Intelligent Microservices. Organizations typically progress stepwise; jumping straight to stage 3 or 4 without stage 1/2 experience is difficult.

Assess where you are. If you are just at traditional automation (stage 0), it might be premature to shoot for full autonomy. Perhaps invest in some analytics and AI monitoring first (stage 1) to build confidence and capability. If you're at stage 1 (lots of data and maybe some predictions), stage 2 might be implementing a pilot AI that actually takes actions (maybe in a non-critical workflow). Stage 3 and beyond require solid infrastructure and trust in AI outcomes.

WebkinsTech's journey roughly followed these stages: they had microservices (stage 0), then added analytics dashboards and basic ML insights (stage 1). Next, their intelligent features were initially advisory (stage 2) – e.g., the risk model gave a score, but the final approval could still be done by a manager for high scores. As they saw the AI was accurate, they let it auto-approve low-risk cases (entering stage 3 for that workflow). Now they retrain that model frequently and it's quite autonomous in deciding approvals, edging into stage 4 for that domain.

Conclusion and Key Takeaway

We stand at an inflection point in software architecture. Microservices brought a paradigm shift in how we design scalable systems by breaking down functions. Agentic AI is bringing a paradigm shift in how these systems can operate by breaking down decision-making and infusing learning. Merging the two – microservices and intelligent agents – is how we build the next generation of systems that are not only modular and scalable, but also smart and autonomous.

As you architect your systems, keep in mind that **technology is only part of the equation**. Equally important are the processes you put in place to manage this intelligence and the people who build and maintain it. Start small, think big, and iterate. Use the best of microservices (encapsulation, interfaces, independent deployability) as the canvas, and gradually paint intelligence onto it where it adds value.

One insightful quote from a forward-looking architect at WebkinsTech encapsulated the journey:

"Architecting intelligent microservices isn't just a tech decision — it's an organizational evolution."

In other words, embracing this approach will likely change how your teams collaborate, how you plan projects, and how you think about problems (you might start asking, "Can the system learn this or adapt to this?" for every feature). It truly is an evolution, much like the leap from monolith to microservices was.

By revisiting microservices architecture with an agentic-AI lens, we see that the principles we started with still matter – perhaps even more so – and now we have new tools to address age-old challenges. The hope is that by applying intelligence thoughtfully, we can tame the complexity of large-scale systems and unlock levels of efficiency and responsiveness that were previously unattainable. As you move forward, use this chapter as a reference and inspiration. The field is rapidly evolving, and you, as an architect or leader, have the exciting job of building the future: one where microservices are not just services, but smart services – each a little agent contributing to the greater collective intelligence of your platform.

Chapter 6: Architectural Patterns for Agentic Microservices

In the previous chapters, we explored *why* enterprises like WebkinsTech are infusing AI into their microservices. We learned what **agentic AI** means and saw the promise of microservices that can make autonomous decisions. Now it's time to tackle the *how*. This chapter takes a practical turn: we'll examine **architectural patterns** that guide *where and how* to incorporate intelligence within a microservices ecosystem. In other words, how do we architect "microservices with minds of their own"?

Think of architectural patterns as blueprints for structuring complex systems. In traditional microservice design, you might encounter patterns like **API Gateway**, **Saga**, or **Event Sourcing**. Here our focus is on patterns tailor-made for **intelligent, agentic microservices** – systems where AI agents and services are intertwined. These patterns address questions like: Should our AI logic live in a centralized service or be embedded within each microservice? How do we orchestrate tasks when AI is calling the shots? How can services learn and adapt over time? By the end of this chapter, you'll have answers to these questions and a toolkit of design patterns to choose from.

Before diving in, let's set the stage with an example. Imagine WebkinsTech's architecture team in a strategy huddle. They have a fleet of microservices powering their operations (orders, inventory, shipping, customer service, etc.), and they've identified numerous opportunities to inject intelligence – from smarter fraud detection to automated inventory reordering and personalized customer experiences. The challenge: **there's no one-sizefits-all approach** to adding AI. One architect suggests creating a dedicated AI service that everyone can call. Another proposes embedding AI inside each critical microservice. Yet another envisions a clever orchestrator that coordinates tasks using AI to make decisions. Who's right? In a sense, *all of them*. Each approach represents a valid architectural pattern for agentic microservices. The key is knowing which pattern fits which problem and how to implement it.

Architectural Patterns in an Intelligent Microservice Context

When we talk about **architectural patterns** here, we mean high-level design templates that describe how intelligent components (agents or AI modules) are distributed and interact within a microservices architecture. These patterns are like recipes – each with its own ingredients, benefits, and trade-offs. Crucially, they help architects reason about **where the "intelligence" lives**:

- Is it concentrated in a single service that all others use?
- Is it spread out, with each service having its own built-in smarts?
- Do we let services coordinate via signals (events) rather than direct calls, enabling more emergent behavior?
- Do we continually train and evolve the intelligence as the system runs?

All these approaches can work, but serve different needs. In traditional systems, patterns ensure stability, scalability, and clarity of design. In **agentic microservices**, patterns additionally must ensure that our AI-driven components integrate smoothly with the rest of the system and with each other. They are about placing "brains" in our architecture in a way that aligns with business goals and technical constraints.

Let's briefly outline the **key patterns** we'll explore in depth in this chapter:

- **AI-as-a-Service** – treating AI capabilities as a standalone microservice (a centralized "brain" others can consult).
- **Embedded Agent Pattern** – embedding an AI agent directly inside a microservice (giving each service its own localized "brain").
- **Intelligent Orchestration** – using an AI-driven orchestrator to control and coordinate workflows across multiple services.
- **Event-Driven Agents** – leveraging an event-driven architecture where autonomous agents react to events and trigger outcomes (a more decentralized, emergent approach).
- **Online Learning Loops** – designing the system to continuously learn and improve by feeding outcomes back into the AI (closing the loop between decisions and learning).

Each pattern addresses a different scenario in WebkinsTech's journey toward intelligence. We'll examine how each one works, dive into its **design approach**, highlight the **benefits and trade-offs**, and illustrate it with a mini case study from fictional WebkinsTech. Along the way, we'll use simple diagrams (using arrows and text) to visualize the architectures, and include tips, examples, and warnings from the field. By weaving in narratives and analogies, we aim to keep things engaging and grounded.

Ready to navigate the architectural landscape of agentic microservices? Let's start with the most straightforward pattern: **AI-as-a-Service**.

AI-as-a-Service Pattern

What if you could add intelligence to your system the same way you add a new microservice for, say, payments or search? The **AI-as-a-Service** pattern does exactly that – it packages AI capabilities into a dedicated service that other microservices call when they need some smarts. In essence, AI becomes *its own microservice*.

Imagine this pattern as **hiring an expert consultant** within your architecture. Instead of every team member learning a complex skill, they all go ask the consultant when needed. In architectural terms, the "consultant" is an AI service. Whenever a microservice needs to make an intelligent decision or prediction, it sends a request to the AI service and gets back a result. This could be anything: a fraud score for an order, a personalized product recommendation, a predicted shipment date, you name it.

Design Approach

In the AI-as-a-Service pattern, the **design approach** is to **decouple AI/ML logic into its own component**. This typically means creating a microservice (or a set of microservices) whose sole job is to handle AI tasks – for example, running machine learning models or complex decision algorithms. All other services that require those AI capabilities will call this service via a well-defined API (e.g., REST or gRPC).

The AI service often encapsulates one or more ML models. It might look like:

```
Client -> Service A -> [AI Service] -> response -> Service A -> Client
```

In this schematic, Service A offloads an AI task to the **AI Service** and awaits the result. Multiple services can reuse the same AI service. For instance, both the **Order Service** and the **Fraud Monitoring Service** could call a central *Fraud Detection AI Service* to assess risk on transactions. Likewise, a *Recommendation AI Service* could be consulted by the ecommerce frontend, the marketing email service, and the mobile app to get personalized product suggestions.

One real-world manifestation of this pattern is the use of cloud-based AI APIs (like

AWS/GCP/Azure AI services). When WebkinsTech's developers call an external vision API to analyze an image or a NLP API to understand text, they are effectively using AI-as-a-Service (with the service hosted by a cloud provider). In our context, WebkinsTech could also deploy its own AI service internally – for example, a container running a trained PyTorch model exposed via an HTTP endpoint.

How it works: Other microservices send data (features) to the AI service and get back an inference or decision. The communication is usually synchronous (the calling service might pause until it gets a reply) because the result is needed to continue the business logic. For instance, the Order Service might call the AI service to get a fraud score *before* finalizing an order. This synchronous nature means the AI service must be responsive and scalable to handle bursts of requests. In some cases, asynchronous calls or caching might be used to mitigate latency (e.g., precomputing recommendations). But fundamentally, AI-as-a-Service centralizes the "thinking" in one location.

Benefits

- **Simplicity and Encapsulation:** This pattern is often the *easiest to implement* when adding AI to an existing system. You don't have to refactor multiple services – you simply bolt on a new one. The AI logic is encapsulated in one place, making it easier for a specialized team (like WebkinsTech's "AI Task Force") to manage it.

- **Reuse Across Services:** A central AI service can be reused by many parts of the system. This avoids duplicate efforts. WebkinsTech's product recommendation engine, for example, can serve the web app, the mobile app, and even in-store kiosks from the same service, ensuring consistency in recommendations.

- **Independent Deployment & Scaling:** The AI service can be deployed, scaled, and updated independently of other services. If the AI model needs a GPU or more memory, you scale that service vertically or horizontally without affecting others. Conversely, if the AI is causing issues, you can deploy a fix without redeploying the whole system.

- **Model Versioning & Experimentation:** Because the AI is in one place, trying a new model or A/B testing different algorithms is simpler. You can route a small percentage of calls to a new model version running in the AI service, for instance, without touching the calling services. This is great for ML experimentation.

- **Consistency of Decisions:** With a single AI service, you ensure that every part of the system uses the *same logic* for the same task. If we centralize fraud scoring, we won't have one service using an outdated fraud rule and another using a new one – all go through the same service, which can be kept consistent.

Trade-offs and Challenges

- **Latency and Network Overhead:** The biggest downside is that every AI call is a remote call. This adds **latency**. A microservice might need to wait, say 50ms or 100ms, for the AI service to respond, which can add up if chained. If network hiccups occur, it might even timeout. In contrast, an embedded approach (discussed later) might complete in microseconds in-memory. For many applications, a few tens of milliseconds are acceptable, but for ultra-low latency needs, this could be a bottleneck.

- **Reliability and Dependency:** By centralizing, you introduce a **dependency**. If the AI service goes down, many parts of your system might be unable to make decisions. It becomes a critical piece that needs high availability. Techniques like replication, load balancing, and circuit breakers (where callers fall back to simpler logic if the AI service is down) become important. Essentially, you've traded a distributed problem for a potentially centralized point of failure.

- **Scalability Bottleneck:** While the AI service can be scaled, it has to handle all requests from all clients. In peak times, this could be a lot of traffic hitting one service. Scaling horizontally (more instances) can handle more load, but certain models (especially big ML models) don't scale linearly without effort (you might need to shard by user, or use caching for popular queries).

There's also the *thundering herd* risk: e.g., during a sale all orders might simultaneously call the recommendation service, spiking load.

- **Data Transfer and Context:** When a service calls the AI service, it needs to send all necessary data. If the needed context is large (say user profile, transaction history, etc.), you are now serializing and sending that over the network. That can be heavy. In contrast, if AI logic were local, it might query a database or in-memory cache directly. In AI-as-a-Service, you might find yourself duplicating data or precomputing features to

send with the request. Designing efficient request payloads and maybe using asynchronous update (where the AI service has its own data store that's kept in sync) is an extra complexity.

- **Not Always Domain-Specific:** A centralized AI might become somewhat generic if it serves many masters. If each service or domain has unique needs, one AI service could grow complex to handle all variations. For example, an AI service might have to handle both fraud scoring and recommending products – two very different tasks – which could lead to a *bloated* service or the need to split into multiple AI services by domain.

Despite these challenges, AI-as-a-Service remains a popular starting point because it cleanly separates concerns. WebkinsTech often treated their first AI deployments this way: as separate services that could be integrated without overhauling everything.

WebkinsTech Case Study: Centralized Recommendation AI

WebkinsTech decided to enhance customer engagement on their e-commerce platform by adding a **personalized recommendation engine**. Instead of baking this AI logic into the existing Order or Product services, they spun up a dedicated **Recommendation Service** powered by a machine learning model. The Recommendation Service was trained on user browsing data and purchase history. Now, when a user views a product, the Product service calls out to the Recommendation Service (with the user's ID or session info) and gets back a list of "You might also like" items. These suggestions are then displayed to the user.

Benefits Realized: This AI-as-a-Service approach meant the core product catalog service remained simple – it just fetched data and delegated the personalization task to the AI service. Different parts of WebkinsTech's ecosystem could reuse recommendations: the marketing team's email microservice began calling the Recommendation Service to include tailored product suggestions in newsletters, and the mobile app did the same for its homepage. WebkinsTech's data science team could update the recommendation model (for example, include a new algorithm for seasonal trends) by deploying a new version of the Recommendation Service, without any changes needed in the other services.

Challenges Faced: Initially, there was a slight impact on page load times because each page view triggered an extra network call to the Recommendation Service. The team mitigated this by implementing caching – the first time recommendations were fetched for a user-product combo, they stored it in a fast in-memory cache. Subsequent calls (within a short window) returned instantly from cache, reducing load on the AI service. They also put a circuit breaker in place: if the Recommendation Service ever became unresponsive, the Product service would simply show a few top-selling products as a fallback, ensuring the page still functioned.

In the end, the AI-as-a-Service pattern allowed WebkinsTech to **inject intelligence incrementally**. They could add more AI services (fraud detection, inventory prediction, etc.) one by one, each time keeping the impact on the overall system isolated. This proved to be a gentle learning curve for the organization's move toward agentic microservices.

Design Tip: When implementing AI-as-a-Service, treat the AI service as a *black box with a clear contract*. Define input and output schemas rigorously (e.g., what fields represent an order for fraud scoring, what output risk levels mean). This not only helps consumers use the service correctly, it also allows the AI internals to evolve (you can swap models or tweak algorithms) without breaking other services – as long as the contract is stable. Also, consider adding instrumentation to the AI service: log the requests and responses (sans sensitive data) and the model's decisions. This creates a feedback loop for improving the model later and provides traceability when debugging system behavior ("Why did the AI recommend *that* product?").

With a solid understanding of AI-as-a-Service, you might wonder: is there an alternative where we don't make that extra network hop at all? What if each microservice could carry its own intelligence internally? This brings us to the **Embedded Agent Pattern**.

Embedded Agent Pattern

The Embedded Agent Pattern takes a more decentralized approach: instead of a single "brain in a jar" that all services talk to, **each microservice gets its own mini-brain**. In practice, this means the AI or decision-making logic is **embedded within the microservice** itself (or very close to it). The microservice isn't just performing its traditional role (like processing orders or managing inventory); it also has an *agentic component* that observes, decides, and acts within that service's scope.

If AI-as-a-Service was like consulting a central expert, the Embedded Agent pattern is like **giving each team member (service) a personal assistant**. Each microservice can autonomously handle certain intelligent tasks without asking a central authority. For example, an Order service might have an embedded agent that automatically offers a discount to a high-value customer if an order is about to be abandoned, all on its own. Or an Inventory service might embed an agent that predicts stock outages and proactively reorders items.

The key distinction: the intelligence lives **inside** the service's boundary.

Design Approach

In this pattern, the design approach is to **integrate the agent logic into the microservice's process or deployment unit**. There are a couple of ways this can happen:

- **In-process integration:** The microservice's code includes modules or components

(like libraries or classes) that implement the AI behavior. For example, WebkinsTech's Inventory microservice might load a trained ML model (say a time-series forecasting model) into memory at startup. When inventory levels change, the code calls that model to predict future demand. It's all happening within the same process that serves inventory API calls.

- **Sidecar or local companion:** Alternatively, the agent might run as a sidecar service or thread alongside the main service. It's still considered "embedded" because it's deployed together with the service (perhaps in the same pod if using Kubernetes) and scoped to that service's data. The main service and the agent sidecar communicate via localhost or shared memory.

This can be useful if the agent is complex or resourceintensive, so you separate concerns but still keep them co-located.

However it's implemented, the microservice and its agent act as a unit. The agent typically has access to the service's state or database and can make decisions in real-time as events occur locally.

For example, consider **Pricing Service** at WebkinsTech. If using Embedded Agent pattern, the Pricing Service might include an agent that monitors market conditions and sales data. When a request comes in to calculate a price, this internal agent could adjust the price based on dynamic rules or a model (maybe offering a discount if sales are slow, or increasing price if demand is high). There's no external call – the logic is right there inside the Pricing Service.

Let's illustrate the structure:

```
Service A (with embedded AI agent)
Service B (with embedded AI agent)
```

Here, *Service A* and *Service B* each have their own embedded agent. If they need to make a decision, they don't call an external service; they consult their internal logic.

Communication pattern doesn't change outwardly – to other services or to clients, these intelligent services still look like normal microservices exposing APIs or events. But internally, there's an extra brain at work.

Benefits

- **Low Latency, Real-Time Decisions:** Since the intelligence is on-board, there's **no overhead of a network call** to get a decision. This makes it ideal for scenarios that require real-time or near-instant responses. For instance, if WebkinsTech's website wants to adjust the price for a flash sale in real-time as the user adds items to cart, an embedded pricing agent can do that in milliseconds. The decision is made right where the data is, providing speed that's hard to beat.

- **Autonomy and Resilience:** Each service can operate more independently. If the central systems are down, an embedded agent might still function (as long as its service is running). There's no single point whose failure cripples decision-making across the board. In a way, you eliminate the dependency on a central brain, potentially making the system *more resilient*. Each microservice still works (and thinks) even in partial outages or network partitions.

- **Contextual Awareness:** An embedded agent has full access to the service's context and data. This can be richer than what it would send to a central service. For example, an Order service's embedded agent can directly query the Order database or combine multiple pieces of internal state to make a decision. It doesn't need to serialize all that state and send it over – it's already next to it. This closeness can enable more nuanced decisions. The agent is basically a part of the service's internal logic, so it can hook into events and data streams that might not even leave the service boundary otherwise.

- **Localized Adaptation:** You can tailor each agent to the specific needs of its service. The Inventory agent might use a forecasting model suited to inventory dynamics, while the Pricing agent uses a reinforcement learning model suited to pricing strategy. They don't have to be one-size-fits-all. This specialization can yield better results because each agent is optimized for its domain. In contrast, a central AI service might have to juggle many domains or use more generalized logic.

- **Scalability (of a sort):** When you scale out microservice instances, you automatically scale out the agents with them. For example, if you run 10 instances of the Inventory service, you have 10 little inventory agents working in parallel on different data shards (or just redundantly handling load). The intelligence capacity scales with the service.

And if one instance is overwhelmed, you add more instances and thus more agents. This is horizontal scaling of intelligence, aligned exactly with the scaling of the service itself.

Trade-offs and Challenges

- **Consistency and Coordination:** A big challenge with multiple agents is **keeping them consistent** if needed. If each instance or each service has its own model, how do you ensure they all behave coherently? For instance, if each Pricing service instance is learning or adapting separately, one might start offering different prices than another if you're not careful. This could confuse customers. To manage this, you might need a mechanism to periodically sync models or share learning across agents (which starts to add complexity and edges toward needing an orchestrator or central coordinator anyway).

- **Operational Overhead:** Instead of one AI service to manage, you now have many distributed pieces of AI logic. Updating a model means deploying new code or new model files to potentially every microservice that carries that agent. If WebkinsTech's ML team comes up with a better algorithm for stock forecasting, they have to update the Inventory service's agent in all environments. This is manageable with good DevOps (CI/CD pipelines can handle it), but it's certainly more moving parts than updating one central service. Testing also becomes trickier – you need to test each service+agent combo in context.

- **Resource Duplication:** Embedding AI means each service might consume more CPU, memory, maybe even GPU if needed. If you had a central AI service with a beefy machine, that might be more resource-efficient than every service carrying a smaller copy of the model. There can be duplication of models in memory across many instances. For small models this is fine; for very large models, it might be prohibitive to load it in every instance. There is a trade-off between the efficiency of centralized processing vs the distribution of computation to where the data is.

- **Complexity in Microservice Design:** One of the appeals of microservices is each service is fairly simple and focused on a business capability. By adding an agent, you're increasing the complexity *inside* a service. Developers working on that service now need to handle not just business logic but also understand the embedded AI logic. The service might need machine learning libraries, model files, etc. This blurs the separation of concerns. It can be mitigated by good modularization (e.g., treat the agent as a separate module that only a few people touch), but the boundary of the microservice is not as clear-cut as before. Essentially, you've made the microservice a bit heavier and more sophisticated.

- **When Agents Interact:** This pattern by itself doesn't dictate how agents communicate, but often embedded agents might still need to talk to something beyond their service (maybe other agents via events, or a central coordinator for learning). If not carefully designed, you can end up with a swarm of agents stepping on each other's toes or conflicting. We'll touch on coordination later (in fact, next chapters dive into that), but it's a consideration: distributed intelligence can introduce emergent effects, for better or worse.

It's worth noting that the Embedded Agent pattern and the AI-as-a-Service pattern are not mutually exclusive in a large system. WebkinsTech might use AI-as-a-Service for some capabilities and embedded agents for others, depending on the use case. For instance, a heavy image-processing AI might remain a central service (since embedding it everywhere is too costly), but a lightweight rule-based agent for dynamic UI customization might live in each relevant service.

Decentralizing the "Brain"

"Our central AI service was turning into a know-it-all bottleneck. We realized not every decision should go through one brain; sometimes you need intelligence right at the source," a WebkinsTech architect recalled. This insight led them to push certain AI capabilities down into individual services, trading a bit of central control for local autonomy.

WebkinsTech Case Study: Inventory Replenishment Agent

WebkinsTech's Inventory microservice manages stock levels for hundreds of products.

Initially, it only kept counts and alerted humans when stock was low. Seeking to automate this, WebkinsTech embedded an **Inventory Replenishment Agent** into the

Inventory service. Here's how it works: each time inventory is updated (say, an item is sold or restocked), the agent wakes up and analyzes the trend. It uses a simple predictive model to estimate how quickly that product is selling and when it might run out. If the forecasted depletion date is within a threshold (for example, it predicts only 3 days of stock left for a high-demand item), the agent automatically creates a restock order with WebkinsTech's supplier via an API call. This all happens inside the Inventory service's process.

Outcome: The embedded agent can make restocking decisions **immediately** as data changes, without needing to ask a central brain. It knows the context (the specific product's sales velocity, current promotional campaigns, etc.) because it's part of the Inventory service. Over time, it even learned seasonal patterns by updating its model with new data (engineers would periodically feed it historical sales and tweak the model).

Benefits Seen: Stockouts (running completely out of an item) dropped by 30% because the agent was quicker and more proactive than the previous manual process. Since the decision to reorder was made locally, there was virtually no delay – the moment stock dipped below threshold, the order was placed. Also, because each Inventory service instance had this logic, it scaled seamlessly; as WebkinsTech added more warehouse locations (each handled by a separate instance of the Inventory service), each one had its own replenishment brain tuned to its local demand.

Challenges: At one point, WebkinsTech introduced a new rule – for expensive electronics, they didn't want to auto-reorder without manager approval (to avoid tying up capital). Making this change meant updating the embedded agent's logic in the Inventory service code. With many instances across regions, the team had to ensure all deployments were updated in sync so the policy would be applied everywhere. This was a DevOps exercise: they rolled out a new version of the Inventory service with the updated agent rules gradually, using feature flags to switch the behavior. It highlighted that governance of these distributed agents needed careful planning. Documentation was key – they maintained a clear description of what decisions the Inventory agent could make on its own, so there were no surprises.

In sum, the Embedded Agent pattern gave WebkinsTech's Inventory service a degree of **self-management**. The service wasn't just passively storing data; it was actively monitoring and taking action. This autonomy reduced latency in decision-making and freed up the operations team from constantly firefighting stock issues. The trade-off was an increase in complexity within that service – one that WebkinsTech managed through disciplined deployment practices.

With each microservice now potentially harboring its own AI, our system starts to resemble a **team of intelligent agents**, each handling its domain. But what about processes that **span multiple services**? How do we coordinate a workflow that involves many parts, especially if we want the coordination itself to be smart? Enter **Intelligent Orchestration**.

Intelligent Orchestration

Traditional microservices architecture often debates *orchestration vs. choreography* for workflows. Orchestration means having a central controller dictating the process (Service A calls Service B, then C, etc.), whereas choreography means services react to events (like a dance where each dancer knows their part without a single director). Both approaches have merits. Here, with **Intelligent Orchestration**, we focus on scenarios where a central **orchestrator** not only coordinates the steps, but does so intelligently, with decision-making capabilities.

Think of Intelligent Orchestration as having a **smart conductor or project manager** in your system. This orchestrator service (or agent) has a view of the overall business process and can make real-time decisions about that process. Unlike a static workflow engine that follows predetermined steps, an intelligent orchestrator can adapt: it might skip a step if it's not needed, retry or compensate differently on failures, or choose different paths based on predictions or learned strategies. It essentially brings AI into the role of workflow coordination.

Design Approach

The design centers on an **Orchestrator Agent** – a service responsible for a specific end-toend process that involves multiple microservices. This orchestrator will call other microservices (much like a conductor cueing different instruments) but with added logic at each juncture.

Key characteristics of the design:

- The orchestrator has **workflow logic with embedded AI**. For example, it might use a decision tree or ML model to decide which service to invoke next or how to handle an exception.
- Communication is often **synchronous** between the orchestrator and individual services (since it's actively directing the sequence), but it could also dispatch parallel calls or use asynchronous waits depending on the process.
- The orchestrator typically holds state about the process (e.g., an order fulfillment state machine, or a case file if we think of an insurance claim process) and updates it as steps complete.
- The "intelligence" can range from simple rules (if X, skip step Y) to complex logic (like optimization or planning algorithms that decide on the fly what the best sequence is).

One way to picture it:

```
[Orchestrator Agent] --> Service 1 (Step A)
      | (decide next)
      --> Service 2 (Step B)

      | (branch or loop based on result)
      --> Service 3 (Step C)
```

The Orchestrator calls Service 1, gets a result, *decides* what to do next (maybe calls Service 2 or skips to Service 3 depending on that result or external conditions), and so on, until the workflow is complete. Throughout, it can apply AI at decision points. Perhaps after Service 1, it uses a predictive model to determine if Service 2 is necessary, etc.

Contrast with normal orchestration: Normally, an orchestrator would have a fixed flow (maybe defined in BPMN or code) – e.g., always call 1, then 2, then 3, with perhaps some hardcoded conditions. An *intelligent* orchestrator might dynamically alter the flow. For instance, consider WebkinsTech's **vendor onboarding process** (which we touched on in earlier chapters): a static orchestrator might always wait 7 days for vendor documents and then send a reminder email. An intelligent orchestrator agent could adjust that wait time based on the vendor's risk profile or past behavior (maybe some vendors usually comply quickly, so it waits longer patiently; for others, it nudges sooner). It could even decide to bypass certain verification steps for low-risk vendors (fast-tracking them) while adding extra checks for high-risk ones.

Benefits

- **Global Optimization:** With a central view, the orchestrator agent can optimize the entire workflow for objectives like time, cost, or customer satisfaction. It's not limited to local decisions. For example, if WebkinsTech's order orchestrator sees that the warehouse that would normally ship an item is overloaded, it could decide to split an order or route it to a secondary warehouse, even if that wasn't pre-defined. It's looking at the *big picture*. This can make processes more efficient than each service acting in isolation.

- **Adaptive Error Handling:** Orchestrators are great for handling failures in a coordinated way (think of the Saga pattern for transactions). An intelligent orchestrator can be even better: it could dynamically figure out compensation or retries. If a payment fails, a smart orchestrator might analyze why (using, say, an AI model predicting that a retry in 5 minutes might succeed if it was likely a temporary issue). Or it might choose an alternate payment processor on the fly if the primary is down, rather than simply failing the order.

- **Simplified Client Interaction:** From the perspective of a client (say the UI or an API consumer), having an orchestrator means they only need to talk to one service to accomplish a complex operation. The orchestrator abstracts the multi-step process. This is true of any orchestration, but the intelligent part means the client doesn't need to embed complex decision logic; the system will handle it. For instance, a customer placing an order doesn't have to worry about anything except sending the order – the intelligent orchestrator coordinates inventory check, payment, fraud screening, shipping arrangement, and maybe even chooses the shipping speed based on customer profile (faster for VIPs, etc.), all automatically.

- **Policy Enforcement and Governance:** A central agent is a convenient place to enforce policies. If WebkinsTech wants to ensure that *every* high-value transaction gets an extra verification step, the orchestrator agent can enforce that no matter how the microservices beneath are implemented. If those policies need AI (like determining what counts as "high-value" dynamically), the orchestrator can encapsulate that logic. Essentially, it's a control point that can be imbued with smarts to enforce business rules and adaptations consistently.

- **Visibility and Monitoring:** Having a single orchestrator for a workflow can improve **observability**. It can log the decisions it made, the path taken, and outcomes. This is helpful for debugging and improving the AI. If an agent orchestrator tries a novel path and it fails, you have a trace of what happened. In purely event-driven systems, by contrast, reconstructing the sequence of events across many services can be challenging. So there's a clarity in having the orchestration centralized, even if the decisions were AI-driven.

Trade-offs and Challenges

- **Centralization Drawbacks:** The orchestrator, intelligent or not, becomes a central component that everything flows through for that process. This can be a **scaling bottleneck** if not scaled out (though you can have multiple orchestrator instances handling different workflows or sharded by some key).

It's also a single point of failure for that workflow – if the orchestrator is down, the whole process halts. So you have to invest in making it robust (think clustering, failover).

- **Complexity in Orchestrator Logic:** Adding AI to orchestration can make the orchestrator's code quite complex. It's now mixing concerns: process management and decision-making. If the logic gets too opaque (say the orchestrator uses a machine learning model to decide steps), it might be harder for developers or operators to predict what it will do. Testing all possible paths becomes complicated because the path isn't fixed; it depends on the model's outputs or learned behavior. This requires careful validation of the orchestrator's AI component, maybe simulation of scenarios to ensure it won't do something crazy in production.

- **Slower Adaptation than Local Agents:** Because orchestration is inherently a more top-down approach, it might not react as quickly as a local agent to certain local conditions. For example, if a particular service has an issue that only that service's agent would detect quickly, a central orchestrator might only find out indirectly or after a delay. Intelligent orchestration shines for cross-cutting decisions, but very granular optimizations might still need local intelligence. It's a balancing act between what the orchestrator decides and what individual services decide. If the orchestrator tries to do too much, it can stifle the local initiative of embedded agents. Conversely, if it does too little, it might not achieve the global optimization we want.

- **Development and Organizational Overhead:** Not every team is ready to build an AIdriven orchestrator. It often requires a good understanding of both the business process and AI techniques like decision optimization or predictive modeling. In WebkinsTech's

case, introducing an intelligent orchestrator for vendor onboarding meant the operations team and data scientists had to work closely to encode domain knowledge and train models for the agent's decisions (like risk scoring vendors to decide fast-tracking). This cross-functional development can be challenging. It's certainly a more advanced pattern, likely adopted after some simpler AI integrations are already in place (e.g., maybe after they've comfortable with AI-as-a-Service or embedded agents in simpler contexts).

- **Potential Overreach:** There's a subtle risk: because an orchestrator can, in theory, control everything, one might be tempted to funnel too many decisions to it, even ones better left to individual services or humans. Over-orchestration by an AI agent could lead to a kind of *centralized AI monolith* if taken to an extreme. The architecture should remain modular – maybe have multiple orchestrators for different bounded contexts, rather than one mega-brain for the whole company's operations. Keeping scope clearly defined is important so the orchestrator agent doesn't become an unwieldy monster.

WebkinsTech Case Study: Smart Vendor Onboarding Workflow

WebkinsTech's vendor onboarding process (bringing new suppliers onto their platform) used to be orchestrated by a simple workflow engine: collect vendor info, run compliance checks, wait for document uploads, send reminders, finalize approval. Elena, the Director of Operations, championed upgrading this to an **Intelligent Orchestrator Agent** when she saw opportunities to expedite trustworthy vendors and scrutinize risky ones.

How it works now: The orchestrator agent coordinates multiple microservices involved in onboarding (Document Service, Compliance Service, Email Notification Service, etc.). As a new vendor goes through the steps, the agent maintains a state for that vendor's onboarding case. Crucially, it uses an **AI risk model** to adapt the process:

Right after a vendor submits initial info, the agent's embedded model generates a *risk score* (based on factors like vendor type, country, initial compliance data, etc.).

If the risk score is low (vendor seems trustworthy), the orchestrator might **skip** certain steps or make them asynchronous. For instance, it may provision the vendor account immediately and only do deep compliance checks in the background, effectively fast-tracking the vendor to start business sooner.

If the risk score is high, the agent adds extra steps: it might request additional documents (like proofs of certification) via the Document Service or schedule an interview with the vendor. It also sets tighter deadlines: a high-risk vendor not completing steps quickly triggers a faster escalation (the agent will send a reminder email after 2 days instead of 7, and if no response, flag a human officer).

The agent orchestrator monitors progress. If a vendor stalls at a step, instead of always waiting a fixed time, it decides when to intervene. For example, in early trials the agent learned that tech startups often drag on legal document uploads, so it proactively schedules a reminder sooner for them, whereas large enterprise vendors usually complete without nudges, so it waits longer patiently.

Benefits: This intelligent orchestration led to **faster onboarding for good vendors** – WebkinsTech reported that low-risk vendors were going live 30% faster than before, giving them a head-start on the platform. At the same time, the orchestrator caught potential problematic cases more reliably, because it dynamically allocated more scrutiny to them. It was as if the process became *self-adjusting*: simple cases flowed quickly, complex cases got the attention they needed. The system also reduced manual oversight; the orchestrator only involved a human manager when its logic determined something truly out-of-normal, rather than on every single case.

Insights and Challenges: During development, the team had to carefully simulate various scenarios to train and fine-tune the risk model. In one early test, the agent was too lenient and auto-approved a vendor who later turned out to have compliance issues. This prompted adding a rule: no matter how low-risk someone appears, certain critical checks (like sanction list screening) *cannot* be skipped – a guardrail in the orchestrator's logic. This mix of AI and rule-based safeguards proved important. They also set up monitoring: each quarter, Elena's team reviews a sample of cases to see if the agent's decisions aligned with outcomes (this is a human-in-the-loop approach to ensure the AI doesn't drift into bad habits).

The intelligent orchestrator became a **central brain** for vendor onboarding at WebkinsTech. By orchestrating with insight, it demonstrated how a central agent can coordinate multiple microservices in a smart way, achieving both efficiency and control. The trade-off was the complexity in that agent – effectively, a lot of domain knowledge and AI logic ended up in one place. WebkinsTech managed this by keeping the orchestrator's scope limited to onboarding (they have separate orchestrators for other areas), and by thoroughly documenting its decision policies. The success of this pattern gave them confidence to attempt similar orchestrators in other processes.

Orchestration Best Practice: When implementing Intelligent Orchestration, treat the orchestrator as a *specialized microservice* with a well-defined bounded context. Don't let it become an all-knowing, all-doing entity for unrelated concerns. For example, WebkinsTech's onboarding orchestrator doesn't handle order processing or customer support – those are separate domains possibly with their own coordinators. Keeping scope narrow makes it easier to train and maintain the orchestrator's AI logic. Additionally, always have clear fallback paths. An intelligent orchestrator might try an optimized path, but it should know how to gracefully fall back to a default flow if the AI is unsure or if external conditions change. This ensures that a glitch in the AI's decision doesn't break the whole workflow. Basically, give your orchestrator agent *the ability to be boring when needed* – sometimes the smartest move is to do the standard thing, and that's okay!

Having a smart conductor can dramatically streamline complex workflows. But in distributed systems, not everything needs or has a central brain. Often, systems lean towards a more **event-driven, decentralized** approach – especially as scale grows. This is where **EventDriven Agents** come into play, essentially the choreographed dance of intelligent services.

Event-Driven Agents

In an **event-driven architecture**, services communicate by emitting and reacting to events rather than direct calls. For example, Service A publishes an event "OrderPlaced" and other services subscribe and react to that event in their own way. When we infuse intelligence into this model, we get **Event-Driven Agents**: autonomous agents (which could be entire microservices or embedded agents within services) that **perceive events**, apply some decision-making, and then take actions (which often result in new events).

This pattern is akin to an **ecosystem of smart actors** all listening and responding to the environment (the stream of events). There's no single boss; instead, coordination happens through the sequence of events flowing in the system. If Intelligent Orchestration was a wellmanaged team project with a leader, Event-Driven Agents are more like a spontaneous block party where everyone dances when their favorite song comes on – coordination emerges from shared signals (music) rather than commands.

Design Approach

The design for Event-Driven Agents revolves around an **Event Bus or Message Broker** (like Kafka, RabbitMQ, etc.) that routes events between services. Each agent service or component:

- **Subscribes** to certain event types (topics, streams).
- **Acts** when those events occur, possibly making decisions or triggering its own processes.
- **Emits** new events as a result of its actions, which in turn may be consumed by others.

The agents themselves can be standalone microservices dedicated to an intelligent function, or they can be embedded logic within existing services triggered by events. The key is that **events are the trigger and the glue**.

Let's visualize a scenario at WebkinsTech in a simplified diagram:

```
Order Service --(OrderPlaced event)--> Event Bus ---> Fraud Agent Service
                                                  |-----> Shipping Agent Service
                                                 |-----> Marketing Agent Service
```

1. The **Order Service** publishes an OrderPlaced event when a customer places a new order.
2. The **Order Service** publishes an OrderPlaced event when a customer places a new order.
3. A **Fraud Detection Agent** (which could be an AI service specialized in fraud) is subscribed to OrderPlaced. When it sees the event, it wakes up, evaluates the order (maybe using an ML model), and if it detects something suspicious, it might emit an event like OrderFlaggedFraudulent.
4. Separately, a **Shipping Service/Agent** is also subscribed to OrderPlaced. It picks a warehouse and carrier for the order and emits OrderShipped or ShippingScheduled event once done.
5. Meanwhile, a **Marketing Agent** listens for OrderPlaced too. For each order, it might trigger a follow-up – say scheduling a personalized thank-you email or updating the customer's profile. It could also listen for events like OrderDelivered later to know when to send a feedback survey.
6. These agents might not even know about each other. The Fraud Agent doesn't directly tell Shipping what to do; it just raises a flag event. The Shipping agent might subscribe to fraud outcome events as well,

e.g., if an order is flagged fraudulent, the Shipping agent could hold off on processing it.

This decoupling means each agent is **focused on its own role**. They communicate implicitly through events.

Crucially, the intelligence comes in how each agent decides to act on the events:

- The Fraud Agent might use an AI model (say, anomaly detection on order data) to decide if OrderFlaggedFraudulent should be emitted or not.
- The Shipping Agent might use an optimization algorithm to decide from which warehouse to ship based on current inventory and backlog (maybe it's even a learning agent trying to minimize delivery times).
- The Marketing Agent could use rules about customer value to decide if this order merits sending a coupon for future purchases.

None of these decisions are centrally coordinated; they happen in parallel, driven by the event stream.

Benefits

- **Loose Coupling & Flexibility:** Event-driven systems are highly **decoupled**.

Services/agents don't directly call each other, so adding a new agent or modifying one usually doesn't require changing others. For WebkinsTech, this meant they could introduce a new **Analytics Agent** that listens to the same order events to gather metrics, without touching any existing code. This flexibility is gold for evolving systems – you can plug intelligence in and out like Lego pieces.

- **Scalability and Parallelism:** Because agents operate independently, the system can handle a lot of work in parallel. In the above example, Fraud, Shipping, and Marketing agents all process the OrderPlaced event in parallel. If one is slow (say the Marketing Agent takes a while), it doesn't hold up the others. Also, each agent can scale out (multiple instances consuming the event stream) as needed. This pattern aligns well with *cloud scalability*: you can scale each stream consumer based on demand. It can handle spikes elegantly – e.g., during a sale, a flood of OrderPlaced events will just enqueue and be processed by however many instances of each agent are available; you can add more if needed.

- **Emergent Behavior & Innovation:** This decentralized pattern often leads to **emergent behavior**, where interesting outcomes arise from the combination of agents. For example, if WebkinsTech's Fraud Agent flags an order, and an event triggers a Customer Service Agent to immediately text the customer for verification, and the customer confirms legit, then an event triggers the order to resume – you get a whole adaptive fraud handling loop without a central brain explicitly coding "if fraud then do X, Y, Z". Each part just knows its role on certain events. New combinations of events can yield new functionality. Teams can innovate within their microservice/agent – like improving their response to events – without needing a system-wide overhaul.

- **Resilience through Isolation:** If one agent goes down or has issues, others can still function (perhaps with some degradation). For instance, if the Marketing Agent is down for a while, orders still get processed and shipped; only the follow-up emails might be delayed (they could even be processed later when it comes back up if events are durable). There's no single orchestrator whose failure stops everything. The event bus can buffer events until consumers are ready (depending on the messaging tech), adding resilience. Each agent can also be updated or restarted independently.

- **Real-time Streaming of Decisions:** Event-driven approach aligns well with **stream processing**. Agents can continuously learn or make decisions on streaming data. For example, an **Online Learning Agent** (tying into our next pattern) might listen to events of user interactions to update a model in real-time. The event stream serves as a realtime data source for learning algorithms, enabling on-the-fly adjustments. This is harder in a request-response model where you don't capture all interactions as easily.

Trade-offs and Challenges

- **Complexity of Observability:** The loose coupling means the overall flow of events (the cause-and-effect chain) can be hard to trace. If something unexpected happens (e.g., an order never got shipped), you might have to dig through event logs across multiple services to reconstruct what happened: OrderPlaced -> FraudFlagged -> (maybe Shipping agent decided not to ship because of fraud, but when fraud cleared did we emit an event to resume shipping? etc.). Debugging requires good tooling: distributed tracing or event logging systems. It's not as straightforward as stepping through a single orchestrator's logic.

- **Event Storm and Handling:** Improper design can lead to a glut of events or **chattiness** that becomes hard to manage. If every tiny action emits an event and triggers multiple agents, you could flood the system or create feedback loops. For example, consider two agents unintentionally bouncing events off each other (Agent A emits event X, Agent B reacts and emits event Y, Agent A also subscribed to Y and emits X again... uh-oh!). Designing idempotent handlers (so processing an event twice doesn't cause double side effects) and using correlation IDs to track cycles is important.

There's a whole discipline to managing event-driven systems to avoid chaos. Introducing AI into the mix (where an agent's reaction might not be 100% predictable) means you must be extra cautious to avoid unintended infinite loops or oscillations.

- **Consistency and Ordering:** In event-driven systems, especially distributed ones, maintaining consistency is tricky. If multiple agents update different parts of the system in response to an event, ensuring that the overall state remains consistent can require careful design (sometimes using patterns like **Saga** or **event sourcing** to reconcile eventual consistency). For instance, the Fraud Agent might cancel an order by emitting an OrderCanceled event. But what if the Shipping Agent already emitted OrderShipped because it processed the original event slightly later? Now you have a race condition – order got shipped even though it was canceled due to fraud. Resolving these kinds of issues might require designing the events and processes to accommodate out-of-order processing (maybe shipping listens for a potential fraud flag for a short window, or cancellation events have higher priority). This is not trivial.

- **Learning Curve and Governance:** Adopting an event-driven approach with multiple intelligent agents requires a **cultural shift** in how you design software. Teams need to embrace asynchronous thinking. It can be hard for folks used to straightforward request-response flows to reason about a flurry of events and partial information. Also, who "owns" the overall outcome? In orchestration, you had an orchestrator to tweak if outcomes weren't as desired. In choreography, it might be unclear which agent's behavior to change to fix a systemic issue. It needs strong **observability and sometimes simulation**:

you might simulate event flows in a staging environment to see how new agents play before unleashing them in production.

- **Potential Redundancy:** Sometimes multiple agents might do similar analysis on the same event (like both Fraud and Marketing might analyze user profile info, duplicating some logic). In a tightly orchestrated system, you might consolidate such checks. In an event system, it's more "let everyone do their own thing." This is usually fine, but can lead to minor inefficiencies. It's a trade-off for decoupling. Teams must ensure critical business rules that need consistency are not inadvertently implemented differently by two agents without coordination.

Despite these challenges, WebkinsTech found the event-driven agent approach very powerful as their system grew. It allowed different departments' concerns to be handled by different intelligent services all under a unified event bus.

WebkinsTech Case Study: Order Fulfillment via Event-Driven Agents

As WebkinsTech expanded globally, they refactored their order processing to be more event-driven, creating a set of specialized intelligent agents:

Fraud Agent Service: Listens to OrderPlaced events. Uses an AI model trained on historical fraud patterns to score each order. If high risk, emits OrderFlagged (with details like risk level).

Payment Service: Also listens to OrderPlaced. It initiates payment processing (charging the customer's card). On success or failure, it emits PaymentSuccessful or PaymentFailed event.

Inventory Allocation Service: Subscribed to OrderPlaced. It reserves items from inventory for the order and emits InventoryReserved or InventoryBackordered if stock isn't sufficient.

Shipping Agent Service: Subscribed to InventoryReserved (meaning items are ready) *and* to OrderFlagged and PaymentFailed. This agent waits until it knows inventory is reserved and payment is successful (and also checks that the order isn't flagged fraud). Once conditions are good, it decides from which warehouse to ship and which carrier, using a route optimization algorithm (it's "intelligent" in choosing fastest delivery vs cost based on customer segment). It then emits OrderShipped with tracking info. If it sees an

OrderFlagged or PaymentFailed for that order, it halts or aborts shipping by not proceeding, possibly emitting an OrderOnHold event.

Customer Comms Service: Subscribed to various events (OrderPlaced, PaymentFailed, OrderShipped, etc.). It sends out emails or SMS accordingly: confirmation email when order placed, alert if payment failed needing customer action, shipping notification when shipped, etc. It uses some basic AI to tailor messaging tone based on customer profile (e.g., very apologetic and with a coupon if a VIP customer's payment failed due to a glitch).

Flow: When an order is placed, multiple events ripple out. The Fraud Agent might flag it (or not) roughly concurrently with Payment and Inventory doing their thing. The Shipping agent effectively **waits for the go-ahead** indicated by events: only when it sees that inventory is reserved and payment succeeded and no fraud flag within, say, a few seconds, it proceeds to ship. (They implemented this wait using an event correlation and a tiny delay to gather signals.) If fraud is flagged, the Shipping agent instead emits OrderOnHold and doesn't ship. If later the Fraud Agent clears the order (perhaps a manual review flipped it back), an OrderCleared event could prompt shipping to resume.

Outcome: This event-driven setup was highly **extensible**. Later, WebkinsTech added a **Recommendation Agent** that listened to OrderPlaced and, if the order was high-value, emitted an event to trigger an upsell email with accessories. They didn't have to modify the Order service or the orchestrations – just added a new subscriber to an existing event. Similarly, a **Metrics Logging Service** subscribed to all these events to update dashboards for managers in real-time. The system handled holiday sale traffic gracefully: when orders spiked, they simply added more instances of these agent services behind the scenes. Each type of agent scaled independently – e.g., Fraud Agent was CPUbound (running models) so they scaled it up, whereas the Shipping agent was I/O bound with external carriers so fewer instances sufficed.

Challenges: During testing, they encountered a scenario where an order got shipped even though the payment failed – a classic race condition. The Shipping agent had a small bug in how it listened for PaymentFailed event. They fixed this by using a **transactional event** approach: the Order service only emitted OrderPlaced after ensuring payment was at least initiated. Also, Shipping agent now double-checks payment status via a query if it hasn't heard anything within a timeout. These kinds of eventual consistency issues required careful handling. Logging was another challenge – initially it was hard to trace an order's journey. WebkinsTech adopted a correlation ID (the order ID) that each event carried, and built a logging view where you could see all events related to that ID in timeline order. This became the team's favorite debugging tool.

Once ironed out, the **event-driven agents** approach proved its worth. It gave

WebkinsTech a *responsive, scalable, and modular* architecture. Different AI agents could be developed by different teams (fraud by security team, recommendations by marketing team, etc.) and plugged into the system. The loose coupling meant one agent's failure rarely brought the whole system down; for example, even if the Recommendation agent crashed, orders still went through and shipped fine – only the upsell emails might be missing. The business folks appreciated that new capabilities could be added by simply introducing new events and agents, rather than disrupting the core order pipeline. For WebkinsTech, this pattern unlocked a kind of **innovation ecosystem** where various intelligent services collaborated via events to handle the business.

Caution: Designing event-driven systems with intelligent agents requires carefully thinking about the worst-case interactions. Always ask, "What if two agents make opposite decisions on the same event?" or "What if this event is missed or processed twice?" In WebkinsTech's case, they introduced governance for events: a central schema registry for events (so everyone knows the structure and meaning of events like OrderPlaced), and a review process whenever a team wanted to introduce a new event or agent that could impact others. This helped catch potential conflicts early. Additionally, they implemented idempotency for consumers – for example, the Shipping agent can receive the same InventoryReserved event twice and will ensure it doesn't ship the order twice (it keeps track if it already handled that order). These defensive practices are crucial because once you have many moving parts, you want to avoid snowballing errors. Testing not just individual agents, but the *system of agents* via end-to-end scenarios, was key to gaining confidence.

By now, we have seen architectures with static or pre-trained intelligence. The agents and AI services we discussed so far have models or rules that, while possibly updated periodically, don't fundamentally change on their own at runtime. But what if we want our microservices to **learn continuously, on the fly**? That's where the next pattern, **Online Learning Loops**, comes into play, closing the gap between acting and learning.

Online Learning Loops

One of the most exciting aspects of agentic systems is the ability to not just make decisions, but to **improve those decisions over time through learning**. The **Online Learning Loops** pattern is about architecting your microservices environment to support continuous learning and adaptation. In other words, the system doesn't rely solely on offline-trained models that get deployed occasionally; instead, it **feeds data from operations back into the AI**, updating its knowledge or model as new information comes in, and does so in a cyclic ongoing manner.

This pattern is inspired by the classic **sense–think–act** cycle in robotics or control systems, with an added twist: after acting, the system observes the *outcome* of that action and *learns* from it to adjust future behavior. It's a closed feedback loop of *Action -> Feedback -> Learning -> Updated Action*.

Design Approach

Implementing online learning loops typically involves integrating components for **data collection, model training (or updating), and model deployment** into the live system. Key elements of the design:

- **Observation/Feedback Collection:** As the microservices operate, they generate data about their performance or the outcomes of their decisions. This could be explicit feedback (user clicked a recommended item or not, order was successful or chargeback happened, system metric improved or worsened after an agent's action, etc.). The architecture needs to capture these events or results. Often an *event log or data stream* is used to accumulate learning data.

- **Learning Process:** There's a component (could be an internal thread, a separate service, or an external pipeline) that consumes the above data and updates the AI model or strategy. This might be incremental learning (updating model weights continuously), mini-batch learning (periodically retraining with new data), or even a reinforcement learning loop (where the agent gets reward signals and updates its policy). The key is this is automated and happens on an ongoing basis.

- **Deployment/Update of Model:** Once new insights or model parameters are learned, the system needs to apply them to the running services. In some cases, the agents themselves might be able to update in-memory models on the fly. In others, a new model version might be deployed (perhaps by swapping a model file or through a model-serving service).

- **Repeat Continuously:** The loop runs continuously or at defined intervals. Unlike traditional offline retraining (say retraining a model nightly and deploying next day, which is more like a **Batch Learning** pattern), online learning suggests more immediacy – the system is learning *as it goes*.

A simple depiction of this loop:

```
(Agentic Service) --action--> Environment
          <-- outcome feedback --
(Agentic Service) [updates its model]
-- improved action next time -->
```

Or to break the components out:

```
Agent makes decision -> outcome occurs -> outcome data logged ->  model
update process consumes new data -> model parameters adjusted ->  agent
uses updated model for next decision.
```

For a concrete example, consider WebkinsTech's **dynamic pricing agent**:

- It sets a price for a product (action).

- The outcome is observed: did the product sell at that price? How quickly? Did the customer show interest or abandon the cart?

- Those outcomes feed into a learning algorithm (maybe a reinforcement learning scheme that tries to maximize revenue).

- The pricing policy is updated (perhaps the agent becomes a bit more aggressive or conservative based on results).

- Next time, the agent adjusts prices slightly differently, gradually honing in on optimal pricing.

Another example: a **recommendation service** that personalizes in real-time. It shows some recommendations (action), then tracks what the user clicks (feedback), and immediately updates the user's profile model so the next set of recommendations is better tailored (learning and updating).

Benefits

- **Adaptability to Changing Conditions:** This is the big one. Online learning loops allow the system to **adapt to drift** and evolving patterns. In business, data today might not look like data from a month ago (seasonal trends, new competitor, changed user preferences). An online learning system will continuously update itself to remain

effective. WebkinsTech saw this when their model for product recommendations started adapting to a new trend within hours instead of the team noticing a drop in engagement days later and manually retraining. The system stays *fresh*.

- **Personalization and Fine-Tuning:** Online learning enables a high degree of personalization. For example, an agent could effectively learn a unique model per user or per context gradually. Rather than one-size-fits-all behavior, it tailors itself. Imagine a support chatbot that gradually learns a particular user's style or needs after a few interactions, improving its responses for that user specifically – that's achievable with continuous learning.

- **Discovery of Novel Strategies:** Especially with reinforcement learning agents, sometimes the agent can discover strategies that humans didn't program in. By trying things and seeing outcomes, it might stumble on a solution that wasn't obvious. For instance, an automated cloud resource manager at WebkinsTech might learn to reboot a server preemptively at a certain time that tends to have issues, thereby avoiding incidents – something humans might not have established as a rule. This emergent improvement can give competitive advantages.

- **Reduced Need for Manual Model Retraining:** Although online doesn't eliminate the need for oversight, it can lessen the engineering burden of constantly extracting data, retraining models offline, and redeploying. The system handles some of that automatically. It can be more responsive than the typical cycle of "collect data for a week -> retrain -> deploy next week" which has inherent lag.

- **Continuous Improvement:** In ideal cases, the more the system runs, the better it gets. It's learning from every additional data point. This can lead to a virtuous cycle especially in high-volume scenarios – if you're processing millions of events, your model could be adjusting minute by minute, potentially achieving high performance and keeping it.

- **Handles Unpredictable Scenarios:** In static systems, when something unforeseen happens, the system might fail or perform poorly until humans intervene. An online learning agent might detect the anomaly and adjust behavior somewhat to handle it. For example, if users suddenly start using a new slang with a chatbot, an online learning NLP model might gradually incorporate those terms into its understanding, whereas a static model would stay clueless until updated.

Trade-offs and Challenges

- **Complexity and Stability:** Online learning architectures are among the **most complex** to get right. You're essentially adding a whole layer of ML Ops into the runtime. Ensuring the learning process itself doesn't introduce instability is a big challenge. A model that is learning can also *mislearn*. If not properly guided, it might overfit to recent noise or chase a wrong objective. For instance, a learning agent might pick up a shortterm pattern that's actually a fluke and change its behavior detrimentally. There's a risk of the system's performance oscillating or even diverging if the learning isn't carefully regularized. Many teams find they need to put a lot of safety checks (e.g., "don't change the model too drastically too fast," or "if new model underperforms old model, revert quickly").

- **Observability and Debugging:** Now not only do you have to debug the decisions of an agent, you have to debug the *evolution* of those decisions over time. It can be very hard to trace why the agent is doing X today when it used to do Y last week – because it changed its internal model incrementally thousands of times since then. We need robust logging of the learning process: when did it update the model, with what data, what metrics did it see? Tools like **model versioning** and experiment tracking become important even in production. Essentially, some practices from data science and ML (like tracking experiments) need to blend into your microservice monitoring.

- **Resource Intensiveness:** Learning continuously can be heavy. Training ML models can be CPU/GPU intensive. If you're doing it as part of your live system, you have to allocate resources for it. Sometimes the learning can be offloaded to a separate service or even a special infrastructure (like a streaming processing job). For example, you might use Kafka to stream events to a Spark or Flink job that updates a model, then push new model parameters to the microservice. That's additional infrastructure and complexity. Also, online algorithms typically simplify models to allow incremental updates (e.g., using linear models or small neural nets that support partial fit). Using very large deep learning models in true online fashion is still an area of research and advanced engineering (it might be done in big companies with custom systems).

- **Data Quality and Feedback Loops:** The effectiveness of online learning is only as good as the feedback you provide. If the feedback is sparse or delayed (e.g., you only know if a recommendation was good when a purchase happens much later), it complicates the loop. You might need to engineer proxy feedback (like click = good for now, purchase = even better signal later). There's also the danger of feedback loops: the system might bias itself. For instance, if the recommendation agent quickly learns to push one type of product, users see mostly that and keep clicking it, reinforcing the model's belief that that product is best – meanwhile, it might have missed other opportunities because it self-confirmed its bias. This is a known issue (exploration vs exploitation trade-off). Handling it might involve injecting some randomness or exploration factor so the agent occasionally tries something different to truly learn if it's better. This again is advanced design to avoid local maxima traps.

- **Safety and Ethics:** With continuously learning agents, you have less human oversight over each change in behavior. This can raise concerns in sensitive domains. For example, an agent might unknowingly learn a strategy that is effective but violates a business rule or fairness criterion. E.g., an online learning pricing agent might discover it can charge certain customers more based on some demographic proxy – that could be seen as discrimination if not caught. Or it might take actions that marginally increase a metric but hurt user experience (classic example: a learning-based content recommendation might find sensationalist content gets more clicks, learning to push it more, leading to a bad outcome overall). Having **guardrails** and periodically resetting or reviewing the agent's policies is often necessary. Many online systems still involve humans in the loop at intervals, to ensure the AI hasn't gone off the rails.

- **Not Always Necessary:** A note of caution is that not every application gains a lot from online learning. Sometimes simpler periodic retraining is enough if your environment doesn't change that rapidly. Online learning adds a lot of complexity. WebkinsTech chose it for specific cases where the extra adaptability was worth it – like dynamic pricing during volatile demand, or their cloud management agent which had to react to load changes on the fly. But for things like image recognition or static fraud rules, they were fine with offline updates. It's important to identify where online loops are truly beneficial, to avoid over-engineering.

WebkinsTech Case Study: Adaptive Cloud Manager Agent

WebkinsTech runs many services in the cloud, and managing costs vs performance is a constant battle. They built an **Autonomous Cloud Manager Agent** (introduced in earlier chapters) that can scale services up or down, schedule maintenance, and even handle certain incidents (like restarting a crashed service). Initially, this agent worked on fixed rules and thresholds. Later, WebkinsTech infused it with an online learning loop to make it more adaptive.

How it learns: The agent continuously monitors metrics like CPU usage, response times, error rates, and cloud costs. It makes decisions (actions) such as: *scale out service X by 2 instances*, or *reboot server Y*, or *throttle down a feature to reduce load*. Each action has outcomes – some positive (issue resolved, performance improved), some negative or neutral (no effect, or made things worse). The agent receives feedback by looking at subsequent metrics. Did errors go down after a reboot? Did cost shoot up after adding instances, and was it worth it in performance gain? These outcomes are turned into a **reward signal** for a reinforcement learning algorithm within the agent. Essentially, the agent gives itself a score based on the impact of its actions (WebkinsTech defined a reward function balancing high uptime, low latency, and low cost).

The agent runs this cycle continuously. It uses a form of Q-learning (a reinforcement learning approach) to update its policy: basically, over time it figures out which actions tend to lead to better long-term reward in various situations. For example, it might learn that *scaling up early when CPU crosses 70%* prevents worse problems later, yielding a higher reward than waiting until 90% when the service might already be struggling. It might also learn subtle things, like if a particular service's memory usage pattern usually spikes at midnight due to a batch job, it's better to preemptively allocate more memory at 11:50 PM. No human explicitly coded these nuances; the agent discovered them by trial and error in the live environment (within safe bounds of course).

Architectural notes: To make this work, WebkinsTech had the agent emit events and log data for every scaling decision and the subsequent system state. They had a separate feedback loop job that aggregated these events to compute rewards and fed back into the agent's learning module. Initially, the agent was conservative – it followed baseline rules 90% of the time and only experimented 10% (to avoid doing harm). As it gained confidence (proving certain actions are beneficial), it gradually took more liberty. The learning was continuous but with safeguards: if performance dipped too much, the agent would temporarily fall back to default behavior (this was a safety trigger). They also kept a human override: ops engineers got alerts of any unusual actions and could step in if needed, especially in early phases.

Results: Over a period of weeks, the cloud manager agent started showing measurable improvements. It managed to reduce cloud costs by ~15% by identifying overprovisioned times and scaling down more aggressively when usage was low (something the static rules were too cautious about). It also improved incident response times – certain types of errors were resolved by the agent 30% faster than before, because it learned patterns (like when service Q throws error XYZ, a restart fixes it 80% of the time, so it would just do it). The ops team reported that they spent less time manually tuning auto-scaling parameters and more time on strategic improvements, because the agent was handling the tactical adjustments itself.

Challenges observed: There were a few hiccups. At one point, the agent learned a "trick" to minimize cost that wasn't desirable: it started delaying scaling-up actions a bit too long, which kept costs low but occasionally caused slow response times for users (basically gaming the reward function since the cost savings slightly outweighed the performance penalty in its calculation).

The team caught this by noticing user complaints. They adjusted the reward weighting to penalize latency more heavily, and the agent's behavior corrected. This illustrated the importance of a well-designed reward function and oversight. They also encountered the complexity of testing: they had simulated the agent's learning in a sandbox on historical data, but real-world behavior had surprises (like unusual traffic spikes the agent had never seen). Gradually, with careful monitoring, they built trust in the agent.

This **online learning loop** example shows how an agentic microservice (here, an infrastructure management service) can evolve its strategy in real-time. It achieved a level of automation that would be hard-coded in static logic otherwise. The trade-off was significant complexity in design, requiring expertise in reinforcement learning and a robust monitoring setup. WebkinsTech deemed it worth it for this use case. In other cases (like their recommendation engine), they chose a simpler approach of nightly retraining because it was sufficient. They learned to apply online learning loops *selectively*, where the domain demanded continuous adaptation.

Be Careful with Continuous Learners: Always implement safety nets for online learning agents in production. At WebkinsTech, any automatically updated model is first validated against recent historical data (a short sliding window) to ensure it's not totally off-kilter before the agent starts using it. They also have a mechanism to roll back to the last known good model if the new behavior starts causing errors or user dissatisfaction. Essentially, an online learning agent should have a "pause" or "rollback" switch. Think of it as an autonomous car with a steering wheel that a human can grab if needed.

Additionally, log the agent's decisions and the reasons (e.g., key features or signals) if possible. This logging helps in auditing why the agent did something crazy at 3 AM. Continuous learning means the system is somewhat a moving target – without logs, you're debugging a shadow that changes shape. Finally, start with conservative learning rates: let the agent make small incremental updates. Rapid learners can overshoot and instabilize the system. It's often wise to blend new learned policy with old policy (e.g., use a weighted sum so it changes gradually) – essentially letting the agent *ease into* new behaviors rather than jerk the wheel suddenly. With these precautions, you can reap the benefits of adaptation while minimizing the risk.

We've now covered five major patterns for architecting intelligent microservices, from a centralized brain to distributed minds, and from static intelligence to continuously learning systems. It's evident that no single pattern is "the best" in all cases – each has its sweet spot. In fact, a large system might employ several of these patterns simultaneously.

To help summarize and compare, let's look at a quick comparison of these patterns side by side on a few key aspects:

Comparison of Architectural Patterns

The table below highlights differences across **AI-as-a-Service**, **Embedded Agent**, **Intelligent Orchestration**, **Event-Driven Agents**, and **Online Learning Loops** in terms of complexity, ideal use cases, and scalability considerations.

Pattern	Implementation Complexity	Ideal Use Case Fit	Scalability Considerations
AI-as-a-Service	*Low-Medium:* Fairly straightforward to implement by exposing an AI model via an API. Some added complexity for maintaining the service and handling latency.	- Adding AI to existing microservices with minimal changes.- Reuse of a common AI capability (e.g., recommendation, fraud check) across many services.- When you have a dedicated team for ML that can own the service.	- Scales by scaling the AI service horizontally for more throughput.- Can become a bottleneck if many calls – ensure load balancing, caching where possible.- Needs robust infrastructure to avoid being single point of failure.

Pattern	Implementation Complexity	Ideal Use Case Fit	Scalability Considerations
Embedded Agent	*Medium:* Requires integrating AI logic into each service. Increases perservice complexity and operational overhead for model updates.	- Low-latency or realtime decision needs (intelligence on the spot).- Domainspecific AI where each service benefits from its own tailored logic.- When	- Scales naturally with microservice instances (each instance carries the agent).- Watch out for resource usage on each instance (may need more CPU/memory per service).- Updating
		microservices are domain-bound and can encapsulate their intelligence fully.	models requires redeploying services, which at scale needs automation.

Intelligent Orchestration	*Medium-High:* Building an AIdriven workflow engine is complex. Involves both workflow management and AI decision logic. Testing all paths is involved.	- End-to-end processes that cross multiple services where central oversight can optimize outcomes (e.g., complex business workflows).- When global policy enforcement or optimization is needed (AI can tweak the process flow for efficiency or personalization).- Scenarios with clear start/finish (transactional flows) that benefit from adaptive logic.	- Orchestrator itself may need scaling or partitioning (e.g., one orchestrator instance per active workflow or use a cluster for many workflows).- If orchestrator is overwhelmed, it could slow entire workflows – design for concurrency and failover.- Still a somewhat centralized element, so ensure high availability (activepassive or sharded orchestration).

Event-Driven Agents	*High:* The architecture is distributed and requires careful design of events and handlers. Complexity in ensuring eventual consistency and tracing.	- Large, complex systems where decoupling and independent evolution of services is paramount.- When you want to enable **emergent behaviors** and allow multiple intelligent services to interact flexibly.- Use cases with naturally asynchronous processes or where many things happen in response to one event (fan-out).	- Highly scalable: each agent service can scale independently based on workload.- The event broker must handle high throughput (design for partitioning topics if needed).- Need to manage event growth (storage/retention) and ensure agents can keep up; otherwise implement backpressure or streaming processing techniques.
Online Learning Loops	*Very High:* Adds ML training into the runtime. Complex ML Ops, careful tuning, and	- Environments that change rapidly or require continuous adaptation (e.g., realtime bidding, adaptive	- Can be resourceheavy: ensure you have scalable compute for training (possibly separate from serving
Pattern	**Implementation Complexity**	**Ideal Use Case Fit**	**Scalability Considerations**

	safety measures are needed.	systems in unpredictable environments).- Use cases where each data point is valuable to immediately improve the model (personalization, adaptive control systems).- Organizations with strong ML engineering maturity (since this is cuttingedge practice).	instances).- Might offload learning to a streaming platform or distributed training setup – which must scale with data rates.- Scalability also in terms of data: need data pipelines that can ingest and process feedback at high velocity (consideration for big data infra).- In terms of performance, the serving part still needs to respond fast; design so learning updates don't stall inference (e.g., decouple training from serving threads).

Notes: Complexity is rated relative to each other (assuming a baseline of a functioning microservice architecture). "Low" means easier relative to others, not trivial per se. Use case fit is where each pattern tends to shine, though they can be applied outside those areas with trade-offs. Scalability includes both scaling the pattern and the implications on system scalability.

As you can see, simpler patterns like AI-as-a-Service are easier to implement but might introduce some bottlenecks, whereas Event-Driven Agents and Online Learning offer powerful capabilities but at the cost of higher complexity and skill requirements. Many realworld systems start on the left side of this table (centralized, simpler) and gradually move right (distributed, adaptive) as they grow in needs and sophistication.

Choosing the Right Pattern: A Readiness Checklist

With multiple patterns in our toolkit, how do we decide which one(s) to use for a given project or component? It often depends on the specific context, constraints, and goals. Below is a **checklist of considerations** to help choose an architectural pattern for agentic microservices:

Pattern Selection Checklist

- **Identify the Decision Scope:** Is the AI decision localized to a single service's concerns (e.g., inventory predictions within Inventory service) or does it span

  multiple services (e.g., an approval decision that touches orders, payments, and notifications)?

  - *Localized:* Consider **Embedded Agent** (for autonomy) or **AI-as-a-Service** (if you prefer centralized maintenance).

  - *Cross-service:* Consider **Intelligent Orchestration** (if a clear process can oversee it) or **Event-Driven Agents** (if it's more asynchronous or you want loose coupling).

- **Latency Requirements:** How fast must the decision/response be?

- *Needs real-time (milliseconds):* **Embedded Agent** is favorable (no network call). Also Event-Driven can be real-time but ensure events are low-latency; AIas-a-Service might be okay if model inference is quick and network overhead is minimal.

- *Tolerant of slight delays:* **AI-as-a-Service** is usually fine (tens of ms added). **Intelligent Orchestrator** may add coordination overhead, so only use if needed for logic, not just for fun.

- **Team and Infrastructure Maturity:** Consider your team's expertise and tooling.

  - *New to AI/ML or small team:* Start simpler with **AI-as-a-Service** or leveraging existing cloud AI APIs. **Embedded Agent** if the integration is straightforward in one service. Avoid jumping into online learning until basics are mastered.

  - *Strong ML Ops and distributed systems experience:* You can confidently explore **Event-Driven Agents** and **Online Learning Loops**. These require good logging, monitoring, continuous deployment pipelines for models, etc.

- **Frequency of Model Updates:** How often do you need to update the AI logic?
  - *Rarely (months):* Any pattern works. **Embedded** is fine since redeploying occasionally isn't a big deal.
  - *Occasionally (weeks):* **AI-as-a-Service** shines (update model in one place). Or an orchestrator agent that can load new rules easily.
  - *Very often or continuously:* **Online Learning Loops** or at least an automated pipeline for updates. If multiple services need frequent updates, centralize that logic (AI-as-service or a centralized embedded library) to avoid operational headache.
- **Coupling and Autonomy:** Are you aiming for maximum independence of services or is a little coupling okay?
  - *Want independent, decoupled services:* **Event-Driven Agents** pattern is aligned to that philosophy, also **Embedded Agent** keeps intelligence local to each service (which is decoupled but duplicates logic if many need same AI).
  - *Okay with some coupling for simplicity:* **AI-as-a-Service** introduces a dependency but simplifies others. **Intelligent Orchestration** couples the process to a controller but yields coordinated behavior.
- **Global vs Local Optimization:** Do you need a global view for the AI to be effective?

- *Global optimization needed:* (e.g., optimizing overall system throughput, multistep workflows) **Intelligent Orchestration** might be needed, or a carefully designed event-driven mechanism that still achieves a cohesive outcome (but that can be harder).
- *Local decisions suffice:* (each part can optimize its piece) **Embedded Agent** or independent **AI services** for each domain might suffice and keep things simple.

- **Failure Impact:** If the AI component fails or is wrong, what's the impact?
  - *High impact (e.g., could disrupt core flows):* Lean towards patterns where you can easily fall back or intervene. For instance, orchestrator can escalate to human or use default logic if AI fails; AI-as-a-Service can return a default response if down (with circuit breaker). Embedded agents might be harder to disable on the fly. Event-driven agents can be individually turned off (stop consuming events) if they misbehave, which is nice.
  - *Lower impact (AI is augmenting, not critical):* You can be more adventurous with patterns. If an online learning agent for recommendations goes awry, worst case users see slightly odd suggestions for a bit. That's tolerable, so pushing boundaries there is fine.

- **Example Fit:** Sometimes it helps to match your scenario with known examples:

- If building a **chatbot service**, AI-as-a-Service makes sense (one service handles the dialogue model).

- If adding **smart features to an IoT device fleet**, Embedded Agent (each device's microservice has local intelligence) might be ideal for autonomy when offline.

- If creating an **automated loan approval system** that involves multiple checks, an Intelligent Orchestrator agent could manage the end-to-end process and use AI at decision points.

- If running a **marketplace platform** with many independent processes (orders, notifications, analytics) that all react to events, Event-Driven Agents keep it flexible and scalable.

- If implementing a **self-optimizing system** (like our cloud manager or a trading system), Online Learning Loops might give that adaptive edge.

- **Combine if Needed:** Remember, these patterns can complement each other. Ask if a hybrid might serve you best:

- e.g., Use AI-as-a-Service for common predictions but call it within an Intelligent Orchestrator – so orchestrator handles flow, AI service handles the model.
- e.g., Use Event-Driven Agents but one of those agents is an orchestrator for a subset of tasks.
- e.g., Implement an Embedded Agent that also publishes events for other services (mixing local decisions with broader signals).
- It's not either/or; it's often a mix, as long as the architecture remains understandable.

By running through this checklist, you can gauge which pattern aligns with your needs and constraints. It's about balancing **strategic goals** (like flexibility, scalability, time-to-market) with **technical feasibility**.

Before we conclude, an important point on continuity: these patterns don't exist in isolation within an enterprise. WebkinsTech's journey, for example, started with simpler automation (earlier chapters) and incrementally adopted patterns as they grew confidence and identified needs. It's perfectly normal to start with AI-as-a-Service for a first AI feature, then later decide to embed some intelligence for latency reasons, and maybe introduce an event bus as the system grows, and eventually dabble in online learning for a cutting-edge advantage. Architecture is evolutionary. The patterns we covered provide a roadmap of options at different stages of that evolution.

Conclusion and Next Steps

In this chapter, we explored **Architectural Patterns for Agentic Microservices**, essentially mapping out how to weave AI-driven intelligence into the fabric of a microservice architecture. We started with the straightforward **AI-as-a-Service** approach, centralizing intelligence for easy reuse, then moved to **Embedded Agents** that decentralize decisions for speed and autonomy. We discussed **Intelligent Orchestration**, putting an AI brain in charge of orchestrating multi-step workflows, and contrasted that with

Event-Driven Agents where intelligent services collaborate through an organic flow of events. We also ventured into **Online Learning Loops**, enabling systems that not only act smart but *get smarter with each action*. Along the way, WebkinsTech's fictional case studies grounded these patterns in reality – from recommendation services and inventory agents to smart onboarding workflows and adaptive cloud managers.

A key takeaway is that **no single pattern is universally superior**. Each pattern involves trade-offs between simplicity, flexibility, performance, and maintainability. As architects and tech leaders, the goal is to **match the pattern to the problem**. It's much like choosing the right tool for a job: sometimes a hammer (simple service) is all you need, other times you need the Swiss army knife (a combination of patterns). Importantly, these patterns can coexist; a complex platform might employ all of them in different corners.

Throughout earlier chapters, we built up the conceptual foundation (the *why* and *what* of agentic AI). In this chapter, we delivered on a big part of the *how* at the architectural level. By now, you should be equipped with a mental catalog of design approaches to try as you infuse intelligence into microservices. We encourage you to think creatively – these patterns are templates, not strict recipes. Real systems might blend them or come up with novel variations.

Let's foreshadow where we go from here. Now that our microservices have brains, the next challenge is getting them to **work together cohesively**. In the coming chapters, we'll zoom into the mechanisms of **communication and coordination** among intelligent microservices. How do these AI-powered services talk to each other? What protocols or messaging patterns suit agent-based interactions best? How do we prevent chaos when multiple autonomous agents are in the mix, ensuring they coordinate towards business goals rather than conflict? We will examine techniques ranging from smart APIs to shared knowledge repositories and even emergent coordination strategies observed in multi-agent systems.

We'll also dive into **state management and data flows** in an agentic context – ensuring that agents have the information they need when they need it, and dealing with the complexity of distributed state and memory. Topics like event sourcing (hinted at in our event-driven discussion), state synchronization, and the role of data platforms will be on the menu.

Finally, looking further ahead, we'll confront the fascinating concept of **emergent behavior** in multi-agent ecosystems. When you have many intelligent microservices interacting, sometimes the overall system exhibits behaviors or capabilities that weren't explicitly programmed into any single agent – *emergence*. This can be powerful (like traffic flows optimizing themselves) or problematic (feedback loops causing instability). We'll prepare you to recognize and harness emergent effects, as well as mitigate unwanted ones.

In summary, we've architected individual intelligent services (the **components**); next, we architect the **collective**. By the end of the subsequent chapters, you'll see how these smart microservices can communicate, coordinate, and even **collaborate to produce emergent intelligence** greater than the sum of its parts. The journey from basic automation to a responsive, intelligent swarm of services is an exciting one, and we're right in the middle of it.

As one WebkinsTech engineer put it:

"We've taught our microservices to think – now we have to teach them to talk to each other."

In the next chapter, we'll do exactly that, diving into communication patterns and coordination mechanisms for our agentic microservices. Get ready to explore how these agents negotiate, synchronize, and work as a team, setting the stage for the richer, sometimes surprising, dynamics of multi-agent systems. The evolution continues – onward to building intelligent conversations and collaborations between services!

Chapter 7: Communication and Coordination

Introduction

In the previous chapter, we established architectural patterns for building intelligent microservices at WebkinsTech. Each microservice was endowed with a degree of autonomy and AI-driven decision-making, essentially becoming an *agent* capable of acting on its own.

However, even the smartest agents cannot operate in a vacuum. Just like a well-rounded team, these microservices need to **communicate** with each other and **coordinate** their actions to achieve broader business goals. Without effective communication and coordination, an ensemble of intelligent services can quickly turn into a disjointed collection of silos or, worse, agents working at cross-purposes.

Communication and coordination are the twin pillars that transform individual agents into a cohesive system. Imagine a group of autonomous drones surveying a large farm: if each drone never shares its findings or flight path, they might all scan the same area redundantly while leaving other regions untouched. Or consider two AI-driven services in a data center – one aims to optimize cost by shutting down idle servers, while another ramps up servers to ensure high performance. If they don't **coordinate**, they'd constantly undo each other's efforts, creating chaos instead of efficiency. These simple scenarios illustrate why our intelligent microservices must talk to each other and align their behaviors.

In this chapter, we take a deep dive into how agentic microservices **talk** and **work together**. We will explore communication mechanisms, from direct request-response calls to distributed event streams, and examine how agents coordinate their actions through patterns like orchestration, choreography, negotiation, and consensus. Along the way, we'll use realworld metaphors – from team collaboration in an office to traffic control in a city – to ground these concepts. We'll also look at technologies like service meshes that provide the connective tissue for reliable inter-service conversations, and discuss how agents share knowledge to build collective intelligence. Finally, we'll wrap up with a practical checklist to help you design robust communication and coordination in your own systems. By the end of this chapter, you'll have a comprehensive understanding of how to make a diverse set of intelligent microservices operate in harmony.

Communication Mechanisms

When designing interactions between intelligent microservices, one of the first considerations is **how** they will communicate. Broadly, communication falls into two styles: **synchronous** (direct request-response) and **asynchronous** (message or event-driven). Each has its uses, and often an architecture will employ both. Let's explore these mechanisms and how they apply to agentic microservices.

Synchronous Communication (Request-Response)

In synchronous communication, one service calls another and waits for a reply before moving on. This is the classic *request-response* model. It's akin to making a phone call: you dial someone and stay on the line until you get an answer. In microservices, this typically means using APIs – often RESTful HTTP calls or a gRPC invocation – where Service A sends a request to Service B and expects an immediate response. The calling service is blocked (waiting) until the response arrives.

Synchronous calls are straightforward. They feel like calling a function on another service – indeed, in a monolithic application, a function call within the same process was immediate. With microservices, that call goes over the network, but the programming model can remain similar: call a method (or API) and get a result. For example, an **Order Service** might make a REST API call to the **Inventory Service** like GET /inventory/item/123 to check stock, and the Inventory Service returns the data right away. Using gRPC, the Order Service could call a method like CheckStock(item_id) on the Inventory service's gRPC interface, which under the hood sends a network request and waits for a reply.

REST vs gRPC: Both REST (often with JSON over HTTP) and gRPC (with binary Protocol Buffers over HTTP/2) are common choices for synchronous agent communication. REST is human-readable and ubiquitous, while gRPC offers better performance and a strongly-typed contract. In practice, an agent might expose a REST API for general use and a gRPC API for internal service-to-service calls where efficiency matters. Either way, the key property is the same: the caller expects an answer immediately.

One advantage of synchronous communication is simplicity. It's easy to understand the flow: call -> response -> next step. This is often used for *querying information* or when an action must be completed immediately. However, there are trade-offs:

- The services are **time-wise coupled** – the caller is stuck waiting on the callee. If Service B is slow or down, Service A is stalled.

- It can introduce **latency** in a chain. If an agent needs to call several others sequentially, those network round-trips add up.

- There's a risk of cascading failures: if one service in a call chain fails, it can affect the others waiting on it.

To mitigate these issues, architects employ techniques like **timeouts**, **retries**, and **circuit breakers** on synchronous calls. For instance, an agent might only wait 2 seconds for a response and then fall back to a default behavior if no reply (preventing an infinite hang). Despite the challenges, synchronous calls remain essential for operations that require an immediate answer (e.g., validating a password, fetching a current price, etc.).

Synchronous Pitfall

Be cautious of chaining too many synchronous calls across services. It can create a brittle, slow "distributed monolith" where a single request fans out to dozens of calls. Aim to keep request-response chains shallow, and use asynchronous patterns for non-critical or long-running interactions to avoid high latency and tight coupling.

Asynchronous Communication (Messaging and Events)

Asynchronous communication means a service sends a message or event and moves on without waiting for an immediate reply. It's like sending an email or dropping a letter in the mail – you send your message and carry on; the recipient will process it at their own pace. In microservice architectures, this is enabled by message brokers and event streams.

With asynchronous messaging, **Service A** can publish an event or message (for example, "Order Placed") to a **message broker** (such as Apache Kafka, RabbitMQ, or a cloud messaging service). **Service B** (and C, D, etc.) can subscribe to that event. When Service A publishes the message, it doesn't stop to wait; the broker will eventually deliver the message to any interested services. Service B processes the message when it gets it, independently of A's timeline.

This decoupling has powerful benefits:

- **Loose coupling:** Service A doesn't need to know about who receives the message or how many services are listening. It just emits an event and forgets about it. Services can be added or removed without changing the sender.

- **Resilience:** If Service B is momentarily down or busy, the message is buffered in the broker. A doesn't have to fail; it continues normally. B will catch up when it can. This way, a temporary outage in one agent doesn't bring others down.

- **Scalability:** Consumers (like Service B) can scale out to handle a high volume of messages, and the publisher doesn't need to change anything. Multiple instances of B could process events in parallel.

However, asynchronous communication introduces its own challenges. Since there's no immediate response, **feedback loops** are more complex. The system becomes **eventually consistent** – there might be a lag between an action and all services knowing about that action. Debugging also requires tracing messages through brokers, and error handling might involve dead-letter queues or compensating actions if something goes wrong asynchronously.

Despite these complexities, event-driven communication is a backbone for intelligent microservices that need to react to events and share information broadly. A good strategy is to use async messaging for updates, notifications, and tasks that can be done "out of band," while reserving synchronous calls for requests that absolutely need a direct answer.

WebkinsTech: Embracing Events

In WebkinsTech's platform, the initial design had the Order Service call the Inventory and Shipping services synchronously to update stock and schedule delivery. This worked, but as traffic grew, the Order Service became slow whenever Inventory or Shipping was under heavy load. The team refactored the process using an event-driven approach. Now, when an order is placed, the Order Service simply publishes an OrderPlaced event to a Kafka topic. The

Inventory Service, Shipping Service, and Notification Service each subscribe to this event. Inventory Service deducts the stock, Shipping Service arranges shipment, and the Notification Service sends the customer a confirmation email – all in parallel, on their own timelines. The Order Service immediately responds to the customer without waiting for those other actions. This asynchronous design made the system more responsive and resilient: if the Inventory Service is momentarily down, the order still goes through and the stock update will happen once it recovers (thanks to the queued event).

Most real-world systems blend synchronous and asynchronous communication. For example, an agent might do a quick synchronous check for user authentication, then fire off asynchronous events for logging and analytics. The art of architecting these systems lies in deciding which interactions truly need to be immediate and which can be eventual.

Below is a simple ASCII diagram contrasting synchronous vs. asynchronous communication between two agents:

```
// Synchronous request-response:
Agent A --> Agent B : [Request]
Agent A <-- Agent B : [Response]
(Agent A waits for Agent B to finish and return) the result

--

// Asynchronous messaging via broker:
Agent A --> [Message Broker] : (Publish Event)
Agent A (continues immediately, doesn't wait)
[Message Broker] --> Agent B : (Deliver Event
Agent B (processes event on arrival, independent of Agent A)
```

In summary, synchronous communication is like a direct conversation, whereas asynchronous communication is like exchanging letters or messages through a post office. Both modes are essential: one for immediate, tightly-coupled interactions, and the other for decoupled, flexible flows. Understanding when to use each is a key skill in designing intelligent microservice ecosystems.

Coordination Techniques Between Agents

Communication (exchanging messages) is just the starting point. **Coordination** is about what agents *do* with those messages – how they align their actions towards a goal. In an intelligent microservice system, coordination ensures that agents aren't working at crosspurposes or duplicating effort. It answers questions like: Who should do what? When should an agent act, and when should it yield? How do agents reach agreement or make joint decisions?

There are several coordination patterns and techniques, each suited to different scenarios.

We'll examine a few key ones:

Orchestration vs. Choreography

These two terms often come up in discussions of microservice workflows. They represent opposite approaches to coordinating a group of services:

- **Orchestration** involves a central **orchestrator** that commands and controls the interactions. One service (or a dedicated coordinator component) acts as the "brain," explicitly invoking each participant service in turn, like calling the plays in a game.

- **Choreography** involves each service working autonomously and reacting to events. There is no single controller; instead, services coordinate by **observing and emitting events**, following agreed-upon rules. The workflow emerges from these interactions, rather than being explicitly directed by one entity.

Think of **orchestration** as a conductor leading a symphony. The conductor (orchestrator) decides who plays when: "Violin section, you start; now brass join in." Each musician (service) waits for its cue. If one part is missing, the conductor notices and can pause or adjust. This top-down control ensures a well-defined sequence.

In contrast, **choreography** is like a jazz ensemble **without a conductor**. Each musician listens to the music and the cues from others, deciding on the fly when to improvise or support. There's a general structure (the chord progression or shared understanding), but no one is explicitly telling the saxophone when to take a solo – it happens organically from the interplay.

Similarly, in a choreographed microservice system, the Order Service might emit an event "Order Placed" and various services (Payment, Inventory, Shipping) know how to react in their own way, in parallel, without a central Order orchestrator telling each one to execute.

Both orchestration and choreography have their strengths:

- Orchestration provides **central oversight**. It's easier to implement complex **business workflows** in a single place, and to handle errors in a coordinated way (the orchestrator knows if a step failed and can take compensating action). It can also simplify understanding the overall process flow – just read the orchestrator's logic.

- Choreography provides **maximum autonomy** and **flexibility**. Services remain loosely coupled – each only knows about the events, not the full process. This makes it simpler to add new services or change existing ones: just hook into the event stream. There's also inherent parallelism; multiple services can work at once when an event is published.

However, each has trade-offs:

- Orchestration can become a **bottleneck**. The orchestrator service itself must be highly available and scalable, or it becomes a single point of failure. Also, too much logic in one place might recreate a mini-monolith of workflow.

- Choreography can lead to **chaos** if not designed carefully. With no central control, it's harder to ensure all necessary steps happen in order. You might get emergent behavior that's difficult to trace. Debugging a choreographed flow means piecing together logs

from many services. Also, handling failures or rollbacks in a choreographed saga (sequence of events) can be tricky, since no one service sees the whole picture.

In practice, many systems use a mix. For example, a payment process might be orchestrated (to easily manage the steps of charging a customer, updating balance, confirming order in sequence), while less critical parts of the domain use choreography (services emitting events like "CustomerCreated" that other services simply listen to and handle as needed).

To illustrate, let's revisit WebkinsTech. Initially, WebkinsTech's Order processing was **orchestrated**: an Order Management service would call

Payment, then Inventory, then Shipping services one by one, coordinating the whole checkout process. This made it straightforward to implement the happy path and retries if something failed at a step. But as new services (like a Recommendation Service or Analytics Service) wanted to do things when an order happened, the orchestrator became a chokepoint – every new action had to be added into its code.

WebkinsTech later shifted parts of the flow to **choreography**. Now the Order Service simply emits an event, and downstream services subscribe. The Payment Service still ensures the transaction is atomic, but ancillary actions (inventory update, sending emails, logging analytics) happen via event listeners. This removed load from the orchestrator and made it easier to plug in new behaviors (for example, a new Loyalty Points Service could subscribe to "OrderPlaced" events and award points, without touching the Order Management logic).

We can compare orchestration and choreography side by side:

Aspect	**Orchestration (Central Control)**	**Choreography (Distributed Control)**
Control	Central orchestrator dictates the workflow sequence. One entity knows the whole process.	No central control; each service reacts to events based on local logic. Global order is emergent.
Coupling	Services are coupled to the orchestrator (they must follow the orchestrator's commands/protocol).	Services are only coupled by the event contracts; they don't directly call each other.
Visibility	Easy to see end-to-end flow in orchestrator code. One place to monitor progress.	Harder to trace end-to-end; need aggregated logging/tracing of events across services.
Flexibility	Workflow changes require updating the orchestrator. Adding a new step means coding it in.	Workflow changes by adding/removing event handlers. Services can evolve independently

		as long as they handle events.
Failure Handling	Centralized error handling (orchestrator can react if a step fails and decide on rollback or retry).	Decentralized error handling (each service must handle failures; compensations via additional events, which is complex).
Aspect	**Orchestration (Central Control)**	**Choreography (Distributed Control)**
Use Cases	Complex, linear workflows; transactions that need coordination (e.g., payment orchestration, batch processes).	Broad event-driven architectures; when many things happen concurrently or you want easy extensibility (e.g., publish/subscribe notifications).

"Orchestration is like a conductor cueing each musician in a symphony, whereas choreography is like a jazz ensemble riffing off each other. Both can make beautiful music – the choice depends on the kind of performance you want."

Distributed Negotiation and Cooperation

Not all coordination comes from a predefined flow. In some cases, agents need to **decide among themselves** who will do a task or how to share a resource. This is where negotiation protocols come in. Instead of a fixed plan, agents communicate to reach an agreement.

One common scenario for negotiation is **task allocation**: suppose a new task arrives, and several agents are capable of handling it. Rather than having a pre-assigned agent, they could negotiate to determine who takes it. A classic approach is an **auction** or **bidding** process. For example, an agent (acting as an auctioneer) announces a task: "I have task X, who can take it and at what cost?" Other agents (the bidders) reply with their bids – maybe one can do it in 5 seconds, another will take 10 seconds but has more capacity, etc. The auctioneer agent then awards the task to the most suitable bidder (say, the fastest or cheapest agent for that task).

This kind of negotiation mirrors how humans might delegate work in a team or how ridesharing services assign drivers to passengers (drivers bid or are ranked by proximity and the "best" one gets the ride). In multi-agent systems research, protocols like the **Contract Net Protocol (CNP)** formalize this: one agent announces a job, others propose bids, and one is chosen to execute the job.

Negotiation isn't only about auctions. Agents might also negotiate by **iterative agreement** –

e.g., two agents working on a schedule might trade proposals: "I'll process these items if you take care of those." They continue back-and-forth until they settle on a division of labor that both find acceptable. Negotiations can be bilateral (just between two agents) or involve multiple parties.

The benefit of distributed negotiation is that it allows for **dynamic coordination**. The system can make decisions on the fly about who does what, based on current context (load, capabilities, etc.), rather than sticking to a rigid pre-planned assignment. This makes the group of agents more adaptive and can lead to better overall performance or fairness.

Of course, implementing negotiation protocols adds complexity. Agents need a way to evaluate proposals (e.g., a utility or cost function) and a mechanism to carry out the negotiation (which could be via asynchronous messages, as one agent sends offers and others respond). There's also the risk of longer decision time if agents take many rounds to agree, or potential standoffs if they can't agree at all.

Autonomous Drones Negotiating Tasks

Imagine WebkinsTech uses a fleet of delivery drones (each an intelligent agent) to handle package deliveries. When a new delivery request comes in, there isn't a fixed assignment of which drone will take it. Instead, the drones engage in a quick negotiation. The request is broadcasted: "Package at Warehouse A needs delivery to Customer B." Each drone evaluates if it can do the job (considering its current location, battery life, and load). One drone replies, "I can do it in 10 minutes," another says "I can do it in 8 minutes," and a third is too far and doesn't bid. The system (or the drones themselves) choose the drone with the 8-minute bid to assign the delivery. In effect, the drones held an auction – the task was awarded to the drone that could perform it fastest. This kind of negotiation ensures the work is assigned optimally without a central dispatcher, and if the chosen drone encounters a problem (say, low battery), the task could be re-auctioned to others.

Negotiation and cooperation protocols are powerful for achieving **collective behavior** that isn't hard-coded. They let agents **resolve conflicts** (like two agents wanting the same resource – they can negotiate a sharing agreement) and **share workloads** dynamically. In building such systems, designers often draw from economic models (auctions, markets) or social models (voting, bargaining) as inspiration.

Leader Election and Consensus

Sometimes coordination requires the agents to **agree on one leader or one value** among many. If there's a need for a temporary coordinator (but not a permanent central orchestrator), agents can perform a **leader election**. This is common in distributed systems to avoid split-brain situations – for instance, out of a cluster of peer services, one is elected as the "leader" to make certain decisions or coordinate a process, and if that leader goes down, the remaining agents elect a new leader.

Leader election algorithms ensure that all agents eventually concur on who the leader is, even if some messages are lost or agents join/leave. A simple example is the "bully algorithm" where the agent with the highest ID takes over as leader if the current leader fails. Another approach is used in consensus algorithms like **Raft**, where candidates solicit votes from others to become the leader.

Beyond choosing a leader, agents may need to reach agreement on shared data. This is the realm of **consensus algorithms**. Consensus means *everyone agrees on the same value*.

Classic examples include agreeing on the order of transactions (like in a distributed ledger or a replicated log) or agreeing on a plan of action. In a group of intelligent microservices, consensus might be needed if, say, they all must agree "yes" or "no" to proceed with a global operation (imagine multiple microservices voting on whether to deploy a new ML model – if all agree, the deployment proceeds).

Implementing consensus correctly is notoriously challenging because of the possibility of network failures and partial outages. Protocols like **Paxos** and **Raft** were invented to solve this, ensuring safety (no two agents disagree on the decided value) and liveness (they eventually reach a decision) under certain assumptions. In practice, most systems use existing tools for this need. For example, ZooKeeper or etcd are distributed coordination services that internally use consensus to present a consistent view (they can be used for things like service leader election, configuration storage, or locks that only one agent can hold at a time).

For our agentic services, leader election might be used if we want **one agent to act on behalf of a group** for efficiency. Suppose WebkinsTech has multiple instances of an AI Scheduling Service for redundancy. They might elect one instance as the leader to make schedule decisions so that two instances don't double-book a resource. If the leader goes offline, the remaining instances elect a new leader to take over. Consensus would be needed if those scheduling services need to **agree on a common schedule** (so that all have a consistent view of bookings).

The key point is that consensus and leader election ensure **agreement**, which is a form of coordination that prevents divergent behaviors. They come at the cost of extra communication overhead and complexity, so they are used sparingly – only when it's crucial that agents stay perfectly in sync on something.

Use Proven Tools for Consensus: Building your own leader election or consensus mechanism is challenging and error-prone. Whenever possible, leverage battle-tested libraries or systems (like etcd, ZooKeeper, or cloud services that provide locking and leader election). These tools implement algorithms like Raft under the hood so your agents can simply use an API to coordinate on a leader or share small bits of state. This lets you achieve coordination without reinventing complex protocols from scratch.

Coordination via Shared State (Blackboards and Shared Memory)

Another way for agents to coordinate is through a **shared state or environment** that everyone can read and write. Instead of sending direct messages or explicitly negotiating, agents leave information in a common place – and other agents respond to those changes. This is sometimes called the **blackboard pattern** in AI, or more broadly, **stigmergy** (a term borrowed from biology describing how insects like ants coordinate by modifying their environment).

In a shared-state coordination model, multiple agents contribute to and observe a common data source:

- For example, all agents might use a shared **knowledge base** or database. An agent that completes part of a task writes its results or status to the database. Other agents continuously watch (poll or subscribe to changes) and, seeing the update, know it's their turn to act next.

- Alternatively, there could be a **distributed in-memory data grid** or a **shared cache** that agents use to post their current state or read others' state. Because it's in-memory, updates appear fast to all participants.

This approach can simplify communication in certain complex workflows. Imagine a scenario where five different specialist agents are collectively diagnosing an IT outage (a classic use of blackboard systems in AI). They could each independently analyze the problem and post their findings to a shared "board". One agent posts an observation: "Server A is down." Another agent, seeing that on the board, adds, "Database connections failing." A third agent might combine those and infer a hypothesis, posting, "Likely root cause: Network issue between Server A and DB." Through this shared space, the agents effectively have a group discussion by editing a common document, rather than sending individual messages back and forth.

The real-world metaphor here is a team working via a whiteboard in a meeting. Any member can walk up and write a note or diagram; others in the team watch the board and get new ideas or tasks from what's written. The whiteboard mediates the coordination.

A more concrete example in microservices could be coordinating a **workflow through a database**. Suppose we have a pipeline of data processing steps, each performed by a different service (agent). We might have a table where each record has a status field. Service 1 picks up a record, processes it, then marks its status as "step1 done". Service 2 continuously looks for records with "step1 done", processes those, and marks "step2 done," and so on. Here the database status is the shared state that triggers each service in turn. No direct calls between services – they just pay attention to the shared state.

The benefit of shared state coordination is that it **decouples agents in time and knowledge**. Agents don't need to know who exactly will act on the information they leave, just that whoever needs it can find it. It's inherently asynchronous as well – an agent can drop off information and leave, and another can pick it up later.

That said, this pattern comes with cautions:

- Shared state can become a **bottleneck** or single point of failure if it's all in one database or memory store.

- There may be **consistency issues**: two agents might try to update the same information simultaneously. Mechanisms like transactions or locking are needed, which can reduce concurrency.

- If many agents are polling a shared store, it can be inefficient compared to direct messaging. Using pub/sub features or change streams (when available) can help so agents get notified of changes instead of constant polling.

Despite these challenges, a shared knowledge base can be very powerful for certain kinds of agent coordination, especially in AI-heavy scenarios. It's a natural way to enable **knowledge sharing** (which we'll delve into shortly) and to achieve coordination even when direct communication patterns become too complex to manage.

Ant Colony Analogy – Coordination via Environment

In nature, ant colonies exemplify coordination through shared state, a concept known as *stigmergy*. Individual ants don't message each other directly. Instead, they leave pheromone trails in the environment as signals. For instance, an ant that finds food will drop pheromones on its way back to the colony. Other ants, detecting the pheromone path, know "food is this way" and follow the trail, strengthening it. Over time, the colony collectively converges on efficient paths to the food source, all by reacting to traces in the environment.

Microservice Parallel: Agentic microservices can use a digital equivalent of pheromone trails. One service might write a note in a shared datastore, like "Order #1001 needs approval." Another service looking at this datastore sees that note and takes action to approve the order, then maybe updates the note to "Order #1001 approved." Through this shared medium, the two services coordinate the order approval without ever sending a direct message to each other. The environment (the datastore) facilitated their cooperation. This indirect method can lead to very robust, decoupled systems, as long as the shared environment is reliable.

By choosing the right coordination strategy for the situation – be it orchestration, choreography, negotiation, leader election, or shared-memory collaboration – architects can ensure their swarm of intelligent services works together smoothly rather than stepping on each other's toes.

Service Meshes for AI Agents

As the number of microservices (agents) in a system grows, managing their communication can become complex. Each agent might need to implement retries, timeouts, security, and monitoring for every call it makes – which leads to a lot of duplicated effort. **Service meshes** address this by providing a dedicated infrastructure layer for service-to-service communication. In essence, a service mesh gives you a way to offload communication responsibilities to a set of common utilities, often running alongside your services.

The typical design pattern here is the **sidecar**. Each microservice instance gets a sidecar proxy (a lightweight network proxy running in the same environment, such as a container). All outbound and inbound calls from the microservice go through the sidecar. The collection of these sidecars, managed as a group, is what forms the service mesh. Examples of popular service mesh implementations include Istio (with Envoy proxies as sidecars), Linkerd, and Consul Connect.

What benefits do service meshes bring to agentic microservices?

- **Service Discovery & Routing:** The mesh can automatically handle finding which instance of a service to talk to. Agents don't need hard-coded addresses for others; they just send requests to a logical service name, and the sidecar proxy routes it to an available instance (using load balancing). If a new instance of a service comes up or one goes down, the mesh updates its routing tables – the agent doesn't need to know about these changes.

- **Resilience (Retries, Timeouts, Circuit Breaking):** The mesh can implement smart communication policies. For instance, if an agent's request to another service fails or times out, the sidecar can automatically retry it (perhaps with exponential backoff to avoid flooding). If a particular service is unresponsive, the mesh can trip a *circuit breaker* to stop sending requests temporarily, preventing wasted calls. This way, one flaky service won't cause cascading slowdowns – the mesh isolates the failure. These policies are configurable, so you might set "try 3 times before giving up" or "if 5 calls fail in a row, stop calling that service for 30 seconds" globally, rather than writing that logic in each agent.

- **Observability (Monitoring and Tracing):** With a service mesh, you gain a wealth of telemetry data about inter-service communications out-of-the-box. The sidecars can collect metrics like request counts, latency, and error rates for every service-to-service call. They can also generate distributed trace spans, so you can follow a transaction as it hops through multiple agents (useful in a choreographed system where an event triggers a chain of reactions). This observability is crucial in intelligent microservice ecosystems – if an AI agent is causing delays, you'll see it in the metrics; if a request fails halfway through a workflow, you can trace exactly which service and call caused it.

- **Security (Encryption and Authentication):** A mesh can enforce security consistently. Sidecars often establish mutual TLS (mTLS) connections between each other, so all traffic between services is encrypted and agents can trust the identity of who they're talking to. This means even within your cloud or cluster, service-to-service calls are locked down – an important aspect of zero-trust security. The mesh can also manage service identities and certificates, rotating them as needed. Additionally, you can define policies like "Service A is allowed to talk to Service B but not to Service C" and the mesh proxies will enforce that, acting like a firewall for internal traffic.

- **Operational Consistency:** Because the mesh standardizes how communication is done, you get a consistent set of features for all agent interactions. If you decide to adjust a timeout, you do it in the mesh config and it applies to all services, rather than updating countless client libraries in each microservice. This consistency makes the system easier to tune and evolve.

From the perspective of an agent developer, using a service mesh is like having an **assistant** handle the nitty-gritty of communication. The agent itself can focus on its goals and logic (whether it's making intelligent decisions or processing data), while the sidecar deals with networking vagaries. For example, without a mesh, if Service X calls Service Y, the team building X would have to implement retry logic, add instrumentation to time the call, ensure the connection is secure, etc. With a mesh, much of that is provided by default or via simple configuration, not custom code.

It's worth noting that service meshes add their own complexity – they are another moving part in the system. But for large-scale systems with many microservices, the benefits usually outweigh the costs. Meshes are particularly useful in AI/machine-learning microservice environments where you might have many small model-serving components and data services interacting, and you need to monitor and secure all that traffic.

WebkinsTech's Mesh Upgrade

As WebkinsTech's ecosystem grew to dozens of intelligent services, they adopted a service mesh (Istio) to streamline communication concerns. When the mesh was deployed, each microservice instance got an Envoy sidecar proxy. Immediately, the team noticed improved visibility: they could see a live dashboard of how services were talking

– for instance, that the Recommendation Service was making 100 req/sec to the User Profile Service with a 99th percentile latency of 120ms. This helped them identify a bottleneck in user profile lookup that they hadn't realized before. The mesh also made the system more robust: once, when the Inventory Agent started failing intermittently, the mesh's retry policy automatically retried calls to it, and a circuit breaker kicked in to prevent overloading it – all without the Order Service having to implement those safeguards itself. On the security side, WebkinsTech turned on mTLS in the mesh, so now all inter-service calls are encrypted. They no longer worried about internal snooping, and they could even prove to auditors that service A cannot talk to service B except through approved APIs (enforced by mesh policy). Overall, the service mesh became the invisible backbone that supported their agents, letting the developers focus on intelligence and functionality rather than low-level networking details.

In summary, a service mesh is a powerful tool to consider when your agent-based microservices reach a certain scale or complexity. It provides a uniform layer for communication management – handling discovery, routing, resilience, monitoring, and security – which is especially beneficial in an environment where services need to communicate frequently and reliably. By using a mesh, you ensure that your intelligent agents can talk to each other

safely, **efficiently**, and with **observability**, without each agent having to reinvent those wheels.

Knowledge Sharing and Distributed Intelligence

Communication and coordination enable agents to work together on tasks, but **knowledge sharing** takes collaboration to the next level. In an intelligent microservice ecosystem, each agent accumulates knowledge – whether it's learning from data, observing events, or caching results. If that knowledge stays locked within one agent, the overall system might miss out. The idea of *distributed intelligence* is that the whole can be smarter than the sum of its parts, as long as the parts actively share what they learn.

Imagine an organization where each department keeps its research secret. The marketing team discovers a new customer trend, but doesn't tell the sales team; the support team finds a workaround to a common issue, but product development never hears about it. Clearly, that company wouldn't be very efficient. The same applies to intelligent services: we want to avoid **knowledge silos** between agents.

Breaking Down Knowledge Silos

One key principle is to ensure that important insights gained by any agent are accessible to others. There are several strategies to facilitate this:

- **Event-driven sharing:** Agents can publish events not just for actions, but for insights. For example, a Fraud Detection service might emit an event "FraudPatternDiscovered" when it learns about a new fraudulent behavior. Other services (like a Payment service or a Login service) can listen and update their own rules or models accordingly.

- **Central knowledge repository:** Establish a service or database that acts as a collective memory or knowledge base. Agents can contribute to this repository with new information, and query it when they need context beyond their own scope. This could be a shared database of facts, a distributed cache of learned features, or even a knowledge graph that accumulates relationships discovered by various agents. For instance, WebkinsTech created an internal knowledge base where the Customer Support agent records frequent issues and resolutions; the Product Management agent periodically mines this to identify areas for improvement, and the Digital Assistant agent queries it when answering user questions.

- **Model and data artifact sharing:** In an agentic AI scenario, one microservice might train an AI model or develop a heuristic that could benefit others. By sharing the model (or its parameters) through a common model registry or service, other agents can incorporate that knowledge instead of learning it from scratch. Think of a recommendation engine agent that learns user preferences – it could export those preferences to a profile store that other agents (like a marketing agent deciding on promotions) can use.

- **Cross-agent learning sessions:** In some advanced setups, agents actually engage in joint training or federated learning. Each agent might learn from its local data and periodically share its learned parameters or insights with peers, who merge that information with their own. This way, the group as a whole converges on better knowledge without centralizing all the raw data. For example, multiple edge devices (agents) could train a machine learning model on local user data and then aggregate their model updates to form a combined model that's better than any individual's.

The goal is that whenever one agent uncovers something valuable, it doesn't remain that agent's exclusive knowledge. By sharing it, the rest of the system can adapt and improve.

Don't Let Agents Become Isolated Geniuses

One pitfall to guard against is each intelligent microservice acting as an isolated genius – brilliant on its own data, but clueless about what its peers have discovered. Without sharing knowledge, two agents might solve the same problem in parallel or, worse, work with conflicting assumptions. Always design channels for agents to exchange key learnings, whether through events, a shared database, or scheduled sync-ups. A win for one agent should become a win for all.

Enabling Collective Intelligence

When agents share knowledge effectively, you get emergent **collective intelligence**. Patterns that no single agent could see alone start to surface when data is pooled or insights are compared. For example, one microservice might notice a spike in user activity at a certain hour, and another microservice separately notices that its recommendation model's accuracy drops during that hour. By sharing data, they realize it's the same cause: a surge of new users with different behavior. Together, they adjust – the first service informs the second to update its model for that time window, improving overall system performance.

In practice, enabling this means paying attention to **data flows** and **data ownership**. It may require a shift in mindset: microservices traditionally encapsulate their data (for good reasons of decoupling and security), but intelligent microservices benefit from selectively exposing data or insights. Techniques like **data federation** or **event sourcing** can help provide a unified view without violating autonomy. For instance, WebkinsTech set up a secure data lake where each service drops anonymized logs of important events. Then they use an AI Ops service (an analytical agent) that crunches this combined data to find correlations and feeds results back to individual agents (like telling an e-commerce search agent that certain products are trending today, which it then highlights in search results).

Another aspect is maintaining a shared **context** during interactions. Agents often form a pipeline handling a user's request; sharing context (state) about that request ensures each step has the knowledge it needs. If a user's intent was interpreted by Agent A, it might pass along that interpretation to Agent B so that B doesn't have to infer it from scratch. This is a form of real-time knowledge sharing specific to a transaction, usually implemented by attaching metadata or a context object as messages pass between services. We'll dive more into how agents manage and pass state in the next chapter, but it's worth noting here as part of knowledge flow.

To truly unlock distributed intelligence, teams should encourage a culture of collaboration among services, not just humans. This may involve:

- Defining **data contracts** or schemas for shared knowledge (so that what one agent writes, another can understand).
- Regularly reviewing where information bottlenecks exist (are some services not exposing useful data they have?).
- Using tooling that supports discoverability of data and models (so agents "know" that certain info is available and how to get it, akin to documentation for APIs but for data/events).

The payoff is a system where learning in one corner can propagate system-wide. Over time, the ensemble of agents becomes more than just a disjoint set of smart services – it becomes a **learning organization** of services, each feeding the others with information and adapting in concert.

Coordination Readiness Checklist

Before deploying a swarm of intelligent microservices, run through this checklist to ensure you've covered the bases for communication and coordination:

- **Clear Communication Patterns:** Identify which interactions should be **synchronous** (request/response) and which should be **asynchronous** (event or message-driven).

Make sure every pair or group of services knows how they will communicate.

- **Timeouts and Retries Set:** For any synchronous calls between agents, configure timeouts and retry policies. No service call should wait indefinitely on another. Implement circuit breakers to stop cascading failures if a downstream agent is unresponsive.

- **Event Infrastructure in Place:** If using asynchronous messaging, have a robust message broker or event stream in place (e.g., Kafka, RabbitMQ). Define clear **event schemas** and topics. Ensure consumers are idempotent (able to handle duplicate messages) and that you have a plan for undelivered messages (dead-letter queues or error handling).

- **Workflow Strategy Chosen:** Decide on **orchestration vs. choreography** for each business workflow. If orchestrated, implement or leverage an orchestrator (or workflow engine) and handle errors/compensations centrally. If choreographed, ensure all services publish and subscribe to the correct events and that there's visibility into the overall process via logging or tracing.

- **Distributed Coordination Tools:** If your agents need to elect leaders or agree on shared state, choose a proven solution rather than ad-hoc methods. This might mean using a coordination service (like ZooKeeper/etcd) or a library for leader election. Don't leave consensus to chance—design it in.

- **Knowledge Sharing Enabled:** Put mechanisms in place for agents to share important context or learnings. This could be a shared database, a cache, a knowledge graph, or simply a set of events that agents emit when they learn something new. Verify that no single agent holds critical information that others couldn't access if needed.

- **Service Mesh or Equivalent:** Consider using a service mesh (or at least common communication libraries) to standardize things like service discovery, encryption, and retries. Ensure that telemetry (metrics/traces) from inter-service calls is being collected, either via the mesh or instrumentation, so you can observe coordination in action.

- **Observability and Logging:** Implement distributed tracing and centralized logging across your microservices. Use correlation IDs for requests that span multiple agents. This will help you troubleshoot when an orchestrated sequence or choreographed event chain doesn't produce the expected outcome.

- **Failure Simulations Done:** Test how the system behaves when an agent fails, is slow, or produces bad data. Disable a service instance or introduce high latency and see if others handle it gracefully (e.g., do events queue up? Does a fallback agent take over?

Does the orchestrator skip or retry the step?). Refine your design based on these drills.

- **Security and Access Controls:** Ensure that agents only communicate through approved channels. Use authentication between services (tokens or mTLS certificates) so that only legitimate services participate in the system. Also enforce authorization: an agent should only be able to access data and endpoints that it needs for its role. This prevents a compromised agent from wreaking havoc on the whole network of services.

By checking off these items, you can be confident that your intelligent microservices are set up to communicate and coordinate effectively, even under stress. This checklist helps catch common oversights before they become problems in production.

Conclusion

Communication and coordination are the lifeblood of any multi-agent system. In this chapter, we saw how intelligent microservices, acting as autonomous agents, can successfully collaborate: by using the right mix of synchronous calls and asynchronous messaging, by orchestrating when centralized control is needed or choreographing for flexibility, by negotiating and reaching consensus in distributed decisions, and by sharing knowledge to become collectively smarter. With robust communication channels and well-designed coordination patterns, a group of microservice agents can achieve outcomes far beyond the reach of any individual service.

At WebkinsTech, and in our analogies from drone swarms to jazz bands, we observed a common theme: **teamwork amplifies intelligence**. The technical strategies we discussed – service meshes, events, orchestration engines, consensus protocols, and knowledge hubs – all serve one purpose: to enable effective teamwork among intelligent services. By applying these patterns, you ensure that your microservices system behaves not as isolated silos but as a cohesive, adaptive whole.

As we move forward, it's important to recognize that communication and coordination alone aren't the full story. Underneath these interactions lies the matter of **state and memory**.

Each agent maintains its own knowledge and context, and how it manages that internal state (and when it shares state externally, as we touched on) is critical to the system's overall intelligence. In **Chapter 8**, we will delve into state management and memory for agentic microservices – exploring how agents remember, how they store and retrieve information, and how state ties into the continuity of long-running goals.

By mastering communication and coordination (the "conversation" among services) and then state and memory (the "mind" of each service), you'll be well-equipped to design truly intelligent microservice ecosystems that are both cooperative and context-aware. Our journey continues into the inner workings of agent memory and state in the next chapter.

Chapter 8: State Management and Data Flows

When WebkinsTech disassembled its monolithic platform into microservices, the team breathed a sigh of relief. Each service was now a nimble racer instead of part of a lumbering freight train. But soon a new challenge emerged: **state and data flows**. How would these independent services remember things, share information, and keep in sync as the business grew smarter? In this chapter, we explore that challenge in depth – from *traditional* state handling approaches to *intelligent, AI-driven* strategies that turn data flows into learning opportunities. We'll journey through event sourcing, CQRS, real-time streaming, adaptive schemas, and more, always anchoring concepts in WebkinsTech's ongoing evolution. By the end, you'll see how managing state is about more than just storing data – it's about understanding and harnessing it. Let's dive in!

Traditional State Handling in Microservices

Microservices were originally conceived to be as **stateless** as possible. In a stateless service, each request is handled independently – the service does **not** retain information (state) from previous interactions. This means after a request is processed, the service "forgets" about it. *In other words, stateless microservices are "forgetful" and carry no memories between calls* . The benefit is simplicity: any instance of the service can handle any request, making horizontal scaling and load balancing trivial. If one instance goes down, it doesn't take any critical session data with it. On the other hand, a **stateful** microservice *does* carry some context or session info across requests, using that "memory" to inform future actions . For example, an online shopping cart service might keep a user's ongoing cart in memory. Stateful services can feel more straightforward for certain tasks but introduce complexity when scaling (you might need the same instance to handle all requests from a given user) and recovering from failures.

Most real-world microservice architectures end up **externalizing state**. Rather than holding important data in memory long-term, microservices rely on **external databases or caches** to persist state. Each service often has its own database (the **"database per service"** pattern), which decouples data between services and avoids global shared databases. This isolation improves independence but makes **multi-service transactions** tricky (we'll get to that with sagas). A classic approach is to keep the services themselves stateless while using an external **session store** (like Redis) or a database to maintain any needed state. For instance, WebkinsTech's early microservices (after the monolith breakup) all used their own datastores: the Order Service had a orders table in a SQL database, the User Service had a user profile store in MongoDB, etc. Services did not cache much in memory; they fetched data from these stores as needed. This ensured any service instance could handle a request, but it also meant a lot of back-and-forth calls to databases.

Stateless by Design

Whenever possible, design microservices to be **externalize their state**
This makes it easier to scale and recover services, since any
request without worrying about

However, statelessness has its cost: *if every request requires fetching data from an external store, performance can suffer*. In fact, accessing state on disk or over the network is far slower than in-memory access – one source notes a disk read can be ~80 times slower than an in-memory lookup . To mitigate this, teams introduce **caching**. A microservice might cache recent results or frequently accessed data in memory (or in a fast store like Redis) to avoid constant database hits. This introduces a bit of statefulness (the cache is a form of state), but it's usually a bounded, ephemeral state that can be regenerated if lost. For example, WebkinsTech's Product Catalog Service caches product details in memory. If an instance restarts, it can warm the cache from the persistent DB, but during operation the cache saves time. The service remains *effectively stateless* from the clients' perspective (clients don't need to always hit the same instance), yet internally it manages state for efficiency.

Another traditional approach is using **sticky sessions** at the load balancer – ensuring a user's requests go to the same service instance that holds their session data in memory. WebkinsTech briefly tried this for a legacy session management service, but it quickly showed limitations: if that one instance crashed or was removed during scaling, the user's session was lost. The modern practice is to avoid sticky sessions and use a distributed session store if you must maintain session state.

In summary, the traditional state handling in microservices emphasizes *stateless services with external state storage*. Each service has its own database or uses shared infrastructure like caches or object storage, and any in-memory state is typically transient (used for a single request or quickly rebuildable). This model fits well with cloud platforms and tools like AWS Lambda (which by design are stateless functions). **AWS Lambda**, for example, gives you ephemeral compute that *cannot assume any state will be remembered between invocations*. If you need to keep state across Lambda invocations, you might store it in **DynamoDB** or another external store. Many serverless microservices follow this pattern:

e.g., a Lambda function processes an event and reads/writes state to DynamoDB (a fullymanaged NoSQL database) as needed. DynamoDB's high throughput and schema-flexible design make it a popular choice for serverless state storage.

WebkinsTech's initial microservice deployment reflected these traditional practices. After breaking the monolith, they had dozens of small services each with its own data. The **Customer Service** stored customer profiles in a PostgreSQL database. The **Order Service** kept order records in its own database. Communication between services was done through APIs – for example, whenever the Shipping Service needed customer info, it would call the Customer Service, which would fetch from its DB. This worked, but as calls increased, the team noticed something: *a lot of internal traffic and repeated data fetching*. Data flows were growing increasingly chatty. It was clear that simply doing database lookups on each request might not scale smoothly.

Event Sourcing and Intelligent State Changes

To address some limitations of the traditional approach, WebkinsTech's architects began exploring **Event Sourcing**. Event sourcing is a fundamentally different way to handle state: instead of storing just the latest state of an entity, you store **all the events** that lead to that state. Every change (an "event") is appended to an **event log** (or event store). The current state can always be derived by *replaying* these events. For example, instead of an Order record that simply says "status: shipped", you would have a series of events: OrderCreated, InventoryReserved, PaymentProcessed, OrderShipped, etc. The latest state (shipped) is obtained by applying all these events in order.

Why do this? One big reason is that you get a **100% reliable audit log** of all changes . Nothing is lost – every state transition is recorded. This not only helps with debugging and compliance (you can answer “how did we get here?” with certainty) but also enables powerful data intelligence. With a history of events, you can ask questions like “what sequence of actions tends to precede a failure?” or “how have user preferences changed over time?” that would be impossible if you only kept the latest state. *Event sourcing makes temporal queries possible – you can determine the state of an entity at any point in time by replaying events up to that moment .*

In a traditional system, publishing events is often an afterthought or an additional mechanism on top of state changes. With event sourcing, **events are the state changes**. This naturally facilitates an **event-driven architecture**. Whenever something happens, an event is recorded and can be published to interested parties. In fact, event sourcing “solves one of the key problems in implementing an event-driven architecture” by ensuring every state change is captured as an event at its source . There’s no risk of missing an update because the event *is* the update.

However, event sourcing introduces complexity: how do we reconstruct state efficiently from potentially long event streams? Common solutions include **snapshots** (periodically save a snapshot of the state so you don’t have to replay from the beginning each time) and **projections** (build derived views for quick reads – this ties into CQRS, our next topic). Also, with events flowing, the system typically becomes **eventually consistent**. That is, when one service emits an event and others react, there might be a short delay before all services’ states reflect that new event. You trade off immediate consistency for flexibility and auditability.

Eventual Consistency Ahead

Embracing event sourcing means accepting *eventual* consistency. There will be moments when the latest event is stored but not yet reflected in a query or a different service’s view. Design your user experience and workflows to tolerate slight delays or to explicitly refresh data when needed . Strong consistency (as in traditional ACID transactions) is usually limited to within a single service’s event handling.

From an **intelligent microservices** perspective, the event sourcing approach is a goldmine. Instead of just storing final data, we’re storing the raw story of what happened – and stories can be learned from. An AI agent or intelligent service can subscribe to the event stream and interpret patterns or anomalies in real-time. For example, WebkinsTech's **Order Service** was redesigned using event sourcing. Every state change in an order's life (created, paid, packed, shipped, delivered, etc.) is recorded as an event in a Kafka topic acting as an **event store**. Now, not only can the Order Service reconstruct an order's state by replaying its events, but other services can listen to these events too. The **Customer Service** listens for "OrderCreated" and "OrderDelivered" events to update customer purchase history. The **Inventory Service** listens for "OrderCreated" to reserve stock. And interestingly, WebkinsTech added an **AI-driven Fraud Detection Agent** to the event stream. This agent consumes events like "PaymentProcessed" and "OrderCreated" (along with user activity events from elsewhere) to predict fraud risk. Because it sees every event, it can spot subtle signals (e.g., many high-value orders in a short time from the same account) and flag an order before it's shipped. In the traditional model, such an agent would be blind to events or reliant on afterthe-fact database states.

Let's visualize how event sourcing works in practice. Consider a simplified flow of an eventsourced **Seat Reservation Service** (to draw on a generic example):

Illustration of event sourcing in a seat reservation system: the Presentation triggers a command, which loads past events from the Event Store (step 2) to decide if two seats are available, then writes a new "SeatReserved" event (step 4) to the store and via a queue/topic for any subscribers (step 5). The Event Store thus becomes the source of truth for all state changes.

In the diagram above, the sequence is: **(1)** the user action comes in (reserve 2 seats). **(2)** The command handler fetches past events from the Event Store (like existing reservations) to build the current state (how many seats are free). **(3)** The logic decides the seats can be reserved and creates a "SeatsReserved" event. **(4)** That event is appended to the Event Store (so state is updated via a new event, not by altering a "seats" record). **(5)** The event is also published (via a queue) to any other part of the system that cares (maybe a billing service to charge for the reservation, or an analytics service tracking booking rate). If any step fails, no new event is recorded, and the state remains as before. This approach scales well and provides a clear history of what happened and when.

WebkinsTech's adoption of event sourcing started with the Order domain and was so successful that they expanded it to other areas. One interesting use was in their **User Preferences Service**. Instead of storing just a blob of the latest preferences, they kept an event log of changes ("PreferenceChanged" events). This allowed an *intelligent recommendation engine* to not only see the current preferences, but also how those preferences evolved (e.g., a user repeatedly increases their budget range – could indicate they're willing to spend more now, a signal for marketing). The recommendation engine (an AI microservice) learned from these event trends to adjust its suggestions in a personalized way.

WebkinsTech Event Sourcing in Action

To illustrate, let's say WebkinsTech's **Inventory Service** switches to an event-sourced model. Initially, they tracked product stock with a simple number in a database. Now, every stock change (StockAdded, StockReserved, StockReleased, StockSold) is an event. After deploying this, the Inventory team noticed an added benefit: they could feed these events to a **forecasting agent**. This intelligent agent analyzes the stream of StockSold events to predict when a product will run out of stock, allowing proactive replenishment. In the past, they might run a daily batch job on the inventory DB to do this. With streaming events, the agent updates its forecasts continuously. One afternoon, the agent alerted that a new gadget was selling out unusually fast (spiking

StockReserved events) – the team increased the purchase orders in real-time. Event sourcing turned their inventory data flow into a loop of continuous learning rather than just record-keeping.

In summary, event sourcing rethinks state as a *chronicle of events*. It pairs perfectly with intelligent services, since that chronicle can be mined for insights. The drawbacks – added complexity in reconstructing state and eventual consistency – are real, but patterns and tools (snapshots, Kafka/Kinesis, Event Store databases) exist to manage them. In fact, many frameworks (like **Eventuate** or **Axon Framework**) provide out-of-the-box support for event sourcing with microservices, often combining it with CQRS, which we'll discuss next.

CQRS: Command Query Responsibility Segregation

While event sourcing deals with *how state changes are stored*, **CQRS** deals with *how reads and writes are separated*. CQRS stands for *Command Query Responsibility Segregation*. The core idea is simple: **split the system's model into two parts – one for commands (writes/updates) and one for queries (reads)** . In a traditional architecture, the same data model (e.g., a relational schema) is used to update data and to query data. This can lead to conflicts: the model might be optimized for neither, or one use case suffers (complex queries slow down writes or vice versa). CQRS says: let's handle commands and queries differently, even use separate stores if needed.

In a CQRS architecture, *commands* are the operations that **change state** – e.g., "PlaceOrder", "UpdateCustomerInfo". These go through the write model. *Queries* are operations that **retrieve state** – e.g., "GetOrderDetails", "ListOrdersForCustomer". These hit the read model. The two models may have different database schemas, or entirely different types of data stores, suited to their tasks. For instance, the write side could use a normalized OLTP database or an event store (if combined with event sourcing), while the read side could use a highly denormalized view (maybe a cached view, or a search index, or a graph, depending on the query needs).

One immediate benefit is **performance and scaling**: each side can scale independently. If you have far more reads than writes (common in many systems), you can scale out the read replicas or use a fast in-memory cache for reads without affecting write performance. It also improves security and separation – you can ensure write logic is very strict and focused, while the read side can be more flexible for various queries . Another benefit is *simpler models*: because a query doesn't have to use the same objects or tables as a command, you might avoid complex join logic. The query side could present the data already in the shape needed by the UI or consumers (sometimes called a **materialized view**).

CQRS often goes hand-in-hand with event sourcing. Typically, the **write side** in such a system is implemented with event sourcing: every command results in events that are stored. Then the **read side** subscribes to those events to update its own state (for example, updating a view in a NoSQL store that is optimized for queries). Microsoft's architecture guide notes that *event sourcing is commonly combined with CQRS by using events to update the read model views* . This combination allows the write side to be purely focused on business logic and integrity (storing events), while one or many read-side projections handle queries.

From an **AI and intelligent microservice** perspective, CQRS enables some creative architectures. Since the read model is separate, you could incorporate **ML models directly into the query side**. Imagine a Customer Support microservice that has a query "GetLikelyIssueResolution(customerId)". Under the hood, that query could run a machine learning model (using features from various data sources) to predict the best resolution, instead of a static lookup. In a traditional single-model approach, mixing that kind of logic with your writes would be messy. With CQRS, the "query side" is free to call ML models, aggregate data from multiple sources, etc., because it doesn't worry about muddling the transactional state. The command side stays clean – it might record feedback or outcomes as events, which later could even retrain the model. This separation can thus foster an **online learning loop**: the commands generate data (events), and the query side (or a parallel ML pipeline) uses those to continuously improve responses.

WebkinsTech's architecture team decided to apply CQRS to their **Order Management** context as they grew. They kept the **Order Write Model** in an event-sourced store (all the order events). For the **Order Read Model**, they created a specialized read database that combined data from orders, customers, and products to answer queries like "show order with full customer and item details". Instead of the Order Service querying three different services (and databases) to compose the response each time, the read model had a denormalized table (or ElasticSearch index) updated whenever new order events come in. This drastically sped up read queries used by their customer-facing portal. Importantly, it also provided a clean integration point for analytics: the analytics team's microservice subscribed to the same order events to maintain a data warehouse and to train recommendation models (like "customers who bought X often also buy Y"). *CQRS decoupled the operational reads from the analytics/ML feeds.* Each service focused on its responsibility, but thanks to the event stream, nothing fell through the cracks.

Let's look at a diagram of a CQRS setup in a microservice system:

A CQRS architecture example: Order Microservice and Customer Microservice each have separate Command (write) and Query (read) components and data models. Commands (writes) go to the left side (which could publish events to an Event Store), and Queries go to the right side (which read from dedicated read storage). A messaging system (event bus) connects the write side to updating the read side. In this way, read and write workloads are isolated per service .

In the diagram, notice that both the Order and Customer microservices have distinct command and query services inside. For instance, the Order Microservice has a **Command Service** with its own data (perhaps an event store or an OLTP DB for orders) and a **Query Service** with a separate read-optimized database. The event sourcing element is hinted by the "Event Store" on each side and the **Event Publisher/Consumer** linking them via a messaging system. When a new Order is placed (Command), an event might be emitted (via an Event Publisher) to the messaging system. The Order's own Query side (or perhaps a separate Reporting service) is an **Event Consumer** of that event and updates the read storage accordingly. Similarly, the Customer service might do the same with customer updates.

This segregation proved beneficial at WebkinsTech. One vignette: the **Customer Service** had an AI component that predicted customer lifetime value (LTV) based on their orders and activities. By using CQRS, the team fed the *write side events* (like "OrderPlaced",

"CustomerComplaintLogged", etc.) into an **LTV Model Trainer** service. This trainer updated an ML model continuously. Meanwhile, the **read side** of Customer Service had an endpoint GET /customer/{id}/value which simply read the latest predicted LTV from a fast lookup store (populated whenever the ML model produced a new prediction). Without CQRS and an event-driven approach, this would have been hard to pull off in real-time – they might have resorted to periodic batch jobs. Instead, the architecture naturally accommodated a real-time learning loop: events -> update model -> write prediction to read DB -> serve via query.

CQRS at WebkinsTech – A Tale of Two Models

WebkinsTech's **Product Catalog** team faced a dilemma: their service was being hit with heavy read traffic (browsing products, searching, filtering) and also had frequent updates (price changes, new inventory, etc.). Using one database for both was causing lock contentions and slow queries . They adopted CQRS. They kept an authoritative product write database (ensuring all updates, like price changes, went through properly) and spun up a separate

search index read store (using Elasticsearch). Whenever a product update happened (e.g., price change event), a small piece of code would update the Elasticsearch index. Now, read queries (searches, product listings) hit the fast index, while writes didn't compete with those reads. As a bonus, the search index could be tuned for keyword querying and faceted search – something that would have been clunky in the normalized SQL schema. The product microservice exposed its own query API that essentially delegated to Elasticsearch, giving clients a rich search capability. The result was a snappier user experience on the storefront and a more robust service. The intelligent part? The team integrated a **recommendation model** that also listened to product view and purchase events (which are part of the overall event stream) and precomputed "related products" into the read model. So the query side not only fetched products quickly but could also directly return a list of "customers also bought…" suggestions, courtesy of an ML model's output being embedded in the read database. This would have been much harder to achieve cleanly without the separation that CQRS provided.

In essence, CQRS encourages us to **think about data flows** in two lanes: one for **changes** and one for **queries**. In an intelligent microservice ecosystem, this means we have a clear pipeline for feeding learning algorithms (the change events are like a stream of experiences the system can learn from) and a clear pathway for serving insights or calculated data (the query side can incorporate those insights). It adds complexity – you now manage multiple data stores and the plumbing between them – but yields a more flexible system. Many cloud services and open-source tools assist with this pattern: for example, **Apache Kafka** (or AWS Kinesis) can carry events from command side to query side consumers; **Read-Replica databases or NoSQL stores** can serve the query side; and frameworks like **Debezium** can even tap into DB changes and propagate events (if you're not implementing event sourcing from scratch).

Learning from Streaming Data (Real-Time Model Adaptation)

One of the exciting opportunities in intelligent microservices is the ability to learn and adapt *in real time* as data flows through the system. Traditionally, data science and model training have been batch-oriented – you accumulate data, train a model offline, and periodically deploy a new model. But modern systems, especially those aiming to be responsive and continuously improving, increasingly rely on **streaming data** and online learning. Streaming data means data that is continuously generated and moving through the system (e.g., user clicks, sensor readings, transactions) rather than static historical datasets . Microservices often communicate or log events in streams (for example, via Kafka topics). Tapping into these streams can enable on-the-fly learning.

There are a few facets to learning from streaming data:

- **Online Prediction:** Using streaming data to make immediate predictions or decisions. For instance, a recommendation service might update its suggestions for a user within seconds of that user's latest clicks. This often means having a pipeline where events (clicks) are fed into a lightweight model or rule engine that adjusts recommendations continuously.

- **Continual Learning:** Updating the model itself incrementally as new data comes. Rather than retraining from scratch once a week, a model might incorporate new information as it arrives (if the algorithm supports online training or by periodically retraining on a sliding window of recent data). This helps the system adapt to changes (concept drift) quickly.

- **Real-Time Feedback Loops:** In intelligent microservices, decisions made by one component can feed back as data for another. For example, if an AI agent automatically adjusts pricing on the website, the resulting sales (or lack thereof) stream back and can immediately tell the agent if the price change worked, leading it to refine its strategy.

To harness streaming data, a robust infrastructure is needed. Tools like **Apache Kafka**, **Apache Pulsar**, or cloud services like **Amazon Kinesis** serve as the bloodstream of data in a modern microservices system. They allow services to publish events and others to subscribe in real-time. On top of these, stream processing frameworks like **Apache Flink**, **Kafka Streams**, or **Apache Spark Streaming** can be used to perform computations on the fly (aggregating events, computing statistics over windows, etc.). For a microservice team, sometimes using these directly is overkill – often simpler approaches work, such as using Kafka consumers within the microservice to update some internal state or model.

WebkinsTech invested in a Kafka-based **event hub** early on, which proved invaluable for streaming analytics. Consider their **User Behavior Analytics Service**. This service subscribed to a stream of website click events, product views, and cart actions. Initially, it was logging these for offline analysis. But as the company moved toward real-time intelligence, the service evolved to do more: it began computing certain features in real-time, such as "number of products viewed in last 5 minutes" for each user. These kinds of features are crucial for on-the-fly recommendations or promotions (if a user has looked at 10 products in 5 minutes, maybe show a discount popup to convert the sale). The analytics service would emit a new event like UserEngagementMetric(userId, last5MinViews=10) to a special topic. Downstream, the **Personalization Service** (which shows recommendations and offers) would consume this and, if the number is high, trigger an AI model to pick a tailored coupon.

All of this happens within seconds, forming a feedback loop.

Another example is **real-time model adaptation** for anomaly detection. WebkinsTech's infrastructure monitoring microservice uses an anomaly detection model to catch unusual spikes or drops in transactions (potentially indicating issues or fraud). Instead of training this model once and static thresholds, they implemented an online learning algorithm. As every minute's worth of data comes in from various services (metrics, counts, etc.), the model updates a little bit. When the pattern of traffic changes (say, due to a holiday surge), the model's parameters adjust in near-real-time to reduce false alarms. This is streaming data learning in action: the model doesn't remain fixed until the next day's retraining; it *adapts with the stream*.

One key enabler for real-time learning is the **Feature Store**. We introduced feature stores earlier as a technology – they act as a centralized place to collect and serve features for ML models. Feature stores often have two components: an **offline store** (historical, big data) and an **online store** (real-time, quick access). For streaming adaptation, the online feature store is crucial. It provides fresh features to the model serving in production. According to Hopsworks, *feature stores provide history and context to stateless online models – enriching models with recent data like a user's latest behavior or what's trending .* In WebkinsTech's case, they set up an online feature store using Redis (for fast retrieval) that stored things like

"user's 1-day purchase count" or "current loyalty points". Whenever the Recommendation Service needed to personalize content, it would pull the latest features from this store, combine with the model, and deliver results. The features in that store were kept up-to-date by consuming events (each purchase event updated the user's purchase count feature, etc.). The result: the recommendations reflected not just last week's training data, but what the user did *minutes ago*. This dramatically improved click-through rates on recommendations.

It's worth noting that not every model or scenario is amenable to true online learning. Some ML algorithms are batch-oriented. But even in those cases, streaming can be used for **micro-batching** (e.g., retrain every hour with the last hour of data) to approximate continuous learning. And streaming data is definitely useful for real-time decisions with *fixed* models (feeding fresh data into models that were trained offline). The key shift is cultural and architectural: thinking of data as continuously flowing and integrating immediate feedback, rather than thinking in day-old snapshots.

To facilitate streaming data flows, WebkinsTech also leveraged **serverless functions** for lightweight processing. For instance, they set up AWS Lambda functions triggered by certain Kafka topics (using an AWS Lambda consumer integration). One Lambda would trigger whenever a new *customer sign-up event* occurred; it would immediately invoke a personalized welcome email service that used an AI model to tailor the email content based on the customer's inferred preferences (from their sign-up data or referral source). This pipeline from event -> Lambda -> model -> action took maybe a couple of seconds, giving customers an immediate, personalized touch. Without a streaming mindset, that welcome email might have been a generic one or sent in a batch hours later.

To sum up, **learning from streaming data** turns your microservices architecture into a living, adaptive system. Data flows in continuously, and intelligence flows out continuously. The technologies that make it possible (Kafka, feature stores, stream processors, etc.) are the plumbing; the design principle is to shorten the loop between observing something and reacting to it. As we prepare to discuss adaptive data schemas next, keep in mind that streaming often forces us to handle evolving data formats on the fly – an area where adaptability is crucial.

Adaptive Data Models and Schemas

In any long-lived system, change is inevitable – especially change in the **data model**. New business requirements lead to new data fields, changing relationships, and evolving schemas. In a monolithic architecture, schema changes are often painful (e.g., big ALTER TABLE migrations, coordinating downtime). Microservices mitigate some of this by isolating data – each service can evolve its database schema independently. But even within a service, if the service is meant to be intelligent, it might need to adapt its data model frequently as it learns new things or incorporates new data sources.

Adaptive data models refer to designing your data storage and schemas in a way that can accommodate change without extensive rework. Part of this is choosing the right technology: for instance, schema-less or schema-flexible databases (like JSON document stores, MongoDB, DynamoDB, etc.) allow adding new fields easily. If WebkinsTech's Profile Service suddenly needs to store a new attribute for users (say, a "preferred communication channel"), a document database can start storing that field in new records without an immediate migration of old records. The service can handle the logic: if the field is absent (old user), treat it as default. This kind of flexibility is a simple example of adaptability.

But adaptive schemas go beyond just using NoSQL. It's also about versioning and evolution practices. One pattern is **schema versioning in events**. If you're using event sourcing or event-driven interactions, you will inevitably need to change the event structure as features evolve. The rule of thumb is to make changes in a backward-compatible way. For instance, if you add a new field to an event, consumers that don't know about it should ignore it. Version your event (embed a version number or use schema registry tools). The Azure guidelines for event versioning note that the event handlers should ideally handle both old and new versions of events, or you might need to implement translators . An intelligent system might even employ an **agent to assist with schema evolution** – think of a smart data pipeline that can detect when a field is consistently coming in null and suggest deprecating it, or detect a new data pattern and suggest splitting an entity.

WebkinsTech encountered a schema evolution challenge in their **Analytics Data Lake**.

Initially, their events didn't include a particular field that an ML model later required (specifically, they wanted to add "customer segment" to each OrderPlaced event for better training the model). They decided to evolve the schema by adding this field. Using Kafka with a Schema Registry (like Confluent Schema Registry), they updated the event schema version. Intelligent consumers (like their machine learning feature pipeline) were built to handle cases where "customer segment" might be missing (for older events) by looking it up from the Customer Service on the fly. Over time, as more events had the field, the lookups became unnecessary. This adaptive approach – designing consumers to handle both old and new data gracefully – kept the systems functioning throughout the transition. It's a mindset of *forwards and backwards compatibility*.

Another facet is **polyglot and purpose-built data models**. An intelligent microservice might decide that one kind of database no longer fits its evolving usage. Perhaps you started with a relational model but now you find a graph database would model relationships better for a recommendation algorithm. In a traditional system, switching databases is enormous. In a microservice, you could introduce a new data store just for the new functionality, gradually move relevant data or dual-write, and then deprecate the old usage. Essentially, the microservice can adapt by becoming **polyglot persistent** – using multiple data stores internally optimized for different subdomains. The key is to manage the data flows between them (often event-driven sync or periodic sync). WebkinsTech's **Social Feedback Service** did this: it began as a simple SQL store of feedback messages. As they built an AI to analyze customer feedback sentiment, they wanted to link feedback with the order data and customer profiles more fluidly. They introduced a graph database (Neo4j) to store a graph of customers, orders, and feedback nodes. They didn't replace the SQL store (which was still used for transactionally recording a feedback submission), but they emitted events that were ingested by a process that updated the graph DB. The schema of the graph (nodes and relationships) could evolve as they added new types of nodes (e.g., a "Product" node to relate feedback directly to product nodes for product-level sentiment). This is an example of *schema evolution by extending the model rather than altering the existing one*. The microservice ended up with an **augmented schema** – relational for the core record, graph for the AI analysis – adapting to the needs of new intelligent features.

From a more automated angle, one could imagine **AI-assisted schema management**. There are emerging ideas where AI systems can observe queries and suggest new indexes or denormalizations. For example, if an intelligent microservice notices that a particular query is slow and frequently executed, it might automatically create a cache or a new materialized view for it. This is related to database tuning, but in a microservices context, the service itself could have an autonomous component (an agent) that manages its data performance. While WebkinsTech hasn't fully automated this, their DBAs (database admins) are working closely with the service teams; one team even built a small tool that analyzes query logs and recommends indexes. One can view that tool as a simple expert system – a precursor to more advanced AI that might directly implement changes with confidence thresholds.

Finally, adaptive data models must deal with **distributed schema consistency**. In a distributed saga (as we will discuss soon), multiple services might each store a piece of a workflow. If the definition of that workflow changes (say a new step is added, involving a new microservice), the overall data flow changes. A robust design uses *version identifiers for workflows* so that in-flight transactions of the old type can complete, while new ones use the new flow. This is akin to schema evolution but for **process state** across microservices. An intelligent orchestrator could route transactions based on version, or even migrate running instances if possible. This is an advanced scenario but highlights that state adaptability is not just at the data storage level, but also at the process level.

To keep things practical: teams adopting intelligent microservices should embrace **schema evolution as a constant**, not a one-time migration. Use JSON or flexible data formats in events. Use migration scripts that can run in a zero-downtime way (like expanding a schema without blocking writes). Test services with mixed old/new data. Consider using **Feature Flags** to handle new data fields – e.g., start producing a new field but have consumers ignore it until you flip a flag to start using it when enough coverage is there. These approaches ensure your data model can grow and learn just as your services do.

Design for Evolvability

When designing microservice data models or events, ask "what if we need to change this later?" For each field or data contract, try to design it such that adding new fields or values won't break things. Prefer additive changes over mutative ones. By making evolvability a upfront concern, you save yourself from a lot of pain when the business inevitably wants something new.

Technologies and Tools for Intelligent State Management

Throughout this chapter, we've mentioned various technologies. Let's briefly summarize some real-world tools that enable the state management and data flow patterns we've discussed, and how they fit in:

- **Apache Kafka:** Kafka is a distributed event streaming platform. It acts as a durable, scalable **log** for events. In our context, Kafka often serves as the **event backbone** – storing event-sourced data (as topics), transporting events for CQRS (commands to query updaters), and feeding streaming data to learning algorithms. It's high-throughput and can retain events indefinitely (acting as an event store). WebkinsTech's architecture uses Kafka for almost all internal event publishing. Kafka's ability to partition topics also helps scale state handling – e.g., partitioning by entity ID ensures events for a given entity are ordered.

- **Redis:** Redis is an in-memory data store, often used as a cache or fast datastore. In state management, Redis shines for **caching state** (to reduce database hits) and for **distributed locks or counters** in microservices. It's also used as a simple message broker in some cases (Pub/Sub) and can act as a lightweight feature store for real-time features. When microservices need quick state retrieval (like user session data, feature flags, or computed metrics), Redis provides millisecond responses. For intelligent services, Redis can even store ML features or model outputs that need to be accessed extremely fast (for example, storing the latest recommendations for each user, updated by a background service, and read by the web service with minimal latency).

- **Amazon DynamoDB:** DynamoDB is a NoSQL key-value and document database known for its scalability and performance. It is schemaless (each item can have different attributes), which makes it adaptable. DynamoDB is often used in microservices that need flexible schemas or massive scale with reliable performance. It also integrates well with AWS Lambda (serverless patterns) – e.g., a Lambda function writes state to DynamoDB because it can't hold it in memory between invocations. For event sourcing, one could even use DynamoDB to store events (partitioned by aggregate ID, for instance) and use Dynamo streams to publish changes. DynamoDB's **TTL (time-to-live)** feature can help manage ephemeral state (like auto-expiring old session entries).

- **AWS Lambda (Serverless):** Lambda represents the epitome of stateless compute – your code runs on demand and has no inherent memory of prior invocations. This forces good state management discipline: use external stores like those above. But Lambda combined with other services (e.g., Step Functions for orchestrating a saga, or S3/DynamoDB for state) can implement fairly complex flows. Intelligent microservices might use Lambda for glue tasks or lightweight event processing as we saw. One advantage for state flows: AWS Step Functions can maintain a state between Lambdas (in a state machine context), effectively enabling a saga orchestrator with minimal code. WebkinsTech uses Step Functions to orchestrate some background workflows (like data pipeline tasks), which is basically a managed way of doing a saga with rollback steps if needed, and all state of the workflow is managed by the platform.

- **Feature Stores:** Specialized feature store solutions (like **Feast**, **Hopsworks Feature**

Store, or cloud offerings like **SageMaker Feature Store** on AWS) are becoming key in AI-centric architectures. They ensure that the features used to train models are available consistently for serving in production. They handle the ingestion of streaming data into feature values, the backfilling from historical data, and provide low-latency retrieval. In microservices, a feature store might be a shared component that multiple intelligent services use to get the latest user or product attributes. This prevents each service from having to independently compute those and risk inconsistency. Essentially, feature stores help manage the **state of the world as seen by ML models** – which is a parallel concern to the state the microservice databases hold. It bridges the gap between data engineering and microservices by providing a real-time view of important derived state.

- **Service Mesh & Observability Tools:** (While not directly state storage, they are relevant in intelligent flows.) Tools like **Istio** or **Linkerd** can trace data flows and enforce policies. They can be augmented with intelligence – e.g., adaptive routing based on performance. Observability platforms (like **Dynatrace, New Relic, or OpenTelemetry** stack) track state changes and events through distributed traces. Some even apply AI (AIOps) to detect anomalies in those flows (which loops back into state management – if a service's state goes unhealthy, detect and maybe trigger self-healing).

In practice, adopting these technologies requires **architectural vision**. It's easy to be overwhelmed by tools; the key is to match the tool to the problem. Need event streaming and replay? Kafka or Pulsar. Need fast ephemeral state? Redis. Need flexible scaling state store? DynamoDB or similar. The combination of these in intelligent microservices often looks like this: use Kafka as the central nervous system (events everywhere), use specialized stores (Redis, Dynamo, SQL, etc.) at each service for their needs, and weave in ML pipelines via feature stores and stream processors. The good news is that cloud providers and open source ecosystems now provide managed versions of many of these (managed Kafka, managed Redis, etc.), so teams can focus on wiring the logic rather than running clusters.

Below is a quick comparison table summarizing **Traditional vs Intelligent state management** approaches to highlight how these tools and patterns manifest differently:

Traditional vs Intelligent State Management

Aspect	Traditional Microservices Approach	Intelligent (AI-augmented) Microservices Approach
State Storage	Each service stores current state in its own DB (relational or simple NoSQL). Stateless where possible, with external session stores.	Event-sourced stores capturing all changes. Combination of databases and distributed logs (Kafka) to retain history. Use of caches/NoSQL for derived state.
Consistency Model	Often strives for strong consistency within a service (ACID transactions in one DB). Cross-service consistency handled via sagas with simple retry logic.	Embraces eventual consistency. Uses sagas with intelligent orchestrators or compensations. Agents predict or detect inconsistencies and resolve them.

Data Flow Design	Request/response flows with synchronous calls for data. Batch jobs for moving data to analytics. Static pipelines for data ETL.	Event-driven flows with asynchronous messaging. Streaming data pipelines feeding ML in realtime. Dynamic data routing (agents decide where data should flow next based on context).
Reaction to State Changes	Explicit logic coded for known conditions (e.g., if order fails, send email). Mostly reactive programming by humandesigned rules.	Intelligent reactions using AI: e.g., anomaly detection on streams triggers alerts, reinforcement learning agents decide how to adjust system parameters on state changes.
Schema Evolution	Manual migrations, versioned APIs. Significant planning to update DB schemas or APIs; risk of downtime if not careful.	Flexible schemas (JSON, Avro with schema registry). Services built to handle unknown fields gracefully. AI assistance in migrating or autotuning schemas (e.g., auto-indexing).
Scaling State	Scale by sharding databases or adding cache layers. Often	State is distributed by default (events partitioned by key, multiple
Aspect	**Traditional Microservices Approach**	**Intelligent (AI-augmented) Microservices Approach**

	requires manual re-partitioning of data. Stateless services scale easily, but stateful components (DBs) are bottlenecks.	materialized views). Leverages cloud auto-scaling databases, and can even predict scaling needs (AI ops) based on usage patterns. State caches warm up proactively using learned access patterns.
Tooling	Relational DBs (MySQL, PostgreSQL), REST APIs, batch processing frameworks (ETL). Caching with Redis as needed. Monitoring via logs and simple alerts.	Event streams (Kafka/Pulsar), NoSQL/NewSQL (DynamoDB, Cassandra) for scalable state, feature stores for ML data, stream processors (Flink/Kafka Streams) for real-time computation. AIOps tools for monitoring with anomaly detection.
Knowledge Retention	Only current state kept; historical data archived separately (logs, audit tables) and seldom used in operation. "Forgetting" system – focus on now.	Full history retained in event logs – "Learning" system. Uses past events to inform future decisions (e.g., trend analysis, model training). State is a continuum, not a point-in-time only.
Failure Handling	Simple retries, manual intervention if state gets inconsistent. Relies on database integrity and backups for catastrophic recovery.	Automated compensating actions (sagas) possibly guided by AI (e.g., choose best compensation path). Failure patterns

		recognized by agents (e.g., rollback transaction if anomaly detected). More resilience through learned strategies.

As the table outlines, the **intelligent approach** isn't throwing away traditional wisdom – it builds on it, adding layers of adaptivity and learning. It's not that every intelligent microservice implementation will tick all these boxes, but the trend is clear: incorporate more *history*, more *asynchronicity*, more *automation* in managing state.

Now that we have a good grasp of state management techniques and how data flows can be leveraged, let's tackle the challenge of maintaining consistency across many services performing a single business task – and how intelligence can assist even there.

Sagas, Distributed Transactions, and Intelligent Rollbacks

In a perfect world, a single transaction could neatly encompass all the work of multiple microservices – but distributed systems laugh at our desire for *global ACID transactions*. Two-phase commit (2PC) across services is too slow or fragile for most scenarios, so instead we have the **Saga pattern** for managing distributed transactions. A **Saga** is essentially a **sequence of local transactions**, each in a single microservice, coordinated to achieve an overall outcome . If one step fails, the saga triggers compensating transactions to undo the prior steps, thereby rolling back the overall operation in a manual way.

Let's break that down. Imagine an e-commerce order placement across services: Order Service, Payment Service, Inventory Service, Shipping Service. A saga might go: reserve inventory -> charge payment -> create shipment -> confirm order. If, say, payment fails, the saga must undo the inventory reservation (compensating action). If shipping fails (perhaps address invalid), maybe it triggers a refund (compensating the payment) and release inventory. Each service is only ever handling its own local database changes (e.g., Inventory Service only marks items reserved in its DB, Payment Service only touches its payment records), so each local operation can be ACID. The saga logic ensures either all the steps happen (and then the saga is completed) or some fail and then all completed steps are compensated (rolling back).

There are two main ways to implement sagas:

- **Orchestration:** A central coordinator (orchestrator) tells each service what to do next, and decides on compensations if needed.

- **Choreography:** No central coordinator. Each service listens to events and performs its step, emitting an event for the next step. If something goes wrong, other services listen for failure events and respond accordingly.

In traditional microservices, sagas are usually implemented with orchestration if the flow is complex (maybe using a framework or an orchestrator service), or with choreography if the flow is simpler (just events triggering the next step). Either way, writing compensation logic is tricky – you have to anticipate how to undo a step (which isn't always straightforward; you can't *truly* undo sending an email, for example, you can only send another email saying "sorry, ignore the previous one").

Now, how do we make sagas *intelligent*? There are a few possibilities:

- **Intelligent Orchestration:** Replace or augment the orchestrator with an AI agent. This agent could use rules or learning to decide the best order of steps or how to handle failures. For instance, if the agent learns that when Payment Service is slow it's better to wait a bit longer rather than immediately time out and compensate, it could adapt its timeout. Or if inventory fails due to stock, the agent might decide to automatically try an alternate warehouse (an alternative flow) instead of just failing.

- **Predictive Failure Handling:** Using ML to predict if a step will fail before actually executing it. If the system, based on past data, predicts a high likelihood that Payment will fail (maybe user's card has failed in the past or external service is down), the saga could choose an alternative path (like prompt user for a different payment method immediately). This reduces wasted undo work.

- **Automated Compensation Optimization:** Over time, an intelligent system could analyze past sagas to see which compensations were frequent or costly. It might then suggest changing the business process. For example, if "payment failed -> refund inventory" happens a lot, maybe require payment authorization before reserving inventory (reorder steps). Essentially, learning from the saga outcomes to optimize the saga definition itself.

- **Human-in-the-loop for exceptions:** Sometimes a saga hits a condition that isn't clearly solvable (e.g., all items reserved, payment charged, but shipping says "country embargoed, cannot ship"). An intelligent workflow could recognize this needs human intervention, pause the saga, alert a human (like a customer service agent) with a recommendation ("Do you want to ship from an alternate warehouse or inform customer of cancellation?"). Once human decides, the saga continues. The AI part might generate the recommendation from knowledge base or past similar incidents.

WebkinsTech implemented sagas for critical processes like **Order Fulfillment**. Initially, they used a straightforward orchestrator (built with a lightweight framework) that called each service in turn and handled compensations. After some incidents, they added a bit of smarts:

the orchestrator was extended with simple decision rules. For instance, if the Payment Service didn't respond in 5 seconds, instead of immediately marking the saga failed, the orchestrator would check the status of Payment Service (perhaps via a health check or a status event). If Payment Service was up but just slow (maybe due to high load), the orchestrator would wait longer if historically those payments eventually succeed. If Payment Service was actually down, the orchestrator would immediately trigger a fallback (like route the payment through a backup payments provider if available). These decisions were based on logic the team coded after noticing patterns. One could see future iterations where a learning agent monitors saga executions and automatically adjusts these thresholds or chooses providers.

A particularly tricky part of sagas is ensuring **data consistency** at the end. Even after all compensations, there could be side effects (like emails sent or external systems updated) that are hard to revert fully. An intelligent approach is to incorporate **post-saga verification**. After a saga completes (success or compensated), an automated check (maybe a script or agent) verifies that all services involved have a consistent view (e.g., no order is left in "pending" in Order Service if payment failed – it should be "cancelled"). If it finds an inconsistency, it flags it for repair. WebkinsTech built a simple verification agent for their order saga that cross-checks a few things and raises an alert if something's off (rare, but a good safety net).

Let's visualize a simple saga flow to concrete it:

Basic Saga pattern overview: a sequence of services each performing a local transaction and emitting a message/event to trigger the next. If any step fails, a series of compensating actions (not shown in this simple diagram) are executed to undo prior steps . The saga ensures either all steps succeed or the system is rolled back to initial state through compensations.

In the above schematic, imagine Service 1 is Order Service (creates an order), Service 2 is Payment Service (charges customer), Service 3 is Shipping Service (arranges shipment). A message or event from one leads to the next. If Shipping fails, it would emit a failure event that perhaps the orchestrator (or each prior service's compensating logic if choreographed) listens to, and they perform undo – Payment Service would issue a refund, Order Service would cancel the order. The diagram is linear, but sagas can have branching logic too (which increases complexity).

Saga Pitfalls

Sagas simplify distributed commits, but they come with pitfalls: compensations can become more complex than the forward logic, and if not careful, you can end up with **partial failures** (e.g., compensation itself fails). Always make compensating actions **idempotent** and as failure-proof as possible . Also beware of long-running sagas – they hold system state over time; you might need to implement timeouts or expiry for sagas that hang (and then have a strategy to clean up). Intelligent monitoring can help spot a saga that's stuck or oscillating (constantly compensating) and bring attention to it.

How did adding intelligence help WebkinsTech's sagas? One tangible improvement was in their **Returns process**. A return involves multiple services: initiate return (Order Service), process refund (Payment Service), update inventory (Inventory Service), notify customer (Notification Service). They noticed that if the refund failed (say the payment provider was offline), the saga would compensate by marking the return as "pending refund" and end, relying on a manual follow-up later. This meant someone had to remember to retry that refund. They enhanced this with an intelligent retrier agent. When a return saga ends in that state, an agent kicks in to periodically retry the refund (and only then mark return complete). It uses exponential backoff and checks the payment service status. This removed a manual task and ensured consistency. It's a simple rule-based agent, but it shows the pattern:

identify saga failure modes and automate the handling beyond just a one-time compensation.

As microservices become more autonomous (the theme of agentic AI), one can imagine sagas being managed by a cluster of coordinator agents that negotiate outcomes. For example, rather than a fixed script of steps, the Order Service's agent could dynamically ask, "Inventory, can you fulfill item X?" If Inventory says "no, stock out," the agent might ask an alternate Inventory (if multi-warehouse) or decide to backorder and immediately inform the customer with a coupon for delay. These are decisions that go beyond a static saga definition – they imply optimization and adaptation. This is where state management meets decision-making (and probably where Chapter 9 on observability and Chapter 10 on autonomous decision loops will go!).

At this point, we've covered a lot of ground: from basic stateless design to advanced event streams and sagas with AI in the mix. Designing and implementing intelligent state management is complex, but incredibly rewarding when done right – systems become more robust, more insightful, and can even handle problems automatically that used to wake up on-call engineers at 2 AM. Before we conclude, let's provide a practical checklist for teams venturing on this journey.

Checklist: Adopting Intelligent State Management

For tech professionals and architects looking to infuse intelligence into their microservices' state handling and data flows, here's a practical checklist and readiness guide. Use this to assess your current systems and to plan next steps:

- ☐ **Inventory Your State:** List where all your microservices currently keep state. Databases, caches, in-memory? Do you have any hidden state (like static in-memory singletons carrying info)? Having a clear map is the first step.

- ☐ **Separate Stateless vs Stateful Components:** Identify which services can be made stateless (pure compute or routing) and which inherently manage data. Ensure stateless services truly don't carry state (e.g., remove any sticky session requirements) and that stateful ones are known points for persistence.

- ☐ **Enable Event Logging:** Even if not doing full event sourcing, start logging important business events. Use an event bus (Kafka or others) to publish state changes. This creates an **audit trail** and can later feed ML models. Ask: "Do we capture why and when things change, or only the end result?"

- ☐ **Assess Database per Service Compliance:** If any services still share a database, plan to separate them (prerequisite for independent state management and sagas). Each microservice should own its data to evolve intelligently.

- ☐ **Introduce an Event Broker:** Set up an event streaming platform (Kafka, Pulsar, etc.) if you don't have one. Begin with simple use – e.g., a service publishes an event when something notable happens, another service consumes it to update something else. This decouples data flows.

- ☐ **Identify Candidates for Event Sourcing:** Not all data needs event sourcing. Good candidates are entities where history is important or where changes are frequent and could benefit from auditing. Pick one or two services and pilot an event-sourced approach for them.

- ☐ **Prepare for Eventual Consistency:** Educate the team about eventual consistency. Ensure product stakeholders know that in an async world, some data might not reflect instantly everywhere. Build user-facing features (like showing an "Order is processing" status) to account for that short lag.

- ☐ **Implement a Simple Saga:** Find a process that currently is handled maybe by manual steps or brittle scripts across services. Try implementing it as a saga. Start with orchestration or even using a workflow engine (Cadence, Temporal, Step Functions,

etc.). Make sure to include basic compensations. This will flush out how ready your team and infrastructure are for distributed transactions.

- ☐ **Add Observability for Data Flows:** Equip your system with tracing and monitoring that can track a request or event as it flows through multiple services. This is crucial when you go async – debugging requires visibility into event pipelines and sagas. Tools like OpenTelemetry can help instrument this.

- ☐ **Pilot a Real-Time Analytics Use-Case:** Choose one area where streaming data could provide quick win. For example, real-time user analytics, or dynamic pricing. Implement a small pipeline (maybe using Kafka Streams or a simple consumer) that computes something continuously and see how you integrate that into the app (maybe just log it or make a dashboard initially).

- ☐ **Evaluate Feature Store Needs:** If you are implementing ML models, assess if a feature store makes sense. Do you have features computed offline that could be served online? Are multiple services using the same data for decisions? If yes, design a basic feature store or use an open source one to centralize that.

- ☐ **Update Data Schema Practices:** Institute a policy for schema versioning. For events, perhaps adopt Avro/Proto with a schema registry. For APIs, semantic versioning. Make sure everyone knows how to extend schemas compatibly. If using JSON, at least document the fields clearly and how to ignore unknown ones.

- ☐ **Introduce AI/ML incrementally:** Identify where a simple ML model or rule-based AI could assist in state management. Maybe a prediction of a value (e.g., expected time to complete a saga) or anomaly detection on a metric. Embed that in the system in a noncritical way at first (e.g., just log the predictions or use them for informational alerts).

This builds confidence and capability.

- ☐ **Train the Team:** Adopting these patterns requires new skills. Ensure your team gets training or at least hands-on learning in areas like Kafka (or messaging), eventual consistency strategies, new databases like Dynamo or NoSQL, and basics of machine learning integration. An upskilled team is part of being ready.

- ☐ **Plan for Failure Scenarios:** As you add complexity (events, sagas), plan chaos testing or at least scenario planning. What if Kafka is down? (Do services buffer or degrade gracefully?) What if an event consumer is slow – does the backlog build up? What if your ML model starts giving bad outputs –

can you quickly disable it? Having answers or mitigations for these will save you later.

- ☐ **Governance and Data Quality:** Intelligent systems are only as good as the data. Ensure you have checks for data quality on your streams. If a service starts publishing gibberish events due to a bug, how will you catch it? Implement some validations or sanity checks in the pipeline.

- ☐ **Start with a Hybrid Approach:** You don't have to flip everything to the new way overnight. It's okay (even recommended) to have a mix: some parts event-sourced, some not; some processes with sagas, others still manual or using ACID within one DB.

Gradually expand as confidence grows.

Using this checklist, you can gauge your maturity. If many boxes are unchecked, start with foundational ones (event broker, observability, training). If you have most basics, push into the more advanced (feature stores, AI orchestration). It's a journey – even WebkinsTech's team took it step by step, often running old and new systems in parallel until the new proved its worth.

Conclusion

State management and data flows form the **beating heart of an intelligent microservices architecture**. In this chapter, we journeyed from the basics of stateless vs stateful services to the forefront of event-driven, AI-enhanced systems. We saw how traditional microservice designs externalize state for scalability, but also how they can evolve to capture rich histories through event sourcing, enabling a whole new world of insight. We explored CQRS as a means to separate concerns and empower both high-performance queries and reliable writes – a pattern that naturally feeds into streaming data pipelines and real-time learning. With streaming data and adaptive models, our services can learn and adjust on the fly, making the system more responsive and smarter with each event. We discussed how to adapt data schemas and models gracefully over time, so the system's knowledge can expand without breaking.

Critically, we tackled the challenge of distributed transactions with sagas, acknowledging that while we can't avoid eventual consistency, we can manage it intelligently. By introducing automated decision points and learning from failures, even the process of rolling back can be optimized. All along,

WebkinsTech's story grounded these concepts: from caching product data for speed, to event-sourcing orders for auditability and AI insights, to handling a return saga with a smart retrier. Their evolution mirrors what many organizations face – a need to balance consistency, performance, and now intelligence.

As we conclude this chapter on state and data flows, a clear theme emerges: **an intelligent microservice is one that not only performs logic, but also perceives and adapts to the data it handles**. State is no longer just something to store – it's something to learn from. Data flows are not just plumbing – they are the sensory system of your architecture, carrying information that can drive decisions.

Looking Ahead

"Our microservices don't just store state – they understand it. We've turned our data from a liability into a learning asset," notes WebkinsTech's CTO, reflecting on their journey. By treating each event as a lesson and each data flow as an opportunity, they've gained agility and foresight that traditional systems could only dream of.

This sets the stage for the next adventure: all this intelligence and autonomy in services requires **careful monitoring and supervision**. How do we watch over a system where each microservice might be making its own decisions or adapting constantly? How do we ensure an AI-driven service doesn't drift into unwanted behavior? In **Chapter 9, "Observability and Supervision"**, we will delve into exactly that. We'll learn how to achieve deep visibility into these complex data flows, and how to implement oversight (perhaps even *agentic supervision*) so that our intelligent microservices remain trustworthy, accountable, and aligned with business goals. The journey of intelligence in microservices continues – from managing state to managing the very minds of those services. Onward to observability and keeping these smart services in check!

Chapter 9: Human in the Loop for Review of Critical Outcomes

Integrating human oversight into intelligent microservice architectures isn't just a safety net or an afterthought – it's a core design principle. In this chapter, we'll explore **Human-in-theLoop (HITL)** systems as a strategic approach to building trustworthy, reliable AI-driven microservices. We'll see why even the smartest autonomous agents still need human judgment at critical junctures, especially in high-stakes domains. We'll identify which decisions demand human review, and how to gracefully weave those review steps into a microservice workflow without breaking the flow. Along the way, we'll dive into real-world examples – from financial fraud detection to healthcare diagnostics – showing how domain experts and AI services can collaborate. We'll discuss practical implementation patterns (using tools like Kafka queues, workflow engines, and dashboards) that allow humans and microservices to work hand-in-hand. Importantly, we'll cover how human feedback loops can make the AI smarter over time, and how to avoid pitfalls like alert fatigue or automation complacency. The tone here is conversational and pragmatic, like an expert mentor guiding you through designing HITL systems that are both robust and user-friendly. By the end, you should have not only a clear understanding of why human oversight matters, but also a checklist of best practices to build your own human-in-the-loop capabilities into intelligent microservices. Let's get started!

Why Human-in-the-Loop is Essential

Why bother putting a human in the loop at all? Can't AI just handle everything? The short answer: **trust and accountability.** No matter how advanced an AI system becomes, there are certain decisions we *shouldn't* fully automate. In legally, ethically, or safety-critical domains, human oversight isn't a luxury – it's often a requirement. Think of scenarios like an autonomous car facing an unexpected road hazard, a medical diagnosis system unsure about a rare condition, or a financial AI flagging a multi-million dollar transaction. In these high-stakes moments, having a human in the loop provides a sanity check and a moral compass. Human judgment remains crucial to ensure outcomes align with societal values and legal standards.

Regulators around the world recognize this. For example, recent AI regulations (such as the EU's AI Act and similar proposals) **mandate human oversight for "high-risk" AI applications**, requiring that **operators must be able to intervene in or override AI decisions** when those decisions can significantly affect people . In other words, if an AI system is making decisions about someone's health, safety, employment, or rights, a competent human agent needs to be accountable for the final result. This ensures we don't blindly delegate authority to algorithms without any human check and balance. Even the most advanced AI must remain *"subject to human judgment to ensure ethical and accountable outcomes"* .

Beyond compliance, **human-in-the-loop design builds trust** with users and stakeholders. Knowing that a real person will double-check a critical outcome gives people confidence that the system won't run amok. For instance, doctors are more likely to trust an AI diagnostic aid if they know a physician will review the AI's recommendations. Bank customers feel safer if unusual transactions require a human supervisor's approval. Human oversight acts as a failsafe against AI errors, biases, or unforeseeable situations – it's the safety brake on the AI engine.

Moreover, humans bring common sense and contextual awareness that AI often lacks. AI models are trained on historical data and patterns; if confronted with a novel scenario outside their training, they might misfire. A seasoned human expert can recognize when "this situation just doesn't fit the usual mold" and handle exceptions appropriately. For example, an AI hiring tool might be great at ranking candidates by resume data, but a hiring manager can spot intangible qualities or one-off circumstances (like an otherwise great candidate who took a non-traditional path) that the AI didn't account for. **Human intuition and ethical reasoning fill the gaps** where pure automation might stumble. As one industry article put it, technology can't yet replace the *"intuition and experience of a team of fraud prevention specialists"* . The best results come when we recognize what each does best – let AI handle the grunt work and pattern-crunching at scale, and let humans handle the hard judgment calls, empathy, and oversight.

It's also essential for **AI alignment** – ensuring AI systems act in accordance with human values and intentions. Autonomous, agentic microservices might take initiative and perform complex tasks (which is fantastic for productivity), but we always want a mechanism to rein them in or adjust their course if they start to diverge from what's acceptable. Human-in-theloop provides that alignment check. It's a way of baking our values and policies into the system's operations. In critical applications, *"even the most advanced AI systems must be subject to human judgment to ensure ethical and accountable outcomes,"* as a governance guide stresses . Ultimately, HITL design is about responsibility: AI may execute, but humans remain responsible. If something goes wrong, a human decision-maker in the loop can intervene, correct, and take accountability, rather than blaming an algorithm.

Finally, consider trust from the perspective of *the AI developers and operators*: having human oversight can actually **mitigate risk** for the organization. It's like having a human supervisor for your AI staff. If an AI service misclassifies something minor, maybe it's no big deal; but if it's about to do something catastrophically wrong, the human-in-the-loop step can catch it. This reduces the chance of costly errors or PR nightmares. It's similar to a pilot monitoring an autopilot system – 99% of the time the autopilot flies fine, but the pilot is there to take over for that 1% when something goes awry. The famous "autopilot + human pilot" analogy is very apt: modern aircraft essentially fly themselves during cruise, but a human pilot is *on the loop* for oversight and handles takeoffs, landings, and any emergency. We trust airplanes because we trust that trained pilots are in control if anything abnormal happens. In the same way, we can trust AI-driven microservices when we know humans are in control of the critical outcomes.

In summary, HITL is essential because it marries **machine efficiency with human judgment**. It ensures **accountability, compliance, and confidence** in intelligent systems. We avoid the extremes of blindly trusting automation or clinging to manual processes; instead, we design systems where AI and humans each contribute to their strengths. In the next sections, we'll discuss how to identify *which* decisions need that human touch and how to implement these human checkpoints effectively.

Deciding What Needs Human Review

Not every single decision in a microservice architecture will require a human approval – nor should it. Part of architecting HITL systems is figuring out **which decisions or outcomes warrant a human checkpoint or escalation**. In practice, we introduce human-in-the-loop at specific **critical decision points**. The trick is to identify those points systematically.

A few guiding criteria can help determine when a human should step in:

- **High Risk or High Impact Decisions:** If an automated decision could potentially cause serious harm, financial loss, legal issues, or ethical dilemmas, mark it for human review. For example, in finance, any transaction above a certain dollar amount or flagged for fraud might require a human analyst to approve. In healthcare, an AI-generated diagnosis or treatment recommendation that could be life-altering would likely need a doctor's sign-off. Essentially, when the stakes are high – safety-critical systems, large monetary moves, anything affecting someone's fundamental rights – you want a human in the loop . As one source succinctly noted, *"the higher the stakes, the more critical human oversight becomes,"* especially in fields like healthcare or aviation.

- **Uncertainty or Low Confidence:** AI systems often provide a confidence score or some estimate of certainty with their outputs. When the system isn't sure of its own answer, that's a good time to ask for human help. Suppose an AI vision service is classifying medical images for cancer, and it's 95% confident for most images but occasionally gets a case where confidence drops to, say, 60%. Those uncertain cases should be funneled to a human radiologist for closer inspection. The AI can handle the easy, clear-cut ones, but if it's essentially saying "I'm not too sure about this," a human should make the call. This can dramatically improve overall accuracy and safety, because the AI isn't being forced to guess on the tough cases – it knows when to defer to a person.

- **Ethical and Legal Judgment Calls:** Some decisions involve nuances that go beyond what an algorithm can decide. For instance, content moderation on a platform might

  use AI to flag hate speech or misinformation, but whether something is truly violative can be context-dependent – thus final decisions often go to human moderators. Similarly, an AI might score job applicants, but questions of fairness and diversity might require a human hiring manager to balance the final pool (to avoid biases the AI might inadvertently introduce). If a decision has a moral dimension or could trigger liability, it's a candidate for human oversight.

- **New or Anomalous Situations:** If the system encounters a scenario that wasn't represented in its training or past data – essentially an "unknown

unknown" – it's wise to have a human take a look. AI is great at interpolating within the patterns it knows, but extrapolating to completely novel cases is where humans excel. For example, a legal AI system might review contracts and flag clauses that usually pose issues, but if it finds a clause it has never seen before, it should defer to a legal expert rather than make a blind judgment. Humans are better at reasoning from first principles when something truly new comes up.

- **Regulatory or Policy Requirements:** In some industries, rules explicitly require a human decision for certain actions. For example, a law might state that any AI-made decision that significantly affects an individual (like rejecting a loan application or making a medical diagnosis) must be reviewed and confirmed by a human. Company policies might also dictate that, say, "all customer-facing messages generated by AI must be approved by a human editor" to maintain brand voice and avoid gaffes. Identifying these externally or internally mandated checkpoints is straightforward: wherever the rules say "human must be in the loop," design for it.

So how do we put this into practice? One approach is to **define thresholds and triggers**. You can design your microservices such that whenever an output crosses a certain risk threshold or falls below a confidence threshold, it automatically gets routed for human review.

Use thresholds and rules to trigger human review: As a best practice, configure your system with clear criteria for escalation. For example, *if an AI model's confidence in its decision is below, say, 80%, then route that case to a human for approval.* Similarly, define business rules like "if the amount is over $10,000, require dual control (AI + human approval)." These conditions act as safety valves – they catch uncertain or high-impact cases and ensure a person looks at them. By tuning these thresholds (and revisiting them over time), you can balance efficiency with oversight: routine cases auto-complete, exceptional ones get human eyes.

It's also useful to maintain a **catalog of decision points** in your microservice workflows. Map out the flow of decisions in the system and mark which ones are critical. For each such point, ask "What's the worst that could happen if the AI gets this wrong?" If the answer is "not much, minor inconvenience," you probably don't need a human there. If the answer is "someone could get hurt, or we'd break a law, or lose a lot of money," insert a human checkpoint. In agentic microservices, where services might be making decisions

continuously and autonomously, you might also implement *periodic human audit*. That is, even if the AI makes many decisions autonomously, have a human randomly sample some decisions daily or weekly to verify nothing funky is happening. Those audit points can catch issues that your pre-set rules might miss.

Another angle: consider **user expectations and acceptance**. If the end-users or stakeholders expect a human in the loop, then it's non-negotiable. For instance, in hiring, candidates might feel dehumanized if an algorithm alone rejects them without human consideration. Or in medical contexts, patients often expect that a human doctor is overseeing their care. These expectations matter for adoption – people tend to reject outcomes if they perceive them as "computer says so" with no human validation. So from a product design perspective, identify where having a human touchpoint will make users more comfortable with the system's decision.

In summary, choose human checkpoints based on **risk, uncertainty, and values**. Not every micro-decision needs human review (that would be inefficient), but the **critical outcomes absolutely do.** Use AI's own signals (like confidence levels) and external criteria (like regulations or risk scores) to flag those critical cases. This ensures that when it really counts, an AI-driven microservice always defers to human judgment.

Integrating Human Checkpoints into Workflows

Once we know *which* steps need a human in the loop, the next challenge is *how to integrate that human step into our microservice workflow*. We want to add human oversight without derailing the efficiency of automation. Ideally, the workflow should gracefully hand off to a human when needed, then resume automation afterward. Let's talk about patterns to achieve this in a microservices architecture.

At a high level, there are two ways a human can be involved in a workflow: **synchronously or asynchronously**. A synchronous human-in-the-loop means the system actually pauses and waits for a human to respond before moving on. An asynchronous involvement means the system can hand off the task and move on to other things (or at least not block other independent processes), and incorporate the human's input whenever it comes back.

In microservice terms, a **common pattern** is to use a **message queue or task queue** to offload decisions to humans. For example, imagine a Fraud Detection microservice that processes transactions. When it encounters a suspicious transaction, instead of the service directly deciding “block” or “allow”, it publishes a message to a “Manual Review Queue” – essentially saying “Hey, I need a human to look at transaction #12345.” A separate component (which could be another microservice, say a *Human Review Service*) picks up that message. That Human Review Service might be connected to a UI that a human analyst uses. The analyst sees the details, perhaps in a dashboard, and then makes a decision (approve/deny or some categorization). That decision is then sent back to the system, maybe by publishing another message or calling an API that the Fraud service listens to, like

“HumanDecision for #12345: Deny.” The Fraud microservice then proceeds accordingly (blocks the transaction) and the workflow continues. This asynchronous queue-based approach decouples the human interaction from the main flow – the system isn’t completely frozen; other transactions can be processed in parallel while one is awaiting human review.

A concrete implementation of this pattern uses tools like **Apache Kafka, RabbitMQ, or Azure Service Bus** for the queues, and perhaps a database or cache to track pending tasks. One Microsoft architecture guide recommends exactly this: *“Implement a dedicated queue or decision manager that routes requests to humans asynchronously,”* using robust messaging infrastructure (Kafka, RabbitMQ, etc.) to handle communication . The idea is to have a **Human Task Queue** as a first-class component of your system.

When an AI-driven microservice reaches a point where it needs human input, it enqueues a request describing what’s needed (for example, “Review transaction 12345 for fraud” or “Validate AI-generated medical report for patient X”). The system then notifies the appropriate human(s) about this task. Notification can be via a specialized dashboard, an email, a Slack/Teams message – whatever fits the operational context. For instance, an HR microservice might send a notification to a hiring manager’s dashboard or email: “AI shortlisted 5 candidates, please review.” A legal AI service might ping a lawyer’s task list: “AI drafted a contract clause, please approve or edit.”

While the task is in the human’s court, the system can either wait or do other things. Often, it will **wait for a callback** – essentially pausing that particular workflow instance until the human responds. Modern workflow engines and orchestration frameworks support this pattern well. For example, **Google Cloud Workflows** and AWS Step Functions allow a step that waits for an external event or callback to resume. In Google's terminology, it's a

"callback endpoint" that the human (via some UI action) will call to signal completion . This is how you implement a synchronous wait in a scalable way: the workflow state is saved, it doesn't consume resources while waiting, and when the human completes the task, an event (HTTP callback, message, etc.) triggers the workflow to continue from where it left off.

Consider a simple visual of this flow:

Figure: Example of a human-in-the-loop workflow within a microservice architecture. An automated process produces an outcome and hits a decision point on whether human review is needed. If yes, the process routes to a human review task (through a queue or similar). The human can then either approve and let the process proceed with the action, or intervene and take an alternate action. In both cases, the outcome and the human's feedback are logged for learning. If no human review is required (low risk or high confidence outcomes), the process proceeds autonomously.

In the figure above, notice how the **decision point** gates whether we involve a human. The "Human Review Task" is a distinct step. In implementation, that could correspond to sending a message to a "Human Review" microservice and then waiting for a response. After the human task completes, we have two paths: "proceed with action" (if the human approved or made no changes) or "alternate action/override" (if the human's input leads to a different outcome). In either case, we log the result along with the feedback. This logging is important (we'll talk more about learning from it later).

Now, how to notify and capture input from the human? A **user interface or dashboard** is typically used. This could be a web application that lists all pending human-in-the-loop tasks. For example, a bank might have an "Analyst Workbench" where fraud alerts show up for review. Each alert has details pulled from various microservices (transaction info, account history, the AI's risk score, maybe an explanation from the AI why it flagged it). The analyst can then click "Approve" or "Flag as Fraud" or whatever actions make sense. When they do, behind the scenes the UI calls an API or posts a message that flows back to the core system.

In some setups, instead of a custom UI, notifications might go to existing tools – like sending a case to a ticketing system (JIRA, ServiceNow) or simply emailing a human who then clicks an approval link. The choice depends on the use case. For a high-volume rapid process (like moderating thousands of social media posts), a purpose-built interface is better. For lowvolume corporate approvals (like approving an expense report flagged by AI), an email with an "Approve/Deny" link might suffice.

Crucially, when designing this integration, we need to think about **state management and timeouts**. What if a human doesn't respond in a timely manner? The system shouldn't be left hanging indefinitely. One pattern is to implement a timeout with a **default policy**. For example, if no human responds within 15 minutes, maybe the system automatically takes a conservative action (like deny the transaction by default, or escalate to a higher authority, or send another reminder). In an Azure architecture example, they suggest: *"If the human doesn't respond within X minutes, the agent takes a default action."* . This ensures that the workflow eventually continues and doesn't stall forever due to a missing human input. Designing these defaults is a matter of policy – e.g., in a safety context, the default might be to fail safe (stop the machinery, etc.), whereas in a customer service context, the default might be to allow something to avoid user frustration (depending on the risk).

Another consideration is **concurrency and scaling**. If you have many automated processes potentially sending tasks for human review, you need to handle those tasks in parallel and possibly distribute them among multiple human reviewers or teams. Message queues and event streams (like Kafka) shine here: they naturally buffer and distribute tasks. You might also maintain a "pending decisions" store (in a database or cache) to keep track of all tasks awaiting human input, which user is handling them, and their status (pending, completed, timed-out, etc.) . This state tracking allows multiple services and users to coordinate. For instance, you wouldn't want two different human operators unknowingly reviewing the same task; a quick lookup in the pending store can prevent double assignment.

It's worth mentioning **workflow orchestration tools** as well. In complex microservice systems, you might use orchestration frameworks (like Apache Airflow, or Netflix Conductor, Camunda, or Temporal) to manage multi-step processes. These tools often allow you to model a human task as a step in the workflow graph (commonly called a "user task" in BPMN terminology). For example, an Airflow DAG could have a task that triggers a Slack message asking for approval, and then waits (by pausing the DAG run) until an operator manually triggers the continuation. This is more commonly done in BPM engines like Camunda, where you'd create a user task with an associated form for someone to fill. The key point: integrating a human step can be done in a structured way with these tools, not just via ad-hoc code.

Real-world tools/platforms: We've mentioned a few already. To recap in context:

- **Message Brokers (Kafka, RabbitMQ, Azure Service Bus):** Use these for asynchronous human task queues. They reliably pass the baton from an AI service to a human reviewer service. Kafka, for instance, can handle a firehose of events – you might have a "fraud_alerts" topic that dozens of transaction microservices publish to, and a pool of human analyst services consuming from it in real-time. Kafka's durability ensures no task is lost, and its scalability ensures you can have many consumers if the volume of alerts grows.

- **Serverless Workflows/Functions:** In serverless environments, you might use AWS Step Functions or Google Workflows with Lambda/Cloud Functions. These can orchestrate calling AI services, then using a Step Function "Wait for callback" step to pause until a human responds via a webhook. AWS also has **Amazon Sagemaker Augmented AI (A2I)**, which is a managed service specifically for routing ML results to human reviewers. With A2I, you configure conditions under which a "human review workflow" is triggered, and it handles storing the result, showing a UI to the reviewer (with Amazon's built-in tooling or your own), and collecting the response. It's an example of a platform trying to make human-in-loop easier to plug in.

- **Custom Dashboards and Portals:** Many organizations build internal web apps for analysts, moderators, or experts to handle AI escalations. For instance, a content moderation system might have a Moderation Portal where AI-flagged posts appear in a queue. These UIs are tailored to the task (showing the content, perhaps the AI's reasoning or highlighted words, and action buttons). The microservice architecture around it might be: Content service -> Moderation service (AI) -> if flagged, put in DB -> Portal pulls from DB for human -> human decision via Portal -> Portal calls API to Content service with decision. A simpler but effective approach is also to use existing workflow tools – some companies integrate with things like Trello or Jira, creating a card or ticket for each AI-flagged item that needs human review.

This leverages notification, assignment, and tracking features those platforms already have.

Integrating the human step does introduce some **latency** in processes – that's inevitable. But by designing it well, we can minimize impact. For example, if 95% of cases don't need human review (they sail through automated), the overall system is still fast for those. The 5% that do need human might take minutes or hours, but that's acceptable if those cases truly need careful attention (and usually they do). Also, by handling the human step asynchronously, we keep the system scalable. The microservices can continue doing other work while a case is under review; the human tasks become their own parallel workflow.

Think of this like an assembly line with a quality control station: most items pass through, but if an item looks defective, it's pulled off the line for special inspection. The line keeps moving while the one item is inspected on the side. That's exactly what our queue + callback pattern does for microservices.

Before we move on, let's highlight the **key integration pattern** in a concise form, as recommended by experts:

1. **Route to a Queue:** When AI logic determines human input is needed, package up the context (data, decision needed) and post it to a dedicated queue/topic for human review

. Use a reliable message system (Kafka/Service Bus) to handle volume and persistence.

2. **Notify the Human:** Ensure the relevant human knows about the task. This could be via a real-time dashboard, or sending a notification (email, chat message) with a link. It can be helpful to integrate with communication tools (e.g., post a message to a Slack channel "Ops Team: please review Alert #8472").

3. **Human Decision Captured:** Provide an interface for the human to submit their decision or input. This could be a form, button clicks, etc. Behind the scenes, this triggers a **callback** to your system – either by sending a message to a reply queue or calling an API endpoint (webhook) that the workflow is listening on .

4. **Resume Workflow:** The original process resumes with the human's input incorporated. The microservice that was waiting can now

proceed. For instance, the Fraud service gets "OK" and completes the transaction, or gets "Block" and cancels it and maybe alerts the customer.

5. **Timeouts and Escalation:** If no human responds within a reasonable time, have a fallback. Possibly escalate the alert to a more urgent channel or to a supervisor, or take an automatic action (like disabling an account temporarily pending later review).

6. **Audit Trail:** Log both the AI's decision and the human's decision. Record who the human was and their action, for traceability (this will be useful for compliance and for improving the AI as we'll discuss later).

By following this integration approach, we embed humans into microservice workflows in a robust way. It's like adding a human "microservice" into the architecture – albeit one that isn't software. In fact, some architects humorously talk about *"human microservices"* – treating human workers as services that can be called with a request and will produce a result. While people are not as predictable or fast as code, thinking in those terms helps design the interfaces and contracts for how software hands off tasks to people.

We've now covered the backbone of integrating human-in-the-loop. Next, let's consider the experience of the human operator and how to support them with good interface design and explainability, so that their involvement is effective and not frustrating.

Designing the Human Interface and Experience

Whenever we put a human in the loop, we need to ensure that the **human reviewer has the right tools and information to do their job well**. It's not enough to toss a decision over the wall to a person and say "you deal with it." The system should be designed to *help* that person understand what's going on and make a decision efficiently. This is where **user interface (UI) design, explainable AI (XAI), and traceability** come into play.

First, think about what information the human needs when reviewing an AI output. At minimum, they'll need the **raw details of the case** – e.g., for a financial transaction: the transaction data, customer info, any relevant account history. For a medical AI suggestion: the patient data, the recommended diagnosis or treatment, and maybe the medical images or test results themselves. Essentially, the human should have access to all the context they would have if they were doing the task from scratch without AI. You don't want them blindly following or vetoing the AI; you want them to apply their knowledge to the same facts the AI saw.

Next, and importantly, the human should see **the AI's output and, if possible, an explanation of why the AI produced that output**. This is the realm of **explainable AI (XAI)**. If your microservice's AI component can provide a rationale or highlight certain data as the reason for its decision, present that to the human. For example, if an AI model flags an insurance claim as fraudulent, show the human reviewer *why* – maybe "This claim was flagged because the repair cost was 3 standard deviations above the average for that car model, and the same customer had another claim 2 months ago," or highlight the suspicious fields in the form. In content filtering, an AI might highlight the specific words or phrases in a post that triggered the flag. In image analysis, it might overlay a heatmap on an X-ray to show what region looks like a tumor. These explanations greatly aid the human's efficiency: they know where to focus their attention and how the AI reached its conclusion.

However, caution is needed: we don't want to bias the human unduly with the AI's opinion (a phenomenon known as **automation bias** – where humans may trust the AI too much, assuming it must be correct). One way to mitigate bias is to present the case data first, let the human form an initial impression, and then show the AI's recommendation and reasoning. Another way is to train humans to critically evaluate the AI's suggestion rather than rubber-stamp it. The UI can help by not making the AI's answer appear like a final verdict, but rather as a suggestion. For instance, instead of the interface saying **"AI RESULT: DECLINE LOAN – 60% confidence"** in big red letters, it might say **"AI suggestion: This loan application seems risky (60% confidence). Please review the application details below."** The phrasing and visual design can frame the AI output as assistive, not authoritative.

Traceability is another aspect: everything the human sees and does should be logged. When the human makes a decision, we want to trace it back to the inputs and the AI's outputs. This is critical for audits. If later someone asks, "Why was this transaction blocked?", we should be able to trace: the AI flagged it with X reason, Analyst John looked at it at 3:45pm, he decided to block because he noted Y, and here's a comment he left. Encourage or even require the human reviewer to provide a brief rationale or select a reason code for their decision. This not only creates an audit trail for accountability but also provides valuable data for improving the AI later (e.g., if the human's reason was "customer confirmed not authorized," that's useful feedback).

From a UI design perspective, **make the human's job as easy and clear as possible.** They should not have to dig around into five different systems to gather info. A common mistake is to give the human just an ID or a link and then they have to manually pull up all details from various databases. Instead, the system should aggregate the needed info and display it in one place. If the task is to "approve a high-value transaction," the screen should show the transaction details, account summary, perhaps risk score, and buttons to approve or escalate – all on one page. Reducing cognitive load and context-switching helps prevent errors and speeds up the process.

Also consider **workload management** features: if there are many human review tasks, the UI could sort or prioritize them (maybe by urgency, or by the type of expertise needed). For instance, in a healthcare setting, perhaps certain AI alerts are best reviewed by a cardiologist vs a radiologist. In content moderation, some staff might handle hate speech flags while others handle self-harm content. The system should route tasks to the right people – possibly through an assignment system or separate queues.

Now, let's talk empathy: We must design these systems with empathy for the human operators. These folks could be fraud analysts, doctors, moderators, HR managers – whoever they are, their time and cognitive energy is precious. **Alert fatigue** is a real risk (we'll delve more into that in the challenges section). To keep humans effective, the UI should help them *focus on the truly important alerts* and not overwhelm them with noise. This means your AI's threshold for sending things to humans shouldn't be so low that they're constantly swamped with trivial reviews (if it is, they'll start clicking "approve" on everything without proper scrutiny). It also means within the UI, highlighting which cases need immediate attention (e.g., "urgent" flags or color coding).

For example, a dashboard might have sections: "**Critical – immediate review needed (e.g., safety issues)**", "**High Priority (e.g., large transactions)**", and "**Normal Priority**". This way a human can triage. Additionally, providing feedback to the human like "how many tasks are in queue, how long each typically takes" can help them manage their time and not feel endless. Some systems even implement **gamification or assistance** – like showing how the human's work is contributing ("You caught 3 frauds today, preventing $50k loss!") to keep motivation, or offering an AI-generated summary to speed up decision-making.

Another UI aspect is enabling **human override or adjustment** in a nuanced way when needed. It's not always a binary approve/deny. Sometimes the human might partially override the AI's work or provide additional input. For example, an AI writing assistant drafts an email, and the human reviewer might tweak a few sentences rather than outright accept or reject. The system should capture those edits. In a customer support scenario, AI might draft a reply to a customer complaint; the human agent might edit it for tone and correctness.

The UI should make it easy to edit the AI's suggestion and send the final – and ideally capture the edits for learning (perhaps highlighting what changed to feed back into the model's improvement).

Explainability tools can help here. There are libraries and techniques (like LIME, SHAP for explaining ML models) that can produce human-readable explanations for model outputs. In an intelligent microservice context, you might deploy a separate **explanation service** or include an explanation module as part of the microservice. For instance, a credit scoring microservice could have an endpoint that, given an application, not only returns a score but also an explanation like "Top factors: income too low, credit history short." If a human underwriter is reviewing a borderline loan, seeing those factors can guide their further investigation (maybe they then focus on verifying income or looking at alternative credit data).

One more thing on traceability: **Logging and audit trails** should record the sequence of events around the human loop. Often it's useful to have a centralized log or audit service in a microservice architecture that records "AI decision X at time T, sent to human, human Y responded at time T2 with decision Z". These logs serve two purposes: post-mortem analysis (if something goes wrong, you can see who did what) and compliance (proving that you had a human in the loop as required). Some industries require maintaining such records for years.

To illustrate effective human-in-the-loop UI/UX, imagine an **Admin Dashboard for AI Decisions** in an HR context (hiring decisions for example): The dashboard shows a list of candidates an AI recruiting tool has screened out or ranked. For each candidate, it shows the AI's recommendation ("Recommended" or "Not Recommended") and key reasons (perhaps "Not

Recommended due to lack of required degree, resume score below threshold"). The HR manager can click on a candidate to see full details: resume, AI's analysis, and maybe even predictions like "% fit for role." The manager notices one "Not Recommended" candidate actually has a unique background that the company values (maybe they come from a competitor, bringing valuable experience). The AI didn't consider that because it's not explicitly in the resume text or it weighted the lack of a formal degree too heavily. The manager can override by marking this candidate as "Worth Interviewing". The system logs that override, and perhaps asks the manager for a quick note ("Reason for override?") – the manager writes "Strong relevant experience outweighs missing degree." This note and the fact of override can be used next time the model is retrained (so it might learn to give less weight to formal degrees in cases where experience is exceptional – effectively learning from the human's nuance).

That example shows how the interface facilitates understanding (the manager saw why the AI said no), provides an easy mechanism to override (a simple button and note), and collects feedback for improvement (the note). It also treats the human with respect – the AI is not fully autonomous, the human can inject their expertise readily.

In designing these HITL interfaces, **collaboration between AI developers and UX designers** is key. Sometimes AI folks think only about the model and backend, but if the lastmile presentation to the human is poor, the benefits of the AI can be lost. As an "expert mentor" voice here: always prototype the human review step and get feedback from actual users of that step. If your fraud analysts say "I'm getting too many useless alerts" or "I don't understand the AI's risk score," that's a sign to improve the thresholds or the explanation content. The success of an intelligent microservice with HITL is as much about the human's experience as it is about model accuracy.

Lastly, consider **training and documentation** for the human reviewers. They need to know their role and how the system behaves. Include guidelines in the interface or onboarding like: "The AI will flag roughly 5% of transactions. Please review each flag carefully. The risk score means X. If in doubt, err on the side of caution," etc. When humans and AI are working together, clear communication of what to expect from each side prevents confusion. We want the human to trust the AI just enough to leverage its help, but not so much that they stop paying attention. This is a delicate balance that user interface cues and organizational training both support.

In summary, designing the human interface and experience means **providing context, explanations, and easy controls** to the people in the loop, while logging everything for traceability. The system should be a helpful assistant to the human, not an opaque box. When humans feel empowered by the interface – knowing they have the information and authority to make the right call – they become effective guardians of the system's critical outcomes.

Modes of Human Intervention: Proactive, Reactive, and Corrective

So far we've mostly discussed humans reacting to AI outputs (like reviewing a flagged decision). But human oversight in intelligent systems can happen at different times and in different ways. Let's distinguish three modes of human intervention: **proactive, reactive, and course-corrective (post hoc)**.

1. **Proactive Intervention:** This is about humans *being involved up-front or continuously monitoring*, even before an AI triggers an alert. In proactive mode, human experts help set the rules, parameters, and boundaries for the AI system. For example, domain experts might be called in during design to identify what conditions should always require manual approval (we did that in the "deciding what needs human review" stage). That's a form of proactive involvement – shaping the system's logic with human knowledge. Another example: a human controller might actively watch a live dashboard of the AI's actions and metrics, and if they see something odd, they can intervene even if the AI didn't explicitly ask for it. Picture an air traffic control system run by AI that sequences landings; a human supervisor might be monitoring all and could step in if they notice a potential conflict the AI hasn't flagged yet. Proactive also includes **training-time and design-time interventions**: humans curating training data, setting ethical guidelines (like "the AI should never do X"), and configuring fallbacks. In microservice terms, a proactive human might set certain microservices to "manual mode" under specific conditions. For instance, if a bank's trading AI normally operates autonomously, a human manager could proactively decide to switch it to supervised mode during times of market volatility (before any issue

occurs), effectively saying "for the next hour, every trade will need a human confirmation because the market is too unpredictable right now."

2. **Reactive Intervention:** This is the classic **in-the-loop scenario** we've been focusing on – the AI does something (or is about to do something), and the human reacts by reviewing or approving/overriding. Reactive mode is triggered by the system: e.g., an alert, a request for approval, an alarm when a threshold is crossed. The majority of human-in-loop implementations are reactive; the human is not actively watching everything at once, but rather responding to items queued up for their attention. In microservices, this reactive involvement is event-driven: the system emits an event "need human input" and the human responds to that event. We've covered reactive in depth with our queue and callback discussions. One thing to add: reactive intervention often implies that the human is only involved in *exceptional cases* or at decision gates. The day-to-day normal operation flows without human involvement except when these triggers occur. This mode is efficient because humans aren't wasting time on things the AI can handle; they're only engaged when needed.

3. **Course-corrective (Post hoc) Intervention:** This mode is about **humans correcting or adjusting the system after observing its behavior over time**. Think of it as feedback and learning from mistakes or near-misses. Even with oversight, an AI system might start to drift or exhibit bias gradually, or external conditions change (data drift, new types of inputs). Human domain experts need to periodically review the outcomes and performance of the AI and then intervene at a system level, not case-by-case. For example, suppose a content moderation AI was doing fine, but over the last month humans notice it's been incorrectly flagging a lot of art-related posts as nudity (maybe due to some new style of images). In a course-corrective intervention, the experts would analyze this pattern and decide to update the AI model or add a new rule (like "if context indicates art, be more lenient"). They might even temporarily dial back the AI's autonomy (e.g., have it route more cases to human review) until the issue is fixed. This is a kind of **human-in-the-loop at the improvement stage**. It involves things like reviewing logs, auditing a random sample of AI decisions (even those that were not escalated) to ensure quality, and then tuning the system. If proactive is about setting initial guardrails, and reactive is about handling incidents in real-time,

course-corrective is about **learning and adapting the guardrails** over the long run.

Another example of course correction: an HR AI tool after a while might show a trend of favoring candidates from certain backgrounds. A human HR analyst might spot this bias by looking at aggregate data (maybe noticing all the hires recommended have similar profiles). They then intervene by retraining the model with more diverse training data or explicitly adding constraints to counteract bias. They are correcting the course of the AI to better align with fairness goals. This is often done in scheduled model review meetings or AI governance processes.

These modes often work together. In a well-designed system:

- Proactive humans define the strategy: what's critical, what's not, and set up the workflow (perhaps even deciding how the UI will look for the reactive steps).

- Reactive humans handle the day-to-day exceptions and critical decisions.

Course-corrective humans (who might be the same people in a different mode, or a separate review committee) look at trends and tweak the system periodically.

Let's map this to a real scenario to see all three in action:

Consider a **medical AI diagnostic service** in a hospital.

- *Proactive:* Before deploying, doctors and administrators define which types of diagnoses the AI is allowed to make autonomously and which always need a human doctor confirmation. They might say, "AI can auto-approve normal lab results, but any cancer diagnosis must be confirmed by an oncologist." They also set up the workflow such that any AI suggestion comes with an explanation, and they hold training sessions for doctors on how to use the new system. They might even run the AI in a shadow mode first to see how it performs and proactively adjust thresholds (e.g., if it's oversensitive, raise the threshold for alerts).

- *Reactive:* Now in daily operation, the AI reads, say, radiology scans. Most come back normal – the AI reports "All clear" with high confidence; the radiologist just glances and signs off quickly (or in some cases, they might even skip looking if the AI is known to be very accurate on normals, though ideally they double-check). But whenever the AI says "possible tumor detected," it triggers a human review task. The radiologist examines the scan and either agrees (confirming the finding, maybe ordering further tests) or disagrees (deciding it's a false alarm and marking it as such). Each such case is a reactive intervention where the human is making the final call on a specific case.

- *Course-corrective:* Over a few months, the hospital's chief radiologist and the AI team meet and review statistics: how many false positives, how many false negatives (if any known), how often doctors disagreed with the AI. Suppose they find the AI missed a certain rare type of tumor that a human caught (false negative) – they then decide to retrain the model with those missed cases included. They also find doctors are overruling a certain kind of alert frequently, maybe the AI is too sensitive on benign anomalies; so they adjust the sensitivity down a bit. They might also update the protocols: for example, if an AI flag is borderline (low confidence), now route it to two different doctors independently to double-confirm (a new rule to increase assurance on uncertain cases). All these are course corrections to improve future performance.

By incorporating proactive planning and periodic course corrections, we ensure the humanin-the-loop system *evolves and gets better* rather than being static. It's not just the AI that learns; the **overall sociotechnical system** (humans + AI together) learns and adapts.

In practical terms, you might establish an **AI oversight committee or workflow** in your organization that meets regularly to review outcomes. This can be seen as part of MLOps (Machine Learning Operations) – monitoring model performance in production and involving humans to handle drift or errors. Logging systems and business intelligence tools can help spot where human overrides are happening a lot, which is usually a sign the AI could be improved or the threshold adjusted.

The proactive/reactive/corrective framework is helpful to ensure you don't focus solely on the reactive piece. All three are needed for a robust setup:

- If you only did reactive without proactive, you might deploy an AI that constantly flags too much or too little, because you didn't set it up right (leading to overload or missed issues).

- If you did reactive but not corrective, you'd handle immediate issues but never learn from them, so the AI would make the same mistakes over and over and humans would forever be stuck correcting them – not very efficient in the long run.

- If you did proactive without reactive, you might severely limit the AI or trust it too much based on initial assumptions, which either underutilizes it or leaves risk during operation.

- And if you somehow tried to do only proactive and corrective (no reactive), basically you'd have either fully manual operations or fully automated with periodic audits, but not actually a human *in* the real-time loop – that's dangerous for high stakes.

So blending these modes gives you a continuous cycle: set the rules (design), handle cases (operate), improve the system (learn), and back around.

To wrap up this concept: **domain experts intervene proactively by setting guardrails and monitors, reactively by handling flagged events and making on-the-spot decisions, and course-correctively by analyzing the system's performance and updating it.** A resilient intelligent microservice architecture benefits from all three layers of human involvement, ensuring that the AI remains aligned with human goals and that the human+AI team gets better over time.

Industry Use Cases and Examples

Let's ground these concepts in concrete industry scenarios. Human-in-the-loop is especially valuable in domains like finance, healthcare, legal, and human resources. We'll explore each, seeing how intelligent microservices can work with human experts to deliver better outcomes.

Finance: Fraud Detection and Trading Oversight

In the finance industry, AI and microservices are used for things like fraud detection, credit scoring, algorithmic trading, and loan approvals. These are high-stakes areas dealing with money and often regulated activities, so human oversight is crucial.

Fraud Detection: Consider credit card fraud monitoring. Banks use AI models (as microservices) that score transactions in real-time for fraud likelihood. A typical microservice architecture might have a Transaction Scoring Service that returns a risk score for each transaction. If the score is below a threshold, the transaction goes through instantly. If above a higher threshold, it might be automatically blocked (very high suspicion). But in a middle band of suspicion, the system can flag the transaction for human review. This is a classic HITL scenario: the transaction gets routed to a fraud analyst's queue within seconds. The analyst sees details: location, amount, customer profile, past transactions, plus the AI's reasons (e.g., "unusual purchase pattern, first time in X country, large amount"). The analyst then calls the customer or uses their own judgment to decide if it's fraud or a false alarm, and then instructs the system to either let it go or stop it.

This approach is widely used in banking. As noted earlier, AI can handle the volume – millions of transactions streaming, scoring each and catching the likely fraud – but **it doesn't operate in isolation**. *"The best approach to fraud prevention consists of recognizing which tasks are better left to AI and which aspects humans should focus on,"* and then building a hybrid system . AI excels at scanning for patterns and anomalies across huge datasets (24/7 monitoring, no fatigue). Humans excel at investigating the tricky cases and making judgment calls on ambiguous situations . Many banks, like Mastercard, use such a hybrid approach: AI issues a risk score for transactions, and *fraud specialists then review the transactions with high risk scores to eliminate false positives* . The human confirms frauds or clears legitimate transactions that were mistakenly flagged.

Importantly, this generates a feedback loop: those *manual reviews create new data that further train the machine learning system* to be more accurate . For example, every time a human says "this was a false positive, not fraud," that transaction can be added to the training set as a negative example so the AI learns not to flag similar ones in the future. Over time, the AI improves, and ideally the load on humans may decrease or shift to only the really thorny cases.

Financial compliance is another area: anti-money-laundering (AML) systems scan transactions and accounts for suspicious behavior (like structuring, unusual transfers). These systems produce alerts that compliance officers must review (often a legal requirement). Here, a microservice might be generating alerts using AI rules, but **every alert must be dispositioned by a human** for regulatory reasons. Too many false alerts lead to *alert fatigue* for compliance teams, so they constantly tune the models and thresholds to find the sweet spot (few false negatives while minimizing false positives) – a good example of course-corrective intervention in practice.

Trading and Risk: In algorithmic trading, speed is essential, and AIs often execute trades in split-seconds based on market data. Humans obviously can't react that fast in real-time. So here, human-in-the-loop is often more "on-the-loop" – oversight rather than approving each trade. However, humans set the boundaries: they set limits on how much can be invested, stop-loss thresholds, and if an AI trading bot starts to behave erratically (like losing beyond a threshold or encountering scenarios not seen before), it might trigger a human review or killswitch. Financial firms sometimes implement a "circuit breaker" – if the AI loses more than X amount, the system halts trading and requires a human to intervene to re-enable it. This prevented some disasters, and failure to do so has caused issues like the "Flash Crash" of 2010 where automated trades cascaded; after that, more safeguards were put in place, including human intervention rules during extreme volatility.

In summary for finance, **HITL microservices protect against costly errors and ensure compliance**. AI microservices do the heavy lifting scanning transactions and markets, while humans handle exceptions and accountability. A quote from a fraud prevention resource sums it up: *"AI can issue a risk score for each transaction. Fraud prevention specialists can then review high-risk transactions to eliminate false positives."* The result is a system that catches fraud efficiently without freezing every customer's card on a hunch, because humans filter out the AI's mistakes. And when the AI is working well, the humans can focus more on truly clever fraud attempts, creating new rules and strategies (something AI often can't do creatively) .

Healthcare: AI Diagnostics with Doctor Oversight

Healthcare has seen an influx of AI-powered microservices – from image analysis (radiology, dermatology) to predictive analytics (risk of readmission, disease outbreak predictions) to treatment recommendations. But given the life-and-death stakes and strict regulations, **human oversight is absolutely mandatory** in most cases.

Consider an AI radiology service as a microservice that reads X-rays or CT scans. It might analyze an image and detect anomalies (nodules, fractures, etc.). Hospitals deploy these to help radiologists cope with high workloads and avoid missing things. But no responsible hospital would let the AI alone diagnose and sign the report. Instead, the AI acts as a **second pair of eyes** – it highlights areas of concern and maybe even drafts a preliminary report, but a human radiologist reviews that and makes the final call. This is human-in-theloop in a supervised sense: The radiologist could accept the AI's findings or could disagree (maybe the AI flagged a benign shadow as a tumor – the radiologist knows it's an artifact). The final report that goes to the patient is signed by the human doctor, perhaps noting that AI was used as an aid.

This approach has shown benefits. When humans and AI work together, accuracy can improve. For example, studies have shown that a combination of an AI diagnostic tool and a human clinician can reduce errors significantly – one report noted that nurse practitioners working with an AI reduced diagnostic errors by 54% compared to either alone . The AI might catch things the human missed and vice versa.

Another scenario: **Clinical decision support** for treatment. An AI might recommend treatments or flag drug interactions. Doctors get these recommendations via a microservice integrated into, say, the electronic health record system. The doctor sees "AI suggests Drug A at dosage Y, because it predicts this will have the best outcome given the patient's genetics and history." That's valuable, but the doctor must review it, considering other factors (perhaps the patient mentioned an allergy the AI didn't have on record, or the doctor knows of a newer therapy not in the AI's training data). The doctor can then accept or adjust the recommendation. The system logs the doctor's decision. Over time, if doctors consistently override a particular recommendation in the same way, that's feedback to update the AI's knowledge base.

Safety-critical decisions: in surgery or intensive care, AI might monitor vitals and even control some support systems. Human-on-the-loop is typical here: e.g., an AI might regulate anesthesia dosage but an anesthesiologist is present watching and can take over manual control if needed. Surgical robots can be semi-autonomous but a surgeon is there to intervene if the robot is unsure or if something goes wrong.

Telemedicine triage: Some health systems use AI chatbots as a first line to triage patient symptoms. The microservice asks questions and either gives advice or determines if the patient should seek care. But usually, any serious case is escalated to a human nurse or doctor. The human-in-the-loop ensures that if the chatbot gets confused or detects emergency signs, a real medic steps in (often the chatbot will say "I'm going to connect you with a nurse now"). That transition is crucial; fully automated medical advice can be dangerous if wrong, so there is typically a human fail-safe.

Also, regulations like those from the FDA (for AI-based diagnostic tools) often require that the AI is used "in an assistive manner" and not as a standalone diagnostic unless specifically approved as such. This effectively mandates a doctor in the loop.

One more example: **Pharmaceutical research** uses AI to suggest drug candidates. Researchers (humans) then review those suggestions and run experiments. That's a different stage (R&D rather than live decision), but still HITL in the workflow of discovery.

Healthcare's motto could be *"AI can assist, but a human is ultimately responsible."* It is literally a matter of life or death that the final decisions are vetted by a qualified person. The microservices approach in hospitals often means AI components are modular – e.g., an imaging analysis microservice, a prognosis prediction microservice – and their outputs flow into clinician dashboards. The human clinician synthesizes all that info with their own examination of the patient. The result is hopefully better care: faster diagnoses, fewer misses, but with preserved accountability and empathy (something only a human caregiver can truly provide).

Legal Tech: AI Assistants with Attorney Oversight

In the legal field, we see AI being applied to contract analysis, case law research, predicting case outcomes, and even drafting documents. Lawyers deal with huge amounts of text and data – tasks AIs are increasingly good at – but the interpretation of law and the responsibility to a client or court means a human lawyer must oversee anything an AI does.

Contract analysis and due diligence: Suppose there's an AI microservice that reviews contracts (a common product nowadays). It can flag risky clauses, extract key terms, and even suggest edits to align with company policy. A legal team might run every new contract through this AI service. However, before that contract goes to signing, a human lawyer will review the AI's findings. The AI might flag, for example, "Indemnification clause is missing or limited, which is unusual" or "Non-standard governing law detected." The attorney then closely checks those parts of the contract themselves. They may agree – "Good catch, let's add a stronger indemnity clause" – or they might have context the AI lacks – "Actually, in this deal it's okay to use the partner's standard terms for X reason." The lawyer then finalizes the contract. Here the AI is like a junior assistant that never tires reading fine print, but the senior lawyer makes the judgment calls.

E-discovery in litigation: In lawsuits, there's often a phase of discovery where each side must exchange vast amounts of documents (emails, files) and find relevant ones. AI (especially using NLP) helps by doing what's called Technology-Assisted Review (TAR). It will go through millions of emails to find those likely relevant to, say, a price-fixing scheme or whatever the case is about. But lawyers oversee this by feeding some example relevant docs, verifying samples (to train the AI), and then ultimately producing the documents deemed relevant. If the AI misses something important and it wasn't reviewed by a human, there's a risk of sanctions or losing the case. So attorneys use the AI to drastically reduce the pile (maybe it narrows to 5,000 documents from 1,000,000), but then often a human will at least spot-check those 5,000 to ensure nothing critical is wrongly omitted or included. Some legal jurisdictions even require a lawyer to certify that discovery was done properly, which by implication means they can't just blame an AI if something was missed.

Legal research and outcome prediction: AI services can read case law and suggest relevant precedents or even predict the likelihood of winning a case given past data. A lawyer might ask an AI, "What are my chances if I file this motion in this court?" The AI might return, "Historically 80% of such motions fail in that court, citing these 5 cases." This is useful insight, but a prudent lawyer will review those 5 cases themselves to ensure they truly apply. The final decision – do we file the motion or not – is the lawyer's, informed by AI but not determined solely by it. If the AI said 80% fail but the lawyer knows some nuance (maybe all those cases had a different fact pattern), they might still proceed and argue why this case is different.

Document drafting: Some legal AI can draft simple contracts or fill in templates. A human lawyer must review the output, because legal language is delicate. The wrong word can change meaning and have huge consequences. So a junior lawyer or paralegal might now be an AI – drafting an NDA for example – but the senior lawyer reviews, edits, and approves it. This speeds up routine work but doesn't eliminate the need for oversight.

From an architecture perspective, legal tech companies often provide these AIs as microservice APIs or cloud services that integrate into law firm workflows. For example, there might be a contract management system; when a new contract is uploaded, it calls the AI analysis service and then presents the findings in the UI for a lawyer to go through.

Legal decisions also carry ethical duties. Lawyers have a duty of competence and cannot offload that entirely to an AI. There's also often confidentiality and privilege concerns – using AI means sharing data with a service, which they must do carefully or on-premises. So humans in the loop are important not just for decision quality but for ethical compliance (ensuring the AI didn't hallucinate a case that doesn't exist, etc., which has happened famously with generative AI where a lawyer submitted a brief full of fake citations because they relied on ChatGPT without verification – a cautionary tale that reinforces why HITL is needed).

Judicial use of AI: In some places, judges have looked at AI risk assessment tools (like for bail or sentencing). These are controversial. The general consensus is that such tools, if used, should *inform* but not determine decisions. A judge might see an algorithm's recommendation (e.g., "High risk of reoffending" score) but must use their own judgment in the final ruling, and in some jurisdictions they are required to justify the decision independently of what the algorithm said. That's an example of a human (judge) rightly being fully in control with an AI as an advisor at most.

Overall, in legal tech, **AI is the tireless researcher/analyst, and humans are the ultimate decision-makers and signatories.** The HITL approach ensures no AI suggestion goes into effect without a legally accountable person validating it. This protects against AI errors causing legal disasters, and it aligns with professional responsibility in the field.

Human Resources: AI Hiring with Human Oversight

Human Resources (HR) is another domain where AI microservices are increasingly deployed – for screening resumes, evaluating candidates through video interviews, even monitoring employee sentiment. However, decisions about people – who to hire, who to promote, who might be struggling – carry ethical weight and risk of bias, so humans must stay in the loop to ensure fairness and compliance with employment laws.

Hiring and Recruiting: Many companies receive hundreds or thousands of applications. AI resume screening services (microservices that score or filter resumes) are used to quickly shortlist candidates. For instance, an AI might parse resumes, compare them to job requirements and past hiring data, and rank applicants. But you wouldn't want to blindly trust that ranking without human review, because AIs can inherit biases (maybe inadvertently favoring certain schools or backgrounds that were common in past hires, thereby discriminating). A good HITL process in hiring is: the AI filters out obvious mismatches (e.g., someone without the required certification if it's absolutely needed, or completely irrelevant applications), and perhaps it highlights the top 20%. Then **a human recruiter or hiring manager reviews those AI selections.** They look for anything the AI might have missed – maybe a gem of a candidate in the "middle" tier who has a non-traditional but interesting profile. They also cross-check the top AI picks to ensure they align with things the AI can't measure (like cultural fit gleaned from a cover letter, or diversity considerations to avoid everyone being the same profile).

The human then makes the decision whom to call for an interview. **No candidate should be rejected solely because "the computer said so."** In fact, some jurisdictions are considering or have passed laws requiring explanation and human review for AI-driven hiring decisions (e.g., New York City's law on automated employment decision tools). To comply, companies keep a human in the loop, especially for final cut decisions.

Human in the Loop in Hiring: An AI recruiting service filters candidates by predicting a "fit score" for each based on the job description and the candidates' resumes and assessments. Suppose the AI outputs a ranked list of 100 applicants for a software engineering role, labeling 20 as "highly recommended," 30 as "recommended," and the rest "not recommended." Rather than automatically rejecting the bottom group, the hiring manager reviews the results. They notice one "not recommended" candidate who worked at a small startup in a role very similar to the open position – something the AI's training data (biased towards big company experience) didn't value. Recognizing the relevant experience, the manager decides to move that candidate into the interview pool. Conversely, they see that among the AI's top 20, there is an overwhelming similarity – say, many attended the same handful of universities – raising a red flag about potential AI bias. The manager ensures a couple of candidates with different backgrounds are also included, even if they were ranked slightly lower, to broaden the perspective. During interviews, the human panel uses their judgment on all those candidates. After hiring, they also provide feedback to the AI system: for instance, they label which recommended candidates turned out great and which didn't, and which overlooked candidates got hired and succeeded. This feedback helps retrain the model to improve future rankings (for example, it may learn that startup experience can correlate with success, adjusting its recommendations). Throughout the process, the AI acted as a helpful assistant to deal with volume and point out likely fits, but human managers ensured fairness, diversity, and final quality of hiring decisions.

In the scenario above, the human-in-the-loop prevented a potentially qualified candidate from being unjustly screened out and added qualitative judgment about team fit that the AI wouldn't know. They also spotted a pattern that could indicate bias (everyone the AI liked looked the same) and took corrective action – a good example of human and AI working in tandem to produce a better outcome than either alone.

AI in Interviews: Some companies use AI tools to evaluate recorded video interviews or online tests. These tools might analyze word choice, tone, even facial expressions to score candidates. This is quite controversial. If such tools are used, a best practice is that **humans review the AI's evaluations and the actual interview videos**. Perhaps the AI provides a recommendation or flags something ("Candidate's answer may indicate low customerservice orientation"), but a human interviewer or hiring committee should review those findings and consider them alongside their own impressions. In many cases, companies choose to use these AI suggestions only as supplementary data, not as pass/fail criteria. A human in the loop can override if, say, they see the AI penalized a candidate for a poor internet connection causing awkward pauses (something an AI might misinterpret as the candidate being not confident or something).

Performance and Promotion Decisions: AI might predict who is likely to be a high performer or who is at risk of leaving (attrition models). Those can help managers give attention to employees or decide who might be ready for promotion. But clearly, you wouldn't promote or fire someone just because an algorithm predicted something. The manager reviews the data and uses it as one input. Perhaps the AI says "Employee X is likely disengaged (based on email patterns, missed deadlines, etc.)" – the manager then proactively has a conversation with X to see if they need support. The decision to intervene was guided by AI, but how to handle it is human judgment. If AI suggests who to promote (maybe scoring people against competencies), the final promotion panel should use their own evaluations and the AI's as a reference, checking if there's agreement or if the AI might have overlooked contextual factors (like someone took on a very challenging project and grew a lot, which raw performance numbers might not fully reflect).

Bias and Fairness Oversight: HR is a domain where bias must be carefully monitored. AIs must be trained on representative data and tested for disparate impact. Human oversight includes reviewing the output distribution of AI decisions – e.g., are certain demographics being screened out at higher rates? If yes, humans need to adjust the system. This is proactive oversight in design and continuous monitoring. Some companies have ethics or diversity officers who evaluate AI hiring tools outputs to ensure fairness. If any bias is detected, humans must step in to modify or even discontinue use of that AI until it's fixed.

In HR use cases, a phrase often echoed is **"AI for efficiency, human for empathy."** Hiring and employee management involve human aspirations, fears, and potentials. AI can crunch data and even find patterns we miss (maybe certain experience correlates with success in a role), but treating candidates and employees fairly and with empathy is squarely a human responsibility. HITL in HR ensures that **the humanity of decision-making is preserved**. The AI might provide a recommendation, but a human can ask: "Is this fair? Does this align with our values? How will this decision impact the team's morale?" Those qualitative judgments are why the human is in the loop.

Across all these industry examples – finance, healthcare, legal, HR – a common theme emerges: **AI handles scale and complexity, while humans handle judgment, ethics, and the unusual cases.** Embedding human review at critical points prevents disasters like unjust rejections, dangerous medical errors, or legal blunders. It also builds trust: clients and the public feel better knowing there's a qualified professional involved when it matters.

We've seen how it's done in practice. Now, let's look at how having humans in the loop not only averts problems but also can make the AI itself better over time, through feedback loops.

Closing the Loop: Learning from Human Feedback

One of the powerful benefits of human-in-the-loop systems is the opportunity to create **colearning loops** – where the human not only corrects the AI in the moment, but that correction is fed back to improve the AI's future performance. In other words, every time a human intervenes, the system can *learn* from it. This turns the HITL design into a virtuous cycle: the AI helps the human, and the human helps the AI.

How does this work in practice? It often comes down to **data collection and retraining triggers**. Whenever a human makes a decision or override on an AI output, we capture that as labeled data. Over time, we accumulate a dataset of cases where the AI's initial output was changed by a human. These are incredibly valuable data points, because they represent the boundary conditions or errors of the AI – precisely the areas it needs to get better at. Periodically, the AI model can be retrained or fine-tuned on a dataset that includes those human-corrected examples. The effect: the model slowly incorporates human expertise and adjusts its parameters to align more with human judgment.

For example, consider our earlier fraud detection scenario. Each transaction that was flagged by AI and then decided by a human becomes a training example: the input is the transaction features, the output label is whatever the human ultimately decided (fraud or not fraud). Initially, the AI may have false positives; the human says "not fraud" for some flagged cases. By retraining on that, the model may learn to reduce the weight on whatever pattern triggered those false alarms. Conversely, if the AI missed some fraud that a human caught later (maybe through an investigation), that should definitely be fed in so the AI can catch similar patterns next time.

Many systems implement a **retraining schedule**. For instance, every week or month, they take all new data (with human-verified outcomes) and update the model. Some go further to near real-time learning: after each human decision, they might do an online learning step for the model (though careful with that to avoid drift or overweighting single examples). A safer approach is accumulating and retraining in batch on a regular cadence, with proper validation.

Triggering retraining can also be automated by monitoring performance metrics. Suppose the proportion of cases where humans override the AI starts increasing – that could indicate model drift or that something in the environment changed. That spike can trigger an automated pipeline to retrain the model (or at least alert data scientists to do so). For example, in content filtering, if moderators suddenly overturn a lot of AI decisions (maybe a new slang emerged that the AI flags incorrectly), that's a trigger. Monitoring systems might say "override rate > 10% this week, schedule model retraining job."

In addition to model updates, **rule-based parts of the system can learn**. Sometimes humans will identify new rules to encode. In fraud, analysts might notice a new scam pattern and add a rule "if X AND Y, flag as fraud." That becomes part of the system's logic (and maybe eventually the AI model will learn it, but rules can cover it in the interim). In HR, if managers consistently override the AI to include candidates from a certain background, it might prompt the developers to adjust the scoring criteria to give more weight to that background going forward.

Think of the human interventions as **annotating data** in a real-world setting. This is akin to active learning in machine learning, where the model asks for labels on data points it's unsure about, and those labels improve it. In our microservice architecture, the cases that reach humans are often the ones the model was unsure or got wrong – exactly the cases you want labels for. Over time, the goal is not necessarily to eliminate the human (because as we discussed, some level of oversight will always be desired for high stakes), but to **raise the bar** so that the AI handles more cases correctly on its own and only the truly ambiguous or novel ones require human brainpower.

However, caution: we must ensure that we **close the feedback loop properly**. It's not automatic that human decisions get fed back – we need to design it. This might involve connecting the output of the human's action to a data store that data scientists or automated retraining pipelines use. For instance, all cases and their final outcomes (AI vs human final decision) might go into a "ground truth" database. We also want to store the features or input data associated, so we have a full training row. Privacy and compliance are considerations here too (especially in domains like HR or healthcare – you have to anonymize or get consent if using that data to retrain, etc., depending on context).

Another aspect is **learning not just from outcomes, but from human reasoning**. Sometimes the system can capture *why* the human made a decision (like those reason codes or notes we have the human enter). If, say, a human moderator marks "this is allowed because it's art nudity, not pornographic," that reason could be used to refine the AI's content classification (maybe by having a special category or by adding training data labeled as art). Or an HR manager's note "overrode because candidate has startup experience which we value" can be turned into a feature for the model ("startup_experience=True") in future iterations.

Furthermore, co-learning isn't just the AI learning from the human – **humans learn from the AI too.** Newer AI systems might identify patterns or signals humans weren't aware of. Through the explanations and through seeing many cases, human experts might update their own mental models. For instance, a doctor might notice the AI picks up on subtle imaging features and might start looking for them too in cases the AI didn't flag (improving their own skill). An analyst might realize a certain transaction pattern is risky after seeing the AI consistently flag it, even though they hadn't considered it before. In an ideal HITL arrangement, the human and AI form a **collaborative team** where each learns and improves by working together – the AI learning from human corrections, and the human learning from AI insights.

Companies also set up **feedback dashboards**: showing humans how their inputs have improved the system. This encourages continued quality input. E.g., "Since you started giving feedback, the AI accuracy on your queue improved from 85% to 90%." It reinforces that their interventions matter beyond just that single case.

One must also be mindful of **confirmation bias** in feedback. If the AI rarely tries certain actions because humans initially discouraged it, the AI might never learn if those actions could have been right in some cases (since it never does them, humans never see them). To combat this, sometimes you need to explore occasionally – e.g., let the AI handle a random case without human check (in a safe sandbox) to see what would happen, or periodically review even AI-approved cases to ensure it's not reinforcing a subtle bias. Keeping humans in the loop for a sample of decisions that *wouldn't* have been flagged is a strategy to ensure the model's unchecked outputs are also monitored occasionally.

In technical terms, incorporating human feedback might use techniques like **reinforcement learning from human feedback (RLHF)** – where a reward model is trained based on human preferences (as done in some NLP models like ChatGPT's training). In microservices, RLHF could mean adjusting policies in an AI agent based on human-provided rewards or rankings. For example, if you had an AI agent optimizing logistics but occasionally a human dispatcher adjusts the plan due to on-ground knowledge, you could treat that adjustment as feedback for the agent's planning algorithm.

An interesting case is **continuous deployment of model improvements**. If your microservice can hot-swap models or update frequently (with A/B testing to ensure safety), you might see improvements quickly. However, always monitor after updating – a new model might behave differently and possibly trigger different or more human interventions initially. It's a cycle: deploy improvement, observe human feedback on new model's outputs, improve again.

We should also consider that learning from feedback sometimes means learning **when to ask for help**. The AI might learn better about its own uncertainty calibration. If it notices that whenever it had a certain intermediate state and asked a human, the human always did a particular action, it might incorporate that knowledge and either automate that next time or at least be more confident. Alternatively, it might realize "Whenever I have pattern X, I get overridden, so I should automatically flag pattern X for human instead of suggesting the wrong thing." This is the AI learning its limits – a very desirable trait. It's akin to a junior employee learning when they need to consult their manager versus when they can make the call themselves.

In summary, **closing the loop** means we do not treat the human-in-the-loop step as the end of the story, but rather as part of an ongoing cycle of improvement. The human's correction or confirmation isn't just resolving one case; it's teaching the AI how to handle such cases in the future on its own. Over time, the collaboration ideally leads to higher accuracy, fewer false alarms, and fewer misses – the human workload may concentrate on only truly novel or contentious cases. And even if we always keep a human checkpoint for critical outcomes (for the reasons of trust and accountability), that human's job becomes easier because the AI's suggestions keep getting more spot-on.

We've built our HITL system now that not only uses human oversight for safety, but actually gets smarter thanks to that oversight. The system is aligning more with the human's expertise and values through each iteration.

However, while all this sounds great, we must remain aware of the potential challenges when embedding human oversight. Let's discuss those pitfalls and how to mitigate them.

Balancing Autonomy and Control

Designing human-in-the-loop systems is a balancing act. On one hand, we have automation and autonomy – which give us efficiency and speed. On the other hand, we have human control – which gives us safety, ethics, and context. The goal is to find the **sweet spot** where the AI handles what it can, and humans handle what they must, with neither being underutilized or overburdened.

If we lean too far toward automation, we risk the AI making a critical mistake unchecked. If we lean too far toward human control, we lose the benefits of automation and bog down humans with too many routine tasks. Striking the right balance is both art and science.

A useful way to think about this balance is through analogies:

The Autopilot Analogy: Modern airplanes use autopilot systems extensively – they can fly the plane through most of the journey with minimal input. The human pilots primarily take off, land, and handle unexpected situations or complex decisions (like rerouting around storms). This is a well-tuned balance: the autopilot handles the mundane, precise work of keeping the plane on course, which reduces pilot fatigue and error; the human pilot oversees everything and is there to intervene for safety. However, pilots are trained to always stay alert and not become complacent, because if something goes wrong, they might have only seconds to react.

This maps to our AI systems. We want the AI (autopilot) to run the show in routine conditions – e.g., auto-approve small transactions, auto-filter trivial job applicants, etc. But we want the human (pilot) monitoring, and ready to jump in for anomalies. We also want the human to handle the critical parts (just as pilots handle takeoff/landing) – e.g., final approval on a major decision or an edge case scenario. The tricky part is preventing **automation complacency**, where the human operator might start to get too comfortable and not pay enough attention because "the AI's got it." Pilots face this; there have been incidents in aviation attributed to pilots getting out of the loop due to over-reliance on automation . To mitigate that, systems sometimes incorporate check-ins (pilots regularly cross-check instruments, or the system might sound a chime periodically requiring a response to ensure alertness). In software, this could mean periodically involving humans even if not strictly necessary, just to keep them engaged and aware, or designing UI that encourages situational awareness (for example, showing a live feed of decisions being made so the human sees a stream of activity rather than sitting idle until an alert pops).

The Orchestra Analogy: Imagine an orchestra and a conductor. Each musician (like an individual microservice or AI component) is skilled and can play their part autonomously from sheet music. The conductor (human oversight) doesn't play an instrument during the performance; their role is to guide tempo, ensure harmony, cue entries, and adjust dynamics to interpret the music as a whole. If a section gets a bit off or the tempo starts drifting, the conductor intervenes with a gesture to correct it. In a sense, the conductor is "on the loop" – not dictating each note, but constantly supervising and guiding the overall outcome.

In an intelligent microservice system, the microservices (especially AI ones) are like those musicians: each performing tasks (data processing, making partial decisions, providing recommendations). The human in the loop is like the conductor who oversees the multiagent system at a higher level, ensuring everything stays in sync with business objectives and ethical standards. They might not micromanage each service's output, but they watch the "big picture." If one AI service's outcomes start conflicting with another's or with policy (like one service optimizing speed starts causing another to generate errors), the human conductor notices the discord and steps in to coordinate – maybe adjusting parameters or in worst case pausing part of the system until harmony is restored. This analogy highlights that the human oversight is about **coordination and alignment**, not just intervention on single tasks. It's a holistic control.

Between these analogies, the core message is: **humans and AI should each do what they do best, in a complementary way.** The system should default to automation for scalability, but have humans in control for direction and exceptions.

How do we find that balance in design? A few guidelines:

- **Define clear hand-off points:** There should be well-defined points where control passes from AI to human or vice versa. E.g., "For risk score above X, escalate to human; below, auto-approve." This clarity ensures we know exactly who (AI or human) is responsible for each part of the process. It also helps avoid confusion where both the AI and human think the other is handling something (leading to a gap).

- **Avoid ping-ponging control excessively:** If control switches back and forth too often, it can be chaotic or frustrating. Ideally, group tasks such that the AI does a chunk, then human does a chunk. If the human must intervene, perhaps let them finish that case entirely, rather than giving it back to AI and forth repeatedly. Frequent toggling might also confuse accountability.

- **Provide context on both sides:** Just as we ensure the human sees AI's rationale, if a human does override or adjust something, feed that context back to the AI or system logic if possible, so the AI doesn't operate in ignorance of what the human did. For example, if a human marks a content item as "safe, do not flag again", the system should not flag that item again and perhaps similar items should also be left alone (until model update). This prevents tug-of-war where AI flags and human clears the same thing repeatedly.

- **Gradual autonomy increase:** You might start a system with more human oversight (to build trust and verify AI performance). Over time, as confidence in the AI grows (through proven accuracy), you can dial up automation (maybe raise thresholds so fewer things come to humans). Essentially, you calibrate. But always leave a mechanism to dial it back if conditions change.

- **Fail-safe defaults:** Decide carefully what happens if the human is unavailable or overwhelmed, and what happens if the AI is unavailable. The system should handle both failure modes gracefully. For instance, if AI service is down, perhaps automatically fall back to human-only processing (slower but safe). If human reviewers are all unreachable (say, 3 AM and an urgent decision needed), perhaps fall back to a conservative automated decision or wake an on-call person. Having layered backups ensures continuity with acceptable risk.

Another balancing consideration: **throughput vs quality.** If humans are reviewing too much, they become a bottleneck and throughput suffers. If they review too little, quality might drop or risk rises. So find metrics to monitor: What percentage of cases go to human? How long is the human taking per case? Are we comfortable with that load and turnaround time? If the human queue grows long, maybe the threshold for what requires review can be adjusted dynamically (some systems do this – if humans are busy, temporarily trust the AI a bit more for borderline cases, though this can be risky; or engage more human reviewers in real-time scaling).

Trust calibration between human and AI is part of balance. Humans need to trust the AI enough to let it do its thing most of the time, but not so blindly that they ignore signs of trouble. Conversely, the AI's design should "know" when to trust itself vs defer. Training the human operators about the AI's capabilities and failure modes helps them calibrate trust. If they know "the AI is 99% accurate on common cases but struggles with edge cases involving X," they can trust it on the common and be extra vigilant on X.

One more analogy, or rather a scenario, often used: **Centaur teams (human + AI)**, as seen in chess where human+AI team can outperform either alone. The human brings strategic thinking and risk management, the AI brings calculation and pattern memory. In microservices, think of the human as a team leader and AI services as team members. A good leader delegates tasks (to AI) but stays informed and steps in for critical decisions. If the AI team members are doing well, the leader can step back; if the project goes off course, the leader intervenes to redirect.

The balance might also shift over the lifecycle of a system. Early on, more human oversight while AI is new. Later, maybe less for the parts that are well-proven. And if a new AI feature is added, again initially treat it with more human checks until validated. It's an ongoing tuning process.

In summary, balancing autonomy and control is about **maximizing the strengths of both AI and humans while minimizing their weaknesses.** Let AI run fast, but not so fast that it outruns human understanding. Let humans ensure correctness, but not micromanage to the point of negating the efficiency gains. The best HITL systems feel like a **seamless partnership** – like a well-trained autopilot with a vigilant pilot, or an orchestra playing beautifully under the guidance of a conductor. When things are balanced, the system achieves higher performance and reliability than either the human or AI could alone.

However, maintaining this balance isn't without challenges. Let's look at some pitfalls to avoid when implementing HITL, and how to mitigate them.

Challenges and Pitfalls of Human-in-the-Loop

While human-in-the-loop systems bring many benefits, they also introduce some challenges that designers must address. Being aware of these pitfalls helps in creating a sustainable and effective HITL solution.

1. Alert Fatigue and Human Overload: If the system flags too many items for human review, human operators can quickly become overwhelmed or desensitized. This is known as *alert fatigue*. It's well-documented in fields like cybersecurity and healthcare alarm systems – when people get bombarded with alerts, they start ignoring them or responding mechanically without investigation . In a HITL context, if your AI is over-cautious and sends 50% of cases to human review, the human is essentially doing half the work manually anyway and will likely start rubber-stamping just to get through the queue. That defeats the purpose of oversight because the scrutiny isn't there; it's a box-checking exercise.

To avoid this, **tune the triggers** so that only a manageable fraction of cases get escalated. The threshold might be adjusted to reduce false positives. You can also implement **prioritization** – not all alerts are equal, some can be auto-triaged to low priority (maybe batched for later review rather than immediate). Another technique is to provide supporting info that makes triaging easier, so each human decision takes less effort (reducing cognitive load). We also discussed dynamic adjustment: if humans are falling behind, perhaps throttle or batch lower priority ones. Or add more reviewers temporarily if possible.

Avoid Alert Overload: Over-reliance on human reviews can backfire if too many items are flagged. When humans are inundated with frequent alerts or approval requests, they can experience "alert fatigue" and start rubberstamping decisions without proper scrutiny. To prevent this, ensure the system only escalates truly necessary cases to human reviewers. Calibrate your AI's sensitivity and use thresholds so that minor or routine cases are handled autonomously. The goal is for human attention to be a scarce resource, focused on the most critical or ambiguous cases. By keeping the volume of human-reviewed items reasonable, you maintain high-quality oversight and keep your experts engaged rather than overwhelmed.

2. Automation Complacency and Skill Degradation: On the flip side of overload is the risk that humans become too passive. If the AI is highly accurate and handles most things, the human might start to assume the AI is always right. This *automation complacency* can lead to missing the rare mistakes because the human isn't as sharp or attentive as they would be if they were doing it all themselves. In fields like aviation, it's noted that pilots who rarely fly manually may lose some flying skills or be slower to take over when needed . Similarly, an analyst who mostly clicks "Approve" might lose the edge in spotting subtle fraud patterns.

To combat this, training and periodic practice are important. Occasionally, simulate a scenario where the AI is wrong on purpose and see if the human catches it – kind of like drills. Also emphasize a culture of *"trust but verify"* for the AI's outputs. Some systems implement **challenge-response** dialogues: e.g., every now and then, require the human to explicitly confirm they've reviewed by answering a quick checklist, to ensure they actually looked (though this can annoy and isn't foolproof). Encouraging humans to always consider "Could the AI be wrong here? If so, why?" as they review helps maintain vigilance.

3. Dependency on Human Override: If an AI system is designed with the assumption "a human will fix anything I do wrong," it might not be tuned to be as accurate as it could be.

Developers might be lazy in optimization, thinking the human safety net will catch issues. This can lead to sloppy AI performance. It can also lead to organizational dependency – the system can't function at all without human oversight, which may be fine by design for critical stuff, but for scale you might want the AI to handle more autonomously. There's a balance: we want the AI as good as possible *and* a human safety net, not the AI mediocre because a human is always there to clean up. The mitigation is to continue to improve the AI and treat human overrides as costly (even if not monetary, in time/effort) so that the AI's goal is still to minimize the need for them.

4. Scalability and Cost: Humans in the loop cost money (salaries) and time. As your system scales, can you scale the human workforce at the same rate? If a process volume doubles, can you double the number of reviewers? In some cases yes (if you can hire or crowdsource), in others no (if it requires highly specialized experts like doctors or pilots, those aren't easy to multiply). So be mindful of where the bottlenecks are. Maybe use tiered systems: easier cases could be reviewed by junior staff or even crowd workers (with privacy constraints considered) whereas only the toughest go to your top experts. Or invest in improving AI so that as load increases, the percentage needing human stays manageable. Cost-benefit analysis should justify the human-in-loop: For example, if reviewing 1000 cases costs X in manpower but prevents Y in losses, is it worth it? Often yes in high stakes, but it should be quantified.

5. User Experience Issues: If not designed well, adding human steps can slow down processes in a way that frustrates end-users or customers. Imagine an e-commerce system where an order might be held for manual review – if that's not communicated, a customer's order might mysteriously be delayed. Thus, transparency is key: if a loan application is under human review, inform the applicant "Your application is under additional review" rather than leaving them in the dark. Managing expectations is part of design. Also, because humans don't work 24/7 or as fast as computers, there might be delays. If your microservices operate in milliseconds but a human takes hours, for certain real-time use cases that's unacceptable (you wouldn't put a human in the loop of an airbag deployment decision!). So ensure humans are inserted only where the latency is acceptable or can be mitigated (like parallel processing or interim responses like "pending review").

6. Communication and Handoff Errors: Whenever you have human and automated components interacting, there's potential for miscommunication. Maybe the information presented to the human is incomplete or confusing, leading them to make a wrong decision. Or the human's decision might not be properly recorded due to an interface bug, so the system doesn't act on it. Careful system integration and UI testing is required to prevent these issues. Logging is vital: if something went wrong, you need to audit whether it was a human mistake, AI mistake, or a handoff issue. Ensure that when control passes, the state is fully passed along. For example, if a human approves something with conditions ("approve if X is fixed"), make sure that nuance isn't lost.

7. Blame and Accountability: In a combined human-AI decision process, if a bad outcome occurs, who is responsible? Humans might over-rely on AI, or conversely blame the AI for mistakes they approved. Organizations need to clearly define accountability. Typically, the organization (and the human operators) remain accountable, since AI is a tool. It's wise to document that human reviewers must verify and are responsible for final decisions. This ensures they stay engaged. But also, internally, use the incident to improve AI rather than simply blame the human for "missing" the error – it's a joint responsibility to improve the system. Ethically, one should not use human-in-loop as a way to shove all blame to humans while benefiting from automation. It should be a partnership.

8. Privacy and Security: If humans are reviewing sensitive data (financial transactions, personal information, medical records), you have to ensure those humans are authorized and that data is handled securely. Microservice architecture might log a lot; ensure access controls so that only the necessary people see the sensitive data. Also, if using external crowds (like Mechanical Turk style for some tasks), make sure data is anonymized or low sensitivity for those cases. Human involvement can increase attack surface too (social engineering of the human operator, etc.), so training and safeguards (like two-person approval for really critical actions) might be needed.

9. Maintaining Human Skills: This ties to complacency but even more broadly – if people rely on AI assistance long-term, they might lose domain skills. For instance, junior doctors might rely so much on AI diagnostic suggestions that they don't develop the same diagnostic acumen their predecessors did. Some argue we might create a generation of professionals who are good at vetting AI outputs but not at working from scratch. Is that a problem? It could be if the AI fails or in novel cases. To mitigate, training programs and job rotations could ensure humans still practice core skills. Perhaps sometimes run without AI assistance as an exercise. It's a bit like pilots using manual flight hours to stay sharp even though autopilot can do it.

10. Ethical Considerations of Deference: Sometimes, a human in the loop might start deferring too much to AI because "the data says so," even against their better judgment. This is a subtle psychological effect. For example, a doctor might feel that the AI likely has processed **more** data than they can hold in mind, so even if they have a hunch the AI is wrong, they might second-guess themselves. The challenge here is to encourage human operators to speak up or override when they truly think the AI is mistaken, and to foster confidence in their expertise. It might require organizational culture messaging that "It's okay to override the AI; we want you to if you feel something is off." If humans feel intimidated by the AI ("It's an expensive system, it must be right"), then the oversight is performative. This is partly solved by showing that their overrides are valued and by demonstrating times the AI was indeed wrong and a human save the day (sharing those stories).

After enumerating these pitfalls, it might sound like having humans in the loop is fraught with difficulties. But each of these can be managed with thoughtful design and process.

For instance, to reduce alert fatigue, one can use techniques like **variable thresholding** (only escalate a proportion of borderline cases) and ensure **high precision in flags** so that most escalations are true needs. For automation complacency, keep the human engaged with active monitoring interfaces and perhaps simulate issues to keep them on their toes. For scalability, incorporate cost of human review into ROI calculations and possibly invest in better AI to minimize that cost over time.

It's also useful to do a **post-mortem on any incident**: if a bad outcome happened despite human-in-loop, analyze whether the human saw the alert, if yes why they didn't prevent it, if not why it wasn't alerted. If a false positive flood happened, analyze why and adjust. Continuous improvement applies not just to the AI model but to the human-integration as well.

"Human-in-the-loop isn't a step backward for automation – it's a safety net that makes the whole system smarter and more reliable." – *AI Ethics Advisor.* This quote encapsulates the mindset to maintain: we add humans not to undermine the efficiency of AI, but to ensure AI is used responsibly and to feed human wisdom back into the system. By acknowledging and managing the challenges above, we leverage that safety net effectively without letting it become a bottleneck or a crutch.

In the end, designing HITL systems is about carefully managing the interaction between humans and AI so that one plus one is greater than two, not less. With awareness of these pitfalls, you can put guardrails in place that keep the collaboration on track.

Checklist: Designing a Human-in-the-Loop System

To wrap up our comprehensive look at HITL in microservices, here's a practical **checklist and readiness guide** for designing and implementing human-in-the-loop capabilities. This can serve as a reference to ensure you've covered the critical considerations:

- ☑ **Identify Critical Decision Points:** Map out your workflow and mark which decisions/outcomes are high-risk, complex, or ethically sensitive. These are candidates for human oversight. Ask "What's the impact if the AI is wrong here?" for each step.

- ☑ **Define Escalation Criteria:** For each identified point, define clear rules or thresholds for when to involve a human. This could be a confidence score threshold (e.g., if confidence < 0.8, escalate), specific trigger conditions (e.g., transaction amount > $5,000), or anomaly detection signals. Make these criteria explicit and adjustable.

- ☑ **Choose Integration Pattern:** Decide how the human will be inserted: synchronous pause (workflow waits for approval) vs asynchronous (case queued for later review). Use message queues or callbacks for asynchronous flows. Ensure your architecture can handle a paused state or a branch out-and-back in the workflow.

- ☑ **Design the Human Interface:** Create a dedicated UI or interface for human reviewers. It should present all necessary context and the AI's recommendation/explanation clearly. Include action buttons for the human (approve/deny/modify) and a way to provide feedback or comments. Make sure it's intuitive and efficient, since this directly affects review quality and speed.

- ☑ **Ensure Explainability:** For each AI decision presented to a human, provide an explanation or factors behind it. Use model explainability tools to generate highlights or reasons. This transparency helps humans trust and verify the AI's suggestions.

- ☑ **Implement Notification & Assignment:** Determine how humans are notified of tasks. This could be via a dashboard, email alerts, or integration with task management systems. Also decide how tasks are assigned – round-robin, based on expertise, severity, etc. Avoid situations where an alert goes unnoticed; have a backup notification (or escalation to another person) if not handled in X time.

- ☑ **Set Timeouts and Default Actions:** Define what happens if the human doesn't respond in time. Do you default to allow or block? Maybe low-risk items auto-approve after a timeout, whereas high-risk auto-deny or escalate further. Consider having an oncall rotation for urgent out-of-hours decisions. The system should not hang indefinitely waiting for input without a plan.

- ☑ **Log Everything (Audit Trail):** Implement comprehensive logging for the HITL interactions. Log when an item was flagged, what information was shown to the human, who reviewed it, and what decision was made (with timestamp). This is vital for compliance, debugging, and later learning. Ensure these logs are stored securely (they might contain sensitive data).

- ☑ **Feedback Capture:** Whenever a human makes a decision, capture it as labeled data. If possible, have structured outcomes (like reason codes for overrides, or whether the AI was correct or not in hindsight). Store these for model retraining or rule refinement. A simple approach is to have a database of cases with fields indicating AI output and final human outcome.

- ☑ **Plan Model Improvement Cycles:** Establish a process to regularly retrain or update your AI models using the accumulated human-verified data. Monitor performance metrics (precision, recall, override rates) over time. If you see drift or too many overrides, schedule a model update. Essentially, integrate this into your MLOps pipeline.

- ☑ **Balance Workload:** Monitor the volume of items going to human review. If it's consistently high, consider adjusting thresholds or improving the model to reduce load. Ensure you have enough staff to handle peak volumes (and consider surge strategies like secondary reviewers or delaying non-urgent tasks during peaks). The system should be designed to handle both normal and stress conditions without overwhelming humans or stalling outputs.

- ☑ **Train and Onboard Human Reviewers:** Don't assume people will instinctively know how to interact with the system. Provide training on the AI tool, what the expectations are for their review, how to interpret the AI's explanation, and what to do in edge cases. Make sure they understand they are empowered to override and that their judgment is valued. Also train them on any bias mitigation strategies (e.g., "Be aware of potential AI bias X, and double-check those situations").

- ☑ **Implement Quality Control for Human Decisions:** Just as AI output quality is measured, have some QC for human decisions too. This might mean secondary review of a random subset of decisions, or peer review for highly critical ones (two humans must agree). Ensure humans are consistent and following guidelines. This can catch mistakes a single reviewer might make and also provide feedback to them.

- ☑ **Privacy and Security Checks:** Verify that in bringing humans into the loop, you're not violating any privacy constraints. Mask or limit data where possible (e.g., do human reviewers need full personal details or just relevant info?). Ensure secure channels for any data the humans see. Also, have NDAs or agreements if necessary for those handling sensitive data.

- ☑ **Define Accountability:** Clearly document who is ultimately accountable for decisions. In internal policy, state that the human approver is responsible for final outcomes of reviewed cases. That helps avoid confusion. At the same time, hold the AI developers accountable for the accuracy of cases not flagged. Basically, every decision either automated or human-reviewed has someone accountable. This clarity is important for compliance and for trust in the process.

- ☑ **Failover Paths:** Have a plan for system failures. If the AI part goes down, do you switch to manual processing of everything? If the human interface or queue system goes down, do you halt certain operations or fall back to automation with caution? Define these contingency modes and test them. For instance, you might simulate "no human available" scenario and see that critical functions still operate safely.

- ☑ **Continuous Communication Between Teams:** The developers of the AI system and the human operators (or their managers) should have a channel of communication. Regular meetings or reports where human feedback is shared with the AI team and vice versa. This ensures the system evolves with input from both sides. It also gives humans a voice in improving the AI tools, which increases buy-in.

- ☑ **Test with Real Scenarios:** Before fully deploying, run simulation tests of the full loop. Perhaps dry-run historical data through the system and let humans practice reviewing (without impact) to see how it flows and tweak the UX. Scenario testing like "AI made a mistake here, did the process catch it?" or "Sudden spike in alerts, how did the system cope?" is invaluable to iron out issues.

- ☑ **Monitor and Iterate:** Once live, treat the HITL system as something you continuously monitor and adjust. Have metrics like average time for human decision, percent of AI suggestions accepted vs overridden, outcomes accuracy, etc. Use these to refine either the model, the threshold, or the process. For example, if humans are consistently approving 99% of a certain category of alerts, maybe you can stop bothering them with those and let them auto-approve. If humans are overturning a certain decision type frequently, focus on that in model improvement.

This checklist isn't static; different domains may have extra steps (like regulatory approval for certain processes, or involving legal counsel in defining when AI can be used). But it provides a baseline to ensure you've considered the main facets of embedding human oversight.

Finally, always approach HITL design with a mindset of respect for human expertise and empathy for human limitations. The checklist reflects both technical and human factors. By checking off these items, you'll be well on your way to creating an intelligent microservice system that is not only powerful, but also trustworthy and aligned with human values and needs.

Conclusion

In intelligent microservice architectures, **human-in-the-loop is the bridge between artificial intelligence and human accountability.** By strategically integrating human oversight, we get the best of both worlds: the efficiency and consistency of AI, and the wisdom, intuition, and ethical judgment of humans. Throughout this chapter, we saw that HITL is far more than an operational failsafe; it's a design philosophy that permeates how we build and deploy AI agents within our systems.

We began by understanding *why* human oversight is non-negotiable in critical domains – from preventing costly mistakes to complying with emerging regulations that demand human control over AI decisions. We identified how to pinpoint those make-or-break moments where a human's touch is needed and discussed concrete patterns to insert a "human checkpoint" in a microservice workflow without breaking the flow. We used real tools and platforms in our examples, whether it was message queues like Kafka routing tasks to analysts or workflow engines waiting for a callback from a human-facing app.

Through industry case studies – finance, healthcare, legal, HR – we illustrated that the HITL approach is versatile and universally applicable wherever AI meets consequential decisions. The recurring theme was collaboration: AI systems doing the heavy lifting, humans providing direction and oversight, and each learning from the other. We explored how domain experts intervene not just reactively (responding to flags) but also proactively (setting the system up for success and monitoring) and in a course-corrective way (refining the system over time).

This 360-degree involvement ensures that the AI remains an aid, not a loose cannon.

Explainability and traceability emerged as key enablers for effective human review. If humans can't understand why the AI did something, it's hard for them to trust or correct it. That's why designing good UIs and logs is as important as designing good algorithms. We also emphasized empathy for the human operators: understanding their experience, avoiding drowning them in alerts,We also emphasized empathy for the human operators: understanding their experience, avoiding drowning them in alerts, and designing systems that support and respect their contributions. A human-in-the-loop system should empower its human users, not frustrate them. When done right, humans feel like valuable "collaborators" with the AI, not button-pushers following a machine's orders.

In closing, remember that **human oversight is not a hindrance to intelligent microservices – it's a catalyst for their success**. By embedding human judgment into our AI workflows, we make those workflows more robust, fair, and aligned with real-world needs. We create a feedback loop where humans and AI improve each other continuously. The end result is an AI-powered microservice architecture that stakeholders can trust with critical outcomes, because they know a responsible person is in the loop ensuring everything stays on track.

As you implement these principles, you'll find that your systems not only avoid the pitfalls of unchecked AI, but they actually deliver better performance and user satisfaction. Human-inthe-loop becomes a competitive advantage, enabling innovative uses of AI in areas that would otherwise be too risky to automate.

With a strong foundation of human oversight in place, you're now prepared to explore the next step: actually building these intelligent agents and integrating them as microservices. In the next chapter, we will dive into *"Building Intelligent Agents as Microservices."* We'll take the concepts of oversight and orchestration and apply them to constructing AI-driven services from the ground up. Together, these chapters will equip you to create AI systems that are not only smart and autonomous, but also aligned with human values and capable of earning and retaining user trust.

Moving forward, always remember the essence of this chapter: **intelligence and autonomy flourish best under the guiding hand of human insight.** By keeping a human in the loop for critical outcomes, we ensure our microservices truly serve us – and in doing so, we make both our AI and ourselves a little more intelligent.

Chapter 10: Building Intelligent Agents as Microservices

Introduction

At WebkinsTech, the customer support team found themselves buried under a growing backlog of helpdesk tickets. As the company's user base expanded, so did the volume of support requests – everything from password reset queries to complex technical troubleshooting. The support agents were struggling to keep up, and response times were creeping beyond acceptable levels. This sparked an idea within the engineering group: **could an intelligent software agent help triage and handle some of these tickets automatically?**

WebkinsTech's engineering and data science teams decided to collaborate on building an AIpowered support assistant as a microservice. The goal was ambitious yet focused. This "support agent" microservice would analyze incoming tickets, determine their topic and urgency, and automatically route each ticket to the right team – or even provide an instant answer if it was a common question. By embedding an AI model into a standalone service, they hoped to reduce the load on human support staff and improve response times, all while maintaining accuracy and customer satisfaction. A cross-functional team was assembled, including software architects, ML engineers, and the support department's product manager, to ensure the solution would meet both technical and business needs.

Now, in this first chapter of **Part III: Implementation Strategies**, we bridge theory into practice. We'll walk step-by-step through the process of designing and implementing an intelligent agent as a microservice – from defining the agent's scope and choosing the right AI approach, to gathering data and training a model, designing the service API, integrating the AI into a microservice, and finally testing and deploying it in production. Along the way, we'll introduce major agentic AI frameworks (like LangChain, Haystack, and others) that can accelerate development, and we'll emphasize a modular, cloud-agnostic deployment approach. Throughout this chapter, we'll use WebkinsTech's support agent scenario as a running example to illustrate each step in a concrete way. By the end, you'll see how building an intelligent microservice isn't just about coding an algorithm – it's a **team effort** blending data science, software engineering, and DevOps, guided by clear objectives and careful planning.

Collaboration is Key

Building intelligent microservices is a team sport. Involve diverse roles early – domain experts to define goals, data scientists to build the models, and DevOps engineers to plan deployment. This collaboration ensures the AI agent aligns with business needs and is reliable in production from day one.

Step 1: Define the Service's Scope and Goal

Every successful intelligent agent begins with a clear **purpose**. The first step is to define exactly **what task or decision your agentic microservice will handle**, and what **success** looks like. It's easy to get excited about AI and imagine a service that "does it all," but in practice, narrowly scoping the agent's role leads to a more feasible and effective solution.

Ask yourself (and your stakeholders) some key questions when scoping an intelligent microservice:

- **What specific problem are we solving?** Identify the decision or task the agent will automate or assist with. For example, is it routing support tickets to the correct department, approving or flagging a transaction as fraud, recommending products to a user, or something else?

- **What is the desired outcome or goal?** Clarify how you'll measure success. It might be reducing average ticket response time by 50%, catching at least 90% of fraudulent transactions before they finalize, or improving user engagement with recommendations.

Define clear metrics or KPIs for the agent's performance.

- **What inputs will the agent receive, and what output will it produce?** Determine the information the service will consume (e.g. a ticket's text description, or a stream of transaction events) and what it will return or effect (e.g. a ticket category and priority, or a fraud risk score and alert).

- **What's in scope and out of scope?** Set boundaries. For instance, the agent might handle routine cases autonomously but defer complex or high-risk cases to humans.

Explicitly decide which scenarios the AI will address versus when it should fall back (perhaps emitting a warning or requesting human intervention).

- **What constraints or requirements must be met?** Consider any business rules, ethical constraints, or runtime requirements. Does the decision need to be explainable to a human? Is there a strict latency requirement (e.g. the agent must respond within 200ms)? These factors will influence the design and approach.

By answering these questions, you create a **specification** for your intelligent microservice. In our WebkinsTech support agent example, the scope was defined as follows: the agent would focus on **ticket triage**. It would take each new support ticket as input and output two things – a predicted category (which team should handle it) and an urgency level. The agent's goal was to correctly route a significant portion of tickets (say, 80% of incoming issues) to the right team instantly, and flag the truly urgent ones, thereby cutting down the initial review work for support staff. Crucially, the team decided that **answering** the tickets (providing a solution to the customer) was out of scope for version 1. The agent would not directly reply to customers – it would only sort and prioritize tickets. Complex issues or anything the model wasn't confident about would still be handled by humans as usual. With this clear mission defined, everyone – from product managers to engineers – was on the same page about what they were building and why.

Defining the scope early on also helps you spot feasibility issues. For example, if the goal requires a level of accuracy or fairness that current AI models can't guarantee, you might refine the scope or plan for a human-in-the-loop step (as you read in earlier chapters about human oversight). In WebkinsTech's case, the requirement was that the routing agent shouldn't misdirect critical tickets (like outages or security issues). They decided that any ticket the agent marked as "high urgency" would still be reviewed quickly by a human, at least until the model proved sufficiently trustworthy. By setting these boundaries and expectations up front, the team established a solid foundation for the project.

Step 2: Choose the Appropriate AI Approach

With a clear goal in mind, the next step is to decide **how the agent will make its decisions**. In other words, what AI or machine learning approach is best suited to achieve the service's goal? This involves weighing different techniques – from simple rules to complex machine learning models or even large language models – and choosing one (or a combination) that fits the problem, the available data, and the operational constraints.

Not every intelligent microservice needs a neural network or an advanced AI. Sometimes a straightforward algorithm suffices; in other cases, a learned model is essential. Consider the main options on a spectrum from manual logic to full AI:

- **Rule-based logic** – You can explicitly program the decision rules using if-else conditions or a business rules engine. This works well when the decision criteria are well-understood and can be codified. *Pros:* Rules are easy to explain and quick to implement for simple cases. *Cons:* They don't learn or adapt – any scenario not anticipated by the rules will be handled incorrectly or not at all. Maintenance can become difficult as rules multiply.

- **Classical ML model** – Train a model (such as a decision tree, random forest, or logistic regression) on historical data to make predictions. This is suited for structured data or well-defined classification/regression tasks. *Pros:* Can capture patterns more complex than simple rules and improve as it sees more data. Many libraries (scikit-learn, XGBoost) make this relatively accessible. *Cons:* Requires labeled data for training and careful validation. The model's logic may be less transparent than rules.

- **Deep learning or Large Language Model (LLM)** – For complex data like images, audio, or human language, more sophisticated models might be needed. NLP tasks (like understanding support ticket text) often benefit from pre-trained transformers or LLMs. An LLM (like GPT-style model) could even be used to interpret a request and decide on an action. *Pros:* These models can handle high-dimensional inputs and often achieve state-of-the-art accuracy, leveraging patterns learned from vast datasets. Using a pre-trained model can jump-start your agent's abilities with minimal task-specific data. *Cons:* They are resource-intensive (potentially requiring GPUs and more memory), and may be overkill if your problem is simpler. Also, relying on an external LLM API has implications for latency, cost, and data privacy.

- **Hybrid approaches** – Often the best solution mixes techniques. You might start with rules for absolute must-haves (e.g. "if transaction_amount > $10,000 and user is new, always flag for review" in a fraud system), and use an ML model for the nuanced decisions in between. Or use an ML model to generate a result and then a rule-based layer to vet or adjust those results (for example, a support agent might use an ML model to categorize tickets, but a rule to ensure any ticket mentioning "urgent" is never marked low-priority regardless of model output).

In deciding the approach, consider the nature of your problem and the data available. Is it a classification task (assigning a label or category)? A regression (predicting a numerical value)? A ranking or recommendation? Or perhaps a sequential decision process (which might hint at using reinforcement learning)? For most intelligent microservices dealing with single-step decisions – which is the majority – you'll likely choose between a rules-based system, a supervised ML model, or an LLM-powered solution.

In our WebkinsTech support agent scenario, the team evaluated a few approaches for ticket triage. They could have begun with a **simple rules engine**: for example, if a ticket's text contains the word "billing", assign it to the Billing team; if it contains "error" or "bug", send to Engineering, and so on. This would be easy to implement, but clearly too brittle and limited given the variety of ways customers describe issues. The more promising approach was **supervised learning** – training a text classification model using the company's past support tickets (which were already labeled by issue type from the support workflow). A machine learning model could learn to recognize patterns in phrasing that correspond to each category, beyond obvious keywords. The team also considered leveraging a **large language model**. For instance, a pre-trained LLM could potentially categorize tickets based on its general language understanding, even with minimal fine-tuning. This option was attractive because an LLM might understand subtleties (like "the app is freezing" implies a technical issue). However, relying on a massive external model raised concerns about **cost and latency** for each ticket processed, and data privacy since support tickets might contain sensitive information. In the end, WebkinsTech decided to **train a custom NLP classifier** for the task, using a moderate-sized pre-trained model (a BERT variant) fine-tuned on their ticket dataset. This gave them a balanced solution: the model was smart enough to handle the variability in support queries, but still efficient and fully under their control within the microservice.

There is no one-size-fits-all here. A fintech company building a fraud detection agent might combine a logistic regression model with a few hard rules for known fraud indicators, whereas a company building a chatbot agent might lean heavily on an LLM. The key is to choose an approach that you can implement and maintain given your team's resources, and that meets the requirements defined in Step 1.

Start Simple, Then Iterate

It's often wise to start with the simplest approach as a baseline. Implement basic rules or a simple model first to set an expectation for performance. This baseline not only gives you something to compare against when you introduce a more complex AI, but it might even solve more of the problem than expected. You can always increase complexity later if needed – for example, by upgrading a rules system to a learned model once you have more data or if the rules prove insufficient.

Step 3: Gather Data and Train the Model

If your intelligent microservice will rely on machine learning, data becomes the lifeblood of the project. In this step, the team needs to **gather the right data, prepare it, and train the AI model**, all while keeping an eye on validation to ensure the model will perform well on real-world inputs. This process typically involves several sub-steps: collecting and cleaning data, crafting features or encoding the inputs for the model, actually training the model on historical examples, and evaluating its performance on test data.

Data collection. Start by identifying and aggregating the data relevant to the task. For a supervised learning approach, this means assembling a labeled dataset (inputs paired with correct outputs). In our support ticket example, WebkinsTech pulled two years' worth of support tickets from their issue tracking system, along with the category each ticket was ultimately assigned to by the support team. They also included metadata like the product area and customer account type, thinking these might help the model. It's important to consider **quality** as much as quantity: the team filtered out spam tickets and merged duplicate reports so as not to skew the dataset. In some cases, you may need to generate labels – e.g. having people manually tag a subset of data – if historical labels aren't readily available.

Data preparation. Raw data often needs to be transformed into a form the model can learn from. Text may need cleaning (removing irrelevant parts, normalizing terminology) and conversion to numerical features (such as word embeddings). Numerical data might require scaling or normalization. It's also vital to split the data – typically into a **training set** (for the model to learn from) and a **validation/test set** (to evaluate generalization). WebkinsTech's team, for instance, set aside 20% of their collected tickets as a test set that the model would not see during training. This split ensures that when they later test the model's accuracy on routing tickets, they're evaluating it on unseen examples, simulating how it will perform in production.

Model training. With data in hand, the team can train the chosen model. Using the approach decided in Step 2, they selected a suitable algorithm and library (in this case, a pre-trained BERT model fine-tuned using a Python ML framework). They fed the cleaned ticket texts and their category labels into the training process, adjusting model parameters to minimize error. During training, it's common to monitor metrics on the validation set – for example, checking how accurate the model is at classifying tickets it hasn't trained on – to detect overfitting. After several iterations of training and hyperparameter tuning, the model usually reaches a point of diminishing returns.

Evaluation and validation. Once a model is trained, it must be evaluated objectively. WebkinsTech's support ticket classifier was tested on the reserved test dataset. The team looked at overall accuracy (what percentage of tickets were correctly routed to the right team) and more granular metrics like precision and recall for each category (to ensure, for instance, that it wasn't only good at identifying "easy" categories while failing on others). They also reviewed examples of misclassified tickets to understand failure modes – many of which turned out to be edge-case issues or ambiguously written queries. This kind of evaluation is crucial; it builds confidence that the model will perform adequately when integrated into the live system. If the model's performance isn't meeting the goal defined in Step 1, the team might decide to gather more data, try a different model type, or refine the feature processing and repeat this step.

Training a Fraud Detection Model

To illustrate the data and training process in a different context, consider WebkinsTech's Fraud Detection microservice. This agent's goal is to predict fraudulent transactions. The data science team collected two years of transaction records, each labeled as "fraud" or "legitimate" based on outcomes (chargebacks, investigations). This dataset was highly imbalanced – out of millions of transactions, only a few thousand were fraud cases – so the team used techniques like oversampling and careful cross-validation to ensure the model would learn the rare patterns indicative of fraud. They engineered features for each transaction, such as the transaction amount, the user's account age, device information, and purchase history. A gradient boosted trees model was trained on this feature set, as it handled tabular data well. After training, the model was tested on a holdout set: it correctly caught about 95% of fraudulent transactions (high **recall**), with a small false-positive rate (precision around 90%). The team also checked where the model made mistakes – for instance, it sometimes flagged very large legitimate purchases as fraud. Such insights helped them refine the model (in this case, by adding a feature for whether the user had a history of large purchases). By the end of the process, they had a validated model that gave them confidence to proceed to deployment in the Fraud Detection service.

It's worth noting that not all intelligent agents involve heavy model training. If you opted for a heuristic or rule-based approach, this step might instead involve **knowledge gathering** – for example, compiling a list of known fraud red flags from experts, or assembling an FAQ database for a support chatbot. If you're using a pre-trained AI (like an off-the-shelf vision model or an API like GPT-4), you might perform a light fine-tuning or simply evaluate it on your task with some test queries. The common thread is that you **prepare and vet the "intelligence" of the agent before deployment**. Whether that's through data-driven model training or through curating rules and knowledge, you should verify that the agent's decision mechanism works well on a dry run. Only then are you ready to wrap this intelligence into the microservice itself.

Step 4: Design the Service API and Interface

With a trained model (or decision logic) in hand, attention turns to the **microservice interface** – how this intelligent agent will interact with the rest of the system. Designing a clear API (Application Programming Interface) for your service is crucial so that other services or clients can easily send it data and consume its results. Essentially, you're deciding how to wrap the "brain" you've built inside a networked service boundary.

Start by determining the **communication pattern**. Will other services call this agent synchronously (e.g. via an HTTP request/response), or will it operate asynchronously (e.g. consuming messages from a queue or event bus)? The choice depends on your system's architecture and the nature of the task:

- **Request-response API:** Common in microservice architectures for interactive or immediate needs. For example, a RESTful HTTP API or a gRPC endpoint where a service sends a request (with necessary input data) to the agent service and waits for a response. This is suitable when the agent's result is needed immediately in a workflow (like getting a ticket routing decision while a user is submitting their support request).

- **Event-driven interface:** Suited for more decoupled scenarios. The agent service can subscribe to events (like "NewTicketCreated" or "PaymentProcessed") broadcast on a message broker. The agent does its analysis and then emits its own events or writes to a result topic (e.g. "TicketRouted" or "FraudScoreComputed"). This is useful when the processing can happen in the background or when multiple consumers might benefit from the agent's output.

Next, define the **data contract** – the input and output schema. What information does the agent service need to do its job, and what will it return? Keep it as minimal and structured as possible, but include all necessary context for the decision. In our support agent example, the service needs the text of the ticket and perhaps some metadata (like product area or customer ID if that influences routing). It will return the category and priority. The team decided on a simple REST POST endpoint for their service, accepting JSON. For instance:

```
POST /route_ticket
Request:
{
 "ticket_id": 12345,
 "issue_text": "I can't reset my password, it keeps giving an error.",
 "customer_tier": "Free"
}

Respons
{
 "ticket_id": 12345,
 "predicted_category": "Account Support",
 "predicted_priority": "Normal",
 "confidence": 0.92
}
```

This design makes it clear what the service expects and produces. The ticket_id is sent back in the response for traceability, the agent's predictions are given in fields (predicted_category and predicted_priority), and a confidence score is included so that the calling system could decide to treat low-confidence predictions differently (e.g. trigger a human review) if desired. When designing your API, consider including such a confidence or status indicator, as it provides a mechanism for downstream systems to handle uncertainty gracefully.

For asynchronous designs, the "API" is the event schema. For example, WebkinsTech's Fraud Detection Agent was built to listen to a Kafka topic of transaction events. The schema of those events (fields like transaction_id, amount, user_id, etc.) was its input contract. The agent then published a FraudAlert event with fields like transaction_id and fraud_risk_score if it detected something suspicious. No matter the mechanism, the principle is the same: clearly define what data comes in and what goes out.

It's also important to address **API considerations** typical to microservices: versioning, authentication, and error handling. If this agent will be consumed by multiple teams or external services, version your API from the start (e.g. /v1/route_ticket) so that future changes can be managed. Decide if the service needs to enforce authorization – in many cases, internal AI microservices are called by trusted systems within a network, but if not, you might integrate with your auth system. Plan for how the service signals errors or exceptions: for instance, if the AI model is unavailable or input validation fails, the API should return a clear error response (HTTP 500 or 400 with a message). In asynchronous setups, you might emit a separate error event or send the result event with an error flag.

By the end of this design step, you should have a well-defined interface specification. The team at WebkinsTech wrote down the API contract in the project wiki for the support agent service, including example requests and responses as shown above, so that developers integrating with it (like the team maintaining the front-end support portal or the orchestration service that creates tickets) knew exactly how to interact with the new intelligent microservice. This upfront clarity reduces integration issues later and ensures everyone has the same expectations for the service's behavior.

Step 5: Implement the Microservice and Integrate the Model

With the interface defined, it's time to build the microservice itself – essentially **writing the code that ties everything together**. This involves choosing the technology stack, setting up the basic service structure (endpoints, message handlers, etc.), and integrating the AI model or decision logic into that structure so that when the service receives a request or event, it uses the model to produce a result.

Choose a tech stack that fits both the AI component and your organization's microservice standards. Many teams use Python for AI-centric services because of its rich machine learning ecosystem (TensorFlow, PyTorch, scikit-learn, etc.) and convenient web frameworks like Flask or FastAPI. Others might use Java or C# for consistency with existing services, possibly calling out to a Python module or a model server. There's no one right choice – the goal is to reliably serve your model's predictions. WebkinsTech's team opted for Python and FastAPI for the support agent service, since their model was built in Python and they wanted to minimize any friction in loading and using it.

Integrate the model. Typically, you will load your trained model at service startup, so it stays in memory ready to handle requests. For example, the team packaged their fine-tuned BERT model as a file (say, model.pt) included with the service deployment. On startup, the service code loads this file into memory:

```
# Pseudocode for service setup
from fastapi import FastAPI
import torch
from model import TicketClassifier # your model class

app = FastAPI()

# Load the trained model once at startup
model = TicketClassifier()
model.load_state_dict(torch.load("ticket_classifier.pt"))
model.eval() # set model to inference mode

@app.post("/route_ticket")
def route_ticket(request: dict):
    text = request["issue_text"]
    # Use the model to predict category and priority
    category, priority, confidence = model.predict(text)
    return {"predicted_category": category,
            "predicted_priority": priority,
            "confidence": confidence}
```

In the snippet above (simplified for illustration), the model is loaded and prepared (with model.eval()) when the service starts. The endpoint handler route_ticket then uses model.predict on incoming data to generate a response. In a real service, there would be additional handling – for instance, translating raw model outputs (maybe numeric class IDs or probabilities) into the final labels and confidence. There might also be preprocessing steps (e.g., tokenizing the text before feeding it to BERT) and postprocessing (e.g., if confidence is below 0.5, you might set a field like "needs_human_review": true in the response). All of this logic gets encoded in the service implementation.

If your agent uses an external AI service or model (say, calling an OpenAI API or a remote model server), the integration will involve invoking that API from within your endpoint logic and handling timeouts or errors gracefully. The key is that **the microservice becomes the glue** that connects inputs to the AI decision and then outputs the result via the API we designed in Step 4.

Manage dependencies and resources. Ensure that any external resources the service needs are accessible. For example, if the model requires additional data (like a lookup table or a connection to a feature store), initialize those connections at startup as well. In the fraud detection agent, if we needed a fresh feed of user account status or a cache of high-risk IP addresses, the service would connect to those data sources when it starts. Also consider hardware requirements: does your model inference need a GPU or special library support (like CUDA)? If so, the service will need to run in an environment that provides those. In development, you might use CPU only, but configuration flags or environment variables can be used to enable GPU usage in production.

At this stage, it's wise to do some **basic testing** of the service in isolation. Run it on your local machine or a test server and send example requests (using a tool like curl or a REST client) to verify it's working end-to-end. For event-driven services, you can publish a test message to the queue and see if the service picks it up and responds as expected. WebkinsTech's engineers, for instance, ran the support agent service locally and sent a few sample ticket texts to the /route_ticket endpoint, checking that they got sensible category predictions back. This sanity check helps catch issues like missing model files, misconfigured dependencies, or mistakes in how input data is passed into the model.

Model Loading and Service Readiness

Load your AI model once during service startup, not per request. This avoids repeated initialization overhead and ensures faster responses. It's a good practice to test the model immediately after loading (for example, run a dummy prediction) and use logging or health check endpoints to signal that the service is ready. In a container orchestration environment, you might set a readiness probe that only passes after the model is successfully loaded into memory. This way, no traffic is sent to your service until it's fully prepared to handle requests.

Step 6: Test and Validate the Intelligent Service

Building the service is not the finish line – **rigorous testing** is essential before trusting an intelligent microservice in production. This testing phase has to cover two facets: the conventional testing of software (ensuring the service behaves correctly, handles errors, scales, etc.) and the AI-specific evaluation (ensuring the model's predictions are sound when running in the service and that the system handles edge cases in inputs or model outputs).

Software testing (functional tests). Start by verifying that the service API works as intended. Write unit tests for your service handlers: for instance, feed a sample input to the route_ticket function (perhaps with a stubbed or simplified model) and assert that you get a response with the expected fields and format. Test how the service handles bad inputs – if required fields are missing or if data is malformed, does it return a clear error? If you have non-ML logic (like certain rules or threshold behaviors), write tests for those as well. Also perform integration tests: deploy the service in a staging environment and simulate interactions with other components. For example, have a dummy application send actual HTTP requests to the service, or publish messages to the queue if it's event-driven, to ensure the end-to-end wiring works (network configs, message formats, etc.). These tests are similar to what you'd do for any microservice, ensuring that the intelligent service will not break the system by mis-communicating or crashing.

AI behavior testing. Traditional tests aren't enough because an AI model can be correct on average but still fail in specific ways. So, you need to validate the model's performance within the service context:

- **Reproduce model evaluation:** Confirm that the model still performs as expected when invoked through the service. Using the same test dataset from Step 3, you might call the service's API (perhaps through a test script) for each test instance and verify the responses. This ensures that any data preprocessing in the service or differences in the runtime haven't introduced errors. For example, WebkinsTech's team ran a batch of historical tickets through the running service and confirmed the accuracy was in line with the offline model's results.

- **Test edge cases:** Think of unusual or extreme inputs the service might encounter. What if a support ticket comes in empty, or with an extremely long description? Will the service handle it gracefully (perhaps by returning an appropriate error or truncating the text for the model)? What if the model is very uncertain and produces a low confidence? The service should perhaps flag that scenario. Also consider performance under load: does the service still respond quickly if many requests come in at once? This might involve load testing tools to simulate high traffic and ensure latency stays within acceptable bounds (keeping in mind that heavy model computations can be a bottleneck).

- **Human review and UAT testing:** For AI services, it's often useful to have domain experts do a manual review of the AI outputs on a sample of cases. In our support agent scenario, before fully deploying, the support team leads reviewed dozens of tickets and the categories the agent would have assigned. This qualitative check can catch any egregious mistakes or patterns (e.g., if all billing-related tickets were being misclassified as technical due to jargon the model learned incorrectly).

Shadow testing and gradual rollout. One prudent approach for a brand new intelligent service is to run it in "shadow" mode initially. This means the service operates in parallel to the existing process but its outputs aren't yet relied on; instead, they are logged for analysis. WebkinsTech did exactly this: they deployed the support routing service in shadow alongside the normal manual triage process. Each new ticket that came in was sent to the AI service **as well as** to the human support triage. The AI's recommendation was recorded, but it was the human decision that actually determined the routing. After a few weeks, the team compared the AI service's suggestions to the actual outcomes – this gave a clear picture of the model's real-world performance (e.g., the AI might have correctly routed 85% of tickets, and for the remaining 15% where it disagreed with the humans, the team examined those cases).

This **shadow testing** phase provided confidence that when they did enable the AI to automate routing, it would perform adequately, and it highlighted specific scenarios where the AI was making different choices. Those insights allowed further refinement (for example, adjusting the model or adding a rule to always escalate tickets containing certain keywords).

By the time the service was fully turned on, the team was already aware of its behavior and had mitigated the major risks.

In cases where an older automated system or heuristic is being replaced by a new AI, you might conduct an **A/B test** instead: send a fraction of live traffic to the new service while the rest goes to the old system, and compare metrics like accuracy, response time, or user satisfaction between the two. A/B testing is common for features like recommendation engines or search ranking algorithms – for example, you could direct 10% of users to see recommendations from the new agentic service while 90% see the status quo, and measure engagement differences.

Evaluate in production-like conditions. It's hard to fully predict an AI system's behavior until it sees real production data, which can drift from your training data over time. By testing in an environment that closely mirrors production (or through shadow/A&B tests in production), you can catch issues like the model being confused by a new product line or a seasonal trend that wasn't in the training set. It's also an opportunity to test the service's monitoring (ensuring it logs key metrics, which will be discussed in a later chapter on observability) and failure handling – for instance, if the model crashes or a bug is hit, does the service fail gracefully and alert the team?

Through this thorough testing regimen, the goal is to uncover and address problems before the intelligent microservice takes on a critical role. By the end of Step 6, you should have strong evidence (both quantitative metrics and qualitative assurance) that the service does what it's supposed to do and is robust under expected conditions. Only then is it wise to move on to a full production deployment.

Step 7: Deploy, Monitor, and Maintain the Service in Production

Finally, after thorough testing, the intelligent microservice is ready to be deployed into the production environment. Deployment of an AI-powered service is not very different from any other microservice, but there are a few extra considerations to ensure a smooth launch and the ability to evolve the service (and its model) over time.

Containerization and release pipeline. Most teams containerize microservices using

Docker (or similar) for consistent deployment. The support agent service was bundled into a Docker image that included the service code, the trained model file, and all necessary libraries (including the ML framework). This image can be run anywhere – the same image was used in testing, staging, and production, which eliminates "it works on my machine" issues. Set up a CI/CD pipeline so that whenever you have a new version of the service (for example, after improving the model or fixing a bug), it goes through automated tests and then can be deployed. WebkinsTech's pipeline, for instance, would build the Docker image, run the suite of unit/integration tests from Step 6, and if all passed, push the image to the registry and deploy it to a staging cluster for final verification. With a one-click (or automated trigger) approval, that image could then be rolled out to production. Automation like this ensures that updates to your intelligent service are delivered reliably and repeatably.

Gradual rollout and rollback. When deploying an intelligent microservice, especially one that will directly impact users or critical processes, it's wise not to flip the switch for everyone all at once. Techniques like **blue-green deployments** or **canary releases** are your friends. For example, the team might deploy the new AI routing service but initially route only a small percentage of tickets through it (the rest still follow the old path). They monitor the outcomes, and if all looks good (the agent is routing correctly and the support teams are happy with it), they gradually increase that percentage until the AI handles 100% of tickets. If at any point a serious issue is detected, they can quickly **rollback** to the previous state (in this case, disable the AI routing and revert fully to manual routing, or revert to an older model version if a new model was the issue). Having the ability to roll back easily is crucial – it means keeping the last stable version of your service (and model) readily available. In practice, this might mean not immediately discarding the containers running the previous model until you're confident in the new one, or versioning your model files so you can re-deploy the old model if needed.

Versioning the model. Unlike typical code-only services, an AI microservice has two evolving parts: the code and the model (which encapsulates learned knowledge). It's important to version them both. Use version numbers or IDs for your model artifacts (for example, "TicketClassifier v2.1"). Embed the model version in the service (the service could log the model version at startup and include it in an endpoint like /health or in a response header). This way, when looking at logs or results, you know which model produced them. Model versioning also allows you to deploy updates carefully: you might test "model v2.2" in staging while "v2.1" is in prod, and only switch over when ready. Some organizations use a model registry or store models in a centralized system (like an S3 bucket or MLflow server) – the service might load "the latest approved model" at startup. In any case, treat model updates with the same discipline as code updates: track what changed (data used, training methodology) and have a rollback plan.

Monitoring and logging. Once deployed, continuously monitor the service's health and effectiveness. At a basic level, monitor typical metrics like throughput (requests per second), latency (how quickly it responds), and error rates – these ensure the service is functioning well as software. Additionally, monitor AI-specific metrics if possible. For instance, you might log the model's confidence scores distribution, or the percentage of tickets the model marked as "uncertain." If you have the ability to later see the real outcomes (e.g., which category the support team actually ended up assigning), that can feed into a live accuracy metric. WebkinsTech set up alerts if the service started returning too many errors or if, say, the fraction of tickets categorized as "General" spiked (which could indicate the model is overloading a default category due to encountering unexpected input). Proper logging also helps debug any issues that arise – for example, logging a few examples of inputs and outputs (without sensitive details) can help diagnose where the model might be confused or if the input formatting from upstream is wrong.

Maintaining and updating the model. Deploying the first version of the model is not the end; models may need updates as new data comes in or as the business needs change. Plan for how you will retrain or refine the model and redeploy. Perhaps every quarter the team will retrain the support classifier with the latest labeled tickets to keep it up-to-date with new product terminology or support issues. Each retraining yields a new model file that goes through the same test/deploy pipeline. Over time, you might even develop an automated process for this (an **MLOps** pipeline). The important part is to close the loop: use what the service sees in production to make it better. (We'll explore continuous learning and feedback loops in a later chapter.)

Be Prepared for Model Drift

The accuracy of AI models can change over time. If the characteristics of support tickets evolve (say, after a new product launch, customers start asking about issues the model has never seen), the model's performance may degrade – a phenomenon known as data drift. It's critical to have a strategy to detect this (through monitoring or periodic evaluation) and update the model. Don't assume your current model will remain at peak performance indefinitely. Plan periodic re-evaluation, and retrain the model when needed to keep the service's intelligence fresh and reliable.

By taking these deployment and maintenance steps, WebkinsTech successfully rolled out their intelligent support agent as a reliable microservice. It was containerized, gradually introduced into the workflow, and set up with the necessary observability so the team could trust its operation. Moreover, they laid groundwork for the agent's ongoing evolution – ensuring that updates to the model or service can be made safely in the future. The result is a living, breathing microservice that can adapt and improve, rather than a one-off deliverable.

Checklist: Readiness for Deploying an Intelligent Agent Microservice

Before you roll out your intelligent agent microservice, run through this checklist to ensure you haven't overlooked any crucial steps:

- **Scope & Requirements Defined:** You have a clear definition of the agent's role and objectives, including how success will be measured (e.g. target accuracy or performance metrics).

- **Approach Selected:** You've chosen an AI/ML approach that fits the problem and the data available (whether it's rule-based, a specific ML model type, or an LLM-based method), and documented the rationale.

- **Data Prepared:** The necessary training data (or knowledge base) is collected, cleaned, and annotated. You've handled class imbalances or biases in the data appropriately.

- **Model Trained & Validated:** If using ML, the model is trained and achieves acceptable performance on a validation/test set. You understand its accuracy, and where it might have weaknesses. (For non-ML approaches, you've tested the logic against historical scenarios.)

- **API Interface Designed:** The microservice's input and output schema are specified (e.g. request/response JSON or event formats). Stakeholders consuming the service have agreed these meet their needs.

- **Service Implemented:** The service code is written and integrates the model/logic. The model loads correctly on startup and the service handles requests/events end-to-end in a development environment.

- **Basic Tests Passed:** Unit tests for the service logic are all green. You've done smoke tests with sample inputs to verify the service returns plausible results and handles error cases.

- **Extended Testing Completed:** You've conducted integration tests with other system components, tried edge-case inputs, and perhaps run a shadow test or A/B test trial. The service's performance and scalability (throughput/latency) have been evaluated under expected load.

- **Container & Deployment Ready:** The service is containerized or packaged for your target environment. Infrastructure (servers or Kubernetes, etc.) is prepared, and the deployment scripts or CI/CD pipelines are in place to deploy the service reproducibly.

- **Monitoring Set Up:** Logging and monitoring for the service are configured. You have dashboards or alerts for key metrics (service uptime, error rates, and possibly model quality indicators) to catch issues early in production.

- **Rollback Plan:** There's a strategy to disable or roll back the service if something goes wrong after launch (for example, ability to route traffic back to a previous system or to a prior model version).

- **Iteration Plan:** Post-deployment, you have a plan to gather feedback and data (such as collecting examples of the service's mistakes) and a process scheduled to improve the model or logic periodically.

If you can check off all (or most) of the above, you're in good shape to deploy your intelligent microservice. This checklist can act as a safeguard to make sure the excitement of AI doesn't cause skipping of essential engineering steps. Remember, a thoughtful, methodical approach is key to a successful deployment.

Tools and Frameworks for Agentic AI Microservices

While you can build an intelligent microservice from scratch (as we did step-by-step above), there is a growing ecosystem of frameworks and tools designed to simplify this process. These frameworks can save development time by providing pre-built components for common tasks – such as connecting to language models, managing prompt workflows, coordinating multiple agents, or serving models at scale. It's worth knowing the major options and understanding their strengths:

LangChain

LangChain is a popular open-source framework for building applications powered by large language models (LLMs). It provides a high-level interface to chain together prompts, models, and other components. In essence,

LangChain helps developers orchestrate complex LLM-driven workflows by composing modular pieces (LLM calls, memory/state handling, tool usage) . For example, with LangChain you can easily set up an agent that takes user input, retrieves relevant context from a vector database, and then queries an LLM

– all with just a few lines of code. LangChain has a large community and extensive documentation, making it a go-to for quickly prototyping conversational agents or GPT-based microservices. Its strength lies in flexibility: you can swap out models or data sources with minimal code changes . If your intelligent microservice involves a lot of prompt design, chaining of LLM queries, or needs features like memory (having the AI remember earlier conversations), LangChain provides those abstractions out-of-the-box.

Haystack

Haystack (by Deepset) is another open-source framework, focused on building search and question-answering systems. It excels at **retrieval augmented generation** – meaning it can take a user query, search through a document database, and then use an LLM or other model to generate a precise answer using the found information. Haystack provides pipeline components for things like document indexing, embedding generation, query filtering, and then integrates with generative models for output . If your intelligent microservice needs to answer questions from a knowledge base (e.g. an automated support FAQ agent) or perform semantic search, Haystack offers a production-ready solution. It's designed with performance and customization in mind – you can choose which database or search backend to use (Elasticsearch, FAISS, etc.), which model to use for retrieval and for generation, and it handles the flow between them. Developers praise Haystack for being **enterprise-friendly**: it has features for scaling, monitoring, and is used in many real-world QA applications. In short, LangChain is often about orchestrating LLM interactions (with some retrieval capabilities too), whereas Haystack is retrieval-first and ensures your AI's answers are grounded in actual data sources.

CrewAI

CrewAI is a newer entrant focusing on multi-agent systems. It's a lean Python framework to create and manage teams of AI agents working together.

Unlike LangChain or Haystack which often revolve around one primary LLM agent, CrewAI is built to orchestrate multiple specialized agents (which could be LLM-powered or rule-based) in a coordinated way. The creators describe CrewAI as a "lightning-fast" framework built from scratch, independent of other agent ecosystems . It introduces concepts like a *Crew* (which is like an organization of agents), and it supports event-driven workflows where agents can react to each other. For example, you might have an agent that acts as a Planner, another as a Solver, and another as a Validator, all working in concert on a task – CrewAI provides the scaffolding to set up such a system with roles and communication channels. One of CrewAI's strengths is finegrained control: you can customize how agents interact, and it's not tied to any particular model or tool (you can use any LLM or even non-ML logic within agents). It's also designed to integrate with real-world applications (for instance, connecting agents to databases or APIs). If your microservice's intelligence is naturally decomposed into multiple cooperating parts (say, an "analysis" agent and a "decision" agent), a framework like CrewAI can handle the heavy lifting of messaging, scheduling, and error handling between them.

AutoGen

AutoGen is an open-source framework from Microsoft that also targets multi-agent AI applications. It is particularly geared towards agents that communicate via conversational language to collaborate on tasks . AutoGen provides high-level abstractions to create agents that use LLMs (and can also incorporate human input or tools) and define how these agents converse and coordinate. One of AutoGen's goals is to simplify the programming of complex agent behaviors by treating inter-agent conversation as a first-class programming paradigm. Under the hood, it has an event-driven core (for performance and scalability) , but as a developer you can mostly think in terms of "Agent A says X, then Agent B responds with Y, using tool Z if needed." For instance, you could use AutoGen to implement an assistant agent and a developer agent that talk to each other to write and debug code (a simplified version of a GPT pair programming scenario). AutoGen comes with support for using multiple LLMs simultaneously and integrating external tool APIs, which means each agent could have different capabilities. It's a powerful framework when you need a lot of **agentic conversation structure** – e.g., running a simulation of two AI agents negotiating or brainstorming. Given that it's backed by Microsoft, it also plays well with Azure's ecosystem and is under active development to incorporate new multi-agent patterns. The trade-off is that it can be a bit complex to learn, but it shines in scenarios where an orchestrated dialogue between AI entities (or AI and humans) is central to the application.

LlamaStack

LlamaStack (often stylized as LLaMA Stack) isn't about multi-agent orchestration like the above; instead, it's a toolkit provided by Meta for deploying and managing large language model services, especially those based on Meta's LLaMA models. Think of LlamaStack as a ready-made backend for an LLM-powered microservice. It encompasses components for running inference, handling chat conversations, generating embeddings for searches, and even enforcing safety filters (via something called Llama Guard) . Meta's goal with

LlamaStack is to make it easier for developers to **self-host** advanced language models in a cloud-agnostic way. For example, if you want to run a ChatGPT-like service on your own servers using LLaMA 2 models, LlamaStack provides the building blocks – you can deploy its Docker containers or use their CLI to set up an API that applications can call. It's designed to handle scale (multiple concurrent requests, batching, etc.) and provides a standard API interface to your model. LlamaStack is especially attractive to organizations that prefer not to rely on external APIs for LLMs due to cost or privacy, or that want finegrained control over the model's behavior (via customization and guardrails). It's somewhat analogous to using TensorFlow Serving or TorchServe, but specifically tailored for modern LLM use cases and with an eye towards a full **microservice architecture** (each component like inference, memory, safety is a microservice in the stack). Using LlamaStack, one could build an intelligent microservice that leverages a powerful LLM entirely on their own infrastructure, with Meta's tooling ensuring it's manageable.

Other Tools

The above frameworks are among the most talked-about in 2025 for agentic AI systems. There are, of course, many others: for instance, **LlamaIndex** (formerly GPT Index) which is often used to connect LLMs with external data sources, or orchestration platforms like **LangFlow** that offer a visual interface for designing LLM workflows. If your intelligent microservice is centered on classical ML models (not LLMs), you might look into modelserving frameworks such as **TensorFlow Serving**, **TorchServe**, or **KFServing (KServe)** for deploying models at scale. Additionally, for managing the lifecycle of models (tracking experiments, versions, metrics), tools like **MLflow** can be integrated into your pipeline. While not specific to "agentic" AI, they complement the deployment of intelligent services. The landscape is evolving rapidly – new frameworks continue to emerge. When starting a project, it's wise to check the latest options and community adoption. The right tool can drastically reduce boilerplate and let you focus more on the unique logic of your agent. (Appendix A of this book provides a curated list of tools and frameworks with more detailed comparisons.)

In summary, frameworks like LangChain and Haystack help accelerate development of common patterns (chaining LLM calls or retrieval-based QA), while multi-agent frameworks like CrewAI and AutoGen help structure more complex agent ecosystems. LlamaStack and similar stacks aim to simplify the deployment and serving of powerful models. Choosing to use a framework depends on your specific needs: if your use-case neatly fits what a framework offers, you can save a lot of time; if not, you might build custom. Importantly, these tools are not mutually exclusive – you might use LangChain inside your service for prompt management, and also use LlamaStack to serve the model, for example. Being aware of these capabilities allows you to make informed design decisions and leverage the AI community's collective innovations, rather than reinventing the wheel for your intelligent microservice.

Conclusion and Next Steps

Designing and deploying an intelligent agent as a microservice is a journey that blends the art of machine learning with the rigor of software engineering. We started with an idea – letting an AI agent handle a support ticket backlog – and broke it down into concrete steps: define the goal, pick the right approach, train the model, build the service around it, test it thoroughly, and roll it out carefully. Along the way, we saw the importance of collaboration between roles (data scientists, engineers, product owners) and the need to treat an AI project with the same discipline as any mission-critical software project. The end result is not just a smart model, but a living service that integrates into a business's ecosystem and delivers real value.

Insight from the Field

"Building our first intelligent microservice taught us that models and code must evolve together. It was challenging, but the payoff was worth it."

— Lead ML Engineer at WebkinsTech, after deploying the support agent service

By now, you have a blueprint for implementing a single intelligent microservice. The next challenge is scaling this approach beyond one service. In a real-world system, you may have multiple AI-driven services – and they often need to work in concert. How do these agents coordinate their actions? How do you design the workflows between them? This is where *orchestration* and *choreography* of intelligent microservices come into play. In the next chapter, we'll delve into orchestration patterns for coordinating multiple intelligent agents, exploring how to make a collection of AI services collaborate smoothly to achieve larger goals. Building one intelligent microservice is a big step; orchestrating many of them effectively is the next – and you're about to learn how to master that as well.

Chapter 11: Orchestration and Choreography in Agentic Microservice Architectures

Introduction

Microservices, each enhanced with intelligent, agent-like capabilities, are like talented musicians in an ensemble. Individually, they can do amazing things – but to achieve larger business goals, they must **work in concert**. This raises a crucial question: **who (or what) directs the show?** In traditional microservice design, two architectural styles have emerged for coordinating services: **orchestration** and **choreography**. Orchestration relies on a central "conductor" to tell each service when to act, whereas choreography is more like a free-flowing dance where services react to events and each other without a single point of control.

In this chapter, we dive deep into orchestration vs. choreography within the context of *agentic microservice architectures*. We will define each approach in plain terms (and even a musical analogy or two), compare their differences, and weigh pros and cons. More importantly, we'll explore how the addition of intelligent agents – services with AI-driven behavior – **changes the dynamics** of both orchestrated and choreographed systems. Are our microservices going to follow a scripted symphony, or improvise like jazz players? We'll use real-world scenarios (from a **loan approval workflow** to a **smart power grid**) and a running case study at **WebkinsTech** to illustrate these concepts in action. Along the way, we'll examine the tooling landscape: from orchestration frameworks like Temporal and Argo to event-driven platforms like Kafka and NATS, and how they support (or sometimes challenge) agent-based architectures.

By the end of this chapter, you'll have a firm grasp on when to choose orchestration, when to favor choreography, and how a **hybrid approach** can often marry the best of both. We'll highlight emerging patterns for coordinating intelligent agents (think of self-organizing service "swarms"), and provide practical checklists to guide your decisions. As always, our tone will be engaging yet informative – consider this a friendly masterclass in making your microservices **play together harmoniously**, whether under a strict conductor or in an improvisational dance. And fear not: whichever approach you choose, we'll set the stage for the next challenge after coordination – observing and learning from this complex ecosystem (the focus of our next chapter).

So, let's strike up the band! It's time to explore the art of orchestration and choreography in the realm of intelligent microservices.

Orchestration vs. Choreography: The Basics

When coordinating multiple microservices in a workflow, **two fundamental modes** of interaction emerge:

- **Orchestration** – a **centralized** approach where one service (an *orchestrator* or controller) **commands each step** of the process.
- **Choreography** – a **decentralized** approach where there is **no central coordinator**; instead, each service **listens for events** and reacts autonomously, resulting in an emergent sequence of actions.

Think of orchestration as a well-rehearsed **symphony** with a conductor, and choreography as an energetic **dance** where each performer responds to the music and to each other. Let's break that analogy down:

Orchestra vs. Ballet Analogy: In orchestration, a central conductor cues each musician (service) on when and what to play. In choreography, each dancer (service) knows the choreography and responds to the music and movements around them, without a single director calling the shots.

In practical terms, an orchestrator contains the workflow logic: it invokes Service A, then Service B, then Service C in order, handling responses and errors along the way. By contrast, in choreography, when Service A is done with its part, it doesn't call the next service directly. Instead, **Service A emits an event** (e.g. "Task A completed" or "Order Placed"), and whichever services are interested in that event (Service B, Service C…) will react to it. Those services may in turn emit further events ("Payment Processed", "Inventory Updated") that other services listen to. The **workflow emerges** as a chain reaction of events, rather than being explicitly defined in one place.

Illustration: Coordination Patterns

To visualize the difference, consider a simple workflow of processing an online order through multiple steps (order entry, payment, inventory update, shipping, confirmation email). We can coordinate these steps via orchestration or choreography:

- **Orchestration Example:** A dedicated **Order Orchestrator** service calls each step in sequence:

1. Orchestrator calls **Order Service** – to validate and create the order.
2. Then calls **Payment Service** – to charge the customer's card.
3. Then **Inventory Service** – to allocate or update stock.
4. Then **Shipping Service** – to schedule delivery.
5. Finally **Email Service** – to send an order confirmation message.

```
[Order Orchestrator] -> [Order Service]
                     -> [Payment Service]
                     -> [Inventory Service]
                     -> [Shipping Service]
                     -> [Email Service]
```

The orchestrator waits for each service to finish before moving on, making sure the flow happens in the right order. It's **step-by-step, centrally controlled**.

Diagram: Orchestration – a central orchestrator directs calls to each service in turn.

Choreography Example: There is **no central orchestrator**. Instead, the **Order Service** does its work and then **publishes an event** say, **"OrderPlaced"** on an event bus. All other services subscribe to relevant events:

- The **Payment Service**, upon hearing an "OrderPlaced" event, automatically attempts to process payment for that order.

- The **Inventory Service** also listens for "OrderPlaced" and reserves stock for the order.

- The **Email Service** listens for "OrderPlaced" and sends a confirmation email to the customer.

- Meanwhile, when Payment is done, the Payment Service might emit a **"PaymentProcessed"** event. The Shipping Service could be waiting for that – upon seeing "PaymentProcessed", it initiates shipping since payment succeeded.

Inventory might also listen to adjust stock only after payment confirmed, etc.

In this way, **each service reacts** to events **asynchronously**:

```
[Order Service] -- emits "OrderPlaced" --> (Event Bus
    ├──> [Payment Service] (on OrderPlaced)
    ├──> [Inventory Service] (on OrderPlaced)
    └──> [Email Service] (on OrderPlaced)

[Payment Service] -- emits "PaymentProcessed" --> (Event Bus
   └──> [Shipping Service] (on PaymentProcessed)
```

Diagram: Choreography – services communicate via events on a common bus. There is no single controller, just a web of events connecting independent services.

In the choreographed model, **no single service knows the entire workflow**. Each just knows, "When event X happens, I should do Y, then maybe emit event Z." The overall business process is the sum of these interactions. It's akin to dancers following the music and cues from each other rather than being told explicitly when to move.

Both approaches achieve the goal (in our example, an order gets fully processed), but they do so very differently. Neither is "better" in all cases; each has its **strengths and trade-offs**, which we'll explore next.

Pros and Cons: Central Control vs. Emergent Behavior

Just as a conductor-led symphony and an improvisational dance troupe have different advantages, orchestration and choreography come with distinct **pros and cons**. It's important to understand these trade-offs, especially as we consider adding AI-driven behavior into the mix.

Orchestration – Strengths: Having a central coordinator provides **clarity and control**. The entire workflow is defined in one place (the orchestrator), which makes it easier to understand the big picture. Error handling can be centralized – the orchestrator knows if a step fails and can decide on a compensation or rollback action (much like a conductor noticing a violinist missed a note and signaling to adjust). For developers, orchestration can simplify reasoning: you often have a **single point of truth** for the process. In fact, companies like Netflix built their own orchestration engines (e.g. Netflix Conductor) to manage complex microservice workflows with reliability . Orchestration also shines when **strict order and timing** are important – it enforces sequencing (do A then B then C) by design.

Orchestration – Weaknesses: The flip side of central control is **central dependency**. The orchestrator can become a **single point of failure or bottleneck**. If it goes down, the whole workflow stalls. Under high load, the orchestrator must scale or it could slow everyone down. Orchestrators can also become complex as they grow – that one workflow brain might turn into a spaghetti of conditionals as new paths are added. This can reduce flexibility: any change in the process (like adding a new service or reordering steps) means updating the orchestrator's logic and deploying it. In large organizations, this can become a governance choke point ("all integration must go through the central engine"). In short, orchestration trades some agility for structure.

Choreography – Strengths: Choreography embraces **loose coupling and flexibility**.

Services don't call each other directly; they only emit and listen to events. This makes it easy to **add new services** or change behaviors without touching a central authority. For example, if you want to add a new step to a process, you can simply have a new service subscribe to an existing event – no other service needs to even know. It's *"like adding another dancer who already* *knows the dance, rather than rewriting the conductor's sheet music for a new musician,"* as one engineer quipped. In event-driven systems, services are more autonomous and **scalable** in their own right: they can be developed and deployed independently as long as they follow the agreed-upon event contracts. This decoupling can lead to faster evolution of the system – teams can build and deploy new capabilities (new event consumers or producers) without orchestrator coordination. The overall architecture can handle **dynamic, complex interactions** because it's not locked into a single linear flow.

Choreography – Weaknesses: If orchestration is a monologue, choreography is a cacophony – and it can be **hard to follow the plot**. The biggest challenge is **lack of global visibility**: no single service knows the whole story, which makes understanding and debugging the system difficult. If something goes wrong in a choreographed system (say an order never got shipped), you have to trace through a chain of events across multiple services to find where the breakdown happened. Error handling and transaction management are distributed – there isn't one component orchestrating a rollback; each service must handle failures in isolation, or you need a clever event-based saga pattern to compensate. This can lead to **complex debugging and emergent behaviors** that weren't anticipated . Also, without careful design, an event-driven system can suffer from *event storms* or cycles (where services keep triggering each other in loops). In terms of performance, while each service is decoupled, the **event broker** or message bus becomes critical – it must handle high throughput and deliver events reliably, or the dance falls apart.

Let's summarize some key differences in a quick comparison table:

Aspect	**Orchestration (Centralized)**	**Choreography (Decentralized)**
Control	Central coordinator dictates the workflow steps.	No central controller; each service reacts to events.
Coupling	Tighter coupling of sequence (workflow is explicitly defined in one place).	Loosely coupled – services only know of events, not each other's internals.

Scalability	Orchestrator can be a bottleneck if not scaled; one more component to scale (the orchestrator itself).	Naturally horizontal – services operate independently; event broker must scale to handle load.
Flexibility	Changing the process requires updating the orchestrator's logic. Adding a new step involves orchestrator coordination.	Easier to extend: add new services by subscribing to relevant events, no need to modify a central workflow .
Aspect	**Orchestration (Centralized)**	**Choreography (Decentralized)**
Fault Tolerance	Single point of failure risk at orchestrator (mitigated by HA setups). Centralized error handling (can coordinate rollback/saga logic in one place).	No single brain to fail; however, failure handling is distributed (each service must compensate locally, harder to do global rollbacks).
Monitoring & Debugging	Easier to trace end-to-end: one service (orchestrator) has the whole picture (aided by centralized logs/traces).	Harder to trace – need aggregated tracing across services to reconstruct the story from events. Observability is crucial (as we'll explore next chapter).
Behavior	Predictable, predefined flows. The orchestrator enforces order and logic – no surprises (assuming orchestrator is bugfree!).	Emergent behavior possible. The overall outcome is the sum of many independent decisions – can be adaptive, but also unpredictable if not carefully designed.

Team Autonomy	Orchestrator team controls the big picture; service teams follow the orchestrator's API/commands. Clear owner of process logic.	Service teams have more autonomy to introduce new behavior via events. No central "gatekeeper" of workflow logic, but requires discipline in defining events and reactions.

As the table suggests, **orchestration** offers more control and simplicity in understanding the workflow, whereas **choreography** offers more flexibility and scalability at the cost of complexity in understanding system behavior.

Now, into this mix we add intelligent, agent-like microservices. How do these pros and cons shift when services can make smarter decisions or even modify their behavior over time? Before answering that, let's ground ourselves with a couple of concrete scenarios illustrating orchestration and choreography in action.

Orchestration in Action – Online Loan Approval: Imagine an online loan application system. When a customer applies for a loan, an AI-powered Orchestrator service coordinates a series of checks: it first calls a Credit Score Service, then invokes an Income Verification Service, then an AI Risk Assessment service to evaluate the application. Each step feeds into the next. Finally, the orchestrator aggregates the results (credit score, verified income, risk score) and decides to approve or reject the loan. The central orchestrator ensures these steps occur in order and within set time limits (e.g., if verification takes too long, it triggers a timeout path). This approach makes it easy to audit the process (the orchestrator logs each decision) and to enforce business rules – for instance, if the AI risk model flags high risk then skip straight to rejection without waiting for other steps. The trade-off is that the orchestrator is doing a lot of heavy lifting. If one of the services changes (say the risk model now requires an additional input), the orchestrator's code needs updating. Still, for the loan team, having a single "brain" coordinating the workflow gives confidence that nothing falls through the cracks.

Choreography in Action – Smart Power Grid: Consider a smart electrical grid with many distributed energy resources. In a region, you have smart thermostats, solar panel controllers, battery storage units, and grid supply monitors – each is a microservice (or an IoT agent) managing a piece of the energy puzzle. There is no single controller telling devices what to do. Instead, they operate via event choreography: e.g., each device broadcasts its state and needs (Thermostat: "I need 5kW more cooling", SolarPanel: "I have 3kW excess", Battery: "I can supply 2kW"). All devices listen on the event bus for these messages. A thermostat, upon hearing that the grid is strained, might lower its demand autonomously. Batteries might discharge when they hear that price signals or energy need events are high. The grid balance emerges from these many-to-many interactions. This is choreography on a grand scale – effectively a dance of energy signals. The advantage is adaptive, real-time balancing without requiring a central brain that could be a single point of failure for the whole grid. The downside? It's complex! If something goes wrong (say a subset of batteries mis-report their capacity), understanding the chain of events leading to a failure requires analyzing logs from many devices. The system's global behavior is an emergent property of all these independent agents. This kind of choreographed multi-agent system scales extremely well (it can accommodate thousands of devices joining or leaving), but it demands robust monitoring to ensure the "dance" stays in sync.

These examples highlight how orchestration and choreography play out in different domains. The **loan approval** scenario benefits from orchestration's control (regulatory and audit needs, clear decision points), while the **smart grid** leverages choreography's flexibility (devices self-adjusting to current conditions). Many systems will lie somewhere in between these extremes, often using a bit of both approaches. Next, we'll explore how adding intelligence – turning services into more autonomous agents – influences our choice of coordination and what new patterns emerge.

Intelligent Agents and Workflow Coordination

So far, we've considered services as relatively fixed in behavior – they perform their function when told (or when an event arrives). Now introduce **agentic behavior**: services that incorporate AI models, learning, or decision-making logic that can change over time or based on context. How does this affect orchestration and choreography?

First, it's worth noting that **intelligence can live in the coordinator, in the participants, or both.** In an orchestrated setup, the orchestrator itself could be a smart agent. In a choreographed setup, each service could be an intelligent agent. You can even have a mix (some intelligence centrally, some at the edges).

Smarter Orchestration

In an orchestration model, an **AI-powered orchestrator** might do more than just follow a static script of "Step A then B then C." It could make **dynamic decisions**: for instance, skip a step if it predicts it's not needed, or run steps in parallel if a learning model suggests it's more efficient. The orchestrator could also adjust the workflow in real-time based on outcomes – e.g., if an AI risk assessment returns a high risk, the orchestrator might **insert an extra verification step** or route the case for human review. Essentially, the orchestrator becomes an intelligent **decision agent** that not only coordinates but also optimizes and adapts the process.

One powerful possibility is using AI planning or reinforcement learning for orchestration. Instead of a fixed sequence, the orchestrator agent could have a **goal** (e.g., "fulfill this order successfully") and it can **plan a sequence of service calls** to achieve it. If one path fails, it can explore alternatives. This starts to blur the line between traditional workflow engines and AI planners. It's not science fiction – research and some cutting-edge systems are looking at letting AI figure out workflows on the fly.

However, injecting AI into an orchestrator also inherits the downsides of centralization. If the AI makes a bad call, the whole workflow follows that bad call. For example, if an AI orchestrator incorrectly predicts that a payment service is likely to fail and skips it, you have a serious business error! We gain adaptability but still have that single point of decisionmaking. Thus, **safety nets** and careful validation are needed. One could imagine a future orchestrator that explains its decisions or is monitored by another "watchdog" agent to ensure it doesn't go off the rails.

On the practical side, modern orchestration frameworks are already evolving to handle more dynamic workflows. Temporal, for instance, touts a **"durable execution"** model ideal for long-running AI-related workflows that might involve many services and human interactions over time . Companies like Nvidia have integrated Temporal to coordinate the deployment of AI models and agents across their platform – effectively treating the orchestrator as the backbone for complex AI-driven processes. This shows that orchestration tools are being **actively used to wrangle intelligent microservices** at scale.

Insight: An AI-driven orchestrator can be thought of as a **"central brain" with reflexes** – it can react to inputs (service responses, predictions) and make judgment calls. It's like moving from a conductor who follows a score to a conductor who can improvise based on the audience's mood. The music might be more interesting, but it relies on that conductor's talent and intuition.

Smarter Choreography

In a choreographed system, each service could have its own AI logic, making it an **intelligent agent** in a community of agents. Here, the overall workflow truly becomes **emergent** and potentially self-optimizing. Each agent-service makes local decisions: what events should I emit? How should I react to incoming events? Should I ignore some events under certain conditions? Possibly, agents could even **negotiate** with each other via events (for example, two inventory services might coordinate to decide who will fulfill an order based on current stock and predicted demand).

Choreography with intelligent agents can lead to **remarkable flexibility**. Suppose each service has a machine learning model that allows it to predict certain outcomes or detect anomalies. They might decide not to emit an event if they predict it's not useful, or they might trigger extra events if something seems off. For instance, imagine a set of microservices handling trading in a stock market. If each service (pricing, risk, audit) is smart, they might collectively prevent a catastrophe by individually deciding to halt trading when their local models sense a flash crash pattern, even without a central circuit breaker telling them to.

However, this power comes with **increased unpredictability**. Now you don't just have a complex event mesh to understand, but each node in that mesh has *changing logic*. The behavior of the system can drift as models retrain or as agents learn from new data. Today your services might choreograph perfectly, but tomorrow, after an update to an AI model, you might observe new emergent behavior that wasn't manually programmed. This is both exciting (the system can adapt to new situations) and scary (it might do things no human anticipated).

One big consideration in intelligent choreography is **maintaining alignment with business goals**. Each agent may be optimizing for something (e.g., a Shipping service might be optimizing for cost, an Inventory service for stock efficiency). If their incentives or models aren't aligned, you could get undesirable outcomes. This is analogous to a team where each member is super smart but if they don't share the same game plan, they might work at cross purposes.

Emergent Chaos: When multiple intelligent agents interact through choreography, *small bugs or divergences can amplify*. For example, if one service's ML model starts misclassifying a normal event as an anomaly, it might broadcast a "panic" event. Other agents, reacting to that, could enter fallback modes or also broadcast alerts. Very soon, the whole system could spiral into an unnecessary failsafe mode – a phenomenon one team jokingly called *"the Skynet sneeze."* In choreographed systems, always have robust monitoring and a way to trace event flows – emergent behavior can be hard to debug and even harder to predict beforehand. Logging, distributed tracing, and even simulation testing of multi-agent scenarios are your friends here 【6†L327-334】.

Despite the challenges, intelligent choreography is appealing for systems that need to be extremely adaptive. Nature works in choreographies – think of ant colonies or flocking birds where simple local rules yield complex, resilient group behavior. We're inching towards that with self-driving microservices. But to get there in a reliable way, architects often introduce some hybrid controls or constraints, which leads us to consider mixing orchestration and choreography.

Tools and Frameworks for Service Coordination

Before jumping into hybrid patterns and advanced strategies, let's survey the **technology landscape** that enables orchestration and choreography. Knowing the tools at your disposal helps in designing solutions (and sometimes the tool influences whether you lean toward one approach).

Orchestration Tools (Workflow Engines & Sagas)

Several frameworks and platforms have risen to make orchestrating microservices easier. These typically provide a way to define workflows (either through code or a visual model) and handle concerns like retries, timeouts, and state tracking for long-running processes.

Here are a few notable ones:

- **Temporal:** An open-source workflow orchestration platform that lets you write workflows as code. Temporal (a successor of Uber's Cadence) handles the heavy lifting of state persistence and retries, so you can write an orchestrator function that *looks sequential* (call A, then B, then C), but under the hood Temporal will ensure if a step fails or the process crashes, it can continue where it left off. It's designed for reliability and **"durable execution"**, making it popular for **microservices that interact with databases, queues, and even human tasks over long periods** . Temporal has been used by companies like Stripe and Netflix for things like payment workflows. Notably, it's becoming a backbone for complex AI workflows too – for example, **Nvidia uses Temporal** in its AI model deployment platform . If you have many moving parts and need an orchestrator that **never gives up on a workflow**, Temporal is a strong choice.

- **Argo Workflows:** A Kubernetes-native workflow engine, part of the Argo project. It lets you define workflows in YAML (as DAGs – Directed Acyclic Graphs of steps) and runs them on Kubernetes. Each step can be a container. It's often used for CI/CD pipelines, data processing tasks, and ML pipelines (it's a core of Kubeflow Pipelines for machine learning workflows). Argo is great when your services (or tasks) can be packaged as containers and you want to leverage Kubernetes to run steps at scale. While not originally designed for orchestrating *business microservices* in the request/response sense, it's very powerful for orchestrating batch processes, and can be adapted for long-running service orchestration as well. Because it runs on Kubernetes, it naturally scales out and survives failures by rescheduling steps.

- **Apache Airflow:** A well-known orchestration platform (created at Airbnb) for data workflows. Airflow uses Python code to define DAGs of tasks and has a scheduler that runs those tasks with dependencies. It's more at home in **ETL jobs, analytics pipelines, and periodic workflows** than in high-throughput microservice transactions. That said, one could orchestrate microservice calls via Airflow operators (HTTP calls, etc.), but you'd be using a cannon to shoot a fly if it's a simple real-time flow. Airflow shines when you have complex batch processes (like "every night, do X then Y then Z, and fan-out to N parallel tasks, etc."). In the context of intelligent services, Airflow might be used to orchestrate training workflows or batch scoring jobs rather than user-facing request flows. It provides nice monitoring of workflows and retry/failure logic, but it's not typically in the critical path of user transactions.

- **Camunda Zeebe:** Part of Camunda's platform, Zeebe is an orchestration engine that allows you to design workflows using **BPMN (Business Process Model and Notation)** diagrams. It's tailored to microservices orchestration with an emphasis on **visibility and monitoring of business processes**. Developers can draw a flowchart (with tasks, gateways, etc.), and Zeebe will execute it, calling microservices via adapters or messaging. It's like having a classical BPM system reimagined for cloud microservices. Zeebe (and similar BPMN engines) can be very useful when business stakeholders want to see a visual of the process or when you need to integrate with legacy human workflows. With intelligent services, one could imagine modeling a high-level process (say, an order fulfillment) in BPMN, and at certain steps an AI decision is made (embedded either in the engine via script or by calling an AI service). One advantage here is the **auditability** – the workflow state can be inspected, which helps in debugging and compliance. The trade-off is potentially less development agility (drawing diagrams vs coding) and the need to fit dynamic behavior into a predefined model (though BPMN has ways to handle event-driven actions as well).

- *(Honorable mentions)*: **Netflix Conductor** (an open-source orchestration engine from Netflix, similar in spirit to Temporal and geared towards microservice orchestration at scale), **AWS Step Functions** (a fully managed service for orchestration in the AWS cloud, great for integrating Lambda functions and AWS services into a workflow), and others like **JBoss jBPM** or **Apache Camel** for various orchestration needs. Each comes with its ecosystem and best-use cases. For instance, Step Functions are great if your architecture lives within AWS and you want a quick way to chain services with minimal code. Netflix Conductor is JSON-based and highly extensible, used internally at Netflix for things like media processing pipelines.

How these apply to agentic architectures: When your microservices are intelligent or you need to orchestrate AI-driven tasks, these tools can serve as the "central brain" or backbone. For example, an AI orchestrator built on Temporal might manage a **conversation with a user**, calling different NLU (Natural Language Understanding) and fulfillment microservices, maintaining state over minutes or hours – something Temporal excels at (it won't lose track even if components crash). An Argo workflow could encapsulate a retraining pipeline: when an AI agent determines it needs to retrain a model, it triggers an Argo workflow that spins up jobs for data prep, model training, and deployment in sequence. Zeebe might be used by a bank to orchestrate a loan approval that involves both AI scoring and human review steps, all modeled in one coherent process diagram. The key is that these orchestration tools **simplify building a complex sequence** of actions that might involve decisions, retries, time delays, and state – all things that a naive implementation would struggle with. As AI introduces uncertainty (e.g., variable execution times, alternate branches, learning loops), having a sturdy orchestration framework helps ensure the whole doesn't fall apart. Indeed, as AI-driven microservices take center stage, orchestration platforms like Temporal have become vital to manage the "complex web of integrations" they bring .

Event-Driven Choreography Tools (Messaging & Event Buses)

For choreography, the focus shifts to the **plumbing that allows services to exchange events reliably**. Here we talk about messaging systems, event logs, and pub/sub platforms. The good news is there are mature tools in this space, each with its flavor:

- **Apache Kafka:** The poster child of event-driven architectures. Kafka is a distributed log where services can publish events (to a topic) and others can subscribe or replay from that log. Kafka is designed for **high throughput and persistence**, meaning your events are durably stored (you can even rewind and re-read them, enabling event sourcing patterns). In a microservice choreography context, Kafka acts as the central **event bus** – services communicate by writing and reading Kafka topics rather than calling each other. Kafka's strong ordering and retention capabilities make it popular for systems where event history is important (for example, an **AI training service** might consume historical events to learn patterns). Companies like LinkedIn (which created Kafka), Netflix, Uber, and many others rely on Kafka to decouple services via events. The ecosystem (Kafka Streams, ksqlDB, etc.) allows building real-time data processing on those events as well. The **downside** could be its operational complexity (running a Kafka cluster) and the fact that consumers have to handle events in an idempotent way (since replays or duplicates can happen). In an agentic architecture, Kafka is often the **backbone for event-driven communication**, and intelligent services might use it to publish events like "anomaly detected" or subscribe to events like "user logged in" to trigger some AI-driven action. It provides the **nervous**

system for a choreographed system, ensuring messages get from point A to B.

- **NATS:** A lightweight, high-performance messaging system that supports pub/sub, request/reply, and other patterns. NATS is known for its simplicity and speed – it's a perfect fit when you need messaging with low latency and you don't want the heavy machinery of Kafka. NATS by default doesn't persist messages (though **NATS JetStream** adds persistence and streaming capabilities), which means it's great for communications where you only care about real-time delivery. For microservices choreography, NATS can serve as an **event bus** or simply a broker for commands and events in a fire-and-forget style. Because it's light, it's often used in **cloud-native deployments** (it's a CNCF project) and is easy to embed in distributed systems. For example, an AI-driven IoT system might use NATS to disseminate sensor readings and control signals between device services because of its efficient binary protocol and small footprint. In an agentic microservice world, NATS could be the medium through which a swarm of services chat with each other at high speed (imagine a swarm of AIdriven trading bots all publishing their bids and offers – NATS would handle that with ease). The trade-off is if you need durability or the ability to replay events, plain NATS isn't enough (you'd use JetStream or another storage).

- **Amazon EventBridge (and similar cloud event buses):** EventBridge is a fully managed event bus by AWS that makes it easy to route events from sources to targets without managing infrastructure. It's essentially a serverless pub/sub system with built-

in integration to many AWS services and SaaS providers. In a choreography setup, EventBridge can be the central hub where your microservices (if running in AWS or sending HTTP events to it) emit events and others subscribe via rules or triggers. It's particularly handy for integrating across different systems – e.g., when something happens in Service A, automatically invoke a Lambda, notify another service, and so on, all configured declaratively. Azure Event Grid and Google Cloud Pub/Sub provide similar capabilities in their clouds. For intelligent microservices, these managed buses allow quick wiring of events. For instance, you could have an **AI anomaly detection service** emit an event "AnomalyDetected" to EventBridge; rules can route that event to multiple subscribers: maybe one triggers an alert notification, another kicks off a remediation workflow (perhaps

Step Functions), another calls a Lambda that logs it to a data lake for training future models. The **benefit** is you offload the event routing infrastructure to the cloud provider. The **limitation** is often in customization and throughput – they are typically designed for moderate event volumes and high-level integration, not the firehose of data Kafka can handle. Also, being cloud-specific can tie your architecture to that environment.

- **Other Messaging Systems:** There are plenty, such as **RabbitMQ** (a popular opensource message broker that supports pub/sub and queues – often used for eventdriven microservices especially when you need robust routing and complex routing patterns via exchanges), **Google Pub/Sub** (cloud-managed, very scalable), **MQTT brokers** (for IoT event choreography, e.g., Eclipse Mosquitto), and more specialized ones like **Pulsar** (which combines features of Kafka and traditional messaging). The landscape is rich. Each differs in performance, features (ordering, delivery guarantees, persistence), and ecosystem. The good news is that conceptually they all allow you to do roughly the same thing: decouple services by using an intermediary for messages/events.

How these apply to agentic architectures: In an intelligent system, the event bus becomes the medium through which agents communicate, learn about each other's state, and even **discover new services**. For example, **WebkinsTech's architecture** made heavy use of event streaming – when they refactored their e-commerce platform, they introduced an event bus (Kafka) to broadcast important business events. This enabled them to later add an **AI-driven Fraud Detection Agent** that tapped into the stream: it listens to events like "OrderPlaced" and "PaymentProcessed" and uses that data (along with user activity events) to predict fraud risk in real-time . Notably, they didn't have to modify the order process orchestrator to add fraud checks – the intelligent agent just *joined the dance* via the event bus. This highlights a huge advantage of choreography in agentic systems: you can introduce new intelligent components on the fly. If you have a new ML service that, say, personalizes promotions after an order is placed, you can simply start consuming the "OrderPlaced" event and do its thing, without disrupting the existing flow.

Furthermore, event streams can themselves be analyzed by AI. Tools like Kafka Streams or ksqlDB can embed simple intelligence (aggregations, windowed counts), and more advanced setups might use stream processing frameworks (Flink, Spark Streaming) to do on-the-fly predictions or anomaly detection. These aren't microservices per se, but they often act as *AI agents within the event pipeline* (for example, a Flink job that detects unusual patterns in events and emits an "Anomaly" event). In a way, the choreography approach encourages a whole **ecosystem of event processors** to evolve, some of which are AIdriven.

One challenge to note is that with great power (lots of events) comes great responsibility: monitoring and managing an event-driven system requires tools like log aggregation, distributed tracing (e.g., OpenTelemetry with trace contexts propagating in events), and perhaps an **event registry** to track what events exist and who produces/consumes them. This is something we'll touch on more in the observability chapter, but suffice to say, when intelligent services are flinging events all over, you'll want a clear view of that superhighway of messages.

Hybrid Coordination Approaches

In practice, **most real-world systems use a blend of orchestration and choreography**. It's rarely a strict either/or choice. Think of it as a spectrum: on one end pure central control, on the other pure distributed messaging – but many architectures sit somewhere in between to get the benefits of both. Especially with intelligent microservices, a hybrid approach can provide structure where needed and flexibility elsewhere.

A common pattern is **to orchestrate at a high level**, but let services choreograph among themselves for subtasks. For example, you might have a top-level orchestrator for a customer order process that ensures the primary steps happen in sequence (Payment -> Inventory -> Shipping as a saga). However, within each of those steps, the individual services might publish events for additional side actions. Payment service might emit "PaymentCompleted" which several other services listen to (fraud detection agent, analytics service, etc.) *without* going back through the orchestrator. Similarly, when shipping is scheduled, that might trigger events to billing or notification systems. The orchestrator doesn't micromanage those, it only cares about the main flow. This way, the **critical path** is centrally coordinated (for reliability and clarity), but the **ancillary processes** are decentralized (for extensibility and minimal coupling).

Another hybrid approach is the opposite: use **choreography for the main flow** but introduce **orchestration for specific complex subflows or exception handling**. For instance, you could primarily rely on events to process an insurance claim across services (policy service emits event, assessment service picks up, etc.), but when a claim is flagged as complex, you hand it over to an orchestrator that manages a special review process involving multiple steps and perhaps human intervention. The system largely dances on its own, but calls in a "conductor" for the tricky solos.

The key idea of hybrid is **"fit for purpose"** – use orchestration where strict ordering, consistency or single-point decision making is crucial; use choreography where autonomy, scalability, and evolvability are paramount. They are not mutually exclusive; they can complement each other.

WebkinsTech's journey provides a good example. Recall their **vendor onboarding** process: initially, they built it with a single orchestrator service handling each step (collect documents, verify compliance, set up account). It was simple and linear – orchestrated. As the company grew, they needed more things to happen on vendor onboarding (send data to CRM, update analytics, notify partners). Instead of bloating the orchestrator with all these new steps, they shifted to a hybrid model: the orchestrator still performs the core sequence, but it also **emits an event** "VendorOnboarded" at the end. That event is picked up by various others services (CRM updater, Analytics service, etc.) which do their thing via choreography . In effect, the orchestrator handled the crucial transaction, and choreography handled the add-ons. Over time, they even moved some parts entirely to events (if a step didn't need strict sequencing, they just triggered it asynchronously). This hybrid approach served them well – the process became more extensible without sacrificing the control needed for the core steps.

WebkinsTech Case Study – Finding the Balance: WebkinsTech's e-commerce platform also found a middle ground in their order processing. They had an Order Orchestrator to handle the customer's purchase transaction end-to-end for consistency. However, around this orchestrator they built an event-driven ecosystem. When an order was placed, the orchestrator would initiate payment, inventory, and shipping in sequence (ensuring that if any failed, the others would roll back). But it would also emit high-level events like "OrderPlaced" and "OrderShipped" on a message bus. This allowed additional intelligent services to attach themselves without disturbing the main flow. For example, their Fraud Detection Agent (an AI microservice) subscribed to "OrderPlaced" and "PaymentProcessed" events – if it detected something suspicious, it would emit a "FraudAlert" event. The orchestrator was designed to listen for "FraudAlert" and if received within a short window, it could halt the order or mark it for review. This is a neat hybrid: the orchestrator remains in control *unless* an external intelligent agent (the fraud detector) raises a flag via choreography. Additionally, a Recommendation Service listened for "OrderShipped" events to trigger a personalized email suggesting complementary products to the customer. That service (powered by a recommender AI) operates completely asynchronously – it doesn't affect the order orchestration at all; it's an *observer* pattern. Through this design, WebkinsTech achieved a robust core workflow with opt-in intelligence layered around it. They jokingly referred to it as *"orchestration at the core, choreography at the edges."* The core ensured every order was handled uniformly, while the edges allowed continuous innovation (new services could join the event stream anytime, like adding new dancers around a steadfast conductor).

The WebkinsTech story above underscores that hybrid models can yield the **best of both worlds**. But it also requires careful design: you need to define clearly which parts are orchestrated (and by whom) and which parts are free-floating events. Setting boundaries (for example, "the orchestrator owns anything related to transaction integrity; everything else is event-driven") helps teams understand where to add new logic.

When designing a hybrid system, consider using orchestration for: **sagas/transactions**, **aggregating results**, **compliance-check processes**, or anything that "must happen in order or not at all." Use choreography for: **publish/subscribe notifications**, **updates to independent systems**, **triggering optional or future-looking actions** (like logging, analytics, recommendations).

Design Insight: It's not either-or; in modern microservices orchestration and choreography often coexist. Use orchestration when you need strict control or auditability of a process, and use choreography when you need adaptability and loose coupling for extension. A common pattern is *orchestrate the critical path, choreograph the rest*. This lets you sleep at night knowing the crucial stuff is handled, while still enabling the system to grow and respond to new requirements easily.

Of course, hybrid approaches come with their own challenge: you have both an orchestrator *and* an event bus in play, so you'll need to ensure they don't step on each other's toes. Clear event naming and responsibilities are important (e.g., the orchestrator emits

"OrderComplete" only once to avoid confusion, etc.). Also, when a lot is happening via choreography, you still need ways to monitor those parts (just because they're "optional" doesn't mean you can ignore their failures). Essentially, you adopt the union of orchestration concerns and choreography concerns – careful **design, monitoring, and documentation** are key.

Emerging Patterns for Intelligent Service Coordination

As we venture into the frontier of multi-agent systems and autonomous microservices, architects are experimenting with patterns beyond the classical orchestrator vs. event-bus model. Here we introduce a few **advanced coordination strategies** that are gaining attention in the context of intelligent services. These may not all be mainstream in industry yet, but they're useful to know as they can inspire robust designs for agent-based systems:

- **The Blackboard Pattern:** This comes from the AI domain (blackboard systems in the 1980s) and is essentially a form of mediated choreography. Instead of services firing events at each other ad-hoc, they all interact with a common knowledge repository or "blackboard." Think of the blackboard as a global state or context that agents can read and write to. In a microservice setting, a **Blackboard Service** could hold the state of a problem (say, a complex scheduling problem), and multiple intelligent microservices (agents) post their partial results or observations to it. Each agent sees the evolving state on the blackboard and contributes until the problem is solved. The blackboard is not a coordinator telling agents what to do; it's more like a shared **environment for collaboration**. This pattern can be powerful for scenarios like **incident response** (different diagnostic agents post clues to a shared incident board and collectively

resolve an issue) or **optimization problems** (each service proposes a solution tweak on the board). The coordination is indirect but structured – the blackboard provides a central place to focus. It's similar to choreography in that there's no boss, but unlike pure pub/sub, the interactions are mediated via shared state. One must ensure the blackboard itself is robust and that there are rules for conflicts (e.g., what if two agents post contradictory data? The system needs a way to resolve that).

- **Market-Based (Auction) Coordination:** Inspired by economics, this pattern treats services as **buyers or sellers** of tasks. If a task needs to be done, an "auction" is held among available agents that could do it. For example, imagine multiple translation microservices (each with different cost/speed/accuracy trade-offs) and an incoming request to translate a document. Instead of a fixed orchestrator choosing one, the system could broadcast a request: "Who can translate this document?" Each service bids (maybe one says it can do it in 1 second for $0.01, another in 2 seconds for $0.005, etc.). The best bid "wins" and that service proceeds with the task. This approach is highly decentralized and dynamic – the system can optimize for cost, speed, or other criteria automatically. It's essentially a choreography with a negotiation step. There are protocols from multi-agent systems like the **Contract Net Protocol** that formalize this kind of bidding process. While not common in typical IT systems yet, we see hints of it in things like cloud function marketplaces or edge computing (where, say, you might auction a computation to the nearest edge server). For agentic microservices, this could be useful if you have **redundant capabilities** and want them to self-organize who does what (like a pool of AI models and you want the one that's most confident to handle a given input). It's a fascinating area that brings economic principles into architecture.

- **Direct Negotiation and Peer-to-Peer Consensus:** In some cases, intelligent services might directly coordinate among themselves to reach a decision. This isn't a single pattern but a family of approaches – it includes things like **leader election** (agents vote who should act as temporary orchestrator for a task), **consensus algorithms** (like Raft or Paxos used among intelligent agents to agree on a value), or even simpler **peer-topeer negotiation** (Service A and Service B chat back and forth to decide something). For instance, two AI microservices might negotiate load balancing: if one is overwhelmed, it might ask a peer to take over some tasks, and they strike a deal. This is more on the choreography side (no central brain, just inter-agent communication), but structured as a dialogue or agreement rather than one-way event publish. We see analogies in the animal world, like how honeybees collectively decide on a new nest site by "voting" (waggle dances). In tech, blockchain networks have nodes reaching consensus – similar idea but applied to transactions. In microservices, you might not implement a full-blown blockchain consensus, but lightweight versions can ensure a set of intelligent services stay in sync on a decision *without* a central coordinator.

- **Stigmergy (Environment-mediated coordination):** This is a fancy term from biology (ant colonies, termites) referring to agents influencing each other by modifying their environment. In a software sense, it's similar to the blackboard pattern but even more indirect – agents leave **signals in the environment** that others sense and respond to.

One could implement digital pheromones: e.g., services add metadata or "tags" to certain events or database records as they process them, and other services adjust their behavior upon detecting those tags. A concrete example: imagine an AI-based content filter service scanning messages – if it finds something suspect, instead of directly alerting others, it *tags the message in the database*. Later, a different microservice reading that message sees the tag and knows to handle it with extra care

(or not at all). The two services coordinated via the data store, not by direct communication. This is an exotic pattern, but it's conceptually useful for **very loosely coupled** scenarios. Modern event streams and state stores could allow some form of this (like a service might increase a "priority" field in an event payload if it thinks it's urgent – others read that and act faster). Stigmergy is rarely designed intentionally in enterprise systems, but it sometimes *emerges* (e.g., one service writes a temp record that another periodically checks – a crude form of environment signaling).

Many of these emerging patterns are still areas of active experimentation. They may or may not fit typical enterprise needs, but as **AI-driven agents proliferate**, we might see them employed more. The main takeaway is that beyond the classical orchestrator and event bus, we can draw on **multi-agent system principles** to coordinate intelligent services:

- **Centralized vs Decentralized is a spectrum:** You could have a hierarchy of coordinators (one orchestrator delegating to mini-orchestrators – akin to manager agents and worker agents).

- **Mediators and Data Hubs:** Using shared data (blackboard) or shared environment to coordinate without direct calls.

- **Competitive or Collaborative agents:** Sometimes you let services compete (like auctions) to optimize outcome; sometimes they collaborate (like jointly building a solution on a blackboard).

For most readers building systems today, these may sound theoretical. But smaller examples exist. For instance, in cloud resource management, some systems let microservices "bid" for resources based on their priority – essentially auctioning CPU time. In distributed AI training, workers might coordinate via a parameter server (a sort of blackboard for model parameters). As your architecture grows more complex and more autonomous, don't be afraid to consider these patterns if traditional ones start to strain. Even if you don't use them directly, understanding them can help you reason about system behavior (e.g., recognizing when your event-driven system is acting like a stigmergy system, and thus how to tweak it).

Goal-Driven Service Composition and Discovery

One exciting possibility with agentic microservices is moving towards **goal-driven composition** – where instead of explicitly programming which services call which, you tell the system *what outcome you want*, and it figures out how to marshal the services to achieve it. This is like taking orchestration to the next logical step: **automated orchestration planning**.

Imagine being able to say: "System, whenever a user makes a support request, ensure that their issue is resolved within 2 hours with minimal cost." That's a goal. The system could then dynamically assemble the right microservices to achieve it – perhaps it discovers that it needs to involve the User Profile service to get context, then an AI classifier to route the ticket, then a Support Agent service to actually handle it, and maybe an Email service to follow up. It might create a temporary orchestrator or just send events to trigger each needed action in turn, essentially **planning a workflow on the fly**.

While we're not fully there yet in general, there are pieces of this already. **Service discovery** systems (like Consul, Eureka, or service meshes) allow services to find each other by name, but in goal-driven composition, you'd find services by capability. This leans into ideas from the Semantic Web and AI planning: if services advertised *what they can do* (their *capabilities* and *inputs/outputs*), an AI planner could chain them to fulfill a goal. For example, one service provides flight booking, another hotel booking, another payment processing – a smart agent could string them together to satisfy a goal "book a travel package and pay for it," even if we never hardcoded that entire flow.

In agentic architectures, you might deploy a special kind of service – a **Planner Agent** – whose job is to compose other services. It would take high-level requests or system goals and generate sequences of service calls or event triggers to achieve them. This Planner Agent is effectively an orchestrator, but a highly flexible one: it might consult a knowledge base of service capabilities and maybe use AI search algorithms to find a valid combination.

For instance, suppose WebkinsTech opens up a partner ecosystem and a customer can choose to redeem loyalty points from either WebkinsTech or a partner's system. If a user action triggers a goal "redeem points for this purchase", a Planner could decide: first call WebkinsTech's Loyalty Service; if it returns insufficient points, then call Partner's Loyalty API, then proceed to apply the discount via Billing service. In a traditional system, you'd code that logic or configure it in a workflow. In a goal-driven approach, the system figures it out because it knows the goal (apply points) and knows two services that can provide points. It can even optimize – maybe use the one with more points first.

Another aspect of goal-driven systems is **dynamic substitution**. Let's say a particular microservice is down or too slow; a smart orchestrator or planner might discover an alternative service or approach to still meet the goal. For example, if the usual **Image Recognition Service** is unavailable, an intelligent coordinator might route the request to a fallback service or even an external API to get the job done, because the goal is "recognize this image", not "use Service X specifically". This requires a registry of alternate capabilities and trust in them, but you can see how it adds resilience.

All this sounds complex to implement, and it is – but incremental steps are possible. You can start by enriching your service registry with metadata about capabilities. Even without full AI planning, you can manually code an orchestrator that uses discovery to pick from multiple services based on context (some companies do AB testing or dynamic routing by having orchestrators choose a service variant at runtime using rules). As AI techniques advance, one can imagine dropping in an agent that observes system performance and automatically reconfigures the workflow to meet high-level KPIs (kind of like an auto-tuner for your microservice composition).

A practical example today might be **routing based on ML predictions**: suppose you have a goal to serve a web request with low latency. You have a normal pipeline of microservices that provides a rich, personalized page (which takes, say, 500ms normally). But you also have a cache or a simpler service that can provide a basic page in 100ms. If an AI predicts that the user is on a slow connection or that the main pipeline might exceed your latency budget (maybe due to current load), a goal-oriented orchestrator could decide to use the fast path for this request (sacrificing some personalizations but meeting the latency goal). In doing so, it has "composed" a different workflow on the fly for the same high-level goal "respond to user request within 200ms".

This overlaps with dynamic reconfiguration (which we'll cover next), but it's specifically about choosing and chaining services based on goals. Some research projects combine **reinforcement learning** with service orchestration, where the orchestrator learns policies on which service to call or which path to take to optimize a reward (like throughput or success rate). That essentially trains the system to do goal-driven composition without explicit programming of each scenario.

To sum up, goal-driven service composition is an emerging capability where **intelligent agents (orchestrators or planners)** can **discover** available services and **compose** them as needed to fulfill a desired outcome. It moves us closer to the vision of *self-assembling software*. While it might not be out-of-the-box in most platforms, keep an eye on this trend – especially as companies expose more microservices via APIs, the ability to stitch them together dynamically with AI could become a competitive advantage.

Dynamic Reconfiguration and Adaptation

Microservice architectures are already dynamic – instances scale up or down, services get deployed or retired. When we infuse AI and make systems more agentic, we unlock a further dimension of dynamism: the ability for the **coordination mechanism itself to reconfigure based on context, signals, or predictions**.

Dynamic reconfiguration means the system can **change its coordination pattern or workflow at runtime** in response to certain conditions, *without human intervention*. This could involve switching between orchestration and choreography modes, altering the path of a workflow, or enabling/disabling certain interactions.

Some scenarios where this makes sense:

- **Load or Performance Changes:** Suppose the system detects unusually high load. A dynamic strategy might be to **relax consistency** to maintain throughput. Concretely, if an orchestrator usually waits for a confirmation from Service C, under extreme load an AI agent might decide to skip that confirmation step (maybe handling it later asynchronously) to keep things flowing. In essence, the workflow toggles from a strict orchestration to a more event-driven fire-and-forget mode temporarily. Once load subsides, it reverts. This kind of adaptation could be learned by an AI ops agent observing that "during traffic spikes, the slowest step is X, so skip X to serve more users, and catch up on X's work offline later."

- **Service Degradation or Failure:** If a microservice becomes slow or unreachable, an orchestrator can reconfigure the flow. For example, if the **Recommendation Service** is down, the system might bypass that step entirely rather than retrying endlessly. Or it might substitute a simpler rule-based recommendation service if available. Dynamic reconfig means this substitution can happen automatically. A concrete case: a purchase workflow normally calls an AI personalization service to tailor an upsell. If that call fails, the orchestrator could decide "skip upsell for now and proceed to payment to not delay the purchase." The orchestrator might learn or be configured to know that the upsell is optional. Essentially, it changes the directed graph of the workflow on the fly.

- **Contextual Optimization:** Use context signals (user location, time of day, etc.) to choose different coordination patterns. For instance, in WebkinsTech's system, they might discover that during peak hours, an event-driven approach for inventory updates is causing too much lag in data consistency (lots of events flying around delaying updates). The system could respond by funneling those through a temporary orchestrator or locking mechanism at peak (ensuring consistency), then revert to eventdriven off-peak. Alternatively, a prediction might indicate that a certain region's users have a high fraud risk today; thus the system dynamically **inserts an extra verification** step into the otherwise orchestrated flow for those users. In effect, the workflow modifies itself in certain contexts.

- **Learning and Improvement:** Over time, an intelligent coordinator can analyze which parts of the process are slow or failure-prone and rewire for better performance. This is where observability (next chapter) feeds into action. If data shows that a particular sequence of events often leads to a deadlock or long completion time, the system might re-route the sequence. For example, say we have Services A, B, C normally working via events. If the combination A->B->C takes too long frequently, an AI might decide to orchestrate B and C tightly for a while (or even co-locate them) to speed it up, effectively treating them as a unit. This veers into the territory of *runtime architecture adaptation*.

Enabling dynamic reconfiguration requires a **programmable infrastructure**. Service mesh technologies (like Istio, Linkerd) can help here as they allow some control over routing between services at runtime. For instance, a service mesh plus an AI controller could redirect traffic from one service instance to another if it predicts failure. Likewise, modern workflow engines allow altering running workflows or choosing branches dynamically via code – which an AI can utilize.

One interesting area is **AIOps** – using AI for operations. An AIOps system might detect anomalies and then act (not just alert humans). For coordination, that could mean automatically quarantining a misbehaving service (like stop sending it events), or spinning up a new instance of an orchestrator in a region that suddenly got a lot of traffic (bringing coordination closer to where it's needed).

We can see early forms of this in auto-scaling (scale-out is a form of reconfiguration – adding more service instances changes the system topology) and chaos engineering (where the system is tested for resilience to reconfigurations). The next step is to make it **adaptive**: the system itself decides on changes based on predictive analytics. For example, using reinforcement learning, a system could learn an **optimal policy for switching modes**:

perhaps it learns that at night (low load) a fully event-driven approach saves resources and works fine, but in daytime (high load) a hybrid approach yields fewer errors, so it schedules that change daily.

Another example: **Feature toggles driven by AI**. Many systems have feature flags to turn certain functionality on/off. Imagine an AI ops agent that flips those toggles in real-time to optimize outcomes. In coordination terms, one such toggle could be "centralize coordination or decentralize it". If the AI senses chaos, it could enable a stricter orchestration mode; if it senses idle capacity and need for speed, it could let services off the leash.

It's important to design safeguards – you don't want the system thrashing, constantly reconfiguring and confusing itself. Any dynamic adaptation logic should be tested thoroughly (even learned policies can be validated in simulations or shadow modes). Also, maintain **observability** of when and what reconfigurations happen – otherwise you'll be baffled when the architecture "mysteriously" changes at 3PM and then reverts at 4PM because some model thought it was a good idea.

In summary, dynamic reconfiguration is about making the architecture **self-tuning**. Instead of a static coordination strategy, it becomes a continuum that the system can adjust. This approach is especially powerful in agentic systems because the environment can change in unpredictable ways (due to both external factors and the agents' own behaviors). Having a meta-level agent or policy that tweaks the coordination as needed can keep the whole ensemble performing optimally. It's like having an AI stage manager who, in the middle of the play, whispers changes to the actors or swaps props to make the scene work better for the audience – all while the play is running.

Choosing the Right Approach: A Practical Guide

With a solid understanding of orchestration and choreography (and everything in between), how do you decide what's right for your project? The truth is, **the best approach depends on your specific context** – the nature of your services, your team's expertise, your scalability needs, failure tolerance requirements, and even organizational factors. Here we provide a checklist and some guidance to help make this decision, especially in light of intelligent, agent-like services.

Coordination Strategy Checklist

Consider the following questions and points when choosing or designing your service coordination strategy:

- **Do you need a clear, central authority for business logic?** If having one place to enforce the order of operations and decisions is important (for example, for auditing, compliance, or ease of understanding), lean towards an orchestrator for those parts. If it's okay that logic is distributed and you're comfortable piecing together the story from multiple services, choreography can work.

- **How independent are the services' tasks?** If each service's work can be done mostly independently and in parallel, choreography often fits well (services can run concurrently on events). If each step strongly depends on the outcome of the previous step, orchestration (or request-response chaining) might be simpler to implement correctly.

- **What are your consistency and timing requirements?** For workflows that are essentially transactions (even if eventually consistent), orchestrators can implement saga patterns or coordinated commits more straightforwardly. Event-driven approaches can also do sagas, but they're trickier as the coordination is implicit. If eventual consistency and slight delays are acceptable, events give you more flexibility. If you need near-immediate consistency across steps, an orchestrator or a locking mechanism is needed. *(E.g., payment and inventory deduction: do you allow them to happen out of order via events or must they be sequential?)*

- **How easy is it to add or change a step?** If you anticipate a lot of changes to the workflow (adding new microservices, changing their order, A/B testing different flows), consider how that impacts your design. In orchestration, changes mean updating the orchestrator's logic (code or model). In choreography, changes mean creating new event handlers or services. The latter often allows adding features faster (just plug in a new subscriber) – if your team values quick iteration and you can tolerate the potential increase in complexity, that's a point for choreography. If you prefer controlled, gated changes, orchestration keeps everything in one spot.

- **Team and Ownership:** Do you have a team that will own the orchestrator service or workflow definitions? Sometimes having a dedicated "workflow team" is a luxury; if not, pushing coordination to the edges (each team handles their part via events) might align with how your org is structured. Conversely, if you *do* have folks who love working in Camunda or Temporal, use that strength – let them build orchestrations that encapsulate cross-cutting processes.

- **Tooling and Infrastructure:** What tools are already in place? If you have Kafka up and running and everyone is familiar with it, building on an event-driven paradigm is natural.

If you are a heavy Kubernetes shop, Argo could slide right in. If you're on AWS, Step Functions or EventBridge might integrate well with minimal effort. Sometimes the decision is made easier by what's readily available and reliable in your stack (you don't always need to introduce a new tool – maybe you can achieve a lot with what you have). That said, don't force-fit – just because you have Kafka doesn't mean every process should be choreographed; maybe a small orchestrator service reading/writing Kafka topics can give a structured sequence where needed.

- **Failure handling:** Think through a failure scenario in your workflow. If something in the middle fails, how do other parts know what to do? Orchestration can catch an error and perform compensating actions (like undo earlier steps). In choreography, you'd implement compensations as reactions to a "failure event" (which is doable but requires every service to cooperate). If your workflow has complex rollback logic, orchestration might make that easier. Also, consider timeouts – orchestrators can enforce timeouts on calls, while in an event system you might need a separate watchdog to notice if something hasn't happened in a while.

- **Observability:** Be honest about your ability to monitor the system. If you have a strong observability platform (distributed tracing, centralized logging, alerting) and your team is comfortable with it, you can afford a more complex choreography because you'll catch issues. If not, a simpler orchestrated flow might be safer to start with. (We'll address observability in detail in the next chapter, but it's part of the decision – a choreographed system essentially pushes more burden to your monitoring to understand what's happening).

- **Performance considerations:** Orchestration often introduces a slight performance overhead because everything funnels through one coordinator (which might become a sequential point in an otherwise parallel-capable process). Events allow naturally parallel processing. If latency is critical and parts of the work can happen simultaneously, choreography (or at least an orchestrator that dispatches without waiting when possible) will help. Also, think about throughput: a central orchestrator has to handle all workflow instances, while events distribute that load among services. You can scale orchestrators, but it's an extra layer. On the other hand, if your steps are mostly I/O-bound and quick, the overhead might be negligible and orchestration overhead isn't a big deal.

- **Need for AI/ML integration:** Since we are focusing on intelligent microservices – consider how your AI components fit in. If you have an AI model that needs to be invoked at a specific point and its result determines the next step, that's often a sign to use an orchestrator or a directed flow (call model, get result, then branch). If you have AI agents that continuously run and provide value whenever relevant, they can be attached via events (like our fraud detection example). Also, if your AI needs training data from the workflow, choreography (with an event log) can double as a data source. Orchestration logs are less data-rich (they might just record what happened, not feed into ML easily). So if learning from the workflow itself is a goal, consider event streaming so you have a record of all interactions for analysis.

After answering these questions, you might find that some parts of your system clearly benefit from one approach or the other. **It's perfectly valid to decide on a per-component basis.** For example: "User login flow – use choreography so authentication and logging and notification services can operate independently; Order processing – use orchestration because it's critical to get it right with all steps; Recommendation generation – let it be choreographed (triggered by events) since it's optional and experimental." Designing this way may yield a mix, and that's fine.

Readiness Checklist for Implementation

Before rolling out your chosen approach, especially if it's choreography-heavy or involves complex agents, ensure you have these in place:

- **Event Schema Contracts (if using events):** Clearly define the structure of each event and share that understanding across teams. Using a schema registry (like Confluent Schema Registry for Kafka or even API docs for your events) helps avoid confusion. Intelligent services might add new fields to events – version your events and keep things backward compatible.

- **Idempotency and Deduplication:** If you go event-driven, can your services handle duplicate events or out-of-order arrivals? This is a must to prevent agents from making inconsistent decisions. For orchestration, idempotency matters too (e.g., if the orchestrator retries a step, can the service handle that gracefully?).

- **Monitoring & Tracing:** Set up a way to trace requests through the system. For orchestration, have the orchestrator tag a trace ID on all calls. For events, propagate a correlation ID in message headers or payload. Tools like

Jaeger or Zipkin can help tie together a story from multiple services. You should be able to ask "what happened to this particular business transaction?" and get an answer from your logs or traces.

- **Fallbacks & Circuit Breakers:** Especially for orchestrated flows, implement circuit breakers (using something like Istio/Envoy or a library like Hystrix if using synchronous calls). If a downstream service is unresponsive, the orchestrator shouldn't hang forever. In event systems, consider using a DLQ (Dead Letter Queue) for messages that couldn't be processed, and have a plan for those. Also decide what an acceptable failure looks like (does the orchestrator abort the whole thing? Do services just not react to an event if they can't handle it?).

- **Scalability Testing:** If using an orchestrator, load test it to find bottlenecks. If using an event broker, make sure it's configured with enough throughput and partitions (Kafka) or nodes (NATS) to handle spikes. Intelligent microservices might produce *a lot* more events (think of an AI that emits events for every anomaly it detects), so factor that in. Ensure your infrastructure can scale out – e.g., orchestrator can have multiple instances (maybe via Temporal's clustered workers or similar), and your message broker is clustered.

- **Security Considerations:** A central orchestrator might need access to various services – ensure it handles credentials/tokens securely (maybe via a vault or cloud IAM roles).

With events, be mindful of sensitive data – do you need to encrypt certain fields in events? Who can subscribe to what topics? In a chaotic choreography, you don't want an internal sensitive event leaking to an unauthorized service. Implement access controls on event streams if possible.

- **Documentation and Visibility:** Maintain a living diagram of how things connect. In an orchestrated world, that might be a flowchart of the process (which could be directly from your BPMN or code). In an event-driven world, consider an "event flow diagram" – which service publishes which event, and who subscribes. This map is invaluable as the system grows. WebkinsTech, for example, created an internal wiki listing each event and its producers/consumers, preventing the "nobody knew that service X was reliant on event Y" problem. For agentic behavior, also document where AI is making decisions in the flow.

- **Simulation or Dry-Run (for critical flows):** If you have an AI-driven coordinator or a very distributed setup, it can be helpful to simulate certain scenarios. For orchestrators, maybe write unit tests or integration tests that simulate services failing or returning odd data (to see if your orchestrator logic handles it). For choreography, consider a test harness that publishes events and verifies the right outcomes (a bit harder to do, but possible with contract testing tools). If agents learn over time, consider how to test changes in their behavior – e.g., a new model rollout shouldn't break the choreography assumptions.

By checking off these items, you increase the chance that your chosen approach will succeed in practice. It's often not the approach itself, but the *supporting infrastructure and practices* that determine success. A choreographed system without good monitoring is a nightmare. An orchestrated system without resilience is a ticking time bomb.

Evolving Your Approach

One final piece of advice: **be ready to evolve your coordination strategy** as you learn. You might start with orchestration to get something working (because it's simpler to conceptualize), and later gradually introduce more events to decouple parts that become bottlenecks. Or vice versa, you might start event-driven and realize you need a bit more central control for a particular sequence – so you introduce an orchestrator for that subset of interactions. This evolution is normal. What's important is to keep the overall architecture clean as you evolve (avoid a mess where you have redundant orchestrators and duplicate events flying around). Periodically revisit your coordination map and refactor if needed.

It's also okay to pilot new approaches. For example, perhaps try an AI-driven planner for one slice of your system while the rest uses traditional means, and see how it goes. If it fails, roll back to something simpler. If it works, expand its use.

In essence, **choose the strategy that makes your system simplest without compromising key requirements**, and remember that you can mix modes. The goal is not to be pure to one philosophy, but to build a system that is robust, understandable, and adaptable. And as you add intelligent agents, always ask: does this agent thrive more under a conductor, or dancing freely? Sometimes the answer is "both, in different ways."

Conclusion

In this chapter, we explored the interplay of orchestration and choreography in the context of intelligent microservices. We saw that **orchestration (central control)** and **choreography (distributed interaction)** are two fundamental approaches to coordinating services – akin to choosing between a strict conductor or an improvisational dance for your system. Each has distinct advantages: orchestration offers clarity, single-point management, and easier error handling, whereas choreography provides flexibility, modularity, and natural scalability. When we introduced AI and agentic behavior, these differences became even more pronounced: an **AI-powered orchestrator** can smartly guide a process but needs safeguards, while **swarms of intelligent services** can self-organize in choreography but demand strong observability to ensure they don't veer into chaos.

Real-world systems often blend both approaches, finding a harmony that suits their needs. We learned from scenarios like the loan approval (favoring a central brain) and the smart grid (flourishing under decentralized coordination), and from WebkinsTech's experiences that a **hybrid approach** can yield robust architectures – orchestrating the essentials while choreographing the extensibles. We also peeked into the future with emerging patterns such as blackboard systems and market-based coordination, imagining how truly autonomous services might coordinate without direct oversight. The notion of goal-driven composition hints at a world where you specify *what* you want and intelligent agents figure out *how* to make it happen by assembling microservices on the fly. And through dynamic reconfiguration, we envisioned systems that can **adapt their own coordination strategy** in real-time, responding to predictions and changing conditions to keep things running optimally.

At this point, you might feel like the conductor of a complex orchestra, or perhaps the choreographer of an intricate dance – or both! The key takeaway is that **there's no onesize-fits-all solution**. The art of architecture is in choosing the right approach for the right problem, and often that means a blend. Intelligent microservices push us to design not just for functionality, but for adaptability and learning.

As you implement these coordination strategies, keep empathy for your fellow developers and ops teams: make the system observable and debuggable, document the "playbook" (even if the playbook is distributed in events), and anticipate where things can go wrong. With the guidance and checklists provided, you're well-equipped to make an informed decision and build a coordination mechanism that aligns with your system's goals and your organization's capacities.

Finally, remember that **the dance isn't over once you pick orchestration or choreography** – in fact, that's just the beginning of making a complex distributed system work in the real world. Regardless of how you coordinate the dance of microservices, you need to be able to see and understand what's happening under the hood, and feed that back into improving the system. In the next chapter, we will delve into exactly that: **observability and learning in agentic microservice ecosystems**. We'll learn how to monitor this "orchestra" or "dance troupe" of services, how to gather feedback from their performance, and even how the system can learn and improve from its own data. After all, an intelligent, coordinated system should not only act – it should also **observe, learn, and adapt**.

Transitioning onward, let's turn up the lights and look behind the curtains: it's time to ensure our beautifully coordinated microservices are not operating in the dark. Observability, monitoring, and continuous learning await in Chapter 12, where we'll equip you with the skills to keep the show running smoothly and even make it smarter over time. Onward to understanding and mastering the insight that **"what we can observe, we can improve"** in our intelligent microservice landscapes!

Chapter 12: Observability and Learning

In the last chapter, we coordinated the dance of intelligent microservices. Regardless of how well we choreograph that dance, we still need to **see what's happening under the hood**. That's where *observability* and *learning* come in. Imagine deploying a fleet of AI-powered services that make autonomous decisions – a fraud detection service blocking payments, a health monitoring agent watching patient vitals, or a recommendation engine suggesting products. How do we trust their decisions? How do we debug issues or improve their behavior over time? In this chapter, we'll dive deep into observability and continuous learning in intelligent microservices. We'll explore how logging, metrics, and tracing (the three pillars of observability) are adapted for AI-driven services, and how the data they generate can feed back into making those services smarter.

Expect an engaging journey: we'll mix practical architecture with storytelling. You'll find tips, warnings, and real-world flavoured examples from our fictional friends at FinBank (finance), MediHealth (healthcare), and ShopEase (e-commerce) to illustrate concepts. By the end, you should have a clear understanding of how to make your intelligent microservices not only observable (so you can peek into their "minds") but also capable of learning continuously from their own operations. Let's crack open the black box!

Understanding Observability in Intelligent Microservices

Observability is the ability to understand a system's internal state by looking at its outputs.

In traditional terms, if you can collect enough information (telemetry) from a running system – like logs, metrics, and traces – you can infer what's going on inside without directly peeking at the code or memory. It's like diagnosing a car by the sounds it makes and how it drives, rather than by ripping apart the engine. For a microservices architecture (lots of moving parts talking to each other), observability is absolutely critical. When you add AI components into the mix, the need goes from critical to *crucial*.

Why? Because an **intelligent microservice** isn't a straightforward input-output machine; it has a mind of its own (well, sort of!). It might be making probabilistic decisions, learning from data, or adapting its behavior. If a regular microservice is a black box, an AI-powered microservice can be a black box filled with more black boxes (models, algorithms, data pipelines). Without observability, you're essentially *flying blind* when something goes wrong or when you simply want to know "why did it do *that*?".

Let's clarify the term a bit more by contrasting it with **monitoring**. Monitoring usually means tracking a predefined set of signals – you set up dashboards for things you know to watch (CPU usage, error rates, etc.), and you get alerted when those signals go out of bounds. It's like having a security camera pointed at specific doors. **Observability**, on the other hand, is about having enough rich data that you can explore new questions *even if you didn't anticipate them upfront*. It's the difference between a security camera (monitoring specific spots) and a full forensic lab (where you can analyze any clue to piece together what happened). In dynamic, distributed systems, you often face **"unknown unknowns"** – problems you didn't even know could happen. Observability helps you deal with those by providing the raw materials (logs, metrics, traces, and more) to answer arbitrary questions about system behavior.

Observability in the Context of AI Services

For AI-driven microservices, observability isn't just about uptime or latency; it's also about **decision quality** and **behavior transparency**. It's not enough to know *that* a service responded in 200ms – we might need to know *why* it recommended loan approval for one customer and denial for another, or why our bot decided to route a support ticket to Department X. Observability provides the data for **explainability**. In regulated industries (finance, healthcare), this is often a compliance requirement – you need an audit trail of decisions. Even outside regulation, having insight into the AI's decisions builds trust with users and helps developers improve the system.

In summary, observability means instrumenting our intelligent microservices so thoroughly that every important action leaves a trace. By examining those traces, we can reconstruct events, debug issues, understand decisions, and ultimately create a loop of continuous improvement. To achieve this, we rely on the three pillars of observability: **logging**, **metrics**, and **tracing**. Let's break those down and see how each is applied in an AI context.

The Three Pillars of Observability

In any modern system, observability rests on three kinds of telemetry data:

- **Logs** – Discrete event records that capture what happened in the system at a certain time (e.g., “User X requested loan approval, decision=DENIED due to risk_score=0.85”).

- **Metrics** – Numeric measurements that are aggregated over time (e.g., “service uptime

= 99.9% this week”, “average response time = 120ms”, or domain-specific ones like

“fraud detection precision = 96% today”).

- **Traces** – Distributed transaction traces that capture a single request or process as it flows through multiple services, often visualized as a timeline of spans across services (e.g., a trace of an order through a dozen microservices, including an AI inventory predictor).

Each pillar gives a different vantage point. **Logs** are like a detailed journal or diary – very detailed, sometimes verbose, but crucial for drilling into specific events. **Metrics** are like a health report or dashboard gauge – they abstract many events into trends and counts that are easy to monitor at a glance. **Traces** are like a storyboard of an adventure – they show the path taken by one transaction through the system, useful to pinpoint where things went wrong in a complex path.

In intelligent microservices, we extend these pillars further:

- We log not just technical events, but also **decision details** of AI.
- We track metrics not just for infrastructure health, but for **model performance** and outcomes.
- We trace not only service calls, but also **data flows and decision chains** across services.

To understand how each pillar works in practice, let’s look at them one by one, with an eye toward the unique needs of autonomous, AI-driven services.

Logging: Capturing Intelligent Decisions

Logging is the oldest and most universal form of observability. Simply put, a **log** is a record of something noteworthy that happened in the system. In code, this is typically done via logging frameworks (like log4j in Java, Python's logging module, etc.) where you write lines to a log at various severity levels (INFO, DEBUG, WARN, ERROR, etc.). These log entries often end up in console output, log files, or centralized logging systems.

For intelligent microservices, what constitutes a *noteworthy event*? Of course, you'll still log errors, exceptions, and high-level lifecycle events ("service started", "connected to DB", etc.). But the interesting new bits are **AI decision logs**. Whenever an AI agent or model inside the service makes a decision, it should log **what it decided and why** (if possible). This might include:

- The **inputs** it received (or a summary thereof).
- The **decision or prediction** it made.
- Possibly the **confidence or score** associated with the decision.
- Any **important factors** that influenced it (if our AI can provide that, e.g. feature values or rule triggers).
- Any **resulting action** taken (e.g., "flagged transaction for review", "sent alert to nurse station 5").

For example, imagine a fraud detection microservice at FinBank. A normal log (in a traditional service) might say:

```
INFO 2025-04-23T10:15:30Z Transaction 123456 checked by fraud
result=ALLOW
```

In an **intelligent** fraud service, we'd want richer info:

```
INFO   2025-04-23T10:15:30Z   FraudCheck   transaction=123456
score=0.02
     user_segment=NEW_ACCOUNT
     message="Low risk, no anomaly detected"
```

Notice we log the **score** (0.02 risk score, meaning low risk) and some context (user segment, any rules triggered). This way, if a legitimate transaction was blocked or a fraudulent one slipped through, engineers (or auditors) can later review what the system was "thinking". Logging these details directly contributes to **explainability** – a developer or compliance officer can trace *why* a particular decision was made by reviewing the logs.

Structured Logging: It's a good practice to log in a structured format (like JSON) rather than free-form text, especially for AI decisions. Structured logs make it easier for machines (and log management systems) to parse fields like score or result. For example, logging the above as a JSON object:

```
{
  "timestamp": "2025-04-23T10:15:30Z",
  "event": "FraudCheck",
  "transaction_id": 123456,
  "decision": "ALLOW",
  "score": 0.02,
  "user_segment": "NEW_ACCOUNT",
  "rules_triggered": ["velocity_check"],
  "message": "Low risk, no anomaly detected"
}
```

This structure can be indexed by a logging system (like Elasticsearch, Splunk, etc.), allowing analysts to query: "show me all transactions with decision=BLOCK and score > 0.8" easily, for instance.

In AI systems, logs can also capture **feedback events**. For instance, if a human later reviews a fraud alert and marks it "false alarm", that can be logged as a feedback event tied to the original decision ("transaction 123456 flagged as fraud was later cleared by human").

Such logs are gold for learning – they provide training data for model improvement (more on that in the learning section).

Tip: Implement explanation logging for AI decisions. Whenever possible, log not just the output but also the key reasons or features behind an AI decision. For example, a recommendation service might log: "Recommended *Item42* to User123 – reason: similar users liked this item (collaborative filtering)". These logs will be invaluable when debugging or explaining the AI's behavior to others.

Naturally, we must be mindful of **privacy and volume** when logging. Logging every detail of every decision can produce huge volumes of data (imagine an AI processing thousands of events per second – the logs can be enormous!). We often find a balance by logging at different levels: critical decisions or errors always, sampling routine decisions, and using debug-level logs for deeper info only when needed. Also, sensitive data (personal info, health records) should be either anonymized or excluded from logs to avoid security/privacy issues.

Modern log aggregation tools (the **ELK/EFK stack**: Elasticsearch/OpenSearch,

Logstash/Fluentd, Kibana, or cloud services like Azure Monitor, AWS CloudWatch, etc.) can collect logs from all your microservices and let you search and analyze them in one place. This is essential in microservices, where each service has its own log – you need a central view to trace an event that might span many services. We'll touch more on tooling later, but suffice it to say: setting up centralized logging is a must for observability.

Lastly, logs are crucial during incidents. If at 2 AM your AI service did something weird, the on-call engineer's first move is usually "check the logs!". You want those logs to tell the story of what led to the incident. In an AI system, that story includes not just error stack traces but also things like "the model started outputting unusually high scores for almost everything". With good logs, you can often pinpoint whether the issue was in the data, the model, or in the surrounding business logic.

Metrics: Quantifying Performance and Quality

If logs are a detailed diary, **metrics** are the key performance indicators (KPIs) on the dashboard. Metrics boil down system behavior to numbers that can be tracked over time: counters, gauges, histograms, and so on. They are typically collected at regular intervals (every few seconds or minutes) or updated in real-time by code, and then aggregated. We often visualize metrics on graphs (like a Grafana dashboard) and set alerts on them (e.g., if "error_rate > 5% over 5 minutes", trigger an alert).

For microservices, common metrics include **latency** (how long each request takes), **throughput** (requests per second), **error rate** (percentage of requests failing),

CPU/memory usage, etc. These keep tabs on the health and performance of services. In intelligent microservices, we have **additional metrics** to worry about: metrics about the AI's performance and behavior itself.

Some examples of AI-specific metrics:

- **Accuracy or Precision/Recall:** e.g., "fraud model precision = 98% this week" (meaning 98% of transactions flagged as fraud were indeed fraud – a measure of false positives). Or recall: "of all fraud, our system caught 90%".

- **Drift metrics:** e.g., "distribution shift = 0.1" measuring how much the input data distribution has diverged from the training data distribution (we'll discuss drift detection later).

- **Throughput of decisions:** e.g., "recommendations per minute = 500" or "transactions scored per second".

- **Success rate of AI actions:** e.g., for an AI that handles support tickets, "automation success rate = 85%" (meaning 15% of cases had to be escalated to a human).

- **Number of human overrides:** e.g., "today 5 transactions flagged as fraud were unblocked by humans" – if this number spikes, it's a red flag (the AI is too aggressive or making mistakes).

- **Model confidence average:** e.g., "average confidence of classification = 0.7" – a drop might indicate the model is encountering more uncertain cases than before.

- **Business metrics tied to AI decisions:** e.g., "conversion rate from recommendations = 5%" for an e-commerce recommender (how many recommendations led to a sale).

This is hugely important: it closes the loop to business value.

These metrics help answer the question: "How well is our AI component doing its job?" Not just from a system perspective, but from an AI quality perspective.

To capture metrics in code, you often use a library or agent. For example, if using Prometheus (a popular open-source monitoring system), you might use language-specific client libraries to define metrics and update them. Here's a quick illustration in Python of defining and using a metric:

```
from prometheus_client import Counter

# Define a counter metric for fraud alerts
fraud_alerts = Counter('fraud_alerts_total', 'Number of
flagged as fraud', ['outcome'])

# ... inside the fraud detection logic:
if is_fraud:
    fraud_alerts.labels(outcome="fraudulent").inc()
else:
    fraud_alerts.labels(outcome="legitimate").inc()
```

In this snippet, we define a counter fraud_alerts_total labeled by outcome. Each time the service processes a transaction, it increments either the "fraudulent" bucket or "legitimate" bucket. Prometheus would scrape this metric and we could graph the total fraud flags vs legitimate transactions over time, or compute the percentage flagged. We might derive a precision metric by combining this with how many of those fraud flags were confirmed true fraud later (if that info is available as another metric or log).

Custom Metrics for AI

General metrics (latency, errors) are still very much needed – an AI service can slow down or crash like any other. But **custom AI metrics** provide insight into something that pure technical metrics can't: the effectiveness of the AI's decisions. It's a good practice to brainstorm a few key metrics for each intelligent microservice that reflect its success. For a fraud service, it might be *false positive rate* and *false negative rate*. For a recommender, it's click-through or conversion from recommendations. For a healthcare vital monitor, it could be *alert precision* (how many alerts were real issues vs false alarms).

By tracking these metrics, teams can set **targets** (like "keep precision above 90%") and get alerted if things degrade. For instance, if a new model deployment causes precision to drop to 70%, a metric-based alert can catch that early, possibly before too much damage is done (e.g., too many false alarms or misses).

Define AI-specific metrics: Don't rely only on generic system metrics. Define at least one metric that captures the quality of decisions for each AI service. For example, FinBank's fraud microservice tracks precision – the percentage of flagged transactions that were actually fraudulent. If this metric dips, it signals the model might be flagging too many innocents (false positives) . Likewise, a medical alerting service might track how many alerts were confirmed as true emergencies. These metrics are your early warning for AI performance issues, not just system issues.

Challenge – False Positives at FinBank: Early on, FinBank's fraud detection agent started flagging a lot of transactions as suspicious – many more than expected. It turned out a hefty portion were false positives (legitimate purchases mistakenly flagged as fraud). This overwhelmed the manual review team and annoyed customers. How did the team notice? They had a custom precision metric in place. The dashboard showed that only 40% of flagged transactions were truly fraudulent – far below their 90% target. Armed with this insight, the engineers retrained the model with better data and adjusted its thresholds. Over time, they watched that precision metric climb back up, confirming that the false-positive problem was largely solved. Without that metric, they might have only realized the issue after getting complaints from the business; with it, they caught the drift in behavior within hours and took action.

Metrics are typically aggregated and visualized. Tools like **Grafana** can display metrics from sources like Prometheus, and you can create charts such as "Model accuracy over last 30 days" or "Number of alerts vs. number of real incidents". An interesting pattern is to put system metrics and AI metrics side by side. For instance, you might have one graph showing API latency and errors (to monitor system health), and right next to it another graph showing model accuracy or business KPI (to monitor decision health). Both are needed for a full picture: your service could be fast and error-free yet making terrible recommendations – or conversely, your model could be spot-on but a bug is causing requests to time out.

One more category to consider: **event metrics vs aggregated metrics**. Sometimes rather than count something, you want to record a value distribution. For example, distribution of prediction confidence scores, or the error between a predicted value and actual outcome. Prometheus offers histograms or summaries for that. Other systems might treat these as events that can be statistically analyzed. If you find yourself wanting to analyze distribution of an AI output (say, the model's score spread), you might log events and analyze offline, or use a specialized monitoring tool.

Remember, metrics feed alerting. We often set up **alerts** on metrics via systems like

Prometheus Alertmanager, PagerDuty, CloudWatch Alarms, etc. In an AI microservice context, it's wise to set alerts not just on the usual suspects (high error rate, high latency) but also on AI metrics. For example: *"Alert if fraud precision < 80% for >1 hour"* or *"Alert if recommendation click-through drops by 50% week-over-week"*. These ensure that if the AI starts misbehaving in a subtle way, someone gets notified. This is part of a practice sometimes called **ML monitoring** or **ML observability**, which parallels traditional monitoring but for model performance.

Tracing: Following the Distributed Journey

Distributed **tracing** is perhaps the most magical of the observability pillars when it comes to microservices. A trace follows the path of a single transaction or request as it travels through multiple services, capturing timing and metadata at each step. If logs are like diary entries and metrics are dashboard gauges, a trace is like a **full play-by-play replay** of a single user's request.

In a microservice architecture, a single action (say, a user clicking "Buy" in a shopping app) might involve a dozen services: Shopping Cart service, Inventory service, Payment service, Fraud service (with an AI model), Notification service, etc. If that overall process fails or is slow, how do you pinpoint where things went wrong? Tracing is the answer. Each service, when tracing is enabled, produces trace spans – pieces of the timeline – that get stitched together via an ID. The result is you can see, for example, that the "Buy" request took 3 seconds, with Service A taking 100ms, Service B 2500ms (ah, a bottleneck!), Service C 300ms, etc., and within Service B, perhaps a database call took 2400ms (the root cause).

For intelligent microservices, tracing plays a similar role, but there are some nuances:

- We want to trace **inference calls and data processing** within services. For instance, if Service B above was an AI service that loaded a model and made a prediction, we might instrument the model inference as a span. This would show up in the trace, e.g., "ModelScoringSpan: 2400ms" which explains why service B was slow – the model took a long time to run or maybe was stuck loading.

- Traces can carry **context information** that helps with debugging AI decisions. Modern tracing systems (like OpenTelemetry) allow attaching metadata to spans. We could attach the model's version or the decision result to the trace. Then when viewing a trace, you not only see timings but also that "the fraud service blocked this transaction (decision=BLOCK, model_version=2025-04-01)". This is powerful for correlating issues ("Aha, all slow traces involve model_version=2025-04-01; maybe that new model is heavy computationally?").

- Tracing is essential for **explainability across services**. Suppose something bad happened – a user complains "your system wrongly denied my loan application." To investigate, you might pull up the trace for that loan application request ID. The trace would show the calls from the frontend to the Credit Check microservice to the AI Risk Assessment service and maybe to a Manual Review service. Alongside each, you can dig into logs if needed. But at a high level, the trace will tell you which service made the deny decision. If we instrument well, it might even tell *why* (via attached metadata or by quickly jumping to logs via trace IDs). In other words, tracing connects the dots in a complex graph of microservices.

To implement tracing, we usually rely on instrumentation libraries or frameworks. **OpenTelemetry** has become a standard for instrumentation, supporting traces (and also metrics and logs). Using OpenTelemetry, for example, you might do something like:

```
from opentelemetry import trace

tracer = trace.get_tracer(__name__)

with tracer.start_as_current_span("fraud_check") as span:
    result = fraud_model.predict(transaction_data)
    span.set_attribute("decision", result.decision) # e.g., "ALLOW"
or "BLOCK"
    span.set_attribute("score", result.score)
    span.set_attribute("model_version", MODEL_VERSION)
    # ... proceed with logic based on result
```

What's happening here: we start a new trace span named "fraud_check" every time we do a fraud scoring. We record attributes like the decision, score, and model version on that span. When tracing is configured (e.g., exporting to Jaeger or Zipkin or a tracing backend), each request that involves this fraud check will have these details. If the fraud check is part of a larger trace (say a whole transaction processing flow), this span will appear as one segment of the whole trace.

Often we integrate tracing in a way that spans are created automatically for incoming web requests (middleware does it) and outgoing HTTP calls (HTTP client library does it). For AI internals, we might add custom spans like the one above for the model call or any slow operations (like loading a model or a large dataset).

Correlation of logs and traces: A powerful practice is to include a trace ID in your logs.

Most tracing systems provide a current trace/context ID that you can inject into log records. This way, when you see a log line, you can know which trace (which specific transaction) it was part of. And vice versa: from a trace viewer, you can often jump to logs for that trace. This cross-linking makes debugging extremely efficient. For example, you could search your centralized logs for a specific trace ID to pull up all logs from all services related to that user request. This essentially reconstructs everything that happened for that request – you see the combined story from logs, aligned with the timeline from traces.

In summary, distributed tracing ties everything together, especially in a **distributed intelligent system** where many decisions by different services collectively determine the outcome. It helps answer questions like:

- "Which service or step caused the slow-down in this workflow?"
- "What path did this particular request follow through our myriad of microservices?"
- "Which microservice's decision led to this end result for the user?"

"Are different microservices properly handing off information? Where could something have gotten lost?"

Tracing tools like **Jaeger** or **Zipkin** provide a UI to visualize these traces. In the UI, you typically see a Gantt-chart-like view: time on the X axis, and each service's span on the Y axis, showing how long each took and how they overlapped. Many also let you click on spans to see metadata (like our decision, score attributes).

One more advanced use: If you run an **A/B test** or canary (which we'll discuss later), tracing can sometimes help ensure you're comparing apples to apples. For example, if we have two versions of a recommendation service (A and B), we might tag traces or spans with the variant ("model_A" vs "model_B"). This way, you could filter traces by variant to see performance differences or see any errors unique to one variant.

It's worth noting that there is some overhead to tracing – it can add a small latency and network cost – but in practice, sampling is used (e.g., capture maybe 1% or 10% of all requests as traces) to limit overhead while still getting useful info. For critical user journeys, sometimes 100% tracing is enabled.

To tie together logs, metrics, and traces, many teams use an **observability platform** or a combination of tools. We might use the ELK stack for logs, Prometheus/Grafana for metrics, and Jaeger for tracing. Or use a commercial solution that handles all three. The emerging standard, as mentioned, is OpenTelemetry which aims to standardize the telemetry data so that it can be sent to different backends interchangeably. This is great because you can instrument your code once and then decide later whether to send data to an open source stack or a vendor SaaS.

Before moving on, let's illustrate how a single service contributes to all three pillars:

```
                        +---------------------------+
Request  ----->  |  Intelligent Service X   |  -----> Response
                        +---------------------------+
                           |        |        |
     Logs  <----------+        |        +---------> Traces
     (events, decisions)      |                 (spans across services)
                                    v
                                 Metrics
                        (counters, timings, KPIs)
```

Diagram: An intelligent microservice produces logs (detailed events), metrics (aggregated stats), and traces (distributed call information) as part of its telemetry.

Each service in a microservice ecosystem does similar, resulting in a rich tapestry of data about the system's behavior.

Now that we have a good grasp of the three pillars of observability, let's talk about how we use all this data not just to monitor the system, but to **explain and improve** it.

Observability as the Key to Explainable AI Decisions

One of the big promises (and challenges) of intelligent microservices is that they make autonomous decisions. But autonomy without accountability is a recipe for mistrust. If our AIdriven service cannot explain its actions, users and developers will be uneasy. Observability data – logs, metrics, and traces – become the foundation for **explainability** and building **trust** in these systems.

When something goes wrong in an AI service (say, a biased decision or an unexpected action), observability is how we perform an "autopsy" on that decision. It's very much like a flight recorder (black box) in airplanes – after an incident, investigators pore over the recorded telemetry to understand what the automated systems and pilots did. For our intelligent microservices, the logs and traces are our flight recorder.

But explainability isn't only post-incident. We also want real-time or proactive transparency. For example, a user might ask: "Why was my loan denied?" You might build a feature to answer that – drawing on logged reasons or an explainability component of the AI. Perhaps your AI logs "denied due to debt-to-income too high (DTI=45% > threshold 40%)." That could be surfaced to the user or at least to a customer support agent. In less formal terms, observability data can serve as a basis for generating human-friendly explanations on the fly.

At a team/process level, having observable AI systems fosters a culture of **open-box development** rather than black-box. Teams regularly review **dashboards of AI metrics** to discuss how to improve the model. They investigate anomalies together, tracing through what the AI did. This cross-disciplinary transparency (DevOps engineers, data scientists, domain experts all looking at the same telemetry) helps align the AI's behavior with business goals and ethical standards.

Consider an example in healthcare with MediHealth's patient monitoring service: If an alert fired that wasn't actually an emergency (false alarm), doctors will want to know why the AI thought it was necessary to call a nurse at 3am. With good observability, the system might have logged: "Alert triggered by Agent_HeartRate: patient 42 heart rate 180 bpm (threshold 170), Agent_Oxygen: oxygen stable. **Reason:** Heart rate spiked while oxygen normal -> potential sensor error or anxiety episode." This log (perhaps accompanied by a trace of what each agent reported) can help the medical team and developers understand that it was a borderline case or a sensor glitch. They might adjust the threshold or add a rule to crossverify with more signals, thereby improving the system. Without that info, the AI just looks flaky and people lose trust, possibly leading them to ignore future alerts (crying wolf problem).

Intelligent Monitoring in Healthcare (MediHealth): MediHealth Systems deploys

AI agents to watch ICU patient vitals. One night, an alert is issued for patient John Doe. The on-call doctor checks the system's observability dashboard: a trace of the alert shows that the *HeartRateAgent* flagged a heart rate of 185 bpm at 02:14, while the *OxygenAgent* reported normal oxygen levels. The log details for the alert say: "Anomaly detected: heart rate 185 (>180 threshold) while O_2 stable; possible arrhythmia or sensor error." The doctor quickly examines John – it turns out to be a transient spike (false alarm). But thanks to these logs and traces, the doctor understands why the AI alerted. The development team later reviews the incident: the metrics show that out of 50 alerts that week, 5 were false alarms. They decide to refine the algorithm (perhaps requiring two consecutive high readings before alerting). This example shows how observability makes the AI's decisions explainable and builds trust – doctors see *why* an alert happened and can give informed feedback to improve the system.

In addition to logs and traces, sometimes **metrics** are used for explainability as well. For instance, you might expose a metric like "risk_score" or "similarity_score" for a particular user's recommendation, which can be displayed as a visual (e.g., a gauge "your risk score is 7/10"). That's more for end-user transparency features. Internally, metrics like feature importance can be logged: e.g., an AI might output that feature XYZ contributed 80% to the decision, which you log or track.

Now, it's worth noting that **observability ≠ interpretability** in the strictest sense. There's a whole field of AI called Explainable AI (XAI) which focuses on interpreting complex models (like showing which pixels in an image influenced a classification, or using SHAP values for feature importance). Observability data might not tell you the innards of a neural network's calculation, but it *does* tell you everything that went into and came out of that network, and the context around it. Many times, that's enough to reason about why it did something. And if not, you can at least determine that it *did* behave unexpectedly, and use that data point for an offline analysis with XAI tools.

Another critical aspect is **auditability**. In finance and other regulated fields, you might need to keep logs for every automated decision and even be able to replay them. Observability supports this by providing the logs and traces to reconstruct scenarios. Some teams even log the input features and model output in a way that they can later feed the same inputs to a model to reproduce the decision (even after the model has evolved, they might store the exact model version or parameters with the log). This overlaps with the concept of a **model registry** or dataset/version tracking in MLOps – ensuring you know exactly which model was used for each prediction. Observability comes in when you annotate your telemetry with model version IDs and have that single source of truth for what happened when.

Let's not forget the human element: it's not just about logs and numbers, sometimes we need narrative. Many teams introduce **incident reviews** or **AI ethics reviews** where they go through notable decisions made by the system. Having good observability data turns these reviews from guesswork into science. The team can say, for example: "On July 3, our loan service denied 123 applications. Out of these, 5 were later manually approved on appeal. Let's see what was common in those 5." Then they can query logs: perhaps all 5 had high debt-to-income but also very high credit scores. That could reveal a blind spot in the model's logic. Without logs to identify those cases and metrics to quantify how often it happens, such issues could go undetected.

To sum up, observability is the **senses and memory** of your intelligent system – it's how it tells you what it's thinking and doing. This not only helps in fixing problems but is fundamental to building trust. As one fictional DevOps engineer on our team would say:

Dana, DevOps Lead: "In complex AI-driven systems, if you're not watching, you're flying blind. Observability isn't just nice-to-have; it's how we trust what the AI is doing."

By watching the logs and metrics like a hawk, Dana's team at ShopEase knows that their recommendation engine isn't running amok – they can see its pulse and thought process. This gives them confidence to let it run autonomously, and also the ability to spot when it needs intervention.

Now that we've established how observability provides insight and explainability, it's time to talk about the next step: using those insights to **continuously improve** the system.

Observability data isn't just for humans to look at; it can feed back into the AI itself. Let's explore how we close the loop: from observation to learning.

Closing the Loop: Learning from Observability

Observability doesn't end with understanding; in intelligent microservices it naturally feeds into **learning and adaptation**. Think of observability data as the eyes and ears of your system. What do living organisms do with sensory information? They **learn and adapt** to their environment. In the same way, our microservices can use logs, metrics, and traces not only for debugging but as fuel for their own improvement loops.

This brings us to the idea of **continuous learning** in production. We've deployed an AI model as part of a microservice – great, it's making decisions. But over time, things change: user behavior shifts, fraudsters find new tricks, new products appear, seasons change, you name it. The model that was great in January might be mediocre by June if left untouched. Observability data – essentially *everything happening to and because of our model in the wild* – is exactly what we need to keep the model fresh.

Building Feedback Loops and Continuous Improvement

A **feedback loop** in this context means using the outputs and outcomes of the system (as observed via telemetry) to update and improve the system's internal logic (the model or decision policy). This concept turns a static deployment into a learning, adaptive service.

Let's outline a typical feedback loop for an intelligent microservice:

1. **Collect data from running system:** This includes logs of decisions, metrics of outcomes, and perhaps external data like user feedback. For example, our fraud service logs which transactions it flagged and whether those were actually fraudulent (maybe via a later label). Our recommendation service logs what it recommended and whether the user clicked/bought.

2. **Aggregate and analyze the data:** Maybe nightly or continuously, these logs are processed. We might calculate new training examples from them. E.g., every fraud decision (transaction + model features + outcome) becomes part of a growing dataset.

3. **Retrain or adjust the model/policy:** Using the new data, we train an updated model or tweak rules. Perhaps we do this offline on a powerful machine once a day, or if we're fancy, in an online manner as data streams in.

4. **Test the updated model:** We validate that the new model is better (maybe using a holdout or comparing metrics on recent data).

5. **Deploy the new model into the microservice:** The microservice now uses the updated model for new decisions.

6. **Observe again:** The cycle repeats – as the updated model operates, we collect new data about its performance.

This creates a virtuous cycle: **Production generates data -> data yields a better model -> better model goes to production -> repeat.**

We can represent this loop in a simple ASCII diagram:

```
Production System (AI Microservices)
    |
    | (telemetry: decisions, outcomes)
    v
[ Data Collection & Pipeline ] (e.g., stream processing or batch
logs)
    |
    v
[ Model Training/Update ] (offline retraining or online learning)
    |
    v
[ Deployment of New Model ] (CI/CD for models, or live update)
    └────────> Feeds back into Production (improved decisions)
```

In practice, implementing this loop touches on **MLOps** (Machine Learning Operations). You need infrastructure to grab the logs or events, a data store to accumulate training data, processes to retrain models, and deployment automation to push new models out safely. It's non-trivial, but many organizations are building such pipelines to keep their AI services sharp.

There are different patterns for how continuous learning can be implemented:

- **Offline batch retraining:** e.g., Retrain the model every night on the day's new data, then deploy it if it performs well. This is simpler and easier to validate.

- **Online learning (streaming updates):** e.g., Each new data point (a user clicks, a fraud confirmed) immediately or continuously updates the model's state (perhaps the model is an online algorithm that can learn incrementally, or it's a reinforcement learning agent updating its Q-values or policy with each interaction). This is more complex, as errors can also propagate quickly if not monitored.

- **Periodic human-in-the-loop refresh:** e.g., Every month, data scientists review the collected data, do some feature engineering or investigation, and release a new model version. (This is more manual but often a real scenario in enterprises.)

Regardless of approach, the key is **closing the loop using observations**.

For example, suppose ShopEase notices that their recommendation engine's click-through rate metric has been dropping for the last 2 weeks – maybe user tastes changed or competitors have better recommendations now. Because they have a feedback loop, they automatically incorporate the last 2 weeks of user interaction data and retrain the recommendation model. The new model, hopefully, picks up on the shift in user interest (say, a surge in popularity of eco-friendly products) and improves recommendations accordingly, which the metrics will then reflect by showing an uptick in click-through again. If they didn't have that loop, the model would continue its slow decline in relevance.

Online Learning vs. Periodic Retraining

Let's discuss the two broad extremes on the spectrum of continuous learning:

- **Online Learning:** The model updates continuously or very frequently (nearly every data point or small batch of data causes an update). This could be done with algorithms that naturally support it (like some reinforcement learning, bandit algorithms, or online gradient descent on streaming data). The advantage is the model is always up-to-date, reacting quickly to changes. The downside is it can potentially learn *bad habits* if not carefully regulated – e.g., noise or a temporary anomaly could throw it off. Also, debugging an online-updated model is harder because it's a moving target that changed a bit with each data point. Reproducibility can suffer if you don't log exactly when and how the model changed.

- **Periodic (Batch) Retraining:** The model is retrained on a schedule (hourly, daily, weekly) using a batch of accumulated data. This is easier to manage and allows more rigorous testing of the new model before it replaces the old. It's more stable in that sense. However, it lags behind; if something fundamentally changes right after the last retraining, you may have to wait until the next cycle to catch up, during which performance might dip.

Sometimes a hybrid is used: e.g., a service might do small online updates for certain parameters but retrain fully once in a while to "re-ground" the model.

Let's illustrate some pros and cons in a table for clarity:

Learning Approach	**Pros**	**Cons**	**Use Cases 🔍**
Online Learning	- Adapts very quickly to new data or trends.- Continuously improves without manual intervention.- Can personalize to users in real-time (e.g., news feed learns your preferences as you click).	- Risk of drifting due to noise or malicious input (needs careful monitoring).- Harder to test or reproduce exact model state at a given time.- Concept drift can mislead it if not handled (it might chase short-term fluctuations).	- High-frequency trading algorithms adjusting to market in real-time.- Personalized content recommendation that updates with each user interaction.- Systems where data distribution shifts rapidly and continuously (e.g., realtime IoT anomaly detection).
Learning Approach	**Pros**	**Cons**	**Use Cases 🔍**

Periodic Retraining	- More stability and ability to QA models before deploy.- Easier to implement with standard ML pipelines (train on batch data).- Can accumulate more data for potentially better model (not just most recent).	- Slower to adapt (could be outdated between retrains).- Sudden changes in environment might cause performance to dip until next retrain.- Operational overhead to schedule retrains and manage model versions.	- Most enterprise ML systems (e.g., retrain fraud model nightly or weekly).- Scenarios where data changes, but not instantaneously (e.g., monthly sales trends).- When regulatory validation is needed for new model versions before deployment.

Neither is "better" universally; many systems start with periodic retraining (for simplicity and safety), then maybe move to more online methods as they mature or if the domain demands it.

Crucially, **monitoring is required for both**. If you do online learning, you must monitor metrics to catch if the model goes haywire (maybe due to compounding error). If you do periodic, you should monitor performance to decide *when* to trigger a retrain (maybe sooner than schedule if needed).

This is where earlier metrics we discussed become triggers for learning. For instance, you might set a rule: if model accuracy on recent data (as estimated by comparing predictions to actual outcomes in logs) falls below X, initiate a model retraining job. Essentially the system can *notice its own decline* via telemetry and take action.

Detecting Anomalies and Model Drift

An intelligent service not only outputs decisions – it also continuously receives inputs. Over time, the nature of those inputs can change. This could be **data drift** (the statistical properties of input data change) or **concept drift** (the relationship between inputs and outputs changes, e.g., fraudsters change tactics so the same inputs no longer mean fraud vs not fraud in the same way).

Observability plays a key role in detecting these shifts. How? Through anomaly detection on the streams of telemetry.

Two broad categories of anomalies we care about:

- **System anomalies:** e.g., a surge in errors, latency spikes – basically what traditional APM monitors. These can affect AI too (like if a model serving endpoint is suddenly slow or failing).

- **AI behavior anomalies:** e.g., sudden change in prediction distribution (the model starts predicting "yes" 90% of the time instead of the usual 50%), or downstream impact anomalies (user engagement drops sharply).

Let's focus on the AI behavior anomalies (often called model drift or data drift):

- We can monitor **input data drift** by tracking metrics on input features. For example, average transaction amount per day, or distribution of customer ages. If those metrics deviate significantly from the training data stats (which we know from training time), that's a signal that the model is now getting data that looks "different" than before. Tools like Evidently AI or WhyLabs provide automated ways to compute drift metrics on continuous data.

- We can monitor **output drift** by looking at the model's predictions. If a binary classifier used to output 50% yes, 50% no, and now it's 90% yes – either the world changed or the model is off. Similarly, if confidence scores skyrocket to always high, maybe the model is overconfident on new data.

- We can monitor **performance drift** by measuring how outcomes compare to predictions. This is tricky because you might not know ground truth immediately (e.g., you predicted user will click – you only know later if they did or didn't). But for something like fraud, eventually you get labels (fraud confirmed or not). So you can measure that precision/recall over time – if it's degrading, that's performance drift. Even before ground truth, you might use proxy feedback (like user behavior) as indicators.

Anomaly detection techniques range from simple thresholds to complex statistical models or even AI watching AI (meta-learning). A simple approach might be:

- Set static thresholds: e.g., if mean of feature_X today is outside [min,max] observed in training, flag drift.

- Use standard deviation or percentile-based bands: e.g., alert if metric goes beyond 3σ from recent mean.

- Use change detection tests (like Kolmogorov-Smirnov test for distribution change) on data batches.

- Use an anomaly detection service or algorithm (like an isolation forest, an LSTM anomaly detector, etc.) on the metrics.

Often, anomaly detection for ML is done by specialized tools that plug into your observability data. For instance, some use **ML for observability** (AIOps) to detect anomalies in logs and metrics – which we can apply to our AI metrics.

Watch for Silent Model Drift: Models can degrade silently over time. Unlike a server crash that triggers obvious errors, a model might continue giving outputs that *seem* fine (no exceptions thrown) but whose quality is slipping. Always monitor your AI outcomes, not just system uptime. For example, an ecommerce model might steadily lose accuracy as user trends shift – if you're not measuring recommendation success rates or error rates of predictions, you won't notice until revenue is hit. Set up alerts for when key AI performance metrics (accuracy, precision, etc.) drop below expected levels or deviate sharply . Early detection of drift allows you to retrain or adjust before things get bad.

In our FinBank case, say the fraudsters discover a new loophole that the model wasn't trained on – suddenly a lot more fraud is getting through. How would FinBank notice? Ideally, they have metrics like "frauds caught vs frauds actually happened" (though knowing actual fraud that happened is tricky – often you only know weeks later due to investigations). However, maybe they notice an anomaly that the model's score distribution changed (lots more low scores even as fraud is higher) – that discrepancy could ring alarm bells. Or their downstream business metric "fraud loss" starts climbing; that definitely gets attention.

Anomaly detection can also be used at the system level to detect weird behavior of an AI service. Perhaps the AI service is supposed to be called 1000 times a day, but one day it's called 100,000 times (maybe a buggy infinite loop or an attack). A simple metric alert on request rate anomaly saves the day.

Another angle: **Meta-monitoring with AI** – using an AI to monitor another AI. For example, a reinforcement learning agent controlling some process could have another agent evaluating whether the actions are out of normal bounds. Or a simple rules-based watchdog that says "if agent output deviates too far from expected range, take action (maybe revert to a safe mode or notify humans)."

All these mechanisms ensure our autonomous services don't go rogue or if they do, we catch them quickly.

A/B Testing and Model Experimentation

Continuous learning isn't just about retraining, it's also about **trying out improvements in a safe way**. Whenever we develop a new model or a tweak to logic, we ideally want to test it against the current version to ensure it's truly better (and not just overfit to a subset of data).

This is where **A/B testing** (and its cousin, **canary releases**) comes in for intelligent microservices. In an A/B test for a microservice, you have two versions of the service (or model) running simultaneously:

- **A (control):** The current model or logic.
- **B (treatment):** The new model or logic you think will perform better.

You route a percentage of real traffic to version B (say 10%) and the rest to version A (90%). Both are serving users in parallel, unbeknownst to them. You then compare outcomes using observability data:

- Metrics: Which version has better business metrics (conversion, error rate, whatever you care about)? For AI, does version B have higher accuracy or user engagement?

- Logs: Are there any concerning logs in B (like more error messages or weird decisions) not present in A?

- Traces: Does version B impact performance (maybe it's slower) compared to A?

If B consistently outperforms A on the key metrics and doesn't introduce problems, you can roll it out to everyone (promote B to 100%). If B underperforms or has issues, you discard it – the observability data saved you from a potentially bad deployment.

Experimentation at ShopEase: ShopEase's recommendation microservice has been running Model v1 for a while. The data science team thinks Model v2 could increase sales by making more personalized picks. Rather than flip the switch for all users, the team does an A/B test. 20% of users (randomly selected) will get recommendations from Model v2 (Group B), while 80% continue with Model v1 (Group A). They instrument everything:

Both versions log recommendations and user interactions identically.

Metrics are split by group: click-through rate, conversion rate, average order value for A vs B are displayed on a Grafana dashboard.

They even tag traces with a label indicating which model served the request, to ensure latency is similar.

After two weeks, the results are in: Group B (new model) shows a **15% higher click-through rate** and a modest uptick in sales, with no increase in average response time. Logs show no red flags in Model v2's behavior. Armed with this evidence, ShopEase rolls out Model v2 to 100% of users. Thanks to observability and a careful experiment, they gained confidence in the new model and quantifiably improved their business metrics.

A/B testing can be seen as a form of "learning" too – the organization learns which version is better. Sometimes multiple versions are tested (A/B/C testing or multi-armed bandits which dynamically allocate traffic to better performing versions).

One challenge is that for some metrics you need statistical significance, so you might run the test for a while. Observability plays a role in monitoring the test in real-time though – if the new version was catastrophically bad, you'd see it quickly in metrics and can stop the test.

Conversely, if it's clearly outshining, you might promote early.

Canary releases are a similar idea but often focused on detecting issues rather than comparing improvements. A canary is when you deploy the new version to a small portion (say 5% of traffic or one region) just to ensure it's healthy (no errors, performance is fine). If it passes the canary, you gradually roll out to more. Observability is what you watch during a canary: you compare error rates, latencies, etc., between canary and baseline. For an AI model, you'd also watch AI metrics. If you had a metric for user engagement, you ensure it didn't drop on the canary users.

The continuous learning loop often involves these experiments: you don't just blindly push a new model – you test it. Some more advanced setups even do **shadow testing** where the new model runs in parallel in the background (not affecting actual users, just getting the same inputs and logging what it *would* have done). Then you compare its decisions to the live model's decisions and see which would have been better. This requires being able to measure "better" offline, but for example, in fraud detection, you could shadow a new model and see, after the fact, if it would have caught more fraud or made more false alarms compared to the production model, based on actual outcomes. If it looks promising, then you A/B or promote it.

Centralized Learning Across Services

So far, we discussed feedback loops mostly per service. But what if you have **multiple intelligent microservices** in your system? Could they learn from each other? The answer is often yes – and this is an area sometimes called **centralized knowledge bases** or **crossservice learning**.

Consider an organization with several AI-driven services:

- A fraud detection service for credit cards.
- A login anomaly detector (flagging suspicious logins).
- A recommendation service for cross-selling products.
- A support ticket triage AI.

Each of these produces observability data (decisions, outcomes). If we keep that data siloed, each AI learns only from its own experience. But sometimes, combining data can yield a richer picture. For instance, the fraud detection and login anomaly detector might share signals – a coordinated attack might cause both logins and transactions to be weird. If both sets of data are fed into a central model, it might detect a pattern that individual services miss. Or simply, the company might build a feature store or data lake where all events (login events, transaction events, etc.) are logged, and then use that to train a more holistic model (like a user risk profile model that informs both services).

A concrete example: one service might detect that a user's account is behaving oddly and assign a risk score to the user. Another service (say a money transfer service) could reuse that risk score as an input to its own model. In effect, one microservice's "observation" (user risk) becomes input for another. This is *feedback across services*.

To enable this, you need some centralization:

- **Central telemetry store:** e.g., send all important events to a message queue or analytics database. Kafka is often used to stream events (each service produces events like "fraud_check_performed" or "recommendation_shown") into a central stream, which can be consumed by both monitoring tools and learning pipelines.

- **Feature store:** an emerging pattern in MLOps where processed features (like aggregated signals about users) are stored and shared so that multiple models/services can use the same up-to-date features. E.g., a feature store might have

"user_recent_login_failures" or "user_purchase_history_vector" that any service's model can query.

- **Cross-service analytics:** teams look at combined data to find insights. Perhaps a trend is only visible when you overlay metrics from two services.

However, there are also **cautions**:

- **Data privacy & domain boundaries:** Combining data from different domains might be inappropriate. For instance, MediHealth wouldn't mix patient data with e-commerce click data obviously. Even within a company, maybe certain data is restricted

(compliance or ethical reasons). So you share knowledge where it makes sense and is allowed.

- **Differing contexts:** Each service's data might have different meanings or scales. Blindly pooling can cause nonsense. It often needs feature engineering to be useful across contexts.

- **Complexity:** Centralizing everything can make the system complex and potentially brittle. We add dependencies between services via shared data. So it's a balance between leveraging shared insights and keeping services decoupled. A smart approach is to share only higher-level insights or distilled knowledge (like the risk score example, rather than raw data).

One approach is to have an overarching "learning platform" in the organization: all telemetry flows into a data lake; data scientists can query it to find patterns; if they find a useful insight (say a feature that predicts failures), they deploy that back into the relevant microservices (maybe as a new model or a rule). It's not fully automated perhaps, but it's enabled by the observability data.

In essence, an organization-wide learning culture may emerge, where every incident or odd pattern observed in one corner is analyzed and then generalized improvements are applied elsewhere if relevant. For example, if FinBank's payment service learned something about fraud patterns at midnight, maybe the login service can pre-emptively bolster security at those times too.

The interplay between **centralized vs localized learning** mirrors the debate of federal vs local control. Microservices traditionally favor autonomy (each service encapsulated). But intelligent microservices can benefit from shared brains at times. It's akin to a team of humans: each team member learns from their tasks, but they also have a weekly knowledge-sharing meeting to discuss what they learned so everyone benefits.

To implement centralized feedback:

- Use an **event bus or stream**: All services publish key events (maybe sanitized) to a central stream. For example, whenever any model makes a high-confidence prediction or encounters an anomaly, publish an event like "AlertRaised(service=X, details…)".

Others could subscribe or it's just stored.

- **Reuse models**: One microservice's model might actually be useful for another. If you have an NLP model that understands customer sentiment from support tickets, that could be reused by a marketing service analyzing reviews.

- Ensure good **data cataloging**: know what data is being collected where, so you can combine meaningfully.

Finally, continuous learning isn't just technical – it's also about the **team learning processes**. Many organizations adopt a practice of periodic model reviews, post-mortems for AI mistakes (e.g., "why did the model screw up on case X?"), and feeding those lessons into the next development cycle. Observability data is always the starting point for those discussions.

We've covered a lot: from instrumentation to using that data for explanation and learning.

Now let's get practical and talk about architectures and tools that make this all possible.

Implementing Observability: Architecture and Tools

By now, it's clear that achieving observability and continuous learning requires solid tooling. It's not all going to happen via print statements and hope! In this section, we'll outline a practical architecture for observability and some common tools and technologies you can use. We'll also see how these tools fit into the feedback loops we described.

Telemetry Collection and Aggregation

At the heart of observability is **telemetry data collection**. Each microservice needs to emit logs, metrics, and traces, and we need to gather those centrally.

A common approach:

- **Logging**: Use a logging agent or forwarder (like **Fluentd/Fluent Bit**, **Logstash**, or cloud-specific ones) on each service instance or host. These capture logs (stdout or log files) and ship them to a central logging system. The central system could be an ELK stack (Elasticsearch for storage + Kibana for search UI) or a cloud service like Azure Monitor, GCP's Stackdriver, etc. Elasticsearch/OpenSearch is popular for on-prem or open setups, allowing full-text search on logs. For high-scale, some use Kafka as an intermediate buffer for logs or use log-specific stores like Splunk. The key is, you must be able to query logs across all services, preferably with powerful filtering (e.g., all logs for a specific user ID or trace ID).

- **Metrics**: The de facto standard in cloud-native is **Prometheus** for metrics collection. Each service exposes an HTTP endpoint (usually /metrics) where current metric values can be scraped. Prometheus server polls these on a schedule and stores time-series data. Grafana is then often used to visualize and create dashboards from Prometheus (or other sources). There are alternatives like InfluxDB/Telegraf, Datadog, CloudWatch, etc., but the idea remains: you need a time-series DB for metrics and a dashboard+alerting layer. Prometheus also supports alert rules natively (Alertmanager). For intelligent microservices, you might integrate custom metrics by using Prometheus client libraries in your code (as shown in our Python example) or pushing metrics (there's a Pushgateway for short-lived jobs). Cloud providers have their metric systems (CloudWatch, Azure AppInsights, etc.) if you are in those ecosystems.

- **Tracing**: **OpenTelemetry** provides instrumentation libraries for many languages, which can export traces to various backends. One common open-source backend is **Jaeger** (which stores and lets you query traces), or **Zipkin**. There are also SaaS products like New Relic, Dynatrace, Lightstep, etc., that can ingest OpenTelemetry or have their own agents. To set up tracing, you often run an OpenTelemetry Collector (which receives trace data from services and forwards to the backend). Each service either does autoinstrumentation (some frameworks auto-generate spans for incoming/outgoing requests) or uses an SDK to create spans in code. The trace data includes span timings and any attributes you attach.

- **Correlation**: Ensure your logging and tracing are correlated. In practice, this might mean configuring your logger to include the trace ID from OpenTelemetry in each log entry. Many OpenTelemetry instrumentation libraries can do this automatically (via context propagation hooking into logging frameworks). If not, you may do it manually (pass trace IDs around or use MDC in Java logging).

Unified Telemetry: OpenTelemetry is worth emphasizing because it's one standard that tries to cover logs, metrics, and traces in a unified way. While still evolving (logs support is newer), the idea is to reduce the need for separate instrumentation for each pillar. You instrument once and can export the data to multiple systems. For instance, instrument your code with OpenTelemetry, then have one exporter send metrics to Prometheus, another send traces to Jaeger, and logs to Elasticsearch, for example. This unified approach also helps ensure things like consistent metadata (tags/attributes) across all telemetry.

Many organizations adopt a **"Telemetry sidecar"** or an agent that runs with each application instance to handle sending data out, so the app code can be cleaner. Service mesh technologies (like Istio, Linkerd) also often provide tracing and metrics for all service calls automatically, which can cover some observability without developers writing as much code (though application-level decisions still need app logging logic).

To handle scale, sometimes a **sampling** strategy is used: e.g., only 1 in N requests are traced (but logs for errors are kept at 100%). Metrics by nature are aggregated so they handle scale well. Logging at high volume may require indexing only partial data or using archival storage for older logs.

Monitoring and Visualization

After data collection, we need to **visualize and monitor** the data:

- **Dashboards (Grafana, Kibana, etc.)**: Grafana is frequently used to bring together Prometheus metrics, logs (via Elasticsearch or Loki), and traces (via Jaeger integration). You can create a single dashboard that, say, shows latency and error rate (from metrics), and have a panel that lists recent error log entries, etc. Kibana is specific to Elasticsearch and is great for exploring logs with full text search and creating some visualizations or quick analysis (like "how many errors per day"). Many choose Grafana for metrics and use Kibana for deep log dives.

- **Alerts**: Setting up alerts on metrics/logs is crucial for production. You might configure Prometheus Alertmanager to trigger an alert (email/Slack/PagerDuty) if certain conditions met (as discussed, system or AI metric thresholds). For logs, some use tools like ElastAlert or the log system's alerting if it has (Splunk has robust alerting on log queries, for example). There are also anomaly-detection-based alerting tools (some APM vendors have AI-driven alerting that learns normal patterns and alerts on deviations).

- **Analytics**: Sometimes you'll use Jupyter notebooks or SQL on the observability data for deeper analysis, especially for learning. For instance, you might dump a month of logs to a dataframe and do data science on it (like training an offline model or analyzing patterns). This requires data pipelines to get log data into an analysis-friendly form (maybe in S3 or a database). Tools like **ELT (Extract-Load-Transform)** jobs or streaming consumers can funnel data from the live systems into analytical stores.

Tools at a Glance

Here's a quick reference table of some tools and their roles, to solidify your understanding:

Aspect	Tools / Technologies	Role in Observability & Learning

Logging	*ELK/EFK Stack* (Elasticsearch/OpenSearch + Logstash/Fluentd + Kibana), **Splunk**, **Graylog**, **CloudWatch Logs**, **Azure Monitor**, **Elastic APM**, **Loki** + Grafana	Collect, index, and search log data from services. Enables querying what happened in the system. Kibana/Grafana provide visualization/search for logs.
Metrics	**Prometheus** (scraping metrics), **Grafana** (dashboards), **Datadog** (hosted solution), **CloudWatch Metrics**, **Azure App Insights**, **InfluxDB + Grafana**, **New Relic**	Continuously measure system and model performance. Prometheus collects numeric time-series (like CPU, latency, custom AI metrics). Grafana visualizes trends and comparisons. Alerts can be set on metrics.
Tracing	**Jaeger**, **Zipkin**, **OpenTelemetry** (libraries & Collector), **SkyWalking**, **AWS X-Ray**, **Azure**	Trace requests across services. OpenTelemetry provides the instrumentation standard;
Aspect	**Tools / Technologies**	**Role in Observability & Learning**
	Monitor (Distributed Tracing), **New Relic**, **Dynatrace**	Jaeger/Zipkin store and show trace timelines. Tracing helps pinpoint issues in complex workflows and correlates events.

Combined Observability	**OpenTelemetry** (unified SDK), **Lightstep**, **Splunk Observability**, **Elastic Observability**, **Dynatrace**, **Datadog**	Many commercial platforms unify logs, metrics, traces in one place with AI analytics on top. OpenTelemetry as an open standard makes it easier to send data to any backend or even multiple backends.
ML Monitoring (Modelspecific)	**WhyLabs**, **Arize AI**, **Evidently**, **Fiddler AI**, **Azure ML Monitoring**, **AWS SageMaker Model Monitor**, custom scripts with Pandas/NumPy on log data	These tools/libraries focus on tracking data drift, model performance, bias, etc., specifically for AI models in production. They often plug into your existing observability stack (e.g., ingesting predictions and outcomes) and provide dashboards/alerts for model metrics. For instance, WhyLabs can compute distribution drift over time for features and output alerts.

Data Pipelines for Learning	**Kafka**, **AWS Kinesis**, **Apache Flink/Kafka Streams**, **Spark**, **Airflow** (for batch jobs), **Feast** (feature store)	These aren't observability per se, but are the plumbing to get from observability data to retraining. Kafka can stream events (like logs or telemetry) to storage or trigger live updates. Airflow or similar can schedule retraining workflows. Feature stores like Feast help serve fresh features to models and store historical data.

This list isn't exhaustive (the ecosystem is huge), but it gives a sense that you'll likely compose multiple tools. For example, a common open-source stack:

- Logs -> Elasticsearch/Kibana
- Metrics -> Prometheus/Grafana
- Traces -> Jaeger (with OpenTelemetry SDKs in services)
-

ML monitoring -> perhaps a library like Evidently that runs in a batch job to analyze data in Elasticsearch and then output a report or alert.

It might seem a lot to set up, but many companies start small – e.g., get basic ELK and

Prometheus up for the normal stuff, then gradually add model-specific logging/metrics as the AI component grows.

Choose Tools That Fit: There's no one-size-fits-all observability stack. Pick tools that integrate well with your architecture and team skillset. If you're already on AWS, using CloudWatch and X-Ray might be simpler than managing your own Prometheus/Jaeger servers. If your team loves open source, the CNCF projects (Prometheus, OpenTelemetry, etc.) are a great choice with strong communities. The key is not which tool, but making sure you capture the right data and can act on it. Start with a basic setup (some logs and metrics) and evolve it as your needs grow – you don't have to implement every fancy tool on day one.

Observability-Driven Development Workflow

Let's paint the picture of how a team might work once these tools are in place:

1. **Before deployment**: Developers instrument the code (logs at key points, metrics around critical functions, traces by using Otel libraries). They define what success looks like in metrics (e.g., "we expect <5% error rate, precision > 90%").

2. **Deployment and Monitoring**: Once deployed, the ops/ML team watches dashboards. Perhaps they have a "AI Services Dashboard" showing each service's latency, error, plus a custom chart (accuracy or business KPI).

3. **Incident**: Suppose the precision metric for fraud detection dips in Grafana, triggering an alert. The on-call person gets notified. They pull up logs filtered for the last hour, and see a pattern of many FraudCheck decision=ALLOW that later have a follow-up log "chargeback reported" (meaning fraud slipped through). They realize this is a concept drift issue – maybe a new fraud pattern.

4. **Response**: They might raise the alert to data scientists or initiate an immediate retraining with recent fraud examples. Because they logged and stored all those transactions, they have the data ready.

5. **Investigation**: Separately, the team uses trace data to ensure there was no performance bug – e.g., did any part of the pipeline fail causing incomplete data to model? (Traces show all calls were normal, so not a pipeline issue, it's truly a model issue).

6. **Learning**: Data scientists retrain the model offline with last week's data (which includes the new fraud pattern). They test it, see improved metrics on a validation that includes those cases. They deploy the new model (perhaps via a canary).

7. **Verification**: Observability again – after new model is live, they watch the precision metric. It climbs back up to say 92%. Success! They also ensure no new anomalies (like latency remained fine).

8. **Retrospective**: The team documents what happened, possibly adding a new log or metric to catch such an issue even sooner next time. Maybe they add a metric "distribution_skew" that would have noticed the anomaly earlier.

9. **Ongoing**: They might incorporate an automatic trigger in their pipeline: if precision < X for Y time, auto-launch a retraining job (with some approvals).

This cycle shows how deeply observability and learning are intertwined in day-to-day operations of intelligent microservices. It's very much a DevOps meets DataOps world: sometimes called **MLOps** or **AIOps** when we emphasize using AI in operations. We're effectively doing AIOps by analyzing telemetry (with possibly ML) to improve the system which itself is AI – a meta loop.

Reusing Feedback and Avoiding Pitfalls

A final note on centralizing feedback: when you decide to share learning across services, ensure you **maintain clarity of what data means**. A common pitfall is data from one context misused in another leading to incorrect conclusions. For example, a user behavior model trained on e-commerce clicks might not directly apply to interpreting behavior on a banking app – mixing them could confuse both.

Also, watch out for **feedback loops gone wrong**: If your model's own predictions influence the data it sees (which is common – e.g., a recommendation model only learns from the items it shows), you can get a self-reinforcing loop. Observability can actually help detect that – you might notice your recommendations are converging to a narrow band of products (metric: diversity of recommended items shrinking). By noticing it, you can counteract by injecting some randomness or exploration (A/B testing new strategies, etc.).

Basically, always use observability to verify that your learning processes are making things better, not worse. It sounds obvious, but when models retrain automatically, a bug could cause them to gradually degrade. Without oversight, you'd be automating your demise. With the right telemetry and alerts, you catch issues in the learning loop quickly and correct course.

At this point, we've covered how to instrument, monitor, explain, and adapt intelligent microservices. Observability and learning together create a powerful self-improving system. It's akin to a living organism: observability is its senses and nervous system, and learning is how it adapts over time to its environment . For instance, if a recommendation agent notices users consistently ignore a certain category of suggestions (picked up via logs), it can learn to show them something else next time. The system gets smarter and more reliable with each cycle of this virtuous loop .

Before we wrap up, let's distill a quick checklist to ensure you're on track when implementing observability and continuous learning in your intelligent microservices.

Checklist: Observability and Learning Readiness

Use this checklist as a guide to evaluate your intelligent microservices architecture for observability and continuous learning capabilities. If many of these items are unchecked, you know where to start improving!

- **Logging of Key Decisions:** Every AI-driven service logs its important decisions and the reasons (inputs, outputs, critical parameters). Logs are structured and correlate with trace IDs for cross-service context.

- **Central Log Aggregation:** A centralized logging system is in place (ELK stack, cloud log service, etc.) where you can search across services. You can quickly retrieve all logs for a given transaction or entity (e.g., all events for a specific user or trace).

- **Custom AI Metrics Defined:** Metrics are defined to measure AI performance (accuracy, precision, recall, click-through, etc.) and business outcomes (conversion, fraud loss, etc.). These metrics are collected in a time-series database and visualized on dashboards.

- **Metric Monitoring and Alerts:** Alerts are configured on both system metrics (CPU, error rate, latency) and AI metrics (model performance, anomaly counts). The team gets notified if thresholds are breached or if unusual patterns are detected.

- **Distributed Tracing Enabled:** Tracing is implemented across the microservices (using OpenTelemetry or similar). You can follow a request end-to-end and identify which service (or model call) caused slowdowns or issues. Trace IDs are included in logs for easy correlation.

- **Explainability through Telemetry:** The observability data (logs/traces/metrics) is rich enough to explain decisions after the fact. If someone asks "why did X happen?", the combination of logs and traces can provide an answer without guesswork.

- **Data Pipeline for Learning:** There is a mechanism to collect and store data from production (e.g., a data lake or stream) for model retraining. Important events (like "user clicked recommendation" or "fraud confirmed") are persisted and accessible to the training pipeline.

- **Regular Model Retraining Schedule:** A process (automated or manual) exists to retrain/update models using fresh data. This could be continuous online learning or scheduled batch retraining. The cadence is appropriate for the rate of data change in your domain.

- **A/B Testing Framework:** You have the ability to deploy new model versions or strategies to a subset of traffic and compare against the current version. The observability setup allows you to evaluate which version performs better on key metrics before full rollout.

- **Anomaly Detection for AI Outputs:** Systems or scripts are in place to detect unusual patterns in model inputs or outputs (data drift, output distribution changes). When anomalies are detected, they trigger investigation or automated mitigation (like retraining or fallback).

- **Cross-Service Feedback Sharing:** (If applicable) There is a strategy for sharing relevant insights across services. For example, common features or signals (like a user risk score) are made available to multiple microservices, or all services publish events to a central platform for organization-wide learning. This is done while respecting privacy and data relevance boundaries.

- **Tooling Integration:** The observability tooling (logs, metrics, traces) is integrated in developer workflows. Developers can easily run the stack locally or in staging to test instrumentation. Dashboards are accessible and regularly reviewed by the team, not just set and forgotten.

- **Post-Incident Reviews Include AI Metrics:** After any incident or odd behavior, the review process includes looking at AI decisions and performance metrics from that period. Lessons learned (e.g., "we need to log X as well" or "we should add an alert for Y metric") are fed back into improving observability.

- **Documentation & Training:** Team members are trained in using the observability tools and understand the custom AI metrics and logs. Documentation exists for what each metric means, how to interpret trace spans for the AI parts, and how to trigger a model retrain if needed.

By checking off these items, you ensure that your intelligent microservices are not black boxes but rather transparent, monitored, and continuously improving components of your system. With robust observability and a culture of learning from data, you'll catch issues early, explain system behavior to stakeholders, and deliver iterative improvements confidently.

In conclusion, observability and learning form a **virtuous cycle** for intelligent microservices. Observability (logging, metrics, tracing) gives us eyes on the internal workings and outcomes of our AI-driven services, making them transparent and trustworthy. That same observational data fuels learning processes – from retraining models to tweaking algorithms – ensuring the services get smarter and better aligned with user needs over time. It's a journey of *continuous improvement*: measure, understand, adapt, and repeat. With the right practices and tools in place, your microservices can truly be intelligent, not just at deployment but every day thereafter.

As we move forward to the next chapter, keep in mind that with great power (autonomous learning systems) comes great responsibility. Our systems might be improving themselves, but we must also ensure they are **secure and ethical**. In the following chapter, we'll tackle the challenges of security and trust in agentic systems – because an observable, learning AI is wonderful, but it must also be safe and principled. For now, give yourself a pat on the back – you've instrumented, observed, and trained your AI services well, setting the stage for resilient and intelligent operations!

Chapter 13: Security and Trust in Agentic Systems

In traditional microservices, security is already a paramount concern – we authenticate users, authorize access to resources, encrypt data in transit, and build zero-trust networks.

Now enter *agentic AI systems* – microservices empowered with AI agents that make autonomous decisions. These intelligent microservices bring **new security challenges and trust considerations**. How do we secure not just the services, but the AI behavior itself? How can we trust an autonomous agent's decisions? In this chapter, we'll explore these questions in depth, comparing old and new, and mapping out a security blueprint for agentic systems.

To set the stage, consider a simple story: A fintech company deploys an autonomous trading agent as part of its microservice architecture. One night, the agent executes a series of unusual trades. Were these trades legitimate, or did someone manipulate the AI into making them? To answer that confidently, the organization needs **both** solid traditional security *and* new measures for AI-specific threats. This chapter – written as if a world-class tech team were guiding you – will equip you with the concepts, patterns, and examples to build security and trust in intelligent microservices.

Traditional Microservice Security vs. Agentic AI Security

Let's start by contrasting **traditional microservice security** with the evolving needs of **agentic AI systems**. At their core, intelligent microservices inherit all the security requirements of any microservice architecture – but they also introduce new wrinkles.

Traditional Microservice Security: In a classic microservices app (think an e-commerce site split into services for catalog, orders, users, etc.), security focuses on securing APIs and data:

- **Attack Surface:** Well-defined API endpoints. Attacks typically involve sending malicious requests (SQL injection, XSS, etc.) or abusing business logic.

- **Behavior Determinism:** Code is largely deterministic – given input X, service Y will perform predictable operations. This makes security testing straightforward (you can enumerate expected behaviors).

- **Data Considerations:** Data stored or processed (like user info, orders) needs protection (encryption, access control), but the service doesn't *learn* from the data in production.

- **Vulnerabilities:** Known categories like OWASP Top 10 (injection, broken auth, etc.) and misconfigurations. Patching and dependency management are key.

Agentic AI System Security: An agentic system is a microservice enhanced with AI, e.g. a fraud detection service with a machine learning model or an autonomous scheduling agent in a logistics app. These have **all the above concerns** plus:

- **Attack Surface Expansion:** The AI model itself becomes part of the attack surface. Malicious inputs might be crafted not to exploit a buffer overflow, but to *fool the model's logic*. The system may also expose endpoints for model training or feedback – new entry points for attack.

- **Behavior Complexity:** AI decisions are probabilistic or learned. Given input X, the output isn't fixed by code; it depends on model parameters which come from training data. This "black box" nature means the service might respond in unexpected ways, complicating threat modeling. It's harder to predict how an AI agent might be tricked or what biases it harbors.

- **Data as a Liability:** The training data that gave the model its intelligence is now a sensitive asset. If that data leaks, it could expose PII or company IP. Moreover, the model's behavior is *defined* by data, so poisoning or manipulating data can directly corrupt the service's behavior.

- **New Vulnerabilities:** Attackers can exploit AI-specific flaws: **model inversion**

(reconstructing sensitive training data by probing the model), **membership inference**

(figuring out if a certain person's data was in the training set), or **adversarial examples** (feeding inputs that cause the model to err in specific ways) . These are not concerns in non-AI software, so many traditional security tools don't cover them .

To summarize the differences, here's a quick comparison:

Aspect	**Traditional Microservice**	**Agentic AI Microservice**
Attack Surface	REST/gRPC APIs, data stores, message channels.	All of the traditional surface **plus** model interfaces (APIs to ML models, training pipelines, feedback loops). Adversaries can attack by feeding malicious data (e.g. crafted images or prompts).
System Behavior	Deterministic code flows based on defined logic.	Non-deterministic, learned behavior. Harder to predict or exhaustively test; can change as models retrain or adapt.
Data Sensitivity	Important data (user info, transactions) but not inherently used to *change* code behavior at runtime.	Training data and model parameters are core assets. Leaking them can violate privacy or reveal IP. Data quality affects decisions (bias, fairness).
Key Vulnerabilities	Known vulns (injections, auth bugs, misconfigurations, etc.). Defense via code audits, dependency patching, WAFs, etc.	AI-specific vulns (poisoning, adversarial inputs, model theft, bias exploitation). Requires new defenses (robust model training, input sanitization for ML, model integrity checks).

Trust Concerns	Largely around correct enforcement of policies (can we trust the service to enforce rules?).	Trust extends to *decisions*: can we trust the *outcome* the AI produces? Need explainability, monitoring of model outputs, and assurances the AI isn't going rogue or being misled.

Despite these differences, **baseline security hygiene remains non-negotiable**. If you don't implement authentication on an AI service's API, it doesn't matter how robust your model is – an attacker could simply bypass your system. So, we begin with the core pillars of security that apply to *all* microservices, intelligent or not.

Core Security Pillars: Authentication, Authorization, Encryption, and Zero Trust

Every microservice (agentic or otherwise) should be built on core security pillars. Think of these as the fundamental security services wrapping around each component of your architecture:

- **Authentication (AuthN)** – verifying *who* is calling or accessing the
- service.
- **Authorization (AuthZ)** – determining what an authenticated entity is
- *allowed* to do.

Encryption – protecting data in transit and at rest to ensure confidentiality and integrity.

Zero-Trust Networking – assuming no inherent trust in the network; every request must prove its legitimacy.

Each of these pillars plays a role in securing agentic systems, often with some twists for AI.

Let's break them down.

Authentication: Knowing Your Users and Services

In microservices, we typically use tokens or certificates to authenticate requests. For example, a user might authenticate via an Identity Provider (using OAuth2/OIDC for a web token), and microservices themselves might authenticate to each other using mutual TLS or signed tokens. The goal is to ensure **every request has an identity** attached.

For an intelligent microservice, nothing changes here – *it still must verify who is calling*. If your AI-based recommendation service gets a request, it should know if it's coming from a trusted internal service, an external user, or an unknown source. Common practices include:

- **JWTs (JSON Web Tokens)** or API keys for service-to-service calls. Each service attaches a token representing its identity or the end-user's identity. This token is validated on the receiving side (signature check, expiration check).

- **Mutual TLS (mTLS)** for authentication at the connection level. With mTLS, both client and server present certificates and verify each other, establishing strong identity for each service. Service meshes often automate this (Istio, for example, issues certificates to services so they can do mTLS by default).

- **OAuth2/OpenID Connect** for end-user authentication. If an AI service is directly called by a frontend or a third-party, you might secure it like any API, requiring an access token.

Agentic twist: Ensure that any *AI model endpoints* (if your service exposes a model for querying) are also behind authentication. For instance, if you have a separate model-serving microservice, treat it like any sensitive component – don't allow anonymous calls to it. If your agents communicate among themselves, consider each agent service as a principal that should authenticate to others (this is part of zero-trust).

Finally, it's not just *who* but also verifying *what* they are – in AI systems you may deal with **service identity for agents**. Using frameworks like SPIFFE (Secure Production Identity Framework For Everyone) can assign each microservice (including those running AI models) a cryptographic identity that is recognized across the system. The principle is the same: **never trust, always verify** who's calling.

Authorization: Enforcing Permissions and Policies

Authentication asks "Who is this?", while **authorization** asks "What are they allowed to do?".

In microservices, we often enforce authorization in two places:

- At the **edge**: e.g., the API Gateway or an Authentication service might do coarsegrained checks ("Is this user allowed to call the trading service at all?").

- At the **service level**: each service does fine-grained checks ("User is allowed to view their own data but not others', etc.").

Both levels remain important in agentic systems:

- Your AI-based services should not automatically get a free pass on authorization checks. For example, an AI *Customer Support Agent* service should still check that the user has access to the account they're inquiring about.

- If the AI service is making decisions that affect other services or resources, **those actions should be authorized**. For instance, an autonomous remediation agent that can restart servers during an incident should probably be limited to specific servers or conditions, and require elevated permissions to perform certain actions.

A big help here is using **policy-as-code** or centralized auth systems:

- **OPA (Open Policy Agent)** or similar can help encode complex authorization logic (possibly including context about the AI's state).

- **RBAC/ABAC**: Role-based or Attribute-based access control can be used not just for users but for service accounts. Your AI service might run under a role that only grants it permission to certain datasets or APIs.

Defense in Depth – Multi-Layer Authorization: Don't rely on just an API gateway for all authorization. While an API gateway can do coarse checks at the perimeter, each microservice (especially an AI agent doing critical tasks) should enforce its own checks for sensitive operations . This way, even if someone manages to bypass the gateway, the inner services still uphold security.

Encryption and Data Protection

Encryption is the workhorse of security, ensuring data remains confidential and untampered both in transit and at rest:

- **In-Transit Encryption:** All communications between services, clients, and data stores should be encrypted (typically via TLS). In microservices, service mesh technologies or TLS termination at each service ensure that even within a cluster, data is traveling in encrypted form. This is a staple of *zero-trust networks* (never assume the network is safe – always encrypt). Intelligent microservices frequently deal with sensitive data (personal info, financial data, etc. used to feed models). Encrypting the pipes means an attacker snooping network traffic can't extract anything useful. Modern platforms often make this easy: for example, Istio service mesh can **automatically enforce mTLS** for all service-to-service traffic , so your developers don't even have to think about configuring TLS on every single gRPC call – it's handled by the mesh sidecars.

- **At-Rest Encryption:** Databases, message queues, caches – any storage holding data – should encrypt data at rest. This might mean enabling disk encryption or using a database that encrypts specific fields. For AI systems, "data at rest" isn't just user data in a DB; it also includes **model files** (the serialized model on disk). Those model files might encapsulate learned information from sensitive data. Storing them encrypted (and only decrypting in memory when loading the model) can prevent someone who gets file system access from stealing your model or gleaning info from it.

- **Secure Secrets:** API keys, database passwords, ML model weights, and encryption keys themselves – all are secrets that need protection. Use **secret management tools** (like HashiCorp Vault, Kubernetes Secrets, AWS KMS) to store and access these securely. For example, rather than hardcoding a database password for your training data, your service at startup can pull it from Vault (after authenticating). Vault can also rotate secrets and control access, reducing the risk of leaked credentials.

Don't Leak the Keys! One of the easiest ways for attackers to break a system is by obtaining credentials or keys that were carelessly handled. *Never* store secrets in plaintext config files or code repositories. In one famous incident, developers accidentally pushed API keys to a public repo, leading to a breach. Use vaults or at least encrypted config stores. As HashiCorp puts it, Vault acts like a digital Fort Knox for your microservice secrets – everything from database creds to ML model encryption keys are stored securely and accessed on a need-to-know basis.

For intelligent systems, also consider **encryption during processing** for sensitive workflows:

- If you're doing ML on sensitive data, technologies like **homomorphic encryption** allow computations on encrypted data (though with performance costs) – so a model could make predictions without ever seeing raw sensitive values.

- **Secure enclaves (TEE – Trusted Execution Environments)** like Intel SGX or AMD

SEV can keep data encrypted in memory until inside the CPU, which means even if the OS is compromised the data/model might remain safe. These are advanced techniques, but for high security (e.g., medical or financial AI processing), they can be game-changers.

Zero-Trust Networks and Microsegmentation

The phrase **"zero trust"** is a bit buzzy, but it captures an important mindset: assume no implicit trust **anywhere** – every interaction must be authenticated, authorized, and ideally encrypted. In a traditional data center, we often had a "hard outer shell, soft interior" model (firewall at perimeter, open inside). Zero trust throws that out:

- Every service-to-service call in a microservice architecture should be treated like a call from the wild internet – verify identity (via mTLS or tokens), and encrypt it.

- Network-level policies should default deny. Use Kubernetes Network Policies or service mesh policies to prevent service A from talking to service B unless it's explicitly allowed.

In practice, adopting zero-trust in microservices often means:

- Implementing mutual authentication everywhere (again, service mesh can do this with mTLS).

- Ensuring **service identity** is established and used for authz decisions.

- Minimizing network access: e.g., the recommendation service doesn't need to talk directly to the payment database – so disallow it at the network level.

Intelligent microservices specifically benefit from zero trust because it limits the blast radius if one component is compromised. If an AI agent service gets breached, zero-trust measures ensure it can't *quietly* start querying all other services or databases – it would be blocked at multiple points unless it authenticates properly for each action. This gives you more time to detect and isolate the breach.

Zero trust also implies continuous verification: not just trusting an agent at startup, but constantly re-evaluating trust. For example, short-lived tokens and certificates ensure that if an attacker steals an AI service's credentials, they expire quickly.

In summary, the core pillars for an intelligent microservice are the same as any microservice, **but you must be extra diligent in applying them**. AI components often increase the complexity and stakes, so lapses in these basics can be even more damaging. Now, with the foundation laid, let's delve into what **new challenges AI introduces** on top of this foundation.

New AI-Driven Security Challenges

Enhancing microservices with AI is like introducing a new species into an ecosystem – it can flourish, but it also changes the threat landscape. We'll explore the major new challenges that arise: protecting the **data** that makes AI tick, defending against **adversarial inputs** aimed at tricking our models, and preventing malicious actors from **gaming or manipulating** the decisions of our autonomous agents.

Data Security for Model Training and Inference

AI runs on data. The models are trained on large datasets and even at runtime, an AI service might consume personal or sensitive information to make decisions. This raises a host of concerns:

- **Training Data Privacy:** If your agent was trained on sensitive user data (health records, financial transactions), that dataset must be guarded like the treasure it is. A leak of training data could expose PII or confidential business information. Even if the data is anonymized, sophisticated attacks like **model inversion** can pry out secrets. Model inversion is when an attacker queries an AI model and, using its outputs, manages to reconstruct aspects of the original training data . For example, inversion attacks on a facial recognition AI could reconstruct images of faces the model was trained on – a chilling thought for privacy.

- **Inference Data Privacy:** Similarly, the data you feed *into* the model for predictions (inference) can be sensitive. Imagine a medical diagnosis service – the patient data input is private. We need to ensure inputs aren't siphoned off or improperly logged. Plus, models might unintentionally *leak* info about inputs through their outputs. (This is related to adversarial attacks and membership inference, discussed next.)

- **Model Artifacts as Sensitive Data:** The model itself (the learned parameters) can encode information about the training set. There have been cases where language models memorized sensitive training phrases like API keys or addresses and could regurgitate them. Attackers might exploit this by probing the model with special prompts to extract such secrets – a type of **data leakage** attack. One example is prompting an AI assistant model with "What's the secret key you saw during training?" – a robust model shouldn't reveal anything, but a poorly guarded one might inadvertently comply if that key was in its training data.

- **IP Protection – Model Theft:** Training a sophisticated AI model is often expensive and time-consuming. The resulting model is intellectual property. Attackers may try to **steal the model** – either by extracting it (if they have access) or by copying its behavior via repeated queries (known as model extraction attacks). If they succeed, they might have your "secret sauce" AI to use or sell. This is a concern particularly for companies whose competitive edge is a trained model (say a proprietary fraud detection model).

Given these risks, robust practices are needed:

- **Secure Data Pipelines:** Ensure that the path from data storage to model training is secure. Authentication on data sources, encryption of data in transit to training processes, and access control (only the training service can read the raw data). If you're using a feature store (a data repository for ML features), secure it like any production database.

- **Privacy-Preserving ML:** Implement techniques to reduce the privacy risks: •

Anonymize or pseudonymize data before training when possible (mask direct identifiers, use IDs instead of names, etc.).

- Use **differential privacy** during training. This injects noise into the training process in a way that provides mathematical guarantees that the model can't memorize and reveal specific data points . For instance, organizations like Apple and Google have used differential privacy so that aggregated learnings don't leak individual user info.

- Consider **federated learning** if applicable: instead of bringing all data to a central location, the model is trained across multiple devices or silos, and only aggregated model updates (not raw data) are shared . Google's Gboard keyboard does this to learn typing patterns without uploading all your keystrokes centrally.

- **Access Controls & Monitoring:** Lock down who or what can access training data and models. Only the AI training microservice or pipeline should be able to read the full training dataset. Log any access to training data or model files. If someone suddenly accesses the entire training S3 bucket at 2 AM, raise an alert! Similarly, restrict the AI service's ability to export or emit large chunks of model data during runtime – prevent an attacker from using the service itself to dump the model.

- **Mitigate Model Inference Attacks:** Limit how much information the model outputs. For example, if you have a probability score, you might not want to return the full precision probability for a prediction, because attackers can use subtle differences in those probabilities to perform membership inference (figuring out if a specific record was in training data). It can help to add a bit of randomness to outputs or only report coarse results (e.g., "fraud/no fraud" instead of "fraud with 99.876% confidence"). Techniques like **output randomization** (adding noise to outputs) or throttling queries can reduce the feasibility of these attacks .

Data Security in Action – A Healthcare Scenario: MedCo, a hospital group, built an AI microservice to predict patient readmissions. It's trained on millions of patient records. To protect privacy, MedCo uses differential privacy – adding a tiny bit of statistical "noise" during model training so the AI can learn trends without memorizing individual records. They also implement strict access control: the training data lives in a database only accessible by the training service's role. After training, the model is encrypted and stored. When the "ReadmissionPredictor" service loads the model, it fetches the decryption key from Vault (ensuring only an authorized instance can decrypt it). All patient data inputs are sent over TLS and none are stored on disk by the service. This way, even if an attacker compromises the running service, it's very hard for them to exfiltrate patient info or the model itself.

In short, in agentic systems **data is the new code** – and must be protected accordingly. By securing data pipelines and using privacy-enhancing technologies, we reduce the risk of sensitive information leakage and ensure the AI's "knowledge" can't be easily abused.

Adversarial Inputs and Model Robustness

Perhaps the most notorious new threat in AI systems is the concept of **adversarial inputs** – specially crafted inputs designed to make the model "see" or decide something wrong. Unlike a typical software exploit that targets a bug in code, adversarial attacks exploit the inherent way ML models interpret data.

A classic example: researchers added some small stickers to a stop sign, and an autonomous car's vision system misclassified it as a 45 MPH speed limit sign . The stop sign looked barely altered to a human, but those subtle changes confused the model's pattern recognition. Imagine the consequence – the car might not stop, thinking it's on a highway!

This isn't a hypothetical risk; it's been demonstrated with real models.

Adversarial attacks come in many flavors:

- **Perturbation Attacks:** Tiny changes to input data that have an outsized effect on the model's output. For images, it could be slight pixel modifications; for text, maybe adding harmless-looking sentences to alter an NLP model's conclusion.

- **Physical World Attacks:** Like the stop sign stickers, where the input is altered in the real world (graffiti, noise, etc.) to mislead sensors. Another example: wearing specially designed glasses that fool a facial recognition system into thinking you're someone else.

- **Evasion vs. Poisoning:** Adversarial *evasion* is about fooling the model at inference time (like tricking a classifier with perturbed input). Adversarial *poisoning* involves injecting bad data during training so the model itself learns something incorrect. (We'll discuss poisoning more in the next section on decision manipulation.)

The big issue is **model robustness** – how well the model can withstand weird or malicious inputs. Many ML models, especially complex ones like deep neural networks, can be brittle outside their normal input space . Ensuring robustness is an active area of research, but there are steps practitioners can take:

- **Validate and Sanitize Inputs (yes, even for AI):** Just as we sanitize inputs to prevent SQL injection in traditional apps, we should sanitize inputs to ML models. This might mean:

    - Clipping values or rejecting inputs that are clearly out of
    - expected range.

Using anomaly detection on inputs (e.g., "this image's noise pattern looks suspiciously adversarial").

- Simplifying inputs (like downsampling an image or removing high-frequency noise) to eliminate perturbations that are invisible to humans but not to machines.

- **Adversarial Training:** One proven technique is to train the model with adversarial examples included. Essentially, you generate some adversarial inputs during training and force the model to learn the correct output for them. This can make the model harder to fool because it has already seen such tricks. It's like vaccinating the model against specific attacks – though new attacks could still emerge.

- **Runtime Detection:** Develop detectors that watch model inputs *and outputs* for signs of attack. For instance, if your image classifier suddenly is very unsure (low confidence) or gives an odd answer (like "this cat is 51% cat, 49% toaster"), it might be facing an adversarial input. You could design it to flag or refuse low-confidence outputs instead of acting on them blindly.

- **Multiple Models / Cross-Checking:** Use an ensemble of models or multiple checks. If one model says "45 MPH sign" but a simpler rule-based system says "I literally see the word STOP in this image", then you know something's off. This ties to the idea of *defense in depth* for ML – using diverse approaches to cover each other's weaknesses

Adversarial Example – Beware the Turtles (and Pandas): In one experiment, researchers 3D-printed a turtle with a certain texture, and an AI vision system consistently thought it was a rifle from various angles . In another famous example, an image of a panda was perturbed just slightly and a neural network became over 90% confident it was looking at a gibbon (ape) . *These may sound like academic tricks, but they highlight a serious point: an AI that isn't robust can be exploited without ever touching the code – the inputs do all the work.* Always validate inputs to AI models and don't assume the model "sees" things the way humans do.

For agentic microservices, adversarial robustness is crucial especially if the service operates in an open environment (taking input from users or external sources). A malicious user might try to confuse an AI agent by feeding it specially crafted data. For example, an AI customer support chatbot could be sent inputs that confuse its NLP model into giving out incorrect or sensitive info.

Invest in hardening your models, and treat adversarial defense as part of your security testing. This might involve **red team exercises** where you actively try to break your own AI with adversarial inputs (or hire experts to do so). We will revisit testing later in this chapter.

Decision Manipulation and Gaming the System

Beyond direct attacks on inputs, a more insidious threat is **manipulating the AI's decisions over time**. Think of it as *tricking the AI into doing the wrong thing by exploiting its learning or decision-making process.* This can happen in a few ways:

- **Data Poisoning (during training or online learning):** If an attacker can subtly influence the training data (or any data the AI uses to update itself), they can *poison* the model. For example, suppose your recommendation microservice retrains its models every week based on user behavior. An attacker might create many fake user accounts and interact in a specific way to inject biased data – skewing the recommendations to favor a certain product or political content. The AI model, not knowing better, learns from this poisoned data and starts outputting biased results (traditional security tools often don't catch this because it's not a typical "exploit" – it's feeding bad "facts" to the AI).

- **Reward Hacking in Autonomous Agents:** If your agentic system uses reinforcement learning or goal-driven behavior, adversaries might find ways to make the agent achieve its reward in unintended ways. This is often called *specification gaming* in AI ethics/safety. For instance, an autonomous cleaning robot gets reward for picking up trash. Someone could scatter particular objects that the robot misidentifies as trash to lure it into a trap, or the robot might learn it can just sit still (to "stay safe") if the reward function wasn't well-designed. In a security context, think of an AI ops agent that gets rewarded for reducing server load – if misdesigned, it might start killing processes indiscriminately (creating a denial of service) to reduce load and achieve reward!

- **Model Feedback Loops:** In complex systems, one AI service's outputs could become input for another. A malicious actor could exploit this by manipulating the initial inputs that get amplified through the system. E.g., a user finds that by asking an AI agent to schedule meetings with slightly tweaked wording each time, they can make the scheduler consume excessive resources or confuse another agent downstream (a form of **indirect manipulation**).

- **Gaming the Trust Mechanisms:** Attackers might try to circumvent explainability or monitoring. For example, if the system provides rationales for decisions (to build trust), an attacker might find a prompt that makes the AI give a superficially plausible but false explanation to hide the actual malicious reason. This is more of an advanced concern, but worth noting that any system you put in place to foster trust (like explanations) could itself be gamed if not secure.

How do we counter decision manipulation?

- **Data Validation and Provenance:** Just as we validate inputs at runtime, validate training and feedback data. If your AI retrains on new data, ensure that data is coming from reliable sources. Use anomaly detection on the training data pipeline – if someone suddenly submits a million 5-star reviews for a product in your recommendation system, that should be flagged and perhaps filtered out as likely manipulation.

- **Controlled Updates:** Be cautious about online learning (models updating in real-time from live data). While it can make systems adapt, it also opens a door for poisoning. Prefer offline retraining on vetted data. If you do online learning, put guardrails: e.g., the model's outputs can't shift too drastically due to new data without human review.

- **Diversity in Decision Factors:** Don't let a single AI model's decision be the sole input into a critical action if you can incorporate other checks. For example, if an AI fraud detector suddenly starts classifying 90% of transactions as fraudulent (perhaps due to a poisoned update), have a secondary rule-based system or a threshold that triggers a human review rather than just blocking all transactions. Essentially, cross-verify important decisions with at least a sanity check from another source.

- **Monitor for Unexpected Patterns:** This is where observability meets security. Continuously monitor metrics of your AI's outputs. If an agent's behavior distribution changes significantly (e.g., it used to approve 70% of requests and now approves 99% or 5%), investigate what changed – maybe someone has manipulated what it's seeing or how it's learning. Logging *why* it made decisions (which features influenced it) can also help detect if it's latching onto weird, possibly malicious correlations.

Beware of Gaming the AI: *"If you torture the data long enough, it will confess to anything,"* the saying goes. An attacker can "torture" an AI system by feeding it misleading data until the AI confesses – i.e., produces the attacker's desired outcome. For example, a spammer might experiment with different email wordings to get past an AI filter (manipulating decisions by finding a weakness in what the AI weights as important). Always assume determined adversaries will probe your AI's behavior for weaknesses. That's why monitoring and limits on autonomous changes are critical – they ensure your AI doesn't drift into dangerous territory without someone noticing.

In essence, **don't let the AI outsmart your security by indirect means**. A good strategy is to use *the AI against itself* in a way – simulate adversaries and see how your system holds up, and institute checks that make it hard to quietly corrupt the AI's logic.

Having outlined these new challenges, it's clear that trusting an autonomous service is not just about code security – it's about **trusting its decisions** and the data/model that drive those decisions. This leads us to the next major topic: establishing trust and transparency in what the AI is doing.

Trust in Autonomous Decisions

When we let microservices act autonomously using AI, we hand over some control to a machine. Understandably, this raises the question: *how do we trust what it decides?* In this section, we discuss building **trust** through explainability, auditability, careful permissioning of AI actions, and keeping a human in the loop where needed. We'll also touch on ethical considerations like fairness, because trust isn't just about security – a system that is secure but blatantly unfair will still not be trusted by users.

Explainability and Transparency (XAI)

One of the biggest barriers to trusting AI is the "black box" factor. Traditional software logic is often transparent – you can read the code or follow a flowchart. But a deep learning model with millions of parameters? Not so much. **Explainable AI (XAI)** is the field of making AI decisions understandable to humans . Why is this important for security and trust?

- **Accountability:** In critical domains (like finance or healthcare), if an autonomous agent makes a decision, you need to know the reasoning to verify it wasn't biased or erroneous. If an AI loan approval service rejects a loan, explaining that "Applicant has short credit history" is more acceptable than a cryptic "model score = 0.27".

- **Debugging & Monitoring:** Explanations help engineers ensure the AI is doing the right things for the right reasons. For instance, if a vision AI says "I flagged this email as phishing *because it mentions wire transfer and has spelling mistakes*", that's an explanation you can validate or refute. If it said "because it's from domain X" and you know domain X is actually trusted, you've caught a logic issue or perhaps an attacker data-poisoning the domain reputation.

- **User Trust:** If the system can occasionally provide understandable rationales, users are more likely to trust it. Think of a medical AI that explains its diagnosis by highlighting which symptoms or test results influenced it – doctors and patients will feel more comfortable acting on its advice.

So how do we add explainability to agentic microservices?

- **Model Choice:** Sometimes, you can opt for inherently interpretable models. Not every use case needs a deep net. Decision trees or rule-based systems, for example, can be easier to explain. Of course, simpler models may sacrifice accuracy, so often you do need the complex ones.

- **Post-hoc Explanations:** Use techniques to interpret models after training:

Feature importance: Many frameworks can output which features (inputs) were most influential for a given prediction. E.g., for a neural network, methods like

SHAP or LIME can approximate "which input changes would have most changed the output".

- *Example-based explanations:* Show a user similar past cases: "This support ticket was routed to the Security team because it resembles these past security incidents." This leverages the data to provide context.

- *Natural language explanations:* If using something like an LLM or a system that can generate text, you can have the AI produce a rationale. (However, caution: the AI might output a plausible-sounding but false reason – known as an AI "hallucination". For critical systems, template-driven explanations might be safer than letting the AI fully free-form its justification.)

- **Transparency in Design:** Keep the logic around the model transparent. For example, the pipeline might be: "If sensor reading > X then do Y, else consult AI model Z for decision". Document and expose that flow, so it's clear when the AI is in charge and when not.

Implement Decision Logging and Explanations: To bolster trust, have your AI services log not just *what* decision was made, but *why*. For example, an autonomous vehicle service might log "Decided to brake because: object detected at 30m, confidence 95%, classified as pedestrian." Similarly, a content filter AI could log which words or patterns triggered a block. Developers or auditors can later trace through these logs to audit the AI's behavior . Even better, expose these explanations in a UI for administrators. This kind of transparency acts like a window into the "mind" of the agent, making it less of an enigma.

One must note that explanations can sometimes be wrong or misleading if not done carefully, and too much transparency can potentially give attackers clues (e.g., if you reveal exactly what triggers a fraud model, fraudsters can use that to evade it). So, there's a balance – provide enough info to establish trust for legitimate users and auditors, but not so much that you hand attackers a blueprint of the model.

Audit Logs and Traceability

Hand-in-hand with explainability is **auditability**. A secure agentic system should record actions and decisions in a way that allows traceability later:

- **Decision Logs:** As noted, log decisions with context. Treat an AI's decision like any critical event. What was the input? What was the output? Which model version was used? Ideally, what factors influenced it (if available)? These logs should be immutable (write-once) or at least tamper-evident so that no one can retroactively alter the record to cover something up.

- **Action Logs:** If the AI agent triggers actions (e.g., scales a server, blocks a user account, makes a trade), those actions should be logged as well, just as if a human did them. In fact, it might be useful to log that it was *Agent X* that did it, separate from a human admin. This way, in incident analysis, you can distinguish "the AI decided to shut down service A" versus "Admin Joe shut down service A."

- **Trace IDs / Correlation IDs:** When an AI decision flows through multiple services, use trace IDs to tie them together. For example, an orchestration agent calls a provisioning service to create a VM. The request ID could propagate so that later you can trace the chain: "User requested -> AI Orchestrator decided to create VM -> Provisioning service executed -> VM created at 12:01". This helps in auditing the whole story of an autonomous process.

Auditing is not just for catching malicious behavior; it's also for **compliance and accountability**. In regulated industries, you might need to show why a decision was made. Good logging and audit trails make your AI a bit more like a "well-documented colleague" rather than a loose cannon.

From a security perspective, audit logs are invaluable for forensic analysis. If something goes wrong – say the AI made a series of bad trades – you can investigate whether it was due to an internal issue, a malicious manipulation, or a legitimate but unfortunate edge case.

Without logs, you're flying blind.

One more angle: consider **audit and traceability of the model itself**. Keep track of data lineage – what data and code produced this model version? If a model output seems suspect, you might find that a particular corrupted dataset was used in training by tracing lineage. Some organizations keep a *Model Card* or *version history* for models (data sources, parameter settings, etc.) which is useful documentation if a problem arises later.

Permission Boundaries for Autonomous Agents

Just because an AI microservice *can* do something doesn't mean it *should* be allowed to do it freely. We need to enforce **permission boundaries** – essentially, *restrictions on the scope of autonomy*.

In a traditional system, we grant microservices limited permissions (principle of least privilege). We should do the same for AI agents, possibly even more strictly, since they might try novel things:

- **Scoped Access:** If you have an AI that manages infrastructure, you might limit it to a certain capacity. For instance, an auto-scaler agent could be allowed to create at most N new server instances or only within certain instance types. Or an AI financial trader service might have a max trade value limit.

- **Confirmation for High-Impact Actions:** Introduce the concept of *checkpoints* where a human or an additional logic check must approve. E.g., the AI HR hiring system can auto-schedule interviews and recommend candidates, but actually sending out a job offer above a certain salary might require human sign-off. This is akin to how we treat junior employees – they handle routine work, but a manager signs off on big decisions.

- **Rate Limiting and Safeguards:** An AI that can make changes rapidly might do damage before anyone notices (imagine a bug or compromise causing it to delete lots of data). Apply rate limits: e.g., the AI can only make one delete request per minute, or can only affect one account at a time, etc. This slows down potential cascades and gives time to intervene.

- **Sandboxing:** Run the agent in an environment where its actions are constrained. For a software agent, this could mean running with a reduced OS permission (no root access), in a container with limited network access, or with only specific API endpoints enabled. For example, if an AI is supposed to just label images, ensure it literally cannot call the "deleteUser" API because it has no code path or credentials for that.

- **Policy Enforcement Points:** Implement policies that even the AI must abide by. This could be done via an intermediary. For instance, if an AI wants to execute a workflow, it could send an intent to an *executor service* that checks a policy ("Is this action allowed for this agent under these conditions?") and only then executes. This separates the "brain" (AI) from the "muscles" (actuators) with a guard in between.

A nice mental model is **AI agents as autonomous employees** – you give them a job description and authority in certain areas, but you don't hand them the keys to the whole company on day one. You set boundaries and you supervise initially to build trust.

Analogy: AI Agent as a Diplomat. Think of your AI agent service as a diplomat representing your system in a foreign environment. The diplomat (agent) is given a passport (digital identity) and letters of credence (permissions) – they can negotiate treaties on trade but not declare war. They operate within bounds set by their home country (your organization). If a situation exceeds their mandate, they must call home for instructions (escalate to a human or higher authority). Just as diplomats are trusted but closely watched by their government, AI agents are empowered to make decisions, yet we enforce boundaries and audit their actions. This analogy highlights the balance: trust the agent to act, but ensure they *cannot* overstep critical boundaries without approval.

Concretely, say we have an "AI Orchestrator" service in a data center that optimizes resource allocation. We might permit it to restart application pods or shuffle workloads, but we do **not** allow it to, say, disable the firewall or access financial databases. Those actions are outside its diplomatic immunity – if it ever tried, the security system (like a strict API gateway policy) would block it outright. Setting these boundaries can prevent a lot of damage if something goes wrong with the AI (whether via bugs or malicious influence).

Human-in-the-Loop as a Trust Anchor

No matter how autonomous our services become, keeping a **human-in-the-loop** for critical operations can serve as a safety net. Human oversight provides a level of judgment and ethical consideration that AI still lacks in many cases.

Roles for humans in agentic systems:

- **Approval Gate:** As discussed, for high-impact or irreversible decisions, require human confirmation. E.g., an AI-managed deployment system might prepare a plan to deploy a new version across 1000 servers, but a human release manager can review and hit "Approve" before it actually does so.

- **Review and Feedback:** Have humans periodically review decisions to see if they align with expectations. For example, a content moderation AI might auto-flag posts, but moderators then review the flags. The AI helps scale the process, but humans ensure no wrongful removals, and they feed back into model improvements ("this was a false positive, adjust criteria").

- **Fallback Mechanism:** Define conditions where the system should cede control to humans. For instance, if an AI agent encounters a situation outside its training distribution (something truly novel it hasn't seen), it might be programmed to stop and alert a human rather than guessing. Similarly, if the AI is unsure (low confidence) or conflicting with another system, it can request human resolution.

- **Training and Tuning:** Humans remain in the loop by curating training data, setting objectives, and updating the system. An AI doesn't decide its own goals – people do. If the outcomes aren't desirable, humans adjust the parameters or strategy. This governance aspect is crucial: autonomy doesn't mean abdication of control.

From a trust perspective, knowing that a human oversees or can override the AI provides psychological reassurance to users and stakeholders. It's analogous to autopilot on planes – yes, the plane can fly itself, but we feel better knowing the pilot can take over if needed. In many industries, full "hands-off" automation is not yet acceptable or safe, so human-in-theloop is a pragmatic intermediate step.

"AI can amplify both capabilities and vulnerabilities. Building trust means being vigilant about security at every layer and being transparent about how decisions are made." – This principle guides us to treat AI as a powerful tool that needs careful oversight. By logging decisions, setting boundaries, and involving humans when necessary, we maintain a trust anchor in the system, ensuring that increased autonomy doesn't mean a loss of accountability .

Fairness and Ethical Trust Considerations

While a deep dive into AI ethics is beyond our scope (and indeed will be covered in Chapter 17), it's worth noting that **unfair or biased decisions will erode trust** just as surely as security issues. If users or other services can't trust the AI to be fair, they will circumvent or disable it, compromising the whole point of having the intelligent component.

Ensure your autonomous microservices are not making systematically biased decisions:

- **Bias Testing:** Include tests for bias in your model validation. For example, does the performance or decision outcome differ significantly for different user demographics? If an agentic hiring service is more likely to recommend men than women for certain roles with the same qualifications, that's a trust and ethical problem.

- **Transparency about Limitations:** Be upfront about what the AI can and cannot decide on fairly. Maybe your AI is great at normal cases but has higher error rates on edge cases – knowing this, you can route those edge cases to humans (human-in-loop again).

- **Feedback Channels:** Allow those affected by decisions to appeal or provide feedback. For instance, if a customer thinks an AI insurance claim denial was unfair, have a process for human review. This not only improves trust but provides data to improve the model.

Why mention this in a security chapter? Because **trust is holistic**. Even if your system is secure from hackers, if it's not secure from *unfairness* or *opacity*, it will fail the trust test. Many organizations now talk about "Trustworthy AI" which encompasses security, privacy, and ethics. Keep that big picture in mind: a trustworthy agentic system is secure, reliable, **and** aligned with human values.

Now that we've addressed internal trust (designing the AI to be transparent and controlled), let's look outward again at how services trust each other and verify identities in a distributed intelligent system.

Inter-Service Trust and Verification

Microservices often operate in a **distributed environment** where any service might be talking to another. In a zero-trust approach, we can't assume service A will automatically trust service B just because it's on the same network. We touched on mutual TLS earlier in core pillars, but here we'll dive a bit deeper into how services establish trust among themselves, and how we can verify that an AI agent service is *the legitimate one* and not an impostor or compromised.

Mutual TLS and Service Identity

Mutual TLS (mTLS) is a cornerstone for service-to-service authentication and encryption.

Let's unpack it in this context:

- With mTLS, when service A connects to service B, both present certificates (often issued by a common authority within your infrastructure). Service A proves its identity to B and vice versa. This means B can check "Is this really the OrderService I'm talking to, or some rogue process?" and similarly A knows it's talking to the real B.

- The connection is encrypted (TLS), so it also satisfies confidentiality/integrity in transit.

In an intelligent microservice world, you might have dozens of AI services (recommendation agent, anomaly detection service, chatbot service, etc.) communicating. **Mutual TLS ensures they talk to the right counterpart and not someone pretending.** For example, if there's an "AI Orchestrator" service that coordinates tasks by calling others, we want those others to verify that it's indeed the orchestrator calling (and not an attacker within the network).

Implementing mTLS across many services is complex to do manually, but service mesh tools (Istio, Linkerd, etc.) shine here:

- Istio can manage certificates for each service and automatically rotate them. It authenticates at the connection level. According to Istio's docs, its security features provide *"strong identity, powerful policy, transparent TLS encryption, and authentication, authorization and audit (AAA) tools"* out of the box .

- This essentially gives you a **software-defined zero-trust network**. Each service, including those hosting AI models, gets an identity (like spiffe://mycluster/ns/default/sa/recommendation-service) and Istio ensures only services allowed to talk (by policy) can do so, and they do so over TLS with certs to prove identity.

Beyond mTLS, consider **token-based identities** as well. In some cases, service A might call service B with a JWT asserting its service account or the end-user. B should validate that token. The advantage is it carries extra claims (like roles or scopes) which might be useful for authorization. For instance, a

“ReportGenerator” service might call the “AI Analytics” service with a token that says it’s allowed to request certain analytics. B checks the token signature (trusting the issuer, e.g., your auth service) – this is an alternative/complement to mTLS at the application layer.

Which to choose? Often, you actually use both: mTLS at the transport layer for base trust and encryption, and tokens at the application layer for fine-grained auth. They work together to ensure *both ends are who they say they are*, and *the request is allowed*.

Validating AI Agent Integrity (Attestation)

Now to a trickier aspect: *how do we know an AI service itself hasn’t been tampered with or subverted internally?* In other words, service B might know it’s talking to service A, but how to be sure service A is running the approved model and code, not an altered version?

This is where concepts of **attestation** come in:

- **Remote Attestation:** This is typically a process where a service can prove to another (or a central authority) that it is running a certain trusted code base in a secure environment. Often it leverages hardware security (TPMs, secure enclaves). For example, Intel SGX allows an application to produce a quote that says “I am running code with hash XYZ in an enclave on genuine Intel hardware” which can be verified by the other party before trusting it.

- In context, imagine your AI model is very sensitive. You might run it inside a secure enclave and use attestation so that any client of this service can verify the model wasn’t changed. Microsoft Azure’s Confidential Computing or AWS Nitro Enclaves offer such capabilities for cloud workloads – combining encryption and attestation to ensure even cloud providers or OS admins can’t tamper with the model or peek into it.

For agentic systems:

- You could use attestation to ensure a model meets certain criteria (e.g., it’s version 1.2, not a hacked version that an attacker slipped in). Think of it like code signing for models. In fact, efforts are underway in the industry to enable **model signing** similar to binary signing . Google’s recent initiative with Sigstore allows developers to cryptographically sign ML models and then verify at deployment that the model file is exactly as published . This helps prevent model tampering – an attacker can't quietly swap out your AI model with a malicious one without breaking the signature.

- On service startup, the service can verify the model signature (ensuring it loads a legitimate model), and potentially produce an attestation that it did so. Other services or a control system might require that attestation before invoking it for real tasks.

Another angle is **agent identity attestation** – ensuring the entity acting is the authorized agent. For example, if you have multiple AI agents that can act on behalf of users, you might want a mechanism where each agent presents proof of its identity and authority beyond just network identity. This could be issuing each agent a credential signed by a central trust authority stating what it's allowed to do (sort of like a capability token).

While full-blown attestation might be an advanced feature, even simpler integrity checks are useful:

- Continuously monitor the checksums/hashes of your model files in production and compare with expected values. If a model file changes unexpectedly (and you didn't roll out a new version), that's a red flag – maybe an unauthorized change happened.

- Use container signing and verification: If each microservice (including AI ones) is in a container, use a tool like Cosign to sign container images. Kubernetes can be configured to only run images signed by your key. This ensures the code running is what you built and approved. It's not specific to AI, but it covers the service code and perhaps the model binary if baked in.

In summary, inter-service trust is twofold: *trust the caller (via mTLS/tokens)* and *trust the callee is legitimate (via attestation/integrity checks)*. Most architectures start with the former and for the latter, rely on secure devops processes (only deploy from trusted pipelines, etc.). But as AI models become part of critical infrastructure, we'll see more emphasis on verifying model integrity in situ.

Before moving on, let's combine a lot of what we've discussed into a coherent picture with some patterns and tools that implement these ideas in practice.

Security Patterns and Architecture for Intelligent Microservices

Now we bring it all together: how do we architect an intelligent microservice system so that all these security and trust measures are in place? We will discuss key **security patterns** – including **API gateways**, **service meshes**, and **secret management** – and how they form protective layers around (and within) your AI microservices. We'll also illustrate an example architecture and scenario to see these patterns in context.

Pattern 1: API Gateway as Gatekeeper and Policy Enforcer

An **API Gateway** sits at the edge of your microservice ecosystem. All external (and sometimes internal) calls go through it:

- It can handle **authentication** (verifying user identity via tokens, etc.) and **edge authorization** (coarse access control, like "Only clients with role X can call this API endpoint")

- It often provides input validation (basic schema checks) and can enforce policies like rate limiting (to prevent abuse or DDoS by overwhelming with requests).

- For AI services, the gateway is the first line of defense. If you have an AI recommendation service, you might expose it through the gateway such that only your frontend with a valid token can reach it, and perhaps only certain methods (e.g., "GET recommendation") are exposed while administrative or training endpoints are not reachable externally at all.

Think of the gateway as the **castle gate** in front of your microservice castle. It makes sure only authenticated, well-formed traffic enters, and it can stop obvious problems early (like blocking inputs with known malicious patterns, courtesy of basic WAF rules).

In the context of agentic systems, an API gateway can also do specialized things:

- **Content filtering** for known bad inputs (like if your AI is an LLM service, maybe the gateway does a quick check for extremely large payloads or known exploit strings before passing it to the AI).

- **Downgrade untrusted requests**: If certain clients are not fully trusted, gateway can route them to a safer path (maybe a less privileged instance of the service).

However, as OWASP notes, don't put all authZ logic only in the gateway for complex systems, or it becomes a choke point and single point of failure . Use it for broad strokes and convenience, but back it up with service-level checks.

Pattern 2: Service Mesh for Service-to-Service Security

We've already talked at length about **service mesh** (like Istio, Linkerd). Let's highlight what patterns it enables:

- **Automatic mTLS:** Every service call is encrypted and authenticated. This is huge for internal security. Mesh sidecars or proxies (e.g., Envoy in Istio) handle the heavy lifting. This means even if someone infiltrates your network, they can't eavesdrop or impersonate services easily.

- **Fine-Grained Authorization (Zero-Trust):** Beyond just encryption, meshes allow you to write policies like "Service A can only call Service B on these endpoints" or "Service C cannot call database D directly". For example, Istio has an AuthorizationPolicy CRD where you can specify allowed principals and paths for services. This microsegmentation is core to zero-trust.

- **Traffic Control & Observability:** Not exactly security, but helpful – meshes give you lots of telemetry (which can feed anomaly detection) and allow things like rate limiting, circuit breaking between services. You can prevent a misbehaving AI service from flooding others with requests by configuring the mesh to cut it off after X QPS, for instance.

- **Identity Federation:** Mesh can integrate with external identity systems. E.g., Istio can validate JWTs on incoming calls at the mesh level. This means even internal calls can carry end-user JWTs and be verified.

In summary, the service mesh is like an **automated internal security guard**. It wraps each service with a proxy that enforces the rules of engagement. So you, as a developer, can focus on the AI logic, while the mesh ensures your service only talks to the right peers and over secure channels.

One important pattern enabled by mesh + gateway together is a **separation of concerns**: The gateway handles entry into the cluster and broad auth, the mesh handles once inside, and each service handles the really specific logic. Together they form layered defense.

Pattern 3: Secure Secret Management (Vaults and Configuration)

Secrets and sensitive config should be centrally managed and not hardcoded. **Secret management tools** like HashiCorp Vault, AWS Secrets Manager, Azure Key Vault, etc., provide patterns such as:

- **Dynamic Secrets:** Vault can generate secrets on the fly (like a database credential that is unique and short-lived for each service on demand) . This means even if a secret leaks, it's short-lived and limited in scope.

- **Encryption as a Service:** Vault can also offer encryption/decryption APIs. So a service doesn't need to handle keys; it sends data to Vault to encrypt, which means the service never sees the raw key. This is great for things like encrypting user data or ML features before storage.

- **Auditing Access to Secrets:** Every access to the vault is logged. So if an AI service suddenly tries to read a hundred secrets it never used before, you have an audit trail.

- **Central Rotation:** When you need to update a secret (say an API key), you can update it in vault and all services fetch the latest when needed, rather than redeploying all services with new config.

A common pattern is to have an init container or sidecar that fetches required secrets at startup, or the service calls the vault at runtime when needed (using its authenticated identity). For instance, our earlier example: an AI service loads a model file from secure storage – it asks Vault for the decryption key, uses it, and never stores it on disk.

Another pattern: **config management** systems (like Consul, ConfigMaps with encryption, etc.) for non-secret but sensitive config (like feature flags to turn off AI features quickly if an issue arises).

By using vault patterns, you ensure there isn't a single static secret that if stolen, compromises everything. Instead, each service gets what it needs, when it needs it, and nothing more (principle of least privilege again, applied to secrets).

Pattern 4: Layered Architecture Diagram

Let's synthesize the above in a conceptual architecture. Below is a text-based diagram of a secure intelligent microservice setup, visualizing the layers of security around the services:

```
                    Internet
                       |
                    [ Client ]
                  |  HTTPS (OAuth2 Token)
v
                 +-----------------+                    | API
Gateway     |  <-- AuthN (OIDC), basic AuthZ, rate limiting
                 | (Edge Security) |      input validation, logging
                 +-----------------
+                     |  ^
   mTLS + JWT auth       |   |  (Token passed through)
between services      v   |
   --------------    +------------------+       Secrets via TLS      +---
-----+
   Identity:         ->| Intelligent       |------(secure API)-------> |
Vault    |
   mutual TLS, JWT     | Service A         |        (Auth to Vault)       |
(Secrets|
   --------------       | (AI Agent +      |                             |  &
Keys)|
                        |  Model)          |--\                          +---
-----+
                 +------------------+  \   mTLS + Policy
                           ^          ^           \   (Service Mesh)
                       |       |          \                      +-------
-----------+
   (Service Mesh) mTLS  |       | (calls)     \--> [ Intelligent     |
Database /        |
   + Authorization      |       |              -->   Service B    | Data
Lake (Encrypted) |
                        |       |             /    (AI Agent) ]    |
(Sensitive Data)|
                 +------------------+     /                     +-----
-
------------+
                 |  Auth Service     |   /
                 |  (Identity &      |<-/
                 |   Token Issuer)   |   <-- calls over mTLS
                 +------------------+
```

In the above (described in words):

- The **Client** (could be a user's browser or an external app) calls into the system via the **API Gateway**. The call is HTTPS and includes an OAuth2 access token obtained from the Auth Service.

- The **API Gateway** verifies the token (authenticates the user) and checks authorization (is this user allowed to call this API). If all good, it forwards the request internally.

- Inside, we have a **Service Mesh** (depicted by mTLS lines) connecting the gateway, the intelligent services (A, B, etc.), and other components. The mesh ensures all communications are secure and authenticated (shown as "mTLS + Policy" on lines). It also can inject JWT info to services or validate tokens if configured.

- **Intelligent Service A** is one of our AI microservices – say it's the Orchestrator agent. It receives the request from the gateway. Perhaps it needs a secret (like a cloud API key or a ML model decryption key), so it calls **Vault** securely. Vault authenticates Service A (maybe via a cert or token tied to its identity) and returns the needed secret. That secret was never exposed to the client or stored insecurely; Vault logged the access.

- Service A then might call **Service B** (another AI service) as part of fulfilling the request. When it calls, the service mesh automatically secures the call (mTLS). Service A might also forward the end-user JWT or include its own service JWT so Service B knows *who* initiated this and can do its own auth checks. Service B might be, say, a ML inference service that returns some data.

- Service A possibly accesses a **Database** or Data Lake to fetch or store results (maybe logging the decision). The database has encryption at rest. The connection could also be secured by TLS. Only certain services have credentials to certain databases, often retrieved from Vault or injected as needed.

- Meanwhile, an **Auth Service** stands by to issue tokens (perhaps the client had to talk to Auth to get the initial token, not shown). The Auth Service might also be used by microservices to validate tokens or to get tokens for service-to-service auth if not using mTLS identities (like getting a JWT to call an external API).

All components produce **logs** – gateway logs incoming requests, services log decisions and actions, Vault logs secret accesses, etc. These logs can be aggregated for monitoring and auditing.

This layered architecture embodies **defense-in-depth**:

1. Gateway: first gate, handling external authentication and gross filtering.

2. Service Mesh: second layer, ensuring any communication is authenticated and authorized at a service level, plus encryption in transit.

3. Service internal logic: third layer, the service code itself which does fine-grained checks (e.g., making sure a user only sees their data) and might implement business-specific security like checking for anomalies, using XAI to validate outputs, etc.

4. Secure data and secrets: supporting layer, ensuring that wherever the AI needs data or secrets, those interactions are secured (with Vault and encrypted DB).

5. Observability: cross-cutting layer watching all of the above for unusual patterns or policy violations.

Pattern 5: Continuous Monitoring and Anomaly Detection

In a dynamic AI-driven system, **monitoring** is both for performance and security. Patternwise, you'd set up:

- **Centralized Logging & SIEM:** All those logs (access logs, decision logs) stream to a Security Information and Event Management system or at least a centralized logging solution (ELK stack, Splunk, etc.). Here you can run queries or set alerts, like "alert if any service starts returning error rate > X (possible attack or malfunction)" or "alert if Vault sees >10 failed auth attempts (could be intrusion)".

- **Anomaly Detection Systems:** Ironically, you might use AI to monitor AI. For example, train a model on what normal metrics look like and alert on deviations. Or simpler, use threshold-based alerts and trending. If your recommendation agent usually calls the product database 100 times an hour and suddenly does 1000 in 10 minutes, something's off – flag it.

- **Health and Behavior Dashboards:** Expose metrics like how many decisions made, average confidence, distribution of outcomes, etc. Unusual changes can indicate either a data drift that needs a retrain or a security issue (like someone gaming the system).

- **Penetration Testing & Red Teaming as a process:** This is more operational than architectural, but essentially treat your AI microservices like any app – regularly test them. This might mean having a dedicated red team that tries adversarial inputs, or using tools to simulate attacks. There are frameworks like IBM's Adversarial Robustness Toolbox or Microsoft's Counterfit that can automate some adversarial testing against AI models. Integrate these into your development or CI pipeline so each new model is evaluated not just on accuracy, but also on robustness.

Real-World Tools and Examples Throughout

Let's pepper in some concrete tools that align with these patterns, to make it more tangible:

- **Service Mesh:** Istio and Linkerd are popular. Istio, for instance, has Citadel (now part of istiod) for certificate management, and Envoy proxies that handle mTLS. Linkerd focuses on simplicity and uses its own lightweight proxies.

- **API Gateway:** Options include Kong, Apigee, NGINX (with Lua plugins perhaps), or cloud-managed gateways (AWS API Gateway, Azure APIM). They allow writing custom auth plugins; e.g., Kong has an OpenID Connect plugin for auth, and rate-limiting plugins out of the box.

- **Auth Service/Identity:** Keycloak (open source) can issue tokens, Auth0/Okta are commercial managed ones. They can integrate with your gateway and also provide service tokens.

- **Secrets Management:** HashiCorp Vault we mentioned. It also can act as a Certificate Authority for mTLS if not using a mesh's built-in one. Kubernetes has Secrets, but Vault is more robust for large systems (with leasing and rotation). In cloud, each platform has a secrets manager or you can use Vault in-cloud.

- **Model Security Tools:** If you want to ensure model integrity, you might look at projects like Google's **Model Card Toolkit** for documentation, or **MLFlow** for tracking model versions (not security per se, but helps in traceability). The earlier reference to Sigstore for model signing is cutting-edge (by 2025, it's emerging tech) . That could soon become part of MLOps pipelines.

- **Adversarial Robustness:** IBM's **ART (Adversarial Robustness Toolbox)** is an open library to test and improve models against attacks. There are also specialized companies like HiddenLayer (as cited earlier) offering monitoring for model threats .

- **Monitoring:** Prometheus + Grafana is common for metrics (including custom AI metrics). ELK or Splunk for logs. For AI-specific monitoring (drift, etc.), startups like Fiddler, WhyLabs, or Arize have platforms, though those are more on the ML monitoring side than security specifically.

Having all these patterns and tools, let's solidify understanding with a concrete scenario callout that demonstrates securing an agentic system in practice.

Case Study: Autonomous Trading System Security

Imagine a fintech company "TradeAI" that deploys an autonomous trading microservice. This agent can execute trades on the stock market without human intervention, based on market data and AI predictions.

Security Measures Implemented:

Strict Access Controls: The trading agent service runs under a locked-down role.

It can only call the external brokerage API via a proxy that enforces limits (e.g., max $1M trade value without secondary approval). Other internal services (like data feeds) authenticate to it via mTLS; no open access. Only the RiskManager service is allowed to instruct the agent to pause or halt (a controlled stop mechanism).

Continuous Monitoring: All trades and their reasons are logged. A separate monitoring service analyzes trade patterns – if the agent's trades deviate from historical norms or risk thresholds (say it suddenly trades an unusual stock or volume), it triggers an alert. In fact, an **anomaly detection model** watches the trading agent (AI watching AI) for signs of either error or external manipulation.

Human Fallback: If something seems off, the system automatically falls back to manual mode. For example, if the anomaly detector flags high risk, it requires a human trader to approve further actions that day.

Defense Against Adversarial Market Inputs: The agent's market data feed could be manipulated (attackers have been known to release fake news to sway algorithms). To mitigate this, the agent uses multiple data sources (cross-verifying news sentiment from different providers), and the developers trained it with some known false data scenarios to make it skeptical of singular, extreme news. A [!warning] in the code comments literally says: "If only one source says 'Company X bankrupt' and others don't, treat with caution – potential false signal."

Secret Management: The API keys for trading and data feeds are stored in Vault. The agent fetches them at startup. TradeAI rotates these keys monthly to reduce the window of misuse if they were compromised.

By combining these measures, TradeAI's autonomous trading system is wrapped in multiple layers of security and trust. Even though it operates largely on its own, the company's architects can sleep at night knowing that a rogue algorithm or malicious actor can't easily wreak havoc – and if the AI ever "goes wild," the system will catch it and hand control back to humans.

This fictional scenario illustrates how the patterns (access control, monitoring, fallback, data verification, secrets, etc.) coalesce to secure an autonomous microservice handling something as sensitive as money.

Expanding the Attack Surface: Testing and Validation

We've woven in the notion that AI expands the attack surface. Therefore, our testing and validation must expand too. In final few sections, let's emphasize how to **test, pen-test, and validate** intelligent microservices against the threats we enumerated:

- **Penetration Testing AI Logic:** Traditional pen-testing of an app would involve trying SQLi, XSS, etc. For an AI service, testers will also craft adversarial inputs. This can be manual or using tools. For example, for an image recognition API, a pen-tester might use FGSM or another algorithm to slightly alter an image and see if it misclassifies. For an LLM-based service, they'll try prompt injection attacks ("Ignore previous instructions and do X...") . The security team should have expertise or playbooks for these AIspecific tests. It's a good idea to include AI subject matter experts in threat modeling sessions.

- **Red Team Exercises:** Some organizations conduct red-team vs blue-team simulations focusing on AI. The red team might attempt a data poisoning by submitting malicious data over time, or attempt to reverse-engineer the model by querying it (model extraction attack). The blue team (defenders) need to see if their monitoring catches it, or if their processes (like reviewing training data) would stop it. These exercises can reveal gaps (maybe your team wasn't checking for subtle drift in model behavior, etc.).

- **Adversarial Attack Libraries:** Utilize open-source libraries to automate generating adversarial test cases for your models. This could be part of CI/CD – every time you train a new model, run a battery of adversarial examples through it. If the model is easily fooled by something known, maybe you need to augment training or tweak preprocessing.

- **Anomaly Injection for Resilience:** Similar to chaos engineering, you might simulate weird inputs or partial system failures to see if the AI still behaves and if security holds up. E.g., "What if the AI only gets half the usual data (simulate a sensor outage)? Does it start making crazy outputs that could trigger a wrong action?" If so, maybe implement a rule "if data incomplete, agent should defer decision".

- **Monitoring Validation:** Test your monitoring and alerts by triggering them intentionally. For instance, perform a safe adversarial attack on a test system – does your anomaly detector or logging alert pick it up? If not, improve it. Or run a scenario where an insider tries to gradually poison data – see if your data pipeline governance notices unusual inputs.

- **Model Performance and Robustness Validation:** From a more engineering QA perspective, ensure you evaluate model robustness metrics, not just accuracy. There are metrics like adversarial accuracy (accuracy on perturbed data), or calibration (how well do probabilities reflect reality) which can hint at how trustable the model outputs are. A well-calibrated model is less likely to be overconfident on an adversarial input, for example.

Finally, remember that **security is an ongoing process**. As attackers evolve tactics (and they will, as AI systems become targets), we must evolve defenses. Keep up with research (the OWASP Top 10 for LLMs or similar guidelines for AI vulnerabilities is a living document). What's secure today might be bypassed tomorrow by a clever exploit, especially in the fastmoving AI field.

Conclusion: A Secure Foundation for Resilient, Trustworthy AI Microservices

In this chapter, we journeyed through the fusion of traditional security with new AI-era measures. We started by reinforcing that *the fundamentals still apply* – robust **authentication, authorization, encryption, and zero-trust** networking are the bedrock upon which any distributed system stands. Building atop that, we addressed the novel challenges of agentic systems: securing the **data** that fuels AI, defending against **adversarial attacks** on model inputs, and preventing the **manipulation** of autonomous decisions. We emphasized making AI decisions **explainable and auditable**, so they are not mystical or unchecked, but rather traceable and justifiable. We set **boundaries and oversight** (both automated and human) to ensure autonomous services remain aligned with our intentions and constraints.

Crucially, we discussed enabling services to **trust each other** through mutual authentication and verifying the integrity of the AI agents themselves, foiling potential impostors or tampering. We introduced practical **security patterns** – API gateways, service meshes, vaults, monitoring frameworks – painting a picture of an architecture where security is deeply woven in, not bolted on. And through scenarios and analogies, we saw that while the technology is complex, the core principles often mirror age-old security wisdom (least privilege, defense in depth, trust but verify) applied in new ways.

By implementing the strategies in this chapter, architects and engineers can make intelligent microservices not only **innovative** and **powerful** but also **safe** and **trustworthy**. A system that stakeholders can rely on is one that will be adopted and thrive. Conversely, one major security failure or breach of trust could derail an entire AI initiative.

Before we conclude, it's worth linking this discussion to the next frontier: **resilience and selfhealing** (Chapter 14). Security and resilience are two sides of the same coin in robust systems. A secure system prevents and protects against incidents, while a resilient system *survives and recovers* when incidents (security breaches or otherwise) occur. The principles we applied – isolation, monitoring, rapid response (like fallback to human) – all bolster resilience. If an AI service is attacked or misbehaving, containing it via zero-trust and quickly switching to a safe state is essentially a self-healing response. In the next chapter, we'll explore how intelligent microservices can detect failures or anomalies in themselves and adapt or heal without human intervention. Combining the secure foundation from this chapter with the adaptive resilience of the next will move us closer to autonomous systems that are **not only smart and secure, but also robust and self-reliant** in the face of adversity .

In summary, the best security chapter (we hope) ever written for intelligent distributed systems boils down to this maxim: *Empower your microservices with AI, but surround that power with safeguards, transparency, and trust mechanisms*. Do this, and your agentic system will be an asset, not a liability, paving the way for the confident deployment of autonomous microservices in the real world.

Chapter 14: Resilience and Self-Healing

Microservices live in an unpredictable world. **Failures are inevitable** – networks glitch, servers crash, and dependencies go offline without warning. Resilience is about **designing for failure** so that our applications can withstand these hiccups and recover quickly. In this chapter, we explore how intelligent microservices achieve high resilience and even **selfhealing** capabilities. We'll start by revisiting the fundamental resilience patterns (like circuit breakers, retries, health checks, etc.) that keep distributed systems robust. Then we'll introduce what it means for a system to be *self-healing*, and how adding an intelligent, agent-driven layer can elevate a system from simply *surviving* failures to **anticipating and autonomously fixing** them. Along the way, we'll dive into real-world implementations: Kubernetes health probes and operators, cloud auto-scaling and Lambda-based remediation, and chaos engineering tools (like Chaos Monkey and Gremlin) that deliberately inject failure to make systems stronger. We'll compare traditional manual recovery approaches with agent-driven intelligent recovery, highlighting where "agentic" intelligence adds unique value (such as pattern detection, predictive scaling, and adaptive resource reallocation). By the end of this chapter, you'll see how an intelligent microservice architecture can **bounce back from adversity** – often before anyone even notices a problem – and how that foundation sets the stage for our next topic, intelligent scaling and elasticity.

"Everything fails, all the time." **– Werner Vogels, Amazon CTO**

This famous mantra underpins resilient architecture. Since failures are unavoidable, the goal is not to prevent every failure (an impossible task) but to **minimize the impact** and recover fast when things go wrong. In other words, design your systems to expect failure and handle it gracefully.

Resilience in Microservices: Fundamentals and Patterns

Resilience is the ability of a system to withstand failures and **keep functioning**. In a microservices context, resilience means that even if one service crashes or a network call times out, the overall application continues to run with minimal disruption. You might experience slight delays or reduced functionality, but the system as a whole remains **available and reliable**. It's like having a ship with watertight compartments: if one compartment floods, the ship still stays afloat. In software, we achieve this using a toolbox of resilience patterns that isolate failures and provide fallback behaviors.

Let's recap some **foundational resilience patterns** commonly used in microservice architectures:

- **Redundancy and Failover:** Always have more than one instance of a service running. If one instance dies, the load balancer directs traffic to another. In databases or critical services, use primary/secondary failover (if the primary node fails, a secondary takes over). Redundancy ensures there's no single point of failure.

- **Circuit Breakers:** A circuit breaker is like an electrical fuse for service calls. If Service A is calling Service B and B starts failing repeatedly, the circuit "trips" and breaks the connection for a while. Further calls to B immediately fail or get a fallback response, instead of hanging and causing cascading issues. This prevents one bad apple from freezing all the others.

- **Retries with Exponential Backoff:** Sometimes failures are transient (a network glitch or a momentary spike). Retrying the request can succeed on a second try. However, retries must be done carefully – too many rapid retries can flood a struggling service. **Exponential backoff** means wait a little before the first retry, and increase the wait time for each subsequent retry, to avoid hammering the service.

- **Timeouts:** Never wait forever for a response. If a service call normally takes 50ms, but this time it's been 5 seconds, it's likely not coming. Setting timeouts on calls ensures that you won't hang indefinitely waiting for a dead service. It's better to fail fast (and maybe try a fallback) than to hang the caller.

- **Bulkheads (Isolation):** The term "bulkhead" comes from ship design – sections of a ship are compartmentalized so that a leak in one doesn't sink the whole ship. In microservices, we isolate resources similarly. For example, use separate thread pools or connection pools for different remote calls. If one external dependency is slow and all its threads get stuck, it won't exhaust the threads needed for other tasks. Each component is insulated from others' failures.

- **Graceful Degradation:** Design the system to degrade functionality in case of partial failure, rather than fully crashing. If a recommendation service on an e-commerce site fails, the site can still serve product pages minus the recommendations (perhaps showing a default message or cached recommendations) instead of an error page. The system *adapts* to work in a reduced capacity.

- **Health Checks and Heartbeats:** Each service exposes a health endpoint (for example, /healthz) or periodically sends a heartbeat signal to indicate it's alive and well. Other parts of the system (like an orchestrator or load balancer) use these to detect unhealthy instances. If a service isn't responding to health checks, it can be restarted or removed from rotation.

- **Graceful Shutdown:** Related to health checks, when shutting down or restarting, a service should finish processing current requests and deregister itself (via readiness checks) so that clients stop sending it new requests. This ensures a smoother recovery or deployment with minimal errors.

- **Load Shedding and Rate Limiting:** In extreme load, a service might start queueing requests and bog down. A resilient approach is to *shed load* – deliberately reject some requests (perhaps less important ones) once you're near capacity, rather than trying to serve everyone and failing universally. This might mean returning errors or a polite "please try later" for some requests to save resources for the rest. Rate limiting clients can also prevent any single client from overloading a service.

These patterns collectively help a system **handle failures gracefully**. A resilient microservice architecture often employs many of them. For example, a **travel booking application** might use retries and timeouts for network calls, circuit breakers between services to prevent cascading failures, and bulkheads in the thread pool to isolate external API calls. If the payment service is down, the app might degrade by queueing orders for later processing (graceful degradation) while immediately showing a message to users that their order is received and will be processed soon. All the while, health check pings help detect when the payment service comes back online to resume normal operations.

Design for failure is the mindset behind all these patterns. Rather than assuming everything will work perfectly, we **assume that each component can and will fail** at some point. By anticipating failure modes (like a slow response, an exception, or a full crash), we plan how the system should respond. This might mean having a backup strategy or simply containing the blast radius of the failure. The result is a system that may bend under stress but doesn't break completely.

Analogy – The Emergency Response Team: Think of a resilient system like a city with a good emergency response plan. If one building catches fire (one service fails), the fire department (circuit breaker) quickly contains the fire to that building so it doesn't spread (no cascading failure). The other buildings can continue their business (the rest of the application still runs). Meanwhile, ambulances and backup utilities reroute resources (failover to redundant instances) to keep essential services running for citizens. Life in the city goes on, perhaps with a few detours, but it doesn't grind to a halt. In microservices, our "first responders" are things like health checks, timeouts, and circuit breakers that detect issues and isolate or remediate them immediately.

Circuit Breakers: Cutting Off Failures

One of the most important patterns in a distributed system is the **Circuit Breaker**. In normal operation, a circuit breaker between Service A and Service B is "closed" – meaning traffic flows through it normally. But if calls from A to B start **failing repeatedly** (e.g., B is down or extremely slow), the circuit breaker will **open** after a threshold is reached. "Opening" the breaker means A will **stop calling B for a while**. Instead, calls fail instantly or go to an alternate path (like a fallback service or a cached response). This prevents A from waiting on B and tying up resources, and it prevents a pile-up of threads or requests that could cascade the failure outward.

After a cooldown period, the circuit breaker can move to a **half-open** state: it lets a test request through to see if B has recovered. If that call succeeds, the breaker closes again and normal operation resumes. If it fails, the breaker stays open for another cooldown interval.

Circuit breakers ensure that **localized failures remain local**. If one microservice is struggling, others will quickly isolate it rather than getting bogged down themselves. It's much like an electrical circuit breaker in your house: a short circuit in one appliance trips a breaker so it doesn't burn down the whole house's wiring.

In practice, libraries like Netflix Hystrix (now legacy) and its successor **Resilience4j** implement circuit breaker logic for developers, and service mesh technologies can enforce circuit breaking at the network level. Without a circuit breaker, a single failing service could cause a **domino effect**, where dozens of other services freeze up waiting for responses that will never come.

How an intelligent agent enhances this: A traditional circuit breaker trips based on predefined rules (e.g., 5 failures in a row triggers it, and it stays open for 60 seconds). An intelligent system can make this **more adaptive**. For instance, an AI agent might monitor not just failure counts but also context like current load, time of day, or the criticality of transactions. It could decide to trip the breaker **sooner** if it detects anomaly patterns (e.g., response times from Service B suddenly 10x slower than usual might merit tripping *before* many failures occur). Conversely, if the agent knows a certain spike is temporary, it might allow a bit more traffic through before opening. An agent could also dynamically adjust the **cooldown period**: if the last few attempts to reset the circuit failed, maybe wait longer before the next try, or if the service B is in the middle of a known deployment (info the agent has), it might keep the circuit open until that's done. Essentially, the AI layer can fine-tune the circuit breaker parameters on the fly, whereas a static config is one-size-fits-all. We could even imagine an agent that correlates data across many circuit breakers – e.g., noticing that multiple services are tripping on calls to Service B and then automatically creating a **high-level alert** or taking action on Service B (like restarting it or diverting traffic).

A quick visualization of a circuit breaker in action:

```
Service A --> [Circuit Breaker] --> Service
                 | OPEN (on failures) ↴
                 |----> [Fallback Response] (when circuit open)
```

In the diagram above, Service A's calls to B pass through a circuit breaker. Normally, they go straight to B. But if B is unhealthy and the breaker "opens", further calls from A get a fallback response or error immediately, rather than waiting on B. This saves A from hanging. When B recovers, the breaker closes and A can talk to B normally again. An intelligent monitoring agent might be watching the state of this breaker and logging how often it trips, possibly using that information to predict issues or adjust thresholds.

Retries and Timeouts: Handling Glitches

Not all failures require drastic measures – sometimes, simply **trying again** will succeed. That's where **retries** come in. Microservices commonly implement a retry policy for calls to other services or external APIs. For example, if a payment request times out due to a momentary network blip, the service can automatically retry the request after a short delay. Often this is combined with an **exponential backoff** strategy: e.g., wait 100ms before first retry, if that fails wait 200ms, then 400ms, etc., doubling the wait each time up to some limit.

This approach avoids flooding the target with retry requests, giving it a chance to recover.

However, **retries need caution**. If the failure is due to the remote service being down, aggressive retries from dozens of callers might worsen the situation (thundering herd problem). That's why retries pair well with circuit breakers and timeouts. A **timeout** ensures we don't wait too long on any single attempt, and a circuit breaker can disable retries entirely if the service is confirmed down.

From a resilience standpoint, retries increase the chance that transient issues self-resolve without human intervention or erroring out to the user. Imagine a slight network latency spike – a single failed request might succeed 100ms later on retry, and the user never even notices a hiccup.

How an intelligent agent enhances this: An AI-driven approach could make retry logic smarter. For instance, an agent might **learn the optimal number of retries** under various conditions by analyzing past incidents. Maybe it learns that for Service B, if it doesn't respond after 2 retries, further retries are hopeless and just waste time – so it dynamically lowers the retry count for that service. Or it could do the opposite: if historically a third retry often succeeds at night when a batch job causes intermittent slowdowns, the agent might allow an extra retry during those hours. Agents could also coordinate retries globally: if Service B is down, a central orchestrator agent might tell *all* services "don't bother retrying right now" to reduce noise, and then signal them when B is back up. This kind of **swarm intelligence** could avoid hundreds of unnecessary retry attempts that traditional independent logic would make.

Another way AI can help: dynamically adjusting timeouts. Many systems set static timeout values (say 2 seconds for a call). But an agent could monitor latency patterns and adjust timeouts on the fly – for example, during peak load it might extend timeouts slightly for noncritical requests (to avoid failing too quickly if things are just a bit slower), but tighten timeouts for critical interactions to detect issues faster. Essentially, the agent can balance **responsiveness vs. patience** based on context, whereas a fixed timeout is a blunt instrument.

Health Checks and Heartbeats: Detecting Failures Early

In a microservice ecosystem, it's vital to continuously **monitor the health** of each service.

Health checks and heartbeats are the pulse-monitoring of your system's circulatory system. A **health check** is typically an HTTP endpoint (like /health or /alive) that a service provides, which returns a simple status (healthy or not) sometimes along with meta-information (like load or version). Infrastructure components like load balancers, service registries, or container orchestrators (e.g., Kubernetes) will ping these endpoints regularly.

- **Liveness vs Readiness:** In Kubernetes (and modern platforms), there are typically two types of health probes:

- A **liveness probe** answers "Is the service process alive and functioning?" If a liveness probe fails (meaning the app is unresponsive or in a bad state),

Kubernetes will assume the container is hung or dead and automatically **restart** it. This ensures that if an application gets stuck (e.g., deadlock or out of memory), it won't stay in that broken state forever – the orchestrator will kill and revive it (a basic self-healing action).

- A **readiness probe** answers "Is the service *ready* to handle requests?" A service might be alive but not ready to serve (for example, still warming up, or temporarily overloaded). When readiness fails, Kubernetes will take the pod **out of traffic rotation** (so no new requests are sent to it via the service/load balancer) until it reports ready again. Readiness probes prevent sending traffic to a service that can't handle it, allowing it to recover or finish startup. They also facilitate graceful shutdown: if a pod is going down for maintenance, it can mark itself not ready, so no new traffic comes in while it finishes existing work.

Health checks can be as simple as returning an HTTP 200 OK if everything's fine. They can also include sub-checks (for instance, ping a database and include that status). However, one must be careful: a health check should usually be **fast and reliable**. It's a bad idea to have a health check that itself might hang or fail due to an optional dependency – that could cause false alarms.

A robust health check system, combined with an orchestrator, gives you **automatic fault detection**. For example, if one instance of Service X on a VM or container crashes, the orchestrator sees the failed heartbeat and can launch a new instance of Service X elsewhere. In Kubernetes, this is part of its **self-healing** nature: *"Kubernetes automatically replaces failed containers, reschedules workloads when nodes become unavailable, and ensures that the desired state of the system is maintained."* . Essentially, the system is always watching its own vitals and **resuscitating** parts that stop working.

Kubernetes Health Probes

In Kubernetes, configure liveness and readiness probes for each microservice. For example, you might set a liveness probe to hit /healthz every 15 seconds, with a failure threshold of 3. That way, if the app doesn't respond 3 times in a row (45 seconds of unresponsiveness), Kubernetes will assume it's dead and restart its container. A readiness probe could check something like the app's connection to its database – if the DB is down, the app might mark itself unready (returning an HTTP 500 on the readiness endpoint), signaling to Kubernetes *"don't send me requests right now."* These probes greatly increase resilience by enabling automatic detection and recovery from common failure modes. Always ensure your health endpoints are efficient and accurately reflect the app's ability to function.

How an intelligent agent enhances this: At first glance, health checks are binary and simple. But an AI layer can bring more **nuance**. Consider an intelligent monitoring agent that doesn't just see healthy/unhealthy, but watches **trends** on health metrics. For instance, a service might be *technically* passing its health checks but running with 90% memory usage and climbing – likely to crash soon from an out-of-memory error. A smart agent could detect this pattern (memory usage anomaly) and pre-emptively restart the service or alert it to scale up, **before** the service actually goes unhealthy. In that sense, the agent is performing a kind of *predictive health check*. Similarly, if a service's latency is creeping up release over release, an AI could flag this regression early as a risk to stability.

Another way intelligence helps: think of **composite health**. Microservice A might depend on B and C. Individually, A, B, C all report healthy. But if B can't reach C, maybe A's functionality is impaired even if A itself is "healthy." Agents can reason about these dependencies: *"Service A's health is green, but its critical downstream C is red; therefore, A's effective health for serving real requests is degraded."* A human operator might infer that from dashboards, but an agent could automate that insight and, for example, **redirect traffic** away from A proactively if one of its required dependencies is hard down. Traditional readiness probes might not catch that scenario unless A explicitly checks C, but an external intelligent orchestrator could.

Bulkheads: Isolating Components to Contain the Damage

Bulkheads in software ensure that **failure in one component doesn't sink the whole ship**. In practice, this often means allocating dedicated resources to different tasks or segments of work. For example, a service might use a separate thread pool for handling incoming HTTP requests and another for processing background jobs. If the background jobs hang and all those threads get tied up, it won't affect the pool that handles live requests from users.

In microservices, bulkhead isolation can occur at multiple levels:

- At the *service instance* level: Run multiple instances of a service in separate containers or VMs, potentially even in different availability zones. If one instance or zone has issues, others are unaffected.

- At the *resource* level: Partition the usage of resources like memory, CPU, database connections. Many servers allow setting pool limits (e.g., "use max 50 connections to DB for feature X, and a separate 50 for feature Y"), so a surge in one feature doesn't starve others.

- At the *service call* level: Similar to circuit breakers, you can isolate calls by type. E.g., use one thread pool for calling external Service P and another for Service Q. If P is slow and those threads backlog, the Q threads are still free to call Q.

Bulkheads complement circuit breakers and timeouts. Timeouts and circuit breakers cut off problematic interactions, and bulkheads ensure that even if something slips through, its damage is confined.

Role of intelligence: Bulkhead configuration is often static (decided by developers or ops based on experience). An intelligent system might **dynamically allocate resources** to where they're needed most while still isolating components. Think of it like an adaptive bulkhead: if Service P is failing and basically not serving, the agent could decide to shrink the thread pool allocated to P (since those threads would just be idle or wasted) and expand the pool for Service Q which is under heavy use, *in real-time*. Or in a Kubernetes context, an intelligent controller might move certain less critical services to separate nodes when it detects another critical service is under heavy load, to avoid competition on the same node's resources. These are advanced scenarios, but they illustrate that AI can manage resource pools and isolation levels more fluidly than fixed configs, essentially **reallocating capacity on the fly** to contain failures and optimize overall system health.

Graceful Degradation: Serving Under Reduced Capacity

We touched on **graceful degradation** as a principle – it's worth emphasizing with an example. Suppose you run a mapping service with two components: one that provides basic map images and one that provides fancy traffic overlays. If the traffic overlay component fails, a graceful degradation strategy would still serve the maps without traffic info, rather than failing the entire map request. Users get core functionality, even if some bells and whistles are missing.

This pattern often relies on having sensible **fallbacks or defaults**. For instance, if your AI recommendation engine is down, you might fall back to showing "popular items" or nothing at all in that section, but the rest of the page (product details, purchase button) works fine. Or if a personalization service is unavailable, serve the page with generic content rather than an error.

The key is to prioritize core service quality over auxiliary features. Identify which parts of your system can be temporarily sacrificed if needed, and build in the logic to do so automatically. Feature flags or config can help toggle features off when dependencies fail.

In traditional systems, graceful degradation might be manually triggered – engineers turn off a feature when something breaks. But resilient microservices should handle it automatically: detect the failure of the optional component and **remove it from the user experience** on the fly.

How an agent helps: An intelligent agent could make degradation decisions based on current conditions. For example, if response times are climbing because an optional service is slow, the agent could *instruct services to stop calling that optional service* (perhaps via toggling a feature flag throughout the system), effectively degrading gracefully across the board. Later, when that service recovers, the agent turns the feature back on. This is a coordinated response that goes beyond one service – it's system-wide adaptation. In an e-commerce scenario, an AI orchestrator might notice that the "similar items" component is timing out and causing checkout pages to load slowly. It could then signal all services, "temporarily disable similar-items feature to keep things snappy," and maybe even display a minor note to users or log that a feature was auto-disabled. This kind of **feature-level self-healing** keeps user experience as high as possible, trading off non-essentials during failures.

Summary of Resilience Patterns

So far, we've reviewed the classic resilience patterns and hinted at how intelligence can extend them. In summary, the fundamental goal is **fault tolerance** – the system continues to operate in the presence of failures, through redundancy, smart retries, timeouts, and isolation. These patterns are the *first line of defense*. They are largely reactive (responding to failure signals) and pre-planned (configured by engineers).

What happens when we add an intelligent, agentic twist to these mechanisms? We move from purely reactive to **proactive and adaptive** resilience. Instead of just absorbing failures as they come, an intelligent microservice system can anticipate and mitigate issues **before** they fully manifest. That's where *self-healing* comes into play, which we'll dive into next.

But before moving on, it's worth noting that resilience is a **holistic property**. It's not just one component or pattern, but the combined effect of many design decisions. A resilient system often has layers of safety nets – for example: retries (to handle blips) within a circuit breaker

(to handle persistent issues) running on an orchestrated platform that will restart failed instances, with good monitoring to alert humans if things get truly out of hand. Now, let's explore how we can push this further with systems that heal themselves.

Self-Healing Systems: What Does "Self-Healing" Mean?

Resilience is great – your system can take a punch and stay standing. **Self-healing** takes it a step further: your system can *treat its wounds and get back to full health on its own*. A selfhealing system can **automatically detect problems and fix them without human intervention** (or with minimal intervention). It's the difference between a ship that can compartmentalize a leak (resilient) and a ship that can automatically patch the leak while still at sea (self-healing).

In practice, a self-healing microservice architecture will *detect* a failure or anomaly, *isolate* or mitigate the issue, and then *restore* the system to normal operation – all autonomously. This often involves some orchestrator or set of agents that watch the system's health and perform corrective actions.

Basic example: Kubernetes restarting a crashed container is a form of self-healing. No human needed to intervene; the system noticed a process died and started a new one. That's a simple automated reaction to failure.

Another example: An Auto Scaling Group in AWS sees that an EC2 instance failed its health check. The group automatically terminates that instance and launches a new one to replace it . The system heals by replacing the bad part.

These are straightforward rule-based self-healing actions. Self-healing can also be more complex, involving multi-step runbooks executed automatically. For instance, if a whole service goes down (all instances), a self-healing script might detect this via metrics and then perform a series of actions: clear a config cache, rollback a recent deployment (assuming the new deploy caused the crash), and then bring the service back up.

Analogy – The Human Immune System: The concept of self-healing in software is often compared to how our bodies heal. When you get a cut on your skin, you don't consciously do anything – your body detects the injury and kicks off a healing process (blood clotting, scab formation, tissue repair). If a virus invades, your immune system detects the intruder and attacks it, often eliminating it before you even realize you were sick. Similarly, a self-healing system detects "injuries" (like a crashed service or a misbehaving component) and initiates a recovery process automatically. There's an innate response mechanism that doesn't require a doctor (or in our case, an on-call engineer) for every little scrape or fever. Of course, severe issues still need a doctor – and likewise, severe outages still need human engineers (we'll discuss that later) – but the idea is to handle the routine and expected failure modes internally.

So what does it take for a system to be self-healing? Two key capabilities:

1. **Automated detection** of anomalies, failures, or performance issues. The system needs to know something's wrong. This comes from rich monitoring, health checks, error tracking, and sometimes predictive analysis.

2. **Automated action** to remedy the situation. Once detected, the system should trigger some corrective action – restart a service, failover to backup, clear a queue, downgrade a feature, etc.

In a classic setup, these might be simple if-then rules. E.g., if a service goes down, restart it (K8s does this). If CPU is 100% for 5 minutes, scale out by launching a new instance. If memory leak, kill and restart process. Many cloud systems have these kinds of automated remedies configurable via policies.

What distinguishes *intelligent* self-healing (the agentic approach) is the use of AI or learning to make smarter decisions on both detection and action. Instead of static thresholds or prewritten scripts for every scenario, an intelligent system can learn from past incidents, recognize complex patterns (like a combination of micro-signals that indicate an upcoming failure), and choose from a range of possible actions the one that's most likely to help.

Benefits of self-healing:

- **Reduced Downtime:** Issues are resolved faster (often immediately), which means less time that the system is in a degraded or down state. Users might not even notice a blip.

- **Less On-Call Fatigue:** Engineers aren't woken up at 3 AM for every minor hiccup – the system handles the easy ones. Humans can sleep through the night more often, only alerted for the truly weird or hard-to-fix problems.

- **Scalability of Operations:** As you scale up the number of microservices, manual monitoring and intervention doesn't scale well. Self-healing automation acts as a force multiplier for a small ops team to manage a large, complex system.

- **Consistent Responses:** Automated healing ensures that known problems are handled the same efficient way each time, rather than depending on who is on call and how tired they are. It reduces human error or delayed responses.

Challenges of self-healing:

- **False Positives/Negatives:** Detecting failures isn't foolproof. An automated system might trigger a "healing" action when nothing is wrong (false positive) or fail to catch a subtle issue until it's big (false negative). Tuning the detection is tricky – too sensitive and you flap (e.g., restarting services too often), too lax and you miss real problems.

- **Choosing the Right Action:** Some problems have multiple possible fixes. Rebooting a service might clear an error, but maybe the better fix was to flush a cache or adjust a config. Hard-coding actions can backfire if the system misdiagnoses the issue.

- **Over-correction:** A classic tale is an auto-scaler adding too many instances, or a selfhealing script that keeps restarting something that is failing for an external reason. If not careful, automated actions can sometimes make a problem worse or cause thrashing.

- **Complex Failures:** Some failures are complex or novel, and an automated system might not handle them well, requiring human insight. Self-healing works best for known failure modes or gradual issues, not as much for chaotic, multi-faceted incidents.

Overall, self-healing is an aspiration: **make the system smart enough to manage routine problems by itself**. In the next section, we'll discuss how we can inject intelligence – often in the form of autonomous agents – to improve both detection and reaction in self-healing systems.

Intelligent Agents for Resilience and Recovery

So far, we've looked at resilience patterns and basic self-healing mechanisms mostly in a rule-based sense. Now let's explore the role of **intelligent agents** in taking resilience to the next level. By "intelligent agents," we mean software components (often powered by AI/ML or complex logic) that have a degree of autonomy: they can perceive what's happening (via metrics, logs, events), make decisions (often based on learned patterns or goals), and take actions – all with minimal human guidance. In the context of microservices, think of an intelligent agent as a super-smart ops team member that never sleeps: constantly watching the system, analyzing it, and tweaking things to keep it running optimally.

What can these agents do to improve resilience and enable self-healing? Let's consider a few scenarios and roles such agents can play:

- **Anomaly Detection Agent:** This agent's job is to watch for **early warning signs** in the system. It uses techniques from AI (like machine learning models for anomaly detection) to spot patterns that humans or simple thresholds might miss. For example, it might learn the normal range of request latency for each hour of the day for Service X. If at 2 PM it suddenly sees latency 50% higher than usual (even if still below a static alert threshold), it can flag this as unusual. Perhaps memory usage in a service is climbing steadily release over release – a trend detection algorithm could catch that long before the service actually runs out of memory. By identifying anomalies in metrics (CPU, memory, response times, error rates, etc.), this agent essentially **augments monitoring** with intelligence. Instead of sending an alert to a human, the agent could directly trigger self-healing actions (like restart a service if it's behaving oddly) or at least warn other parts of the system.

- **Predictive Scaling Agent:** This specialized agent tries to forecast the system's needs. Traditional auto-scaling reacts to metrics crossing thresholds (e.g., if CPU > 70% for 5 minutes, add a server). A predictive agent uses historical data and perhaps external signals (like time of day, day of week, special events) to **anticipate load spikes or drops** and adjust capacity *in advance*. For instance, say every weekday at 9 PM there's a spike in traffic to a streaming service when new episodes of a show release. A predictive agent notices this pattern and at 8:45 PM it scales up the necessary services proactively – so when the spike hits, the system is already prepared and no performance degradation occurs. Likewise, it might scale down preemptively after the rush to save cost. By staying **one step ahead**, such an agent prevents the system from ever hitting a breaking point due to load. This is a self-healing angle because it avoids a scenario where services would otherwise become overloaded (essentially preventing a failure through prediction).

- **Auto-Tuning Agent:** Microservices have many knobs – timeouts, cache sizes, thread pool limits, circuit breaker thresholds, etc. These are typically configured based on conservative estimates. An intelligent agent could **dynamically tune** these values based on current conditions or learned behavior. For example, if it observes that a certain downstream service is responding slowly, it could temporarily lengthen the timeout for calls to that service (to reduce the chance of failure), or perhaps tighten the circuit breaker to trip sooner if it suspects a bigger issue. If it sees a surge in error rate with messages indicating queue overload, it might increase the size of a message queue or slow down the producers. This is essentially the system learning to optimize itself in real-time to avoid hitting failure conditions. Think of it like a thermostat adjusting heating/cooling to keep the environment stable – the agent adjusts system parameters to keep everything in the "green zone".

- **Healing Orchestrator Agent:** This is like the **brain of self-healing**. It takes input from anomaly detectors, health checks, and performance data, and decides **what corrective action to run**. Suppose a service goes down. A basic orchestrator (like Kubernetes) will restart it – that's one action. But an intelligent orchestrator could add more contextaware actions: e.g., "Service X went down right after a new deployment – likely the deployment is bad, so roll back to the previous version *before* restarting." That's a smarter sequence than just blindly restarting the bad version repeatedly. Or consider a scenario: the agent detects that Service X memory usage is growing abnormally (maybe a memory leak) but it hasn't crashed yet. It could proactively schedule a rolling restart of Service X instances one by one (so that each gets a fresh memory slate) at a low-traffic time, preventing an outage that would have happened hours later. The healing orchestrator basically encodes operational runbooks and decision logic, possibly enhanced by AI if it learns which actions tend to fix which problems. Over time, such an agent might learn, for example, "If database connections are exhausted, 90% of the time the fix is to recycle the connection pool – I'll just do that automatically now."

- **Network and Traffic Control Agent:** Failures aren't just about processes dying; sometimes parts of the network between services become slow or flaky (e.g., one availability zone's network is having issues). An intelligent agent at the routing layer could detect unusual latency between services and **reroute traffic** through healthier paths. For instance, it might temporarily avoid sending requests through a particular API gateway instance that looks unhealthy, or shift load away from a region that's experiencing high error rates due to an infrastructure issue. This is akin to how the Internet's routing protocols find new paths when a fiber cut happens – here we apply it at the application level with an AI deciding the best routing adjustments.

- **Knowledge Accumulation:** One often overlooked aspect of intelligent systems is that they can **remember and learn from past incidents**. Suppose an agent encountered a particular failure mode last month and tried a few things, eventually a certain sequence fixed it. The next time a similar pattern arises, the agent can recall what worked before and apply it faster. Over time, as the system experiences various faults (and as engineers provide feedback or label causes), the agent builds a knowledge base. This moves towards the concept of *AIOps* (AI for IT Operations) – where machine learning is applied to ops data (logs, incidents, performance metrics) to glean insights and suggested remedies. In a way, your system's self-healing gets **smarter with age**, because it has seen more and more and can generalize that experience.

Let's ground this in a concrete scenario:

Predictive Self-Healing in a Video Streaming Platform

Imagine a company *StreamCo* that runs a video streaming platform. They have a microservice architecture with a **Streaming Service** that delivers video chunks to users, and a **Recommendation Service** that suggests new content. One evening, the Streaming Service experiences a memory leak issue – its memory usage climbs steadily with each hour of uptime. Traditionally, this might go unnoticed until the service crashes

(likely during the peak of a new episode release, causing user-visible downtime). But StreamCo has an intelligent monitoring agent watching. The agent has learned the normal memory usage patterns and detects the anomalous growth. Around 30 minutes before the expected crash, it triggers a proactive **self-healing workflow**: it gracefully spins up a fresh instance of the Streaming Service and starts draining traffic from the leaky instance to the new one. Then it takes the leaky instance out of rotation and restarts it, effectively cleaning the memory. This all happens **just in time** – users didn't experience any outage. Later, the agent also flags the incident in a report for developers to investigate the memory leak bug, but in the meantime, the platform's users were none the wiser. StreamCo's team jokingly calls this agent their "night shift engineer" because it often handles issues in the middle of the night, like when a sudden surge of viewers at 2 AM causes high load. The agent will preemptively scale up nodes and even shed some non-critical processing to keep streams smooth. The engineering on-call gets to sleep through the night, and in the morning they check a summary: *"Memory leak in Streaming*

Service mitigated via proactive restart; no user impact. Please address underlying bug." This is the promise of intelligent self-healing – issues resolved with zero human intervention and zero customer impact.

In this example, the key difference from a normal auto-healing system is the **predictive, almost surgical response** by the agent. A simple rule-based system might only restart after a crash (causing at least some downtime) or restart periodically without knowing if it's needed (which could disrupt users unnecessarily). The intelligent agent found the sweet spot to act.

To integrate agents like this, architects often build or leverage frameworks for **AIOps** – where agents ingest logs, metrics (through observability platforms), events (like deployments), and maybe even external data (user feedback, support tickets) to make informed decisions. Tools and platforms in this space (e.g., IBM Watson AIOps, Dynatrace, Datadog with AI features, etc.) are emerging to provide this kind of functionality out-of-thebox or allow customization.

However, one can also DIY an intelligent agent using a combination of monitoring, scripting, and machine learning models. For instance, you might have a Python service that periodically runs anomaly detection on metrics and calls APIs (like the Kubernetes API or AWS API) to execute fixes (scale up, restart pod, etc.) when certain patterns occur. The complexity can vary widely.

An important consideration: **trust and safety of AI actions**. If an agent can do powerful things (like shutdown services, change traffic routing), you want to ensure it doesn't run amok. Often, there will be **guardrails**: e.g., only allow the agent to do reversible or safe actions automatically, and require human approval for riskier moves (like scaling down a database cluster or triggering a failover that might lose state). In Chapter 9 on human-in-theloop, we discussed having checks and balances. For resilience, one might set up rules like "the agent can restart services and scale up, but it cannot delete data or permanently shutdown something without human sign-off." This ensures that if the agent misfires (false alarm or misdiagnosis), the consequences are limited to things that are easily recoverable (like a momentary extra restart, which is usually fine).

In distributed systems, there's also the notion of **inter-service trust**: one service (or agent) shouldn't blindly trust another's actions. If an orchestrator agent says "Service B is bad, route away from it," individual services or the mesh might verify that claim (maybe by checking their own experience calling B). Zero-trust principles apply even in automated healing – agents need to have clearly defined scopes and authentication, so a compromised or buggy agent doesn't cause system-wide issues. Security is not the main topic here, but it's worth noting that adding AI agents means adding components that themselves must be secure and reliable.

Alright, we've painted a picture of what intelligent agents can do. They can **detect** better (pattern recognition, anomaly detection), **decide** better (learned responses, predictive moves), and **act** faster than traditional systems, all of which contributes to resilience.

Next, let's combine these ideas into **real-world strategies and tools** that implement resilience and self-healing. We'll look at how platforms like Kubernetes and AWS support these goals, and how practices like chaos engineering validate them.

Real-World Self-Healing Strategies and Tools

Concepts are great, but how do we apply resilience and self-healing in practice? In this section, we'll explore concrete tools and techniques used in industry to build and maintain resilient microservices. We'll see how Kubernetes offers self-healing out of the box (and how we can extend it with operators), how cloud platforms like AWS enable automated recovery and scaling, and how **chaos engineering** tools are used to continually test and improve resilience. Throughout, we'll contrast the "traditional" usage of these tools with how an intelligent or automated approach can further enhance them.

Kubernetes: Health Probes and Automated Restarts

Kubernetes has become the de facto platform for running microservices, and it heavily emphasizes resilience. One of Kubernetes' core features is its

self-healing capability for containers. We discussed **liveness and readiness probes** earlier – those are your hooks to inform Kubernetes of an application's health.

- If a **liveness probe** fails, Kubernetes assumes the container is unhealthy and will perform a **restart** of that container. This is typically faster than waiting for a human to notice and restart it. In many cases, a simple restart solves transient issues (like clearing a bad state).

- If a **readiness probe** fails, Kubernetes stops sending traffic to that pod (it's marked as "unready"). The pod can then fix itself or just stay idle until it becomes healthy again, without hurting incoming requests.

Kubernetes Deployments manage a set of replicas for a microservice. If one replica goes down (say the process dies or the node it's on crashes), Kubernetes notices the missing pod and will **create a new pod** to replace it, maintaining the desired number of replicas. This is a classic case of *healing by replacement*. Similarly, if a node (VM) in the cluster goes offline, the pods that were on it are rescheduled to other nodes.

Kubernetes also lets you specify **pod disruption budgets** and configure rolling updates, all in service of keeping the system available even during changes or failures. A disruption budget might say "ensure at least 80% of pods are up"; so if a failure happens, Kubernetes will try to respect that and not kill too many pods at once. All of this contributes to resilience.

Custom Health Checks: Kubernetes' default notion of health is binary (the probe succeeds or not). However, you can implement logic in the probe endpoints to incorporate more sophisticated checks (like checking dependent services, or ensuring a certain workload has drained). Some advanced setups have external controllers that decide to mark a service unhealthy if certain complex conditions are met (via Kubernetes API).

In summary, if you containerize your microservices and run them on Kubernetes with properly configured health probes, you get a lot of self-healing for "free": auto-restarts, autoreplacement of failed pods, and isolation from unhealthy instances (via readiness gating).

Intelligent twist: Even with Kubernetes doing basic healing, you can run an **intelligent agent inside the cluster** (or integrated with it) to enhance this. For example, an agent could monitor application-level metrics and preemptively cycle pods. You might integrate an AIOps tool that tells Kubernetes to cordon off a node (mark as unschedulable) when it predicts that node will fail (perhaps due to hardware signals or kernel errors that an AI model picks up on). Kubernetes is very extensible, which leads us to the next tool: **Operators.**

Kubernetes Operators and Custom Controllers: Automated Repair Crews

A Kubernetes Operator is essentially a custom controller that extends the Kubernetes API to manage not just pods, but entire applications or systems. Operators often encapsulate **expert knowledge of running a specific application**. For example, there are operators for databases like MongoDB or Cassandra, which know how to set up a cluster, reconfigure it, handle node failures, take backups, etc., in an automated way.

How do operators relate to resilience and self-healing? They can be thought of as **domainspecific self-healing agents** within Kubernetes. A few examples:

- **Database Operator:** Suppose we have a MySQL operator managing a MySQL cluster on Kubernetes. If one MySQL instance (pod) crashes, the basic K8s will restart it – but the cluster might need more careful handling. The operator can notice the crash and ensure that when the instance comes back, it rejoins the cluster properly, perhaps by restoring data from a backup or syncing from replicas. If the primary fails, the operator can promote a replica to primary (failover logic), which is more application-specific than Kubernetes would handle on its own.

- **Config Checker Operator:** You could write an operator that watches config maps or environment variables and the pods using them. If an invalid config is applied (say someone sets an illegal value that causes the app to crash on startup), the operator might detect the crash-loop and revert to the last known good config automatically – a self-healing action beyond the generic restart (which would just keep failing).

- **Custom Auto-Scaler:** While Kubernetes has the Horizontal Pod Autoscaler (HPA) for scaling based on metrics (like CPU), one could create a smarter scaler operator. For instance, an operator that scales services based on business metrics (like number of active users or incoming orders), or one that coordinates scaling between a set of services (ensuring if Service A scales up, Service B and C also scale to handle the increased load from A). This goes into intelligent territory by encoding knowledge of how services relate.

- **Workload Recovery Operator:** Imagine an operator for a streaming platform that monitors Kafka queue lengths, processing rates, etc. If it sees a backlog growing dangerously (perhaps one service in the pipeline is slow), it could automatically spin up more consumers or even redirect some partitions of work to a backup service. Essentially, it's watching a workflow and self-healing performance bottlenecks by adding capacity or rerouting – all coordinated through K8s APIs to create/delete pods or tweak configs.

Operators usually run as **controllers inside the cluster** and have permission to create/update/delete resources based on the custom logic. They typically respond to events or periodic sync (reconciliation loop) where they observe the current state and compare it to a desired state, then act to converge the two (this is the Kubernetes controller pattern).

From a resilience standpoint, operators are powerful because they allow **automation of operational tasks** that used to require human intervention. They also can encapsulate complex multi-step processes. Essentially, they are the embodiment of "self-healing" for complex apps. Instead of a person manually doing failover steps or cleaning something, the operator does it immediately.

For intelligent microservices, one could embed AI into an operator – for example, an operator that uses a machine learning model to predict when to scale or when a component is likely going to fail, then take action. There are experimental projects and research on reinforcement learning based controllers for scaling or scheduling.

However, even without heavy AI, operators make your platform more *autonomous*. They're like having a specialized SRE team member dedicated to one problem, running 24/7 in code form.

One thing to be cautious about: Ensure that your operators themselves are reliable and don't introduce bugs or runaway behavior. It's somewhat ironic if the "self-healing" operator crashes and needs healing. So treat operators as first-class components: monitor them and maybe have fallback mechanisms if they fail (for instance, manual scripts or alerts if an operator isn't functioning correctly).

Cloud Auto Scaling and Automated Remediation (AWS focus)

Outside the Kubernetes world (or even within it on cloud), there are many resilience features provided by cloud platforms like AWS, Azure, and Google Cloud. We'll focus on AWS as an example since it's widely used and was mentioned, but the concepts translate.

Auto Scaling Groups (ASGs): In AWS, an Auto Scaling Group manages a collection of EC2 instances (virtual machines) together. You define policies for scaling (increase or decrease the number of instances based on CloudWatch metrics like CPU, network, etc.). Importantly, ASGs perform **health checks** on instances – either the EC2 status checks (like is the VM responsive) or deeper health via an attached Load Balancer's health check. If an instance is marked unhealthy, the ASG will automatically terminate it and launch a new one to replace it . This is analogous to Kubernetes replacing a failed pod, but at the VM level. For example, if an instance gets stuck or the application on it fails such that health checks to the load balancer fail, you get a new fresh instance (hopefully with the app configured to start on boot, etc.). This drastically reduces time to recover from instance-level failures.

- You can tune the health check grace period (giving instances time to fully start before health checks count) and the conditions for health (like based on LB 200 OK responses).

Amazon EC2 Auto Recovery: For single instances not in an ASG, AWS has a feature called EC2 Auto Recovery – if the underlying hardware fails or certain types of failures happen, AWS can automatically move your instance to new hardware (basically a reboot on a healthy host) to recover it. This is more infrastructure level but contributes to resilience (less likely a hardware fault takes you permanently down).

AWS Lambda and Event-driven Remediation: AWS Lambda (serverless functions) combined with CloudWatch Events (or EventBridge) allows you to automate responses to certain events or alarms. A common pattern:

- A CloudWatch alarm triggers when something goes out of bounds (e.g., high error rate, low available memory, queue length high, etc.).

- The alarm can target a Lambda function as a trigger.
- The Lambda function contains the logic to **remediate** the issue. This could be anything from calling an AWS SDK to scale something, restarting a service, or even performing a complex sequence like draining a queue or toggling a feature flag.

For example, suppose you have a legacy service that occasionally hangs and needs a restart. You can set an alarm on “no requests succeeded in last 5 minutes” for that service.

When it triggers, a Lambda could call the ECS (Elastic Container Service) API or SSM

(Systems Manager) to restart the task or instance. That’s self-healing via a Lambda script.

Another example: If a certain log pattern appears (indicative of an error), send it to CloudWatch Logs which triggers a Lambda that automatically creates a Jira ticket and reboots the service and maybe sends a Slack message to the on-call. The possibilities are endless because with Lambda you can script any logic and use any AWS API or even thirdparty API.

AWS Systems Manager Automation: AWS has Systems Manager (SSM) which can run predefined automation documents on your instances. You can trigger these via events or manually. For instance, if an alarm says “disk space low”, you could have an automation that clears caches or increases volume size. It’s a bit more pre-baked approach than writing raw Lambda code, but similar idea: automated runbooks.

Automated Runbooks

A best practice is to take common manual recovery procedures (runbooks) and automate them. If your ops team has a documented step-by-step for “when X happens, do Y, then Z, then check Q,” see if you can encode that into a script or cloud automation. For example, *“If the web server is unresponsive, restart the service. If it’s still unresponsive, reboot the instance. If that fails, roll back the deployment.”* You can implement this logic with a combination of CloudWatch alarms and AWS Lambda or SSM Automation. By doing so, you turn institutional knowledge into code, enabling the system to handle known failure scenarios on its own. Start with safe, easily reversible actions (like service restarts) before automating drastic ones (like database failovers). Over time, build a library of these self-healing scripts. They will significantly reduce the need for 3 AM calls, as the system will often resolve the issue before you even know it occurred.

Auto Scaling Policies – Predictive: AWS has introduced some predictive scaling features where the system can scale based on patterns (using machine learning under the hood) – you enable it, and it looks at historical data to forecast scaling needs (similar to the predictive agent we described, but built-in). This again is bridging into the intelligent territory with outof-the-box services.

Serverless and Managed Services: Another note – using serverless services (like AWS Lambda, DynamoDB, etc.) offloads resilience to the provider. For instance, Lambda will retry failed executions depending on how you configure triggers, and DynamoDB automatically replicates data across availability zones. While not "self-healing" that *you* implement, it's resilience provided as a service. The less you manage, often the more the platform is doing auto-healing behind the scenes (because, say, if a Lambda server fails, AWS runs your function on another transparently).

Cross-Region and Multi-AZ Resilience: AWS allows architectures that are resilient to an entire data center failure. For example, running instances in multiple Availability Zones (AZs) means if one AZ goes down, the others still serve (load balancer will stop sending to the down AZ). Multi-region active-active setups even survive region outages (though that's complex). While this is beyond basic "self-healing" (it's more about disaster recovery), an intelligent system might orchestrate failover between regions. AWS Route 53 (DNS) can do health-based routing – if region A's health check fails, route traffic to region B. An agent could similarly detect a region issue and shift traffic globally (perhaps by manipulating DNS or using anycast or other advanced routing). Netflix's system has elements of this with their global traffic management.

In summary, cloud platforms provide building blocks for resilience:

- Automated instance replacement (ASG health checks),
- Auto-scaling (reactive and some predictive),
- Event-driven remediation (with Lambdas or similar),
- Managed services that inherently auto-heal (like replacing nodes in a managed cluster, etc.),

- Multi-zone redundancy.

By utilizing these, even traditional systems can achieve high resilience. Adding intelligence often means making those triggers smarter or more dynamic: e.g., using ML-based policies instead of static CPU thresholds, or coordinating multiple actions.

One thing to emphasize: **instrument everything and observe**. Whether it's Kubernetes or AWS, you need good monitoring to feed into these automated systems. If you don't have visibility (metrics, logs, traces), your self-healing mechanisms might not know what to do or could act on wrong assumptions. Intelligent resilience is tightly coupled with **observability** (which was likely covered in a previous chapter on monitoring).

Chaos Engineering: Testing and Improving Resilience

As a microservices practitioner, you might have heard of Netflix's Chaos Monkey – a tool that randomly terminates instances in production to test the system's resilience. This is the essence of **chaos engineering**: **deliberately introduce failures** in a controlled way to see if the system can handle them and to uncover weaknesses.

Why do this? Because you don't want the first time a failure happens to be a surprise in a critical moment. If you regularly inject a bit of chaos (like shutting down services, adding network latency, or disabling a dependency), you can verify that your resilience patterns and self-healing mechanisms actually work. It's like a fire drill – practicing what to do in an emergency so that in a real emergency, everyone knows their role and the safety systems (alarms, sprinklers) function correctly.

Chaos Monkey and the Simian Army: Chaos Monkey, created by Netflix, randomly kills virtual machine instances in their auto-scaling groups. If your application is truly resilient, it should not be bothered by one instance disappearing – maybe a momentary blip but no outage. Netflix expanded this to a whole "Simian Army" of tools (Chaos Gorilla, Chaos Kong etc.) that simulate bigger disasters (like an AZ outage or region outage). These tools helped Netflix build confidence that they could survive even major failures by forcing them to regularly deal with simulated ones.

Gremlin: Gremlin is a modern chaos engineering platform (offered as SaaS and on-prem tooling) that lets you define and run chaos experiments. It provides a GUI and API to, for instance, spike CPU on a host, fill up disk space, drop packets on the network, kill processes, etc., in a controlled manner (you target specific hosts or containers, specify duration, etc.). Gremlin essentially industrializes chaos testing – you can schedule experiments or run them ad-hoc. The philosophy is the same: break things on purpose to see if the system heals itself.

Analogy – Vaccines and Immune Training: Chaos engineering is like giving your system small "vaccines" of failure to help it build immunity. By exposing the system to controlled doses of failure (like a service dying or latency injection), you trigger the selfhealing responses frequently. This not only verifies they work, but often the system "learns" from it (especially if you have intelligent agents that adjust based on past incidents). Just as a vaccine teaches your immune system how to fight a virus, a chaos test teaches (or validates) your software and your team on how to handle a failure. Over time, the system becomes **antifragile** – it actually gets stronger and more robust by encountering failures regularly, because you've fixed the weaknesses those failures revealed.

When you incorporate intelligent agents, chaos engineering can become a learning loop for them as well. You can use chaos experiments to **train your AI models** – for example, generate many failure scenarios (like simulate high latency, simulate exceptions in a service) and let your anomaly detection agent see them, so it learns to recognize those patterns in the future. Or let your orchestrator agent try different healing strategies during chaos tests to see which resolves issues fastest, refining its strategy.

Some organizations even implement **automated chaos** in production (with a lot of care!). They might have a rule like "we randomly kill one container somewhere every workday" – ensuring the team's on their toes and the systems are built to handle it. Others do

GameDays – scheduled chaos experiments, often in staging but sometimes in prod during a prepared time window. The key is to have hypotheses (e.g., "if we kill this service, our failover should take < X seconds and users should not see errors") and then run the experiment to validate. If the hypothesis fails (e.g., it actually took 60 seconds and alerts blew up), then you found a resilience gap to fix.

Gremlin & Chaos Monkey usage: In a Kubernetes environment, you might use the opensource Chaos Mesh or LitmusChaos (similar chaos tools). But Gremlin can target K8s pods as well, and Chaos Monkey now has variants for K8s. These can randomly delete pods or simulate node failures.

The intelligent approach to chaos is to possibly let **agents induce chaos when safe**. For example, an agent might detect that the system is currently under low load and everything is healthy, and decide to opportunistically kill a instance to test resilience (like an automated Chaos Monkey that knows not to do it at peak time). Or after each deployment, automatically run a suite of chaos experiments to ensure the new code didn't break any recovery mechanism.

Caution with Chaos

When introducing chaos engineering, start small and in non-critical environments. Make sure you have proper monitoring and alerting in place, and everyone (especially on-call engineers) knows that an experiment is happening. A poorly planned chaos test in production could cause an outage if your system isn't as resilient as you thought. The goal is to learn, not to create unnecessary customer impact. Use incremental steps:

begin with killing stateless, easy-to-recover services before you, say, corrupt a database replica or partition your network. And always have a "kill switch" to stop the experiment if things go wrong. Chaos should be like a fire drill, not an actual fire – controlled and with safety measures in place.

Combining Tools and Intelligence

All these tools – Kubernetes healing, operators, cloud auto-scaling, chaos experiments – can be combined with intelligent agents for maximum effect:

- Kubernetes handles the *simple cases* (restart pods, etc.), while an intelligent operator or external agent handles *complex recovery* (like cross-service coordination or predictive actions).

- Cloud auto-scaling ensures capacity, while an AI agent fine-tunes the scaling or preempts it.

- Chaos engineering provides continuous feedback into the system's resilience, and agents use that feedback to adjust.

Think of it as a layered defense and adaptation system:

1. **Reactive Layer:** Basic health checks, restarts, retries, etc. (Kubernetes, ASG, etc.) – fast and simple reactions.

2. **Adaptive Layer:** More nuanced logic (operators, scripts) that handle known scenarios in a smarter way.

3. **Intelligent Layer:** Agents that learn and predict, handling unknown or future scenarios and optimizing as conditions change.

4. **Validation Layer:** Chaos engineering and monitoring to validate that 1-3 are working and to provide new data to improve them.

By combining these, you aim for a system that *rarely needs human intervention*. However – and this is critical – you should **always keep humans in the loop for oversight**. Let's discuss that briefly.

When Humans Step In: Fallbacks and Transparency

No matter how autonomous our system becomes, we need to acknowledge that some situations will require human decision-making or creative problem solving that AI can't (yet) provide. Part of resilience engineering is knowing **when to alert a human** and how to **fail gracefully to manual control** if needed.

Human-in-the-Loop for Critical Failures

Your self-healing mechanisms should have thresholds or **circuit breakers of their own** to know when to stop and ask for help. For example, if an agent has restarted a service 3 times in 10 minutes and it keeps failing, it might refrain from a 4th restart and instead alert the oncall engineer: something deeper is wrong that isn't being fixed by the usual remedy. This prevents endless cycles of automated action that aren't solving the issue (and potentially making it worse or hiding it). It's similar to how after a couple of automatic retries, you surface an error to the user – here after a couple of self-healing attempts, you surface an incident to an operator.

Notification and Explainability: When an agent takes an action, it should log what it did and why. If it scaled up the service, it might log "Scaled Service X from 3 to 5 instances because CPU was at 90% and rising, exceeding policy threshold." This provides a breadcrumb trail for engineers to follow later. If something goes wrong, these logs help debug why the agent did what it did. In intelligent systems, especially, we want a level of **explainability** – if an AI agent predicted a failure, what evidence did it base that on? Maybe it noticed a pattern in request latencies. Having that explanation helps engineers trust the system and also verify it's not acting on garbage data.

Don't Hide Failures

A self-healing system can be so seamless that failures are fixed before anyone notices – which is awesome, but also dangerous if those failures silently keep occurring. It's important to **surface issues even if they were automatically resolved**. For instance, if a service restarted itself 5 times overnight, there's probably a bug or issue that needs attention, even if users never felt it. The system should at least record and report this to the team. Otherwise, you might be papering over a crack that's widening under the hood. **Transparency** is key: ensure your dashboards, alerts, and reports give insights into what the self-healing system is doing. That way, if the AI can't handle something or if it is handling something repeatedly, the humans can step in to improve the underlying system. Remember, self-healing is not about masking problems to the point of ignorance; it's about managing them in real-time while still enabling continuous improvement.

Fallback to Manual Control: In certain sensitive domains (finance, healthcare), organizations might design the system to fall back to a manual mode if anomalies reach a certain level. For example, an autonomous trading microservice might have a rule: if something truly bizarre is detected (like trades that look highly irregular), the system halts automated trading and notifies humans to make decisions. This is more about *trust* and risk management – you might trust the AI to handle normal failures, but not a once-in-a-bluemoon scenario that could have massive consequences.

Resilience of the AI Itself

One often overlooked aspect: if you have intelligent agents running, you must consider *their* resilience too. What if your monitoring agent crashes? Or the orchestrator agent gets into a bad state? They themselves should ideally be deployed in a redundant, monitored way (e.g., run multiple instances of the agent, perhaps as a Kubernetes Deployment with health checks!). It's somewhat meta, but you don't want your whole resilience strategy to hinge on a single AI component that could fail.

Also, think about scenarios where the AI's *outputs* are faulty. For instance, an ML model that predicts failures might itself start giving incorrect predictions (maybe due to concept drift – the patterns changed – or a bug). The system should detect if the model is now performing poorly (perhaps by feedback – it predicted X but nothing happened, repeatedly). In such cases, a backup strategy could be to fall back to simpler static rules until the model is fixed. Or maintain a "last known good" model: if a newly deployed model starts behaving strangely, automatically roll back to the previous model version (this is analogous to rolling back a bad deployment, but for AI logic).

If your agent takes actions, ensure those actions have some kind of safety check. For example, if an agent script is about to delete all caches, maybe it should ensure that it's not deleting something outside its scope (a sanity check: is the cache directory path as expected?). These little guardrails prevent an AI gone wild from causing harm.

Finally, test your intelligent agents under failure conditions too. Use chaos engineering *on the agents* if you will – e.g., disable the monitoring agent and see if the system's baseline resilience (Kubernetes, etc.) can still hold the fort. You don't want to become overly reliant on the AI to the point that everyone ignores basic best practices.

Traditional vs Intelligent Recovery: A Comparison

Let's take a step back and compare how a traditional microservice system handles failures versus an intelligent, agent-driven system. This will highlight the added value of the agentic approach and also show that it builds upon rather than replaces the fundamentals.

Aspect	Traditional Resilience Approach	Intelligent (Agent-Driven) Approach
Failure Detection	Relies on static thresholds and heartbeats (e.g., CPU > 85% triggers alert, missing heartbeat = down). Alerts are sent to humans or simple scripts.	Uses anomaly detection and predictive analytics. Agents notice subtle pattern deviations (e.g., latency creeping up abnormally) and predict failures before thresholds are hit.

Reaction Time	Reactive, often with human in loop. An alert might wake an engineer who then executes a runbook. Automated reactions (like auto-scaling) kick in only after metrics cross set limits for some time.	Proactive or faster reactive. Agents can trigger actions immediately upon detection (often within seconds), or even beforehand (e.g., scaling up at 8:55 because it predicts 9:00 surge). No need to wait for human approval for routine fixes.
Recovery Actions	Pre-defined scripts or platform automation handle simple cases (restart, replace instance). Complex fixes require manual intervention (engineer debugs, finds issue, applies fix).	Adaptive, multi-step actions by agents. The system might try a sequence: restart service, then clear cache, then roll back deploy, based on what it "thinks" is the likely cause. Agents choose actions based on learned effectiveness. Human called only if those fail.
Configuration & Tuning	Static configuration for timeouts, retries, etc. set during development. Operators might tweak them over time via deployments. The system doesn't change these at runtime on its own.	Dynamic tuning by agents. System parameters adjust on the fly (e.g., lowering circuit breaker threshold when a dependency is under strain, raising it when healthy). The system optimizes itself for current conditions without redeploying code.

Learning from Failures	Post-mortems after major incidents; humans update documentation and maybe improve alerts or code for next	Continuous learning loop. Agents record the outcome of each action. The system can automatically refine its models (for prediction or
Aspect	**Traditional Resilience Approach**	**Intelligent (Agent-Driven) Approach**
	time. The system itself doesn't inherently learn; it's improved via human feedback loop slowly.	decision) based on success/failure of its interventions. It effectively gets smarter and more efficient at recovery over time.
Resource Utilization	Typically conservative overprovisioning to handle spikes and failures (since automation is basic, you play it safe by having extra capacity at all times). Manual scaling is common.	Optimized resource usage. Predictive scaling and adaptive reallocation mean the system can run closer to ideal capacity (neither too much slack nor too little), as it trusts the agents to rapidly adjust to any changes or issues.
Visibility	Dashboard per service, alerts on failure – mostly after the fact awareness. Difficult to track cross-service incident cause without manual analysis.	Holistic system view by agents that correlate signals. They might provide a higher-level incident report like "Service A, B, and C experienced issues due to root cause in Service D's memory leak." Better at connecting the dots automatically.

As the table outlines, the intelligent approach isn't about throwing away traditional wisdom – in fact, it **builds on it**. Circuit breakers, health checks, retries, redundancy: those remain absolutely essential. The AI agents don't replace them; they make them **more effective** by adding adaptability and foresight. You can think of it this way: *traditional resilience is the solid foundation, and agentic intelligence is the flexible, smart roof on top.* Without the foundation, the roof would collapse; without the roof, the foundation alone might not fully protect you from every storm.

It's also worth noting that not every system needs the highest level of intelligence.

Depending on the criticality, you might choose a simpler setup. There's a maturity curve: first ensure you have basic resilience (if you don't even have health checks and restarts, do that before worrying about AI). Then automate known procedures (maybe some scripts or operators). Then add learning/predictive capabilities for extra gains.

Case Study: WebkinsTech's Self-Healing Evolution

To make these ideas more concrete, let's follow a fictional company **WebkinsTech** on its journey toward a resilient, self-healing microservice architecture. WebkinsTech provides an ecommerce platform that started as a monolith and evolved into microservices (we've seen bits of their story in earlier chapters). Now, reliability is paramount – downtime means lost sales and unhappy customers. We'll see how WebkinsTech applied the patterns and intelligent techniques discussed in this chapter, progressing from basic resilience to an automated, agent-driven system.

Stage 1: Baseline Resilience (After the Monolith)

After breaking the monolith into microservices, WebkinsTech's team immediately implemented fundamental resilience patterns. Each microservice had timeouts on outbound calls and retries for idempotent operations. They used Netflix Hystrix for circuit breaking between services (later migrating to Resilience4j). For example, the Order Service wrapped calls to the Payment Service in a circuit breaker – if Payment was slow or down, Orders could quickly fail the payment step and mark the order as pending, rather than blocking the whole order workflow. They also established **graceful degradation**: if the Recommendations Service (which suggests products to users) was unavailable, the UI would just skip that section rather than fail the page.

They containerized all services and deployed them on a Kubernetes cluster. Kubernetes' self-healing took care of restarting crashed pods. WebkinsTech set up **liveness probes** for critical services like the API Gateway and the Inventory Service, so any hang or deadlock would be resolved by a swift restart. They also set resource limits and requests for each pod (a bit of bulkhead in terms of CPU/memory isolation).

At this stage, most failures were handled gracefully, but some still required manual intervention:

- If a new deployment caused issues (e.g., a bug causing high CPU), the on-call engineer would get paged by their monitoring system (Datadog) and they'd manually rollback the deployment.

- If the load spiked beyond what they anticipated, sometimes the cluster ran hot because the auto-scaling was set to scale on CPU only and not very quickly. A human might notice and add more nodes/pods.

Stage 2: Automated Playbooks and Chaos Testing

WebkinsTech's SRE team decided to automate some common ops tasks. They identified that memory leaks were a recurring issue in a couple of services (let's say the Image Processing Service). The typical fix was to restart the service when memory usage got near 90%.

Instead of waiting for it to OOM kill, they wrote a small script and packaged it as a

Kubernetes CronJob that ran every hour: it checked if the Image Processing pods' memory usage was above a threshold (via metrics query) and if so, triggered a rolling restart of those pods. This is a basic form of *self-healing via automation*. Over time, they refined this by instead integrating with the Kubernetes API and eventually created a small **Operator** to handle it more elegantly.

They also set up **AWS Lambda functions for specific alerts**. One Lambda watched for an AWS CloudWatch alarm "Order Queue Length High."

If triggered (meaning orders backing up likely due to a downstream issue), the Lambda would automatically double the number of Order Worker instances processing that queue. This way, a backlog would auto-correct quickly. Another Lambda listened for EC2 instance recovery events; if an instance in their older environment failed a status check, the Lambda would notify the team and attempt to restart the instance or move workloads, acting faster than a human could.

Wanting to be confident in their growing automation, WebkinsTech embraced **chaos engineering**. They started in their staging environment: using Gremlin, they simulated scenarios like:

- Killing one instance of each service randomly during business hours.
- Inducing 500ms latency on the database network link.
- Shutting down an entire Kubernetes node unexpectedly.

These tests revealed some weaknesses:

- When the database was slow, they found that a couple of services didn't have proper timeouts and thus hung, causing user requests to pile up. They fixed this by adding timeouts and fallback logic.

- A random kill of a service instance was usually fine, but occasionally their message consumer service would lose the offset when killed in the middle of processing. That uncovered a bug in how they handled at-least-once processing; they fixed it to be idempotent on restart.

- The chaos tests also validated that Kubernetes indeed restarted things as expected, and their load balancer properly routed around the killed instances – boosting the team's confidence.

They then dared to run Chaos Monkey (the open-source version) in production with very tight scope: it would kill at most one non-critical service instance in a day, and not during peak hours. After a few months, resilience became ingrained in the culture – developers knew their service should tolerate sudden death at any time.

Stage 3: Intelligent Observability and Proactive Agents

WebkinsTech then invested in an AIOps tool (let's call it *InsightAI* for story's sake). InsightAI ingested all their metrics and logs, and started to provide anomaly detection alerts. For example, it learned the normal error rate for each service. One day, it alerted an anomaly: the Checkout Service's error rate was 2% – not high enough to trigger a traditional alert (which was set at 5%), but unusual given it's normally 0.1%. This early notice led the team to discover a misbehaving feature flag rollout before it became a severe incident. This sold them on the value of intelligent monitoring.

Encouraged, they collaborated with the vendor to let the tool not only alert but also trigger actions. For instance, if the anomaly is a memory leak pattern, automatically restart the pod (with a limit of say 1 restart per hour to avoid flaps). If the anomaly is latency rising in a service, automatically spin up one extra instance to see if it alleviates (a safe action). These were essentially *guarded actions* tied to certain anomaly signatures.

Additionally, WebkinsTech built a custom **Load Forecaster Agent**. They had lots of historical data and noticed predictable rhythms: traffic would surge every weekday around lunch (when people shop on their break) and in the late evening. Instead of reacting with auto-scaling each time (which sometimes led to 2-3 minutes of high latency before new pods were fully ready), the forecaster agent would schedule scale-outs 10 minutes before expected peaks. It used a simple ARIMA time-series model and later incorporated promo calendar events (marketing could feed it when a sale was scheduled, so it would anticipate a spike). This resulted in much smoother handling of traffic – customers saw consistent performance, and the ops team saw fewer "scale up" alerts because the system was already scaled.

Stage 4: Full Self-Healing Orchestration

By this stage, WebkinsTech had dozens of microservices, chaos tests running regularly, and a lot of automation. The next step was a unified **orchestration agent** (internally named *"Doc"*, after Doctor, since it heals the system). Doc integrated with Kubernetes, AWS, and their monitoring tools. It acted on a number of triggers:

- If a service crashed repeatedly on deploy, Doc would automatically rollback that service to the last version (it interfaced with their CI/CD pipeline).

- If a certain error pattern was detected in logs (they fed logs to an AI that could recognize specific known failure patterns, like "OutOfMemoryError" or database deadlock traces), Doc would classify the issue and run an appropriate script (like restarting the service for OOM, or cycling a database connection pool for deadlocks).

- If an entire dependency went down (like their third-party tax calculation API), Doc could detect the 100% failure rate on those calls and flag all services to switch to a degraded mode that bypassed tax calculation (using a default tax rate) until the API came back. This was done by toggling a feature flag in their config service, which all relevant services honored in near real-time. Essentially, within a minute of the tax API outage, WebkinsTech's checkout flow degraded gracefully to assume a tax, allowing orders to continue (with a note to possibly adjust tax later in records).

One night, at 2 AM, a power failure hit one of their hosting zones, knocking out a chunk of their infrastructure. Normally, this would be a major incident waking up a whole team. But Doc had been programmed (and trained in chaos tests) for exactly this: it detected multiple services failing heartbeats from that zone, and immediately initiated a failover procedure. It rerouted incoming traffic entirely to the other availability zone using their cloud load balancer API (essentially mimicking what a DNS failover would do, but quicker via API calls), and spun up replacement instances for critical services in the healthy zone to take over the load. By the time the on-call engineer's pager went off (which it did, as a precaution for multi-AZ failure), the system was already stabilizing in the single remaining zone. That engineer checked the status, saw that Doc had things under control, and mostly monitored the situation rather than doing the recovery themselves. They later said, "It was surreal – I was paged for an outage, but when I looked, our AI Ops agent had already done 90% of the mitigation. I just verified and gave a thumbs-up."

Over the next hours, as the power issue was resolved, Doc slowly reverted traffic and workloads back to balance across zones and marked everything green. The team in the morning analyzed the incident and found that the self-healing system saved them from possibly an hour of downtime. Customers saw maybe a few seconds of hiccup and then service continued.

WebkinsTech's journey shows an evolution:

- Start with the basics (solid resilient design, health checks, etc.).
- Gradually automate known fixes (scripts, operators, simple rules).
- Incorporate chaos testing to validate and improve.
- Add intelligent analysis (anomaly detection, predictive scaling).

Finally, orchestrate it with an agent that can make decisions quickly across the system.

At each step, they reduced the frequency and severity of incidents that required human intervention. By the end, humans were mostly focused on improving the system and handling novel situations, while the "boring" everyday failure modes were handled by the system itself.

One could say WebkinsTech's architecture had become like a living organism: it could detect pain, respond with reflexes, heal wounds, and even learn to avoid future pain – all autonomously to a large extent.

Conclusion and Next Steps

Resilience and self-healing are about achieving **robustness** in the face of constant change and failure. In this chapter, we saw how microservices can be designed to not only survive failures (through patterns like circuit breakers, retries, health checks, bulkheads, etc.) but also to **recover automatically** through self-healing mechanisms. By introducing intelligent agents into the mix, we unlock a new level of resilience: systems that can anticipate issues, adapt in real-time, and learn from incidents to get better over time. This results in higher uptime, more efficient operations, and ultimately happier users and developers (who get to sleep more and firefight less).

A key takeaway is that intelligent resilience is **evolutionary, not revolutionary**. You enhance traditional techniques with AI; you don't replace them outright. The smartest healing agent in the world is of little use if your basic architecture is fragile. So, it's about layering intelligence on a strong resilience foundation.

We also learned the importance of **testing resilience continuously** (via chaos engineering) and ensuring transparency – making sure we're aware of failures and how they're handled, even if no one gets paged at 2 AM anymore. As WebkinsTech's story illustrated, the road to a self-healing system is a step-by-step journey with immediate benefits at each step.

At this point, we have a system that can **heal and maintain itself** to a large degree. The natural next question is: how do we handle growth? As our system scales to more users, more data, and more complexity, how do we ensure it remains efficient and performant? Resilience helps keep it running, but we also need to be able to *expand* and *contract* the system intelligently based on demand.

In **Chapter 15: Intelligent Scaling and Elasticity**, we will build upon the resilient foundation and explore how agentic microservices handle scaling. We'll see how the same principles of automation and intelligence can be applied to ensure a system not only survives load but thrives under it – allocating resources when and where needed, optimizing cost, and delivering consistent performance at any scale. Just as we've taught our microservices to heal themselves, next we'll teach them to **grow (and shrink) themselves** in response to the world around them, completing our picture of an autonomous, efficient microservice ecosystem. Get ready to scale up!

Chapter 15: Scaling Intelligent Microservice Ecosystems

Introduction

WebkinsTech – our trusty fictional enterprise – has come a long way. In earlier chapters, we saw how they broke apart a tangled monolith, stood up dozens of **microservices**, and even imbued some of those services with **intelligent agents** that can make autonomous decisions. The result was an ecosystem of smart microservices (or "agentic" services) that collaborate to run the business. But as WebkinsTech's system grows in scope and traffic, a new challenge looms: **scaling** this intelligent microservice ecosystem without letting it dissolve into chaos. How do you keep hundreds of AI-powered services working in harmony as you add more users, more data, and more complexity?

This chapter tackles that question head-on. We'll start with conceptual and architectural insights, using cloud-neutral examples (and plenty of WebkinsTech stories) to illustrate the principles. Gradually, we'll move into real-world implementation guidance. Key themes include:

- **Hierarchical organization of agent collectives:** structuring intelligent microservices (agents) into scalable clusters or layers – from low-level task agents to high-level coordinators – and patterns for managing these groups (federations, swarms, and more).

- **Managing complexity in large-scale deployments:** strategies to maintain control, observability, and adaptability across a massive distributed network of services. We'll discuss principles like bounded contexts, modular design, and clear domain ownership that keep complexity in check even as services proliferate.

- **Performance considerations for AI-powered services:** techniques to scale up model inference workloads, retraining jobs, and data pipelines. We'll explore how to serve ML models at scale (using accelerators like GPUs), handle traffic spikes gracefully, and ensure the whole system stays responsive.

Along the way, we'll include practical checklists, comparison tables, and even a few in-line diagrams (in text format) to make concepts crystal clear. You'll also find tips, examples, quotes, and a running narrative with

WebkinsTech's journey to keep things engaging. By the end of this chapter, you should have a deep understanding of how to grow an intelligent microservice architecture from just a handful of services into a robust, scalable ecosystem.

So, let's step into WebkinsTech's shoes once more. They've taught their microservices to **think** and to **talk**; now they need to teach them to **scale** – without losing their minds (or yours) in the process!

Hierarchical Organization of Agent Collectives

Scaling an ecosystem of intelligent services often starts with **how you organize the agents** (intelligent microservices) within it. When only a few AI-driven services exist, they might communicate in an ad-hoc peer-to-peer fashion. But as their numbers grow, WebkinsTech's engineers found that some structure was necessary – otherwise, it felt like a chaotic meeting where everyone's talking at once and no one's listening. In an unstructured swarm of agents, two services might unknowingly duplicate work or make conflicting decisions. Clearly, some *coordination mechanisms* are needed so that these agents can act collectively towards larger goals.

In essence, we're looking at ways to arrange multiple AI agents into **collaborative groups** that can tackle complex tasks together. Should there be a leader coordinating the others? Or should each agent simply react to its environment (and to others) with no central control? The answers lie in a spectrum of architectural patterns, from strict hierarchies to loose federations to free-form swarms. Let's explore these structural patterns for agent collectives and how they enable scaling.

Structural Patterns: Hierarchies, Federations, and Swarms

One common approach is to organize agents in a **hierarchy** – a layered structure where higher-level agents oversee and coordinate lower-level ones. This is analogous to an army's chain of command or a corporate org chart. In a hierarchical collective, a **top-level coordinator agent** delegates tasks to several **mid-level agents**, each of which may in turn break the work into pieces handled by **lower-level agents**. This tree-like structure allows complex problems to be decomposed into sub-tasks at multiple levels . Higher layers focus on big-picture goals while lower layers handle the detailed execution.

Researchers call this a *Hierarchical Swarm* architecture, and it's great for efficient task distribution and clear responsibility boundaries .

WebkinsTech applied this pattern in certain parts of their system. For example, they created a

"**Global Orchestrator**" agent service that monitors incoming customer orders and overall system load. This top orchestrator doesn't process orders itself; instead, it delegates orderrelated tasks to domain-specific mid-tier agents. One mid-level agent (let's call it the *Inventory Coordinator*) ensures stock is allocated from the right warehouse, while another (the *Payment Coordinator*) confirms payment processing. Those mid-level agents further rely on low-level microservices – for instance, an Inventory service that actually updates stock counts, or a Payment service that interfaces with the credit card gateway. In this hierarchy, each layer plays a role: the top orchestrator sees the whole order pipeline, the mid-level agents handle domain-specific orchestration, and the leaf-level services do the grunt work. **Figure 15.1** illustrates a simplified hierarchy of agents:

Figure 15.1: Hierarchical Agent Collective at WebkinsTech

```
graph LR
  A[Order Orchestrator]<:::top> --> B1[Inventory Coordinator]<:::mid>
  A --> B2[Payment Coordinator]<:::mid>
  B1 --> C1[Inventory Service]<:::low>
  B2 --> C2[Payment Service]<:::low>
```

In this hierarchy, a top-level orchestrator agent delegates tasks to mid-agents (for Inventory and Payment), which in turn invoke lower-layering helps scale coordination by dividing

Hierarchies like this bring **scalability through delegation** – no single agent has to handle everything. WebkinsTech found that by introducing a top-level orchestrator for critical workflows, they prevented their many microservices from "stepping on each other's toes." As one engineer quipped, *"Our AI services were brilliant individually, but without a game plan, it was like a soccer team with no coach – lots of movement, no coordination."* The orchestrator provided that game plan. It ensured, for instance, that two different optimization agents weren't trying to tweak the same inventory levels simultaneously in conflicting ways. Higherlevel agents make **global**

decisions, while lower-level ones execute **local tasks** in parallel – a balance that improved throughput and kept the system organized.

However, a pure hierarchy isn't the only way to scale out agent collectives. Another pattern WebkinsTech considered is the **federated model**. In a federated setup, you have multiple semi-autonomous groups of agents – essentially **clusters of microservices** – that collaborate by sharing information, but **without a single global leader**. Each cluster (or "swarm") operates mostly on its own, handling a subset of tasks or a particular domain, and occasionally communicates results or requests to other clusters. Federated swarms have been described as *"multiple independent swarms that collaborate by sharing information and results"*, where each swarm can contribute to a larger goal while still running autonomously . This is a divide-and-conquer strategy: you break the giant ecosystem into smaller ecosystems that are easier to manage, then let them sync up as needed.

Imagine WebkinsTech divided its platform into regions – North America agent collective, Europe agent collective, Asia agent collective – each with its own hierarchy of services handling local customers. These regional agent groups might only share high-level summaries with a central system (like reporting daily sales or requesting inventory transfers), but otherwise they function independently. The **advantage** here is that each federated group can scale internally without a central bottleneck, and problems in one group (say, a spike in European traffic) can be contained and handled within that group. It's **fault isolating** – an outage in one region's services might not bring down the others – and it allows solutions to be tailored per group. The **trade-off** is that you need well-defined communication protocols between the federated groups; they must agree on how to share data or requests so that the overall system still works towards common goals. In essence, a federated model introduces a **layer of coordination between clusters** instead of a single coordinator between all agents.

At the far end of the spectrum is the **swarm-based design** with **decentralized coordination**. In a pure swarm (or fully distributed) architecture, there is **no explicit hierarchy or central controller**. Each agent communicates with others as needed (often through a common medium like an event bus or shared data space), and complex behavior *emerges* from many local interactions. This is inspired by **swarm intelligence** in nature – think of ants finding the shortest path to food without a general, via simple rules and pheromone trails, or bees collectively deciding on a new nest location. In software terms, this often translates to **event-driven microservices** where any service can emit events and react to events, and global order is achieved through carefully designed protocols rather than a coordinator dictating actions.

WebkinsTech uses swarm-like designs in scenarios where flexibility and resilience are top priorities. For example, their **incident response agents** work in a decentralized way: monitoring agents, diagnostic agents, and remediation agents all post their findings to a shared incident log (a kind of virtual "blackboard"), and any agent can pick up those clues and act. There's no single incident commander; instead, the agents collectively converge on a solution. One agent notes high database latency, another sees unusual error rates in the inventory service, and a third agent puts those together and hypothesizes a root cause – all through the shared context of event data. This *blackboard pattern* is one form of swarm coordination where **agents communicate indirectly** by sharing state, leading to an emergent problem-solving process. The benefit of such a swarm approach is **robustness** – if any one agent fails or is absent, others can still carry on and adapt. There's also no obvious performance bottleneck, since work is distributed and parallelized naturally.

However, pure decentralization has its **challenges**. Without a clear hierarchy, you must be very careful to prevent *conflict* and *redundancy*. If multiple agents can act on the same event, how do you avoid duplicate work or contradictory actions? The system's overall behavior can be harder to predict, because it's not orchestrated in a single place – it emerges from the interplay of many parts. Debugging a swarm can sometimes feel like herding bees; you have to trace which agent did what in response to which signals, which requires excellent observability (a topic we'll cover later). WebkinsTech mitigated these issues by defining simple rules for their swarm agents and having a limited shared vocabulary on the "blackboard" (so agents don't misinterpret data). They also implemented a sort of *gentle referee* mechanism: one agent in the swarm (e.g. the analysis agent) can signal when a hypothesis is confirmed, telling others to stand down on further investigation of that incident.

This isn't a hierarchy per se, but it injects just enough coordination to prevent chaos.

To summarize these patterns, **Table 15.1** compares hierarchical, federated, and swarmbased organizations of agentic services:

Pattern	Description	Pros	Cons	Use Cases
Hierarchical	Tree-like layers of agents (coordinators and workers). Higher-level agents orchestrate lower-level agents .	Clear structure and control; welldefined responsibilities at each layer; easier to reason about each component's role.	Potential bottlenecks or single points of failure at top levels; added latency if many layers deep; requires reliable coordinator(s).	Complex workflows that naturally break into subtasks (e.g. multi-stage business processes, hierarchical decision making).
Federated	Multiple autonomous groups of agents that share information or results. Each group handles a domain or subset, collaborating via agreed interfaces.	Scales by **division of responsibility**; fault isolation between groups; each cluster can use the approach best suited for its domain.	Requires careful design of intergroup communication (to avoid inconsistency); potential duplication of effort across groups; no single global view by default.	Large enterprises with distinct domains or regions; cases where separate agent teams solve local problems, then integrate outcomes (e.g. crossdepartment optimization).

Pattern	Description	Pros	Cons	Use Cases
Decentralized Swarm	Agents communicate peer-to-peer or via a common event hub with no central controller. Global behavior emerges from local interactions.	No single failure point; highly adaptive and **parallel**; naturally loaddistributes work; flexible to changing scenarios.	Harder to predict and debug emergent behavior; needs strong observability; risk of conflicting actions if not carefully governed; each agent must handle partial info.	Adaptive systems requiring extreme resilience (e.g. fault detection networks, collaborative filtering by many small services); scenarios where designing a single coordinator is impossible or undesirable.

Table 15.1: Comparison of agent collective organization patterns for scaling. (*Note*: Many real-world architectures blend elements of these patterns.)

As the table suggests, there's no one-size-fits-all structure. In practice, WebkinsTech's architecture ended up a **hybrid**: mostly decentralized and event-driven for flexibility, but with a few *intentional hierarchies* superimposed to guide critical workflows. For instance, their ecommerce order pipeline is orchestrated (hierarchical) to ensure consistency, but their system monitoring and optimization tasks use a swarm approach to leverage collective intelligence. They also effectively **federated by business domain** – different product lines and departments at WebkinsTech have their own services and even their own AI models, connected through well-defined APIs and event streams rather than one giant mesh of interactions. This domain-based federation is a nice segue into our next topic: managing complexity in a large-scale microservice landscape.

Before we move on, keep in mind a key lesson: *scaling isn't just about adding more instances or more CPU; it's also about **structuring** your system so it can grow gracefully.* By thoughtfully organizing agent collectives – whether via clear hierarchies, federated clusters, swarms, or a combination – you create an architecture that can handle more agents and more interactions without collapsing under its own weight. In the next section, we'll see how to maintain **clarity and control** as the system scales up, diving into complexity management techniques that complement the structural patterns we just explored.

Managing Complexity in Large-Scale Deployments

When WebkinsTech hit roughly a hundred microservices (many with AI capabilities), they encountered a new kind of complexity. It wasn't a single bug or a database bottleneck – it was the **complexity of the system itself**: so many moving parts, interdependencies, data flows, and teams involved. Scaling an intelligent microservice ecosystem isn't just about handling more traffic; it's also about keeping the architecture **comprehensible, controllable, and adaptable** as it grows. In a word, it's about **governance** of complexity. How do you prevent a sprawling collection of services from turning into the dreaded

"distributed big ball of mud"? How do you ensure engineers can still understand and manage the system, and that changes don't have unintended ripple effects?

In this section, we'll explore strategies WebkinsTech and other organizations use to tame complexity at scale. These include **domain-based decomposition (bounded contexts)** to keep services aligned with business areas, strong **observability** to maintain visibility and control, and design approaches that promote **adaptability** so the system can evolve without breaking. Think of these as the architectural practices and operational disciplines that make a large microservice ecosystem sustainable.

Domain-Based Decomposition and Ownership

A fundamental principle for managing complexity is **modularity** – breaking the system into smaller, well-defined pieces. In the world of microservices, this often means aligning services to **bounded contexts** from Domain-Driven Design. A *bounded context* defines a clear boundary within which a particular domain model applies; outside of it, the same terms or concepts might mean something else. In practice, this translates to grouping functionality by business domain and ensuring each microservice has a single, cohesive responsibility within that domain .

WebkinsTech embraced bounded contexts early in their microservices journey (Chapter 2 introduced this with our example company). After breaking up their monolith, they defined services around distinct business capabilities. For instance, they carved out an **Order Service** to handle order processing logic, an **Inventory Service** to manage stock levels, and a **Billing Service** for payments and invoicing . Each of these services corresponds to a bounded context – "Order Management", "Inventory Management", "Billing" – with its own data and business rules. This clear separation paid off: when WebkinsTech's team updated how inventory is tracked (say, adding a new warehouse management algorithm), **only the Inventory Service was affected**, and other services remained untouched . Developers could work in parallel on different services without stepping on each other's toes . In other words, **bounded contexts prevented the system from devolving into an entangled mess**, and they scaled out the organization too – smaller teams could *own* each context end-to-end.

By keeping each microservice focused on a single domain, you reduce mental load. An engineer working on the Shipping service doesn't need to know the internals of the Pricing service. They communicate through well-defined APIs or events. This **loose coupling** and **high cohesion** (each service doing one thing well) is a hallmark of successful microservice architectures at scale. It localizes complexity. Problems in one domain ideally don't cascade to others because cross-service interactions are minimized and controlled.

That said, splitting an application into many pieces introduces its own complexity – primarily in the interactions between services. To avoid simply trading a "big ball of mud" for a "distributed ball of mud," WebkinsTech set some rules for inter-service communication. They treated integrations between bounded contexts as carefully managed contracts. For example, the Order service might emit an "OrderPlaced" event that Inventory and Billing listen to, rather than Order calling Inventory and Billing directly. This asynchronous, eventdriven integration means each context updates on its own terms, and a failure in one doesn't immediately break the others. WebkinsTech's architects deliberately **avoided creating tightly coupled synchronous dependencies whenever possible**, especially across bounded context boundaries (e.g. Order service making a blocking call to Inventory) – because that could reintroduce high coupling and cascade failures (one slow service making others slow) . Instead, they favored patterns like event messaging, or if synchronous calls were needed, they implemented fallback logic and timeouts to keep failures isolated.

Crucially, WebkinsTech assigned **domain ownership** to different teams. Each bounded context (and its related microservice(s)) had a small team responsible for it – developing it, maintaining it, and being on-call for it. This aligns accountability with the service boundaries. It also meant that as the system grew, no single team had to understand everything; each team became experts in their domain. To coordinate, they established cross-team forums and an architecture guild that agreed on system-wide standards (like API styles, event schemas, and observability practices), but day-to-day, teams could work fairly independently. This organizational approach scaled their **human processes** alongside the technology. A big part of managing complexity is making sure the *people structure* mirrors the system structure – a concept sometimes called Conway's Law. WebkinsTech's move to domain-based teams ensured that as the number of services grew, they didn't overwhelm any one group of engineers.

Of course, domain decomposition isn't a one-and-done task. WebkinsTech periodically revisited their service boundaries to ensure they still made sense as the business evolved. They merged a couple of services that had split too finely and were causing too many chatty calls between them, and they split one service that grew too broad. **Boundaries need to be just right:** too broad and you lose the benefits of modularity; too narrow and you get excessive fragmentation and overhead. The guiding question was always: *does this service have a clear purpose and isolated knowledge, or is it entangled with others?* If the latter, refactoring was in order.

In summary, **using bounded contexts to structure the microservices** gave WebkinsTech a fighting chance to keep complexity in check as they scaled. It created natural modules in the system, allowed teams to focus, and reduced the blast radius of changes. Next, we'll see how they maintained *visibility* into this distributed system, because even well-modularized services can misbehave or interact in unexpected ways at runtime.

Observability and Control at Scale

In a large-scale distributed system, **"you can't manage what you can't see."** When you have dozens or hundreds of intelligent microservices handling live traffic, it's vital to have great observability – a window into the internals of the system – so you can understand its behavior and quickly pinpoint issues. Observability is often defined by three pillars: **logging**, **metrics**, and **tracing**, complemented by **alerting** and visualization tools. WebkinsTech invested heavily in this area as their ecosystem grew, because they knew that a minor issue in one corner could be nearly invisible without the right telemetry, yet could snowball into major user-facing problems if left undiagnosed.

Logging: Every microservice at WebkinsTech produces structured logs for key events, errors, and decisions. For intelligent services, logs might include things like "AI model X decided to decline transaction Y due to fraud score 0.98" – providing insight into not just what happened, but *why*. They set up a central log aggregation (using tools like the ELK stack) so that any service's logs can be searched in one place . This way, when an incident occurs, engineers can quickly pull up all related log entries across services within the relevant timeframe. Logging is the most basic form of visibility, and at scale it requires infrastructure to collect and store huge volumes of log data. WebkinsTech managed this by defining log retention policies (to not keep everything forever) and using searchable storage that can handle high write throughput.

Metrics: Each service also emits metrics – numerical measurements like request throughput, latency, CPU/memory usage, number of orders processed, number of AI inferences performed per minute, etc. These metrics are collected in a time-series database (e.g. Prometheus) and visualized on dashboards (e.g. Grafana) . WebkinsTech's operations center has big screens showing live graphs of key business and system metrics. For example, one graph tracks the recommendation engine's response time; another shows the order pipeline's overall success rate per minute. When scaling out, metrics are crucial for capacity planning – by watching trends, they can decide when to add more instances or optimize code. Metrics also feed into **alerting**: WebkinsTech set up alerts for abnormal conditions (like if error rates on any service spike above 5% or if an AI model's inference latency goes beyond, say,

200ms). Using tools like Alertmanager or PagerDuty, these alerts notify the on-call engineers so they can respond . The team also employed **automated anomaly detection** on some metrics – e.g. using machine learning to detect when a metric deviates from its typical pattern – which is quite fitting in an intelligent microservice ecosystem!

Distributed Tracing: One of the trickiest parts of a microservice architecture is that a single user request may hop through many services. For instance, a customer's action on the website might go through API Gateway -> Order service -> Inventory service ->

Recommendation service (AI) -> back to Order -> out to user. If that overall operation is slow or fails, which service was the culprit? To answer this, WebkinsTech uses **distributed tracing** tools (like Jaeger or OpenTelemetry tracers) . These instruments tag each request with a correlation ID and trace it across service boundaries. In practice, this means each service logs (or sends to a tracer system) the incoming request ID, what it did, and passes the ID along to any calls it makes to other services. The tracing system then lets you visualize the entire call graph and timings for a single transaction. WebkinsTech's engineers could pull up a trace and see, for example, that the overall operation took 1.2 seconds, with 1.0 second of that spent in the Recommendation service (maybe waiting on a model prediction). That pinpoints where to focus optimization. Traces help untangle inter-service interactions at runtime and are invaluable for debugging performance issues in complex, scaled systems .

Control and Governance: Observability gives you data, but controlling complexity also means having mechanisms to **act on that data** and to enforce standards. At WebkinsTech, a central **SRE (Site Reliability Engineering) team** acts as guardians of the system's health. They built tooling for chaos testing (like inducing random failures in lower environments to ensure services can handle them), and capacity testing (gradually increasing load to see where things break). This proactive approach catches issues before they happen in production. They also established **governance policies** for new services: e.g. any new microservice must include certain standard metrics (like requests per second, error count) and must integrate with the centralized logging and tracing from day one. By enforcing these rules, even as the number of services grew, each one was born with the hooks for observability and control, rather than being wild rogue elements.

Another important aspect of control at scale is **configuration management**. With so many services, each might have dozens of configuration parameters (thresholds, model versions, feature flags). WebkinsTech introduced a centralized config service and/or used

infrastructure-as-code to manage these systematically. This way, they could change a config (say, turn off an AI feature flag) across all relevant services in one fell swoop, rather than editing config files manually on each service – a process that doesn't scale. They also versioned configurations and used automated rollouts for config changes, treating them with the same caution as code deployments (because a misconfiguration can be just as disruptive as a code bug).

Service discovery was another piece: as new instances of services come and go

(especially with auto-scaling, which we'll discuss later), they needed a way for services to find each other. WebkinsTech employed a service registry (backed by Kubernetes or a tool like Consul) so that any service could dynamically locate the current endpoints of another service by name. This avoids hard-coding addresses and allows the system to **adapt** to scaling events (like if you scale Inventory service from 5 to 8 instances, other services automatically start load-balancing across 8 endpoints). The service mesh concept (e.g., using Istio) can further provide discovery, traffic routing, and uniform telemetry collection across languages – something WebkinsTech considered to simplify cross-cutting concerns for all services.

In sum, managing a large-scale microservice system is like being an air traffic controller: you need radar on every plane (observability) and protocols to direct them safely (governance and control mechanisms). By investing in logs, metrics, tracing, and robust operational practices, WebkinsTech gained the confidence to deploy ever more services and agents. When something went wrong (and at that scale, something *always* eventually goes wrong), they could swiftly identify where and why, often before users even noticed, thanks to proactive alerts. Observability and good practices turned the complex web of services into a glass box rather than a black box.

Adaptability and Continuous Evolution

The only constant in a scaled system is **change** – workloads change, user behavior changes, and the software itself is continuously evolving (new features, updated ML models, etc.). A truly scalable microservice ecosystem must be able to **adapt** on the fly, without requiring a complete overhaul for every change. Adaptability comes in a few flavors for WebkinsTech: **automatic scaling** to handle variable load, **flexible deployment processes** to push frequent updates safely, and even **self-healing** or adjusting behavior based on learning (the "intelligent" part helping run the show).

Auto-Scaling: One of the biggest advantages of microservices is you can scale them independently. At WebkinsTech, each service has an *auto-scaling policy* – rules that tell the cloud or orchestrator when to add more instances or allocate more resources. For example, they set the Inventory service to scale out if CPU usage stays above 70% or if incoming request rate exceeds a certain threshold for a few minutes. In Kubernetes (EKS), this was done with Horizontal Pod Autoscalers; in AWS autoscaling groups for ECS or EC2; in serverless (Lambda), it's inherent as the platform will just run as many Lambdas as needed (with some concurrency limits). By tuning these, the system can **elasticly grow and shrink** with demand. On Black Friday, when traffic spiked 5x, WebkinsTech's services automatically scaled out to handle the load – additional containers spun up to serve the rush – and then scaled back down overnight. This elasticity is critical for cost-effective performance: you don't want to run 100 instances at all times "just in case" when 10 will do most of the day, but you also don't want to be caught short-handed during a surge. Modern cloud platforms even support **predictive scaling**, using historical data and ML to anticipate spikes (e.g., scaling up *before* Black Friday hits) .

Graceful Degradation: Despite best efforts, there could be times when load exceeds capacity (maybe an unprecedented spike or a bug causing slowness). An adaptable system is designed to **degrade gracefully** rather than fail catastrophically. WebkinsTech built in circuit breakers and timeouts on service calls – if Service A is calling Service B and B is overloaded, A will stop waiting after, say, 2 seconds and either try a fallback or at least fail in a controlled way (perhaps returning a partial result to the user). They also identified noncritical features that can be turned off under extreme load. For instance, if the recommendation engine (AI service) is under heavy strain, the site can temporarily stop showing personalized recommendations and instead show best-sellers (a simpler query) – thus relieving the AI service. This kind of adaptability ensures the **core functionality** (like placing orders) still works even if ancillary intelligent features are struggling. It's a bit like how a graceful web application might disable fancy animations if the client device is slow, except here it's at the system architecture level, shedding load to protect the overall user experience.

Continuous Deployment and Evolution: As WebkinsTech's engineering teams deliver new code and new models frequently, the architecture supports rapid, safe deployments. They practice **canary releases** – rolling out a new version of a microservice to a small percentage of users or one region first, monitoring it, and then scaling it out to everyone if all looks good. This minimizes the risk of a bad update taking down the whole system. It's especially useful for AI models: they might deploy a new ML model behind an endpoint serving 5% of traffic and compare its performance to the old model before full rollout. If the new model is slower or less accurate, they can roll it back without most users ever seeing it.

The microservices approach, with its decoupled nature, inherently enables such experimentation: since services communicate via interfaces, you can swap out the internals or even the implementation of a service (e.g., replace a rule-based service with an AI agent or vice versa) as long as the interface stays consistent. WebkinsTech leveraged this by continuously improving their intelligent agents. For example, they had a simple rules-based fraud check service that they gradually replaced with an AI-powered Fraud Agent. During the transition, both ran in parallel on a subset of transactions to validate the AI's decisions. Once confident, they retired the old service. This evolutionary approach kept the system improving without large "big bang" rewrites.

Self-Monitoring and Learning: Since this book is all about intelligent microservices, it's worth noting that those same agents can help manage the system itself. WebkinsTech experimented with an AI Ops agent that watches metrics and adjusts parameters automatically. For instance, a *Resource Optimizer Agent* could observe that Service X's 95th percentile latency is creeping up and decide to preemptively spin up an extra instance (effectively acting as a smart auto-scaler, perhaps even anticipating issues before the standard CPU-based auto-scaler triggers). Another example is a *Self-Healing Agent* that detects when a service instance is misbehaving (error rate shooting up) and automatically restarts or replaces it. These kinds of agents make the ecosystem more **adaptive** by closing the loop: not only are the microservices doing business tasks, but some are dedicated to **observing and modifying the system** itself in real-time. We touched on this in the resilience chapter (Chapter 14), but it's relevant here as a scaling aid – you essentially add meta-level agents to manage the growing complexity and ensure smooth scaling.

Finally, **adaptability** also means scaling *down* when needed, not just up. WebkinsTech was careful to make services stateless where possible (especially for things like Lambda functions or containerized web services) so that instances can be terminated anytime without losing data. State is kept in databases or distributed caches, enabling the infrastructure to remove servers during low load periods. This elasticity saves cost and reduces waste. For long-running stateful components (like perhaps a streaming analytics job), they scheduled scale-down windows (e.g., at midnight) or used retention policies to drop old data so those components don't grow unbounded.

All these practices – auto-scaling, graceful degradation, continuous deployment, and selfhealing – combined to give WebkinsTech a system that could **bend without breaking**. As the ecosystem expanded, they could introduce changes and handle variability with confidence that the architecture would adapt. It transforms the environment from a brittle one (where one unexpected event could cause a serious outage) to a resilient, pliable one that can absorb shocks and keep on running.

Having covered structural scaling patterns and complexity management, let's consolidate these operational insights before diving into raw performance and infrastructure considerations. Here's a quick **readiness checklist** WebkinsTech developed to ensure they were prepared for large-scale deployments:

Scalability Readiness Checklist-Use this to gauge if your microservice ecosystem is prepared to grow gracefully:

[] Clear Boundaries:Microservices are organized around well-defined bounded contexts, with minimal overlapping responsibilities. Each team knows which domain they own.

[] Loose Coupling: Services interact through stable APIsorasynchronous events. No critical functionality hinges on a tightly coupled chain of calls without fallbacks.

Observability Everywhere:Logging, metrics, and tracing are in place for each service. You can answer "What's happening right now?" in the system with dashboards and trace views.

Automated Alerting:Set up alerts for key indicators (errorrates, latency, throughput, resource usage) so issues are caught early. Include anomaly detection for the unexpected.

Auto-Scaling Policies: Each service has rules or schedules for scaling out/in based on load. Test that they actually prevent performance from dipping under heavy load.

Resilience Mechanisms: Timeouts, retries, and circuit breakers are implemented to prevent cascading failures. The system can survive the failure of any one component without total outage.

Continuous Delivery Pipeline: You can deploy updates to services rapidly and safely (canaries, blue-green deployments) to fix issues or add capacity.

Emergency Switches Ability to disable or limit non-essential features (especially AI-heavy ones) if the system is straining, to preserve core functions.

Documentation & Maps: An up-to-date service catalog or architecture diagram exists, so any engineer (or even an automated agent) can discover what services exist and who owns them.

Teams Prepared On-call rotations, runbooks for common scenarios, and cross-training are in place so that human responders can effectively manage incidents in a complex system.

Ticking all the boxes above doesn’t mean you won’t have surprises, but it greatly increases your odds of scaling without major mishaps. WebkinsTech learned some of these the hard way; by the time they had them all in place, they could sleep a lot more soundly – and focus on scaling further!

Now, with the architectural and operational groundwork laid, let’s turn to the raw performance side of scaling. How do we scale the AI and data-heavy workloads that are often at the heart of intelligent microservices? The next section delves into performance considerations: making sure your intelligent services are fast, efficient, and able to handle the volumes of data and requests that come with success.

Performance Considerations for AI-Powered Services

When we talk about scaling intelligent microservice ecosystems, performance is front and center. It's not enough that our architecture is well-structured and observable – each service, especially the AI-driven ones, must be able to handle increasing loads without slowing to a crawl. This is often where the rubber meets the road: can your **ML inference service** handle 10x traffic? How do you retrain models on ever-growing datasets without taking all night? Can your data pipelines ingest and process a surge of events in real time? In this section, we'll explore strategies for scaling the performance of AI-powered microservices, covering inference, training, and data processing aspects.

Key topics include scaling **model inference workloads**, making efficient use of **hardware accelerators (GPUs, TPUs, etc.)**, handling **spikes in traffic**, and planning for **continuous model retraining** in a production environment. We'll also look at some architectural choices like model serving frameworks and asynchronous processing that can dramatically affect throughput and latency.

Scaling Model Inference Workloads

Inference is the act of running data through a trained AI/ML model to get predictions or decisions. In an intelligent microservice, inference might be happening all the time – e.g., a recommendation service generates product suggestions for each user view, or a vision service labels each image uploaded. Scaling inference means ensuring that as requests increase, the service can handle them within acceptable latency.

One of the first things to consider is **workload characteristics**: is the inference workload real-time (interactive) or batch (offline)? Real-time inference (say, a user waiting for a response on a web app) typically has tight latency requirements (often 50-200 milliseconds, or even less for some applications). Batch inference (say, processing last night's logs to flag anomalies) might allow minutes or hours of processing time, as it's not user-facing in the moment. You might design differently for each:

- **Real-Time Inference:** Focus on minimizing latency and supporting high concurrency. You'd likely run the model in a persistent service that's always up (or a pool of instances), so you don't incur startup costs per request. Scaling out horizontally (running multiple instances) is common to handle more concurrent requests. You might also implement **request batching** carefully –

e.g., batch together a few incoming requests and feed them through the model in one go if using a GPU, which can amortize overhead. But batching adds a bit of delay to collect requests, so there's a trade-off. For very tight latency budgets, you may even consider lighter models or approximate methods. WebkinsTech, for instance, had a deep learning model for personalized recommendations that initially took ~300ms per request on CPU – too slow when added to other processing. By deploying it on a GPU and using some optimization (like TensorRT), they brought this down to ~50ms per inference, which was acceptable. They also set an upper limit: if the model didn't return in 100ms, they'd fall back to a simpler cached result, ensuring the user interface was never stuck waiting indefinitely for AI.

- **Batch Inference:** Here the goal is throughput – how many predictions can you churn through per second/minute – rather than per-request latency. It may be fine if a batch job takes 30 minutes, as long as it processes a million data points in that time. To scale batch, you can leverage **parallel processing** heavily: spin up multiple workers (possibly serverless functions or containers) each handling a shard of the dataset. Since no user is waiting on each piece, you can utilize the full capacity of hardware, run jobs in parallel, and even schedule them in off-peak hours to avoid competing with realtime services. WebkinsTech does nightly re-scoring of their entire product catalog for search relevance; they use a Spark job that distributes the work over many executors so the whole task finishes before morning. That job can eat a lot of CPU/GPU, but it's isolated from the live microservices so it doesn't degrade customer-facing performance.

For scaling inference in an online service, a critical architectural decision is whether to use a **dedicated model serving solution** or roll your own within a generic microservice. There are specialized frameworks (like NVIDIA Triton Inference Server, TensorFlow Serving, or TorchServe) designed to efficiently serve ML models with features like dynamic batching, model versioning, and GPU support out-of-the-box. For example, **NVIDIA Triton** can accept incoming requests, automatically batch them to utilize a GPU effectively, and even serve multiple models on one GPU . Using such a framework can save a lot of engineering effort and squeeze more performance from hardware. WebkinsTech initially embedded models directly in their Python microservice code. This was simple but had downsides: each service instance loaded its own model into memory (potentially duplicating a large model N times) and they didn't easily batch requests. Later, they moved to a model serving service for their largest models – essentially treating the model server as a special kind of microservice that multiple other services could call. This allowed them to **centralize heavy models** on a few GPU-enabled instances, rather than every microservice needing a GPU. The trade-off is an extra network call (services must query the model server), but if that model server is welloptimized, it can actually be more efficient overall.

An interesting debate in scaling AI microservices is **monolithic model server vs. many micro model services**: do you host many models in one service or break them into separate services? WebkinsTech faced this when they had a dozen different ML models (recommendations, fraud detection, personalization, forecasting, etc.). One approach is a single "AI Gateway" that contains all models and returns results for any query – this leverages sharing of resources (one GPU can serve multiple models if memory allows) and simplifies client logic (just call the AI Gateway for anything). The other approach is to have each model as its own microservice (Recommendation Service, Fraud Service, etc.), which aligns with bounded contexts and allows independent scaling (the fraud model might need more instances than the recommendation model, for instance). They chose a middle ground: models that were small and specific were deployed with their corresponding microservice (e.g. the Inventory service carries its forecasting model in-process, scaling together), whereas very heavy models or multi-use models were deployed on a central model server cluster. This prevented every service from needing a GPU and avoided "model bloat" in memory across dozens of services . It did introduce a dependency on the model serving layer, but they mitigated risk by making that layer robust (multiple instances, caching results for a few seconds to relieve load, etc.). There's no universal right answer – it depends on model sizes, update frequency, and team organization – but it's a crucial consideration for performance.

Another performance trick: **caching** inference results. If certain predictions are repeated often, a cache can save a full model run. For example, if user A asks for recommendations twice within a minute, you likely don't need to recompute it from scratch the second time – you can cache user A's recommendations for a short period. WebkinsTech caches things like "top N products for category X" since those queries hit frequently and the results don't change too fast. But caching needs careful use in AI systems: you don't want to serve stale or irrelevant results (e.g., caching a stock prediction for too long could be dangerous). Still, at scale, judicious caching (even for a few seconds) can drastically cut the load on a model service under heavy barrage.

Finally, **monitoring performance** of models is key. Through the observability setup, WebkinsTech watches the response times of each AI service. If an inference service's latency starts creeping up as load increases, that's a signal to scale out (if not already auto-scaling) or optimize the model (maybe use a smaller neural network or distill the model) or add hardware. They set SLOs (Service Level Objectives) for critical AI endpoints – e.g., "95% of recommendation responses < 100ms". If that SLO is breached in production, it triggers an internal investigation and response (which might be scaling or might be code changes). This kind of **performance budgeting** ensures that even as they pile more data or more complexity into models, they keep the end-user experience snappy.

Efficient Use of Accelerators (GPUs and More)

For many AI tasks, the secret sauce of performance is hardware acceleration. CPUs can only take you so far, especially with deep learning models. **GPUs** (Graphics Processing Units) and other accelerators (like Google's **TPUs**, FPGAs, or AWS's Inferentia chips) are specialized to perform the math of AI models (linear algebra, matrix ops) much faster via parallel processing. To scale AI services, you often need to leverage these accelerators effectively.

WebkinsTech's infrastructure team had to answer questions like: Do we equip some of our servers with GPUs? Do we use cloud services that provide managed GPU environments? How do we schedule workloads to those scarce (and expensive) resources optimally? Here's how they approached it:

- **Identify GPU-worthy workloads:** Not every microservice needs a GPU. Many of

WebkinsTech's services are simple CRUD or rule-based services that are fine on CPU.

They identified the handful of AI services (like image processing, recommendations, NLP analysis) that would significantly benefit from GPUs for inference. A rule of thumb: if it's a deep neural network or a large matrix computation and is becoming a bottleneck on CPU, consider a GPU. They ran tests – for instance, their vision model that took 500ms on CPU might run in

50ms on a decent GPU, a 10x improvement. That justifies using the GPU for that service.

- **Provisioning and scheduling:** In Kubernetes, they labeled certain nodes as GPU nodes and deployed the heavy services there. In AWS specifically, they could use instance types like p3 (NVIDIA Tesla V100 GPUs) or newer g5 instances for GPU. They also explored AWS **SageMaker endpoints** for model serving, which allows you to deploy a model to a managed cluster of GPU instances and auto-scale it. This was attractive because SageMaker can handle the scaling and provisioning, and even do multi-model endpoints (hosting several models on one instance if they're small). We'll talk more about SageMaker soon in the AWS example, but suffice to say, using a managed service saved them from having to reinvent the wheel on deploying models to GPUs.

- **Dynamic Batching and Parallelism:** As mentioned, GPUs shine when they can perform work in parallel. WebkinsTech tuned some of their GPU-enabled services to batch multiple requests together. For example, the recommendation service might wait 5-10 milliseconds, gather all user recommendation requests that came in during that tiny window (perhaps dozens of them), and send them as one batch to the model on the GPU. The GPU processes them in one go, and then the results are split back out to the respective responses. This tiny delay is usually unnoticeable to users but allowed the GPU to run at high utilization. Idle or underused GPUs are a waste of money – you want them crunching as much as possible once you have them in play. By contrast, a CPU-bound service often handles one request per thread at a time; with GPUs and batching, you handle many at once, achieving much higher throughput.

- **Concurrency and threads:** Even with GPUs, there's an optimal way to feed them work. They used async or multithreading to ensure the CPU can prepare the next batch of data while the GPU is busy, to maximize overlap. This gets low-level, but for performance-critical services they fine-tuned thread pools and memory copies so that the GPU was the only bottleneck.

- **Other Accelerators:** GPUs were the main tool, but they kept an eye on emerging tech. For instance, Amazon's Inferentia chips (accessible via AWS Neuron SDK) are specialized for deep learning inference at lower cost than GPUs in some cases. TPUs are mostly used on Google Cloud; not in WebkinsTech's stack, but conceptually similar purpose. If your platform is cloud-agnostic, you might consider whichever accelerator fits your models best. The key is abstracting your code to be able to run on different hardware. Frameworks like TensorFlow or PyTorch make it relatively easy to switch between CPU/GPU with minimal code changes. WebkinsTech containerized their model servers with all the necessary drivers, so deploying to a GPU node vs. a CPU node was mostly a configuration change.

- **GPU Utilization vs. Cost:** GPUs are expensive. One challenge was ensuring they weren't over-provisioning. In early days, they had a GPU instance running 24/7 for the image analysis service, but traffic to that service was sporadic – at times the GPU sat mostly idle. To fix that, they considered using **serverless GPU** models – for example, running that image analysis as an AWS Lambda with GPU (which AWS supports for certain sizes) or on demand via SageMaker. In the end, they consolidated multiple lowtraffic models onto one GPU server (multi-tenancy) to better utilize it. This is a common practice: pack models together until utilization is high, but be mindful not to overload memory. Tools like NVIDIA MIG (multiple instance GPU) allow slicing a single GPU into separate logical ones for isolation. WebkinsTech didn't need MIG given their scale, but it's worth noting for large deployments where you want guaranteed slices of a GPU for different tasks.

In short, accelerators gave WebkinsTech the raw horsepower needed for their AI-heavy services to scale without strangling on CPU limits. They saw firsthand that using a GPU can dramatically speed up deep learning workloads . The performance boost translates directly into better scalability: one GPU can handle what 10 or 20 CPUs might take, meaning you can serve many more requests per machine. The caveat is to manage those GPUs smartly – they introduced additional complexity in deployment and cost management. But as the saying goes in the ML ops world: **"don't skimp on hardware prematurely – use the right tool for the job."** If the job is heavy AI inference, a GPU is often the right tool.

Handling Spikes and High Traffic Loads

No discussion of scaling would be complete without addressing **traffic spikes** – those sudden surges in requests or data volume that can crush an unprepared system. Spikes might come from external events (a viral moment, a big sale day, a sudden news event driving usage) or internal triggers (a bug causing a service to send way too many requests, for instance). An intelligent microservice ecosystem has to handle not just steady growth but also these bursts gracefully.

WebkinsTech had its share of spikes. Black Friday for their retail side, an unexpected celebrity

endorsement that sent users flocking, even a scenario where their own QA team inadvertently doubled the event stream volume during a test. Here's how the architecture coped:

- **Aggressive Auto-Scaling and Pre-Provisioning:** Earlier we mentioned auto-scaling based on metrics. For known big events (like Black Friday), WebkinsTech didn't wait for the system to notice the load – they proactively **pre-scaled** critical services by 2-3x capacity the morning of, based on predicted load. Cloud providers also offer *scheduled scaling* (e.g., scale up at 8am, down at 9pm on a schedule) which they utilized. For unpredictable spikes, they configured auto-scaling to react quickly: low cooldown periods (don't wait too long to add more instances) and somewhat lower thresholds so that scaling kicks in earlier. They also enabled **burst capacity** where possible – for example, AWS Lambda can burst to a high concurrent execution count in a region (there are limits, but they're high for initial bursts). One of their strategies was to offload some surge traffic to **Lambda functions** even if normally a service ran on ECS. For example, during a spike in image uploads, they had an S3 trigger that launched a Lambda to process images if the main image service queue length grew too large. This acted as a relief valve – Lambda to the rescue with virtually unlimited scaling for that event-driven workload.

- **Load Shedding and Queueing:** If a spike is truly extreme, there comes a point where you may choose to shed load rather than drown. This means intentionally not processing some requests to preserve overall system stability. WebkinsTech set upper bounds on certain queues – e.g., their order ingestion queue had a max length, after which new events might be dropped or pushed to a "dead letter" for later. It's not ideal to drop anything, but consider the alternative: if a spike is 100x normal and you try to process everything, you might backlog so badly that latency goes through the roof for all users. Instead, sometimes it's better to serve, say, 95% of requests normally and fail fast on 5% when overloaded, rather than slow 100% of requests to a crawl. They implemented HTTP 503 responses with retry-after headers for some services if capacity was exceeded, signaling clients (or upstream proxies) to try again later. In event-driven parts, they used buffering with tools like **Amazon Kinesis** or SQS queues: these can absorb bursts of events temporarily, acting as a shock absorber. For instance, incoming clickstream events go into Kinesis; the microservices reading from Kinesis will auto-scale, but if they lag a bit, Kinesis retains the burst of events until they catch up. This decoupling via queues is a classic way to handle spikes – it turns a sudden spike into a manageable backlog that can be processed over a slightly longer time. The users might not even notice a tiny delay in an analytics pipeline, for example.

- **CDN and Edge Offloading:** For certain read-heavy scenarios (like serving AIpersonalized content on a website), WebkinsTech also leveraged CDNs (Content Delivery Networks) to cache responses at the edge. This is more relevant for static content, but even dynamic content that isn't user-specific can be cached. While not directly about microservices scaling, it's part of the holistic strategy: the less load that hits your core services, the better they handle what does hit. If an AI service generates a common piece of content (say, a personalized homepage layout for a popular

segment of users), that could be cached for a short time on a CDN. Then a spike in user traffic doesn't proportionally spike the service load because many users get the cached version. This technique is limited for highly dynamic or user-specific things, but it's worth noting as a tool in the toolbox.

- **Testing and Game Days:** One way WebkinsTech ensured they were ready for spikes was by simulating them. They conducted load tests where they'd artificially generate traffic 2-3x beyond peak to see which part breaks first. Each time, they learned something – maybe a database connection pool limit was too low, or the autoscaler wasn't scaling fast enough, or the network egress limit of a node was hit. By tweaking those and testing again, they gradually raised confidence. They also did "Game Day" exercises (chaos engineering) where an unexpected event was simulated – e.g., "what if Service A goes down during a peak hour?" – and watched how the system adapted. This helped verify that failovers and fallbacks (like the graceful degradation we discussed) actually worked under stress.

- **Backpressure and Throttling:** Within the mesh of microservices, they implemented backpressure mechanisms. For example, if a downstream service (say, the ML model server) was running at max capacity, it could return a signal to upstream to slow down (or the upstream noticed increased latency and started to self-throttle). Throttling could mean queuing requests, or outright rejecting some. They aimed for intelligent throttling – perhaps an AI agent could prioritize certain requests over others. A simple instance: the system might prioritize checkout-related requests over a background data crunch if both compete for resources. So during spikes, non-essential tasks were paused to free up capacity for critical ones. This kind of dynamic prioritization is something an

"intelligent orchestrator" agent could do in real-time, based on defined priorities.

To illustrate, during one Black Friday promotion, their **Checkout Orchestrator** agent noticed the payment gateway was responding slower as volume grew. It automatically signaled the recommendation service to temporarily reduce its activity (the recs service was trying to update recommendations in real-time with purchase data – which was not critical at that moment). It also increased the timeout for the payment service slightly, anticipating slowness, to reduce immediate errors. These little adjustments, done by smart automation, helped the whole system ride through the wave until things normalized.

Overall, handling spikes is about being proactive (scale out, use buffers) and graceful when overwhelmed (shed load intelligently, maintain core functions). Users often forgive slower service or limited features during an extraordinary event, but they won't forgive a total crash. So the motto was, "**bend, don't break**." WebkinsTech's ecosystem certainly bent under some extreme loads, but with these strategies, it did not break.

Continuous Learning and Model Retraining

Performance isn't only about real-time speed; in an AI-centric system, it's also about **keeping the AI models up-to-date and effective** without disrupting the running system. This brings us to the aspect of scaling the **retraining of models** and continuous learning processes. As data grows over time, models may need retraining to improve accuracy or prevent drift. But retraining is often computationally intensive – sometimes even more so than inference – and doing it naively can hog resources or require downtime if the model must be taken offline. Let's see how WebkinsTech approached ongoing learning in a scalable way.

Dedicated Training Pipelines: WebkinsTech set up separate pipelines for model training that run in parallel to the live services. For example, they have a nightly job that retrains the product recommendation model using the latest purchase data. This job runs on SageMaker training clusters or a transient Kubernetes job that grabs a bunch of CPU/GPU resources, trains the model, evaluates it, and saves the new model artifact. By isolating training from inference serving, they ensure that heavy training workloads (which might use 100% CPU/GPU for minutes or hours) don't interfere with the responsiveness of the production services. It's a separation of concerns: serving infrastructure vs. training infrastructure. They even use separate AWS accounts or clusters for some of this to really firewall resources.

Incremental Learning vs. Full Retraining: Not every model needs a full retrain from scratch on the entire dataset frequently. For some, you can do **incremental updates** or online learning (updating the model with new data continuously). WebkinsTech did this for their fraud detection model – it uses an online learning algorithm that ingests new fraud signals and updates its internal state without a full retraining run. This is great for adaptation speed, but you have to ensure the algorithm and system support it (and that it doesn't introduce instability). Other models, like deep neural nets, often require full retraining. In those cases, they scheduled them during off-peak hours and perhaps only daily or weekly.

A/B Testing and Shadow Deploys: When a new model is trained, WebkinsTech doesn't flip the switch immediately for all users. They first run the new model in a **shadow mode** – the new model is deployed alongside the old one, receiving a copy of live traffic but its results are not served to users, they're just logged for analysis. This allows them to compare performance: is the new model faster/slower? More accurate? Does it produce any errors? This "shadow testing" is done at scale on real data, which is the best way to vet a model. If it passes muster (accuracy gains, no latency regression beyond acceptable, etc.), then they promote it to serve production traffic (either gradually through canary or all at once if low risk). This approach ensures that scaling the intelligence (making models more sophisticated) doesn't inadvertently hurt system performance or user experience.

Automating the Pipeline: At scale, you can't have someone manually retrain and deploy models every time. So WebkinsTech invested in **MLOps automation** – when new data arrives or a certain time passes, a pipeline (using AWS Step Functions or Jenkins or a specialized tool like MLflow) kicks off: it trains the model, evaluates it, and if metrics look good, possibly even automatically deploys it. Human oversight is still there for critical models, but a lot of the heavy lifting is automated. This reduces the operational burden and speeds up how quickly they can improve models. It's essentially CI/CD for machine learning. From a performance perspective, this means models stay **fresh** and effective, which indirectly is a performance issue – a stale model might make poor predictions that cause inefficiencies downstream (e.g., recommending wrong inventory which then causes extra processing to fix).

Scaling the Data Pipeline: Feeding the beast (the models) with data is another challenge. WebkinsTech's data engineers ensure that as data volume grows (more users, more events), the data storage and feature processing pipelines can handle it. They use scalable storage like S3 or Snowflake for raw data and Spark/Flink jobs for feature extraction. These tools are built to scale horizontally, processing partitions of data in parallel. When they moved from 1 million records to 100 million, they just added more nodes to the Spark cluster and perhaps optimized the code – the pipeline runtime increased, but remained manageable. They also consider **feature stores** – a centralized service where computed features are stored and served to both training and inference, so you don't recompute them too often. This can cut down on redundant computation and ensure consistency. At scale, if multiple models use the same feature (e.g., "customer lifetime value"), computing it once and reusing is better than each model's code doing it.

Resource Budgeting for Training: A practical tip – always ensure your production inference has the resources it needs before allocating to training. If there's a crunch, it's better for training to take a bit longer than to starve inference and hurt users. WebkinsTech sometimes delayed a retraining job by a few hours because the system was under heavy load serving an ongoing promotion; they waited until things calmed down to run the training. The model might be slightly less up-to-date, but users wouldn't notice that as much as they'd notice a slow site. In cloud environments, one can also provision separate quota for such jobs so they don't consume the same pool as production.

By making model retraining a first-class citizen of their architecture, WebkinsTech ensured the "intelligence" in their microservices stayed sharp as they scaled. The models improve over time, and the system handles the improvement process smoothly. It's a bit meta: you need to *scale the process of getting smarter*, not just scale the outputs. Many companies fail when their

ML pipeline can't keep up with the velocity of data – they end up serving models that are weeks or months out of date relative to the data. WebkinsTech was keen to avoid that, knowing their competitive edge lies partly in quickly adapting their AI to new trends (for example, if a new popular product emerges, their recommendation model should learn that pattern within a day or two, not a month later).

We've now covered a lot: from how to structure and govern a growing mass of microservices, to how to squeeze performance and resilience out of each part. To ground these concepts, let's walk through a concrete real-world style example, using a specific technology stack (AWS in this case) to see how all the pieces can come together for scaling intelligent services.

Real-World Example: A Scalable Agentic Architecture on AWS

To bring everything together, let's look at WebkinsTech's platform (or you can imagine a similar company) running on a **cloud infrastructure (AWS)**, and how various AWS services can be used to achieve scale, intelligence, and resilience. This example will tie the conceptual ideas to tangible components: we'll see how **Amazon SageMaker**, **AWS Lambda**, **Kinesis**, **ECS/EKS**, and other tools fit into the picture of an intelligent microservice ecosystem.

Picture WebkinsTech's system in full swing on AWS. The goal is to handle a user doing something like placing an order on their website, which triggers a cascade of events and AIdriven decisions behind the scenes – all scaled to serve millions of users.

Architecture Overview: When a request comes in (say, user submits an order on the web app), it goes through an API Gateway or Application Load Balancer to reach the backend microservices. WebkinsTech runs most of their long-running microservices in containers on Amazon ECS (Elastic Container Service) or EKS (Elastic Kubernetes Service). These include services like Order Service, Inventory Service, User Profile Service, etc., each in its own ECS task or Kubernetes deployment. AWS's auto-scaling for ECS/EKS ensures these services can scale out by adding more containers when needed. They use Amazon CloudWatch alarms tied to CPU/MEM metrics (or custom metrics) to trigger the scaling policies.

Parallel to that, they have an **event-driven backbone** using Amazon Kinesis (a managed streaming platform, similar to Kafka). Many services emit events into Kinesis streams. For example, when an order is placed, the Order Service not only processes it synchronously (replying to the user that their order is confirmed), but also puts an OrderPlaced event onto a Kinesis stream. This event can be picked up by numerous other pieces: the Inventory Service will consume it (to decrement stock), the Recommendation service might consume it to update its models (hey, this user bought item X, update their profile), the Shipping service will start a shipping process, and so on. Using Kinesis here allows these to happen asynchronously and in parallel, rather than chaining a bunch of calls that would slow the user's checkout. It also naturally buffers any surge of events. Each consuming service can scale its Kinesis consumer apps horizontally to keep up with the stream.

Now, certain tasks are perfectly suited for AWS Lambda – short-lived, event-driven, and needing high concurrency. WebkinsTech uses Lambda for things like sending notifications (a Lambda function triggers to send a confirmation email or SMS when an order is placed, reading from the same Kinesis event), resizing images (on an image upload event from S3, a Lambda function resizes it for thumbnails), and running periodic maintenance tasks (like a Lambda that runs every hour to clean up stale cache entries). Lambda gives them massive concurrency scaling without having to manage servers – if 10,000 orders come in at once, 10,000 Lambda invocations can theoretically run in parallel (within account limits) to send out 10,000 confirmation messages, and they don't have to pre-provision that. This serverless approach is cost-efficient and auto-scales by design, great for spiky workloads.

The **AI components** are front and center in their architecture. One example: the

Recommendation Service, which provides personalized product recommendations. Instead of doing all heavy lifting in the container, WebkinsTech hosts the actual ML model on Amazon SageMaker Endpoints. SageMaker endpoints are a fully managed way to deploy models; you choose the instance types (say, ml.g4dn.xlarge for a GPU), and SageMaker will keep a fleet of these instances up, load your model, and expose it via a web-serving endpoint. The Recommendation Service (running on ECS) calls the SageMaker endpoint (essentially an internal API call) whenever it needs fresh recommendations for a user. SageMaker takes care of scaling the number of instances based on traffic – using the auto-scaling policies you can configure, similar to EC2 auto-scaling but specific to the model endpoint . For instance, if requests per minute double, SageMaker can automatically add another instance of the model to handle it. This way, the WebkinsTech team doesn't worry about directly managing GPUs or servers for that model – they hand off that responsibility. They might have multiple endpoints, e.g., one for the recommendation model, one for a fraud detection model, etc.

Another AI piece: suppose WebkinsTech wants to do real-time fraud checks on each order. They could implement this via a Lambda or a microservice calling a SageMaker endpoint running a fraud-detection model. To minimize latency, maybe they keep a warm Lambda function (using Provisioned Concurrency) or an ECS service in memory with the model. Either way, they likely leverage AWS's **AI services** where possible – for instance, if they needed NLP sentiment analysis, they might even use Amazon Comprehend (an AI API) rather than build their own model.

Let's not forget data streaming. They use **Amazon Kinesis Data Analytics** (or Spark on EMR) to do streaming analytics on the Kinesis events for monitoring and feeding some aggregated data into AI models. For example, they maintain a real-time dashboard of top trending products, which is computed by a Kinesis Data Analytics job scanning the purchase events stream and updating a DynamoDB table that the website reads. This is highly scalable (Kinesis can handle thousands of events per second, and the analytics job can scale in parallel shards).

Speaking of **DynamoDB**, that's another piece: many microservices use DynamoDB for a fast NoSQL datastore (e.g., User profiles, product catalog caching, etc.) because it scales virtually infinitely with usage and requires no manual sharding. They set read/write capacity modes as auto-scaling or on-demand so that the DB throughput scales with access patterns.

Some services still use an RDS (SQL) database for transactional consistency (like the Orders database), scaling that with read replicas and so forth, but for truly high scale or flexible data, DynamoDB or Aurora (which can scale read throughput substantially) is used.

All these components produce a lot of logs and metrics, which CloudWatch collects. They also likely use AWS X-Ray for distributed tracing across AWS services (like from API Gateway to Lambda to DynamoDB, X-Ray can create a trace graph). This mirrors our observability discussion, implemented with cloud-native tools.

Let's visualize a part of this architecture to solidify it. **Figure 15.2** shows a simplified flow for the order placement scenario:

Figure 15.2: WebkinsTech's Scaled Architecture on AWS (Simplified)

```
flowchart LR
    subgraph UserInteraction
      WebClient(Website/App) -->|Order Request| API_GW[(API Gateway)]
end
    API_GW --> OrderSvc[Order Service<br/>(ECS)]
    OrderSvc -- writes event --> Kinesis[(Order Events Stream)]
    OrderSvc -->|Sync call| PaySvc[Payment Service<br/>(ECS)]
    OrderSvc -->|Sync call| SageMakerFraud[(Fraud Model<br/>SageMaker
Endpoint)]
    OrderSvc -->|Response| API_GW
    Kinesis ==> InvProc[Inventory Processor<br/>(Lambda)]:::lambda
    Kinesis ==> RecProc[Recommendation Updater<br/>(ECS)]
    InvProc --> InvSvc[Inventory Service<br/>(ECS)]
    RecProc --> SageMakerRec[(Recommendation Model<br/>SageMaker
Endpoint)]
    InvProc --> ShipSvc[Shipping Service<br/>(ECS)]
    OrderSvc -.-> SNS{Order Notifications}:::event
    SNS --> NotifyLambda(Confirmation Email Sender<br/>(Lambda)):::lambda

    classDef lambda fill:#e6ffe6,stroke:#2ecc71;
classDef event fill:#fdf5e6,stroke:#f1c40f;
```

Figure 15.2: A blend of services on AWS enabling WebkinsTech's order workflow. An order request goes through API Gateway to an Order Service (ECS). The Order Service calls a Payment Service (to charge payment) and a Fraud detection model on SageMaker (to vet the order) synchronously. It then emits an event to a Kinesis stream for downstream processing. A Lambda function consumes that event to update Inventory (via the Inventory Service) and trigger shipping. Another consumer (Recommendation processor) updates the recommendation model using SageMaker. Separately, an SNS topic and Lambda handle user notification. All components are scalable and managed: API Gateway and Lambda auto-scale by design; ECS services auto-scale on metrics;

SageMaker endpoints auto-scale on request load; Kinesis buffers the event flood.

This diagram is a mouthful, but it showcases how different AWS components work in concert. Let's unpack a few salient points in the flow:

- **API Gateway** handles scaling of incoming requests (it can handle thousands of requests per second easily, and you can raise limits as needed). It forwards to Order Service, which is running on ECS behind a load balancer. If the Order Service needs more capacity, ECS will launch more tasks (perhaps on AWS Fargate or EC2 instances).

- The Order Service, to keep the user waiting as little as possible, only does minimal synchronous work: charges the payment (maybe an external API call) and checks for fraud quickly (via SageMakerFraud endpoint). The heavy lifting like adjusting inventory and starting shipment is offloaded to asynchronous flow via Kinesis.

- The use of **Amazon SNS** (Simple Notification Service) for notifications is another eventdriven piece: the Order Service publishes to an SNS topic "OrderPlacedNotifications". This fan-outs to a Lambda that sends an email (and could fan-out to other subscribers like an SMS service, mobile push, etc.). SNS and Lambda both scale automatically – if 1000 orders come in, SNS will queue up messages and 1000 Lambda executions will send emails in parallel.

- Inventory updates are critical but can be slightly decoupled. The Lambda (InvProc) reading from Kinesis ensures inventory is eventually consistent. If needed, they could have chosen to update inventory in the OrderService synchronously (for strong consistency – ensuring no oversell), but maybe they use a reserve system with eventual adjustment for simplicity and speed. (The architecture can be tweaked based on requirements; the point is Kinesis allows it to be async if eventual consistency is acceptable, improving throughput).

- The Recommendation Updater (RecProc) could be another Lambda or a lightweight ECS service that takes events to update user profiles or trigger a retraining. In this diagram, it calls SageMakerRec endpoint – perhaps updating some real-time features for the user's next recommendation.

- **Shipping Service** (not shown deeply) might also consume the event and create a shipment order via an external courier API. It could be triggered by the Inventory Processor or separately from Kinesis.

Notably, **all these components scale independently**. If the bottleneck becomes the SageMakerRec endpoint (say recommendations are heavy), SageMaker will scale it out to more instances under the hood. If Lambdas are the bottleneck, AWS will run more Lambdas (as long as you've allowed it via concurrency limits). If Kinesis throughput is an issue, you increase the number of shards (each shard is like a parallel pipe). ECS services scale by tasks, and DynamoDB behind any of these can scale in IOPS.

WebkinsTech's team also uses AWS CloudWatch ServiceLens or X-Ray to trace an order request through this whole flow, which is amazing to observe: you can see the initial request, the calls to Payment and Fraud model, then the asynchronous trail of events flowing through Lambda and so on.

For completeness, their setup likely also includes **Amazon EKS (Kubernetes)** for some services that require custom scheduling or if they prefer open-source stacks (like running Kafka, etc., if not using Kinesis). They might run a **service mesh (App Mesh or Istio)** for easier observability and reliability features across services. And they probably use infrastructure-as-code (CloudFormation or Terraform) to manage all these moving parts declaratively, because at this scale, you automate everything.

The key takeaway from this example is that a cloud like AWS provides *building blocks for scale*. By using managed services (Lambda, SageMaker, Kinesis, etc.), WebkinsTech offloaded a lot of the heavy lifting of scalability to AWS. They didn't have to implement their own messaging system or design an auto-scaler for models – they leveraged what exists. This allowed them to focus on the unique parts of their application (their business logic and proprietary AI models) rather than re-inventing core infrastructure. It's a cloud-neutral chapter, but realistically, whatever cloud or platform you choose, you'll find similar capabilities – use them!

We've walked through the theory, the practice, and a concrete example. At this point, you might feel like you've traveled through a **vast landscape of scaling techniques**. It's true – scaling intelligent microservice ecosystems is no small feat, but with the right patterns and tools, it's absolutely achievable.

Before we conclude, let's recap the major points and set the stage for what's next.

Summary and Next Steps

Scaling an intelligent microservice ecosystem involves much more than adding servers. It's about thoughtfully **structuring your agents**, rigorously **managing complexity**, and wisely **engineering for performance**. In this chapter, we saw how hierarchical and decentralized patterns can organize a multitude of AI-driven services, how discipline in design (bounded contexts, observability, autonomy) keeps a large system manageable, and how leveraging technologies (from GPUs to AWS managed services) ensures the system runs efficiently under heavy load.

We followed WebkinsTech's journey as they grew their platform. They found that a mix of architectures – part hierarchy, part swarm – gave them the best of both worlds: coordination where needed, freedom where possible. We learned that **clear domain boundaries** and **strong team ownership** prevent a sprawling system from collapsing under its own weight. We emphasized building **observability** in from day one, so that even if you have hundreds of microservices, you maintain a god's-eye view of what's happening. We discussed preparing for the worst (spikes, failures) so the system can bend without breaking. And we dove into **AI performance** – scaling model inference with accelerators and good design, keeping models fresh through automated retraining pipelines, and handling the massive data flows that feed these intelligent services.

A few recurring themes emerged:

- **Structure scales:** A well-structured system (whether via hierarchical layers or federated domains) can grow more easily than an ad-hoc one. Architectural patterns act as force multipliers for scaling, by reducing chaos as you add more pieces.

- **Automation is your friend:** From auto-scaling to self-healing to continuous deployment, automation handles scale far better (and faster) than manual intervention. The more you can encode smart behavior into the platform (even using AI agents to manage AI services!), the more smoothly you'll ride the growth curve.

- **Stay observant and adaptive:** Assumptions that held at 10 services might break at 100. By watching the system and being ready to adapt – re-architect hotspots, tune parameters, split or merge services – you ensure scalability is a continuous process, not a one-time project.

- **Leverage platform capabilities:** Whether on cloud or on-prem, use the tools and services that abstract away the gritty details of scaling infrastructure. This lets you concentrate on higher-level scaling challenges (like algorithms and architecture) rather than reinventing load balancers and task schedulers.

As we wrap up, imagine WebkinsTech's platform a year from now. Perhaps they have five times the users, and their ecosystem has grown to include IoT sensors streaming data, even more AI agents optimizing logistics in real-time, and global deployments across multiple regions. Thanks to the foundations laid and lessons learned, they're not only coping with this scale – they're thriving on it. Their intelligent microservices aren't buckling under pressure; instead, they're **learning and improving**, making the system more efficient and robust over time. This is the ideal end-state: scaling not just in size, but in *wisdom*.

Now, our exploration of intelligent microservices doesn't end here. In fact, it's time to see how all these principles play out in the real world. In **Chapter 16**, we will dive into industry case studies – real companies and projects that have built agentic microservice ecosystems. We'll analyze how they applied (or sometimes discovered the hard way) the very strategies we've discussed. You'll get to compare notes between WebkinsTech's journey and those of major players in finance, e-commerce, healthcare, and more. These case studies will solidify your understanding and perhaps even inspire new ideas for your own scaling adventure.

Scaling up intelligently is both an art and a science – and with the knowledge from this chapter, you're well-equipped to practice that art in your own projects. Onward to the case studies, and the final leg of our journey into the world of Intelligent Microservices!

Chapter 16: Real-world Applications Enterprise Case Studies

Now that we've explored the principles of agentic AI and intelligent microservices, it's time to see how they come to life in real-world scenarios. In this chapter, we dive into **nine in-depth enterprise case studies**, each illustrating how *WebkinsTech*, our fictional enterprise, applied these concepts to solve complex problems. We'll compare WebkinsTech's journey with parallels from industry leaders like Amazon, Netflix, UiPath, Stripe, and others. The case studies span domains from document processing and workflow automation to customer experience and self-healing infrastructure. For each case, we'll discuss the conceptual overview, the technical architecture (with inline diagrams), the tools and tech stack (cloudagnostic, noting AWS examples where relevant), the integration of agentic AI, and the impact measured in metrics like efficiency gains, cost savings, and reduced downtime. Along the way, we include callout tips and real-world analogies to keep things concrete and engaging.

Let's get started on this tour of intelligent microservices in action!

Case Study: Document Processing and Validation

Conceptual Overview

WebkinsTech's finance department was drowning in paperwork. Accounts Payable specialists spent hours manually handling invoices, purchase orders, and compliance documents. The process involved receiving documents (often as PDFs or scans), reading and extracting key information, entering data into systems, and verifying everything for accuracy. It was **slow, error-prone, and costly**. A single contract or invoice could take days to validate if data had to be cross-checked with multiple systems. WebkinsTech knew that automating this **document processing and validation** would save time and reduce human error. Traditional optical character recognition (OCR) solutions helped a bit, but they were rigid and required template-specific rules. The company envisioned an intelligent microservice that could *ingest documents, "understand" their content, and validate them autonomously*, only involving humans for edge cases. This is a classic use case for intelligent document processing (IDP)

– blending OCR, natural language processing, and business rules, wrapped in a microservices architecture.

From a **business perspective**, the goal was to accelerate processing (turnaround in minutes instead of days), improve accuracy in data entry, and free up staff for more valuable work than staring at forms. Conceptually, WebkinsTech designed a solution where an **Agentic Document Processor** microservice takes in any document, uses AI to extract structured data, validates that data against business rules (and other microservices), and decides if the document is approved, rejected, or requires human review. Key challenges included handling a wide variety of document formats (invoices from different vendors, regulatory forms, etc.), ensuring data privacy, and building trust in the AI's outputs. A human-in-the-loop mechanism would be crucial for validation until the AI was sufficiently accurate.

Technical Architecture and Agentic Solution

WebkinsTech's engineering team broke the problem into a set of collaborating microservices, each with a clear role, coordinated by an intelligent agent inside the workflow. The architecture they implemented looked something like this:

```
[Document Intake API] --(new file)--> [Document Processor Service] --
(extracted data)--> [Validation Service] --> [ERP System]
                                   \                            /
                                    \--(low confidence)---> [Human Review
Portal]
```

Diagram: Document Processing Microservice Flow – When a new document arrives

(through a scanning tool or upload portal), the **Document Intake API** passes the file to the **Document Processor Service**. This service is augmented with an AI **documentprocessing agent**. It uses OCR and NLP models to extract text and key fields (e.g. invoice number, dates, line items, totals). It then uses machine learning classifiers to identify document type (invoice, contract, etc.) and to parse relevant sections. Next, the extracted data is sent to a **Validation Service** microservice. The Validation Service contains business rules and calls out to other systems (like the ERP or a vendor database) to cross-check details – for example, verifying that a purchase order number exists and matches the vendor. If everything checks out and the AI is highly confident in the extraction, the validated data is automatically recorded in the ERP system or database. If there are any discrepancies or the

AI's confidence in its understanding is below a threshold, the case is routed to a **Human Review Portal**, where a person (e.g. an AP clerk) can see the document and the AI's suggested data, then correct or approve it. This human feedback is fed back into the learning loop to continuously improve the AI model over time.

Under the hood, this architecture uses a mix of batch and real-time processing. Document files are often placed into an **object storage** (like an S3 bucket in AWS) and a message or event triggers the processing service. The Document Processor Service might be implemented to scale out as needed (many documents in parallel) and uses AI models – which could be custom-trained or calls to cloud AI services – to perform OCR and entity extraction. The **agentic** element lies in how the Document Processor makes decisions: it's not just blindly extracting text, but also running an intelligent validation workflow. We can think of the Document Processor as having an embedded *document agent* that has a goal: "extract accurate data from this document and get it validated." It will decide, based on confidence scores and validation results, whether to mark the document as processed or involve a human or other services. This kind of autonomy (deciding the next step without a human explicitly coding every rule) is what makes it *agentic*. For example, if the agent notices a total amount on an invoice that doesn't match the sum of line items, it can automatically flag an anomaly and route to a human or request clarification from a supplier service – actions beyond a rigid, pre-defined script.

To ensure reliability, WebkinsTech's team included *redundancy and monitoring*. Every step logs the outcomes (extraction confidence levels, validation checks). They started with conservative settings – the agent would only auto-approve when extremely confident – to build trust. Over time, as the models learned and proved their accuracy, more documents could go straight-through without human touch.

Industry Example: In the insurance sector, companies are leveraging similar intelligent document pipelines. For instance, Anthem, a large health insurer, used AI to automate about **80% of its claims-processing workflow**, drastically reducing manual data entry . They achieved this by using an OCR and machine learning service (Amazon Textract) to extract data from claims forms and validate it, expecting to reach over 90% automation .

This real-world success mirrors WebkinsTech's goals: faster processing of forms and fewer human hours spent on rote tasks.

Tools and Technology Stack

To keep the solution cloud-agnostic, WebkinsTech used a mix of open-source tools and modular cloud services. Key components included:

- **OCR and NLP Engines:** They evaluated open-source OCR (like **Tesseract**) and cloud APIs (such as **AWS Textract** or Google's Document AI). In the end, they opted for a combination: a self-hosted OCR for simple documents and AWS Textract for complex forms, taking advantage of Textract's ability to recognize tables and handwriting.

- **Machine Learning Models:** For classification and data extraction, they trained custom models using frameworks like **TensorFlow**. For example, a model to classify document types and another to extract fields (alternatively, they could use services like AWS Comprehend for entity extraction).

- **Microservices & Orchestration:** Each function (intake, processing, validation, etc.) was a separate microservice (developed in Python and Node.js). They communicated via an event bus (Kafka) and REST APIs. AWS **Step Functions** was considered to orchestrate the workflow, but the team decided on a lightweight orchestrator within the Document Processor agent logic for flexibility.

- **Datastores:** Documents were stored in a secure file repository (on-premises storage appliance for sensitive docs, mirrored to cloud storage for processing). Extracted data and validation logs went into a **Document Processing DB** (e.g. a PostgreSQL instance) that tracks each document's state. This also served as a training data repository for improving the AI.

- **Human-in-the-loop Interface:** A simple web portal built with React where humans could review and correct data. This portal connected to the back-end via an API gateway. It was crucial for user experience that this interface be fast and show the original document side-by-side with extracted data.

By keeping components decoupled, WebkinsTech ensured the system was cloud-agnostic. For instance, they could swap out AWS Textract for another OCR service or an on-prem tool without affecting other services, thanks to a clear interface contract (the Document

Processor service abstracts away the specifics of extraction). In fact, a **Design Tip** emerged here: *treat AI services as black boxes with clear input/output contracts*. WebkinsTech defined strict schemas for what the Document Processor sends to the OCR/AI component and what it expects back (e.g., a JSON with fields and confidence scores). This way, whether it's Textract or another library under the hood, the rest of the microservices remain unaffected .

Tip: Implement **explanation logging** for the AI's decisions. WebkinsTech added logging such as: "Extracted total $5,230 with 95% confidence from field X on page 1." This audit trail not only helps debug issues ("Why did the AI read $5230 instead of $5320?") but also builds trust. Over time, these logs became a source of truth to measure model improvements and were used to refine the ML models (feedback loop).

Integration of Agentic AI

The agentic behavior in this case study is embodied by the Document Processing service's AI agent. Unlike a traditional RPA script that follows fixed rules, this agent **perceives** (reads text, parses meanings), **decides** (based on confidence and rules outcomes), and **acts** (routes the document accordingly). It was programmed with goals: maximize accuracy of extracted data and minimize turnaround time, while avoiding false approvals. In practice, this means the agent would:

- dynamically adjust OCR strategies if initial results were low confidence (e.g. trying an alternative OCR engine or requesting a higher-resolution scan from the intake if possible),

- learn from corrections (the agent flags uncertain fields and later re-trains on the confirmed correct values provided by humans),

- and coordinate with other services autonomously. For example, if an invoice's vendor ID isn't found in the vendor service, the agent might query a secondary database or trigger a process to create a new vendor entry (depending on company policy), rather than simply failing.

Initially, the "intelligence" was mostly supervised ML (the models for extraction). Over time, as WebkinsTech grew confident, they introduced more *autonomy*: the agent could handle some exceptions on its own. One experimental feature was an **anomaly detector agent** that watched the stream of processed documents to detect outliers – e.g., if a normally 2-page form came in as 10 pages, flag it. This was done using a simple statistical model (to demonstrate multi-agent coordination, this anomaly detector was a separate microservice agent that would alert the main processor agent).

Crucially, WebkinsTech kept humans in the loop for training and governance. The AI didn't live in a vacuum – it was **augmented intelligence**, working alongside staff. Over months, as the agent handled more volume, the team tuned the threshold for human review (lowering it as confidence improved). The agent's policies were updated with a form of reinforcement: whenever a human corrected a field that the AI was confident about, that scenario was analyzed to adjust either the ML model or the confidence threshold.

Impact and Results

The results were dramatic. After deployment, WebkinsTech saw immediate gains:

- **Throughput:** Document processing time dropped from an average of **2-3 days to a few minutes** per document. In cases where no human review was needed, some documents went from inbox to fully validated in under **30 seconds** (mostly spent in compute time and cross-checking systems).

- **Automation Rate:** Within the first three months, about 65% of incoming documents were processed end-to-end with no human intervention. The remaining required partial or full review. As the models improved, the automation rate climbed. They reached **85% automated processing** on invoices after a year, meaning only 15% needed a person's eyes. This closely mirrored industry benchmarks; for example, insurers like Anthem hit ~80% automation on claim documents with such AI .

- **Accuracy:** Accuracy of data extraction was measured by comparing with humanentered ground truth. Initially, the AI matched or exceeded human accuracy on structured forms (like 99% on government forms), but lagged on unstructured documents. With active learning, the overall accuracy reached **around 95%**, significantly reducing downstream errors. More importantly, certain error types (like typo transcriptions) were virtually eliminated.

- **Efficiency and Cost:** The department calculated that they saved **hundreds of manhours per month**. Staff who used to key in data could focus on exception handling and supplier relationship tasks. The company estimated an annual saving of $1M in labor and error-related costs. They also avoided late payment penalties on invoices since nothing languished in queues. Even considering the cost of the AI infrastructure (which ran mostly on existing servers and some cloud services), the ROI was clear.

- **Employee Experience:** At first there was resistance or fear that the "AI would take our jobs," so WebkinsTech involved the end-users (AP clerks, auditors) in the design of the Human Review Portal. The result was an interface that *made their job easier* – showing them exactly where to look on a document for issues, etc. Over time, employees reported higher satisfaction as they spent less time on drudgery.

Beyond these quantifiable impacts, this case set the stage for **broader intelligent automation** in the company. It proved the concept that microservices infused with AI agents can handle core business processes reliably. After seeing the success in finance,

WebkinsTech's leadership looked to apply similar patterns to other processes, which leads us to the next case study.

Case Study: Intelligent Automation of Complex Business Processes (Order Fulfillment & Claims)

Conceptual Overview

With document processing under control, WebkinsTech turned to automating its *complex business workflows*. Two high-value targets were identified: **order fulfillment** in its ecommerce operations and **claims**

processing in its warranty service. These processes span multiple steps, departments, and decisions – perfect for an intelligent microservices approach.

Order fulfillment at WebkinsTech involved a series of coordinated tasks: receiving a customer order, checking inventory, processing payment, scheduling warehouse picking and packing, shipping via carriers, and handling any exceptions (like item out-of-stock or payment issues). Traditionally, WebkinsTech had semi-automated this with a mix of an ecommerce platform and manual interventions. For example, if an item was out of stock in the nearest warehouse, an employee might manually reroute the order to another location. Similarly, high-priority orders (like expedited shipping) were flagged by rules, but if something went wrong (missed SLA, or an address issue), humans had to jump in. This resulted in delays, especially during peak seasons. The process was ripe for an overhaul: WebkinsTech envisioned an **Intelligent Order Orchestrator** – a brain that could oversee the end-to-end order pipeline, make smart decisions in real-time, and orchestrate microservices and human tasks as needed.

Simultaneously, WebkinsTech's **warranty claims processing** (for its hardware products sold to customers) was another complex workflow. Customers would file a claim if a product failed under warranty. This required verifying the warranty validity, possibly arranging a return or repair, approving or rejecting the claim based on conditions (was the device misused? Is it within the covered period?), and then triggering compensation (replacement unit shipment or refund). This was very much like an insurance claim process, albeit on a smaller scale. WebkinsTech saw parallels with how insurance companies use AI for claims – from fraud detection to automatic approvals for straightforward cases. They decided to incorporate an *agentic AI* to triage and process claims intelligently.

The overarching concept here is **Intelligent Process Automation** – going beyond static rule-based workflows (like those in traditional BPM or RPA systems) to workflows that can *adapt on the fly* using AI decisions. These processes involve multiple microservices (inventory service, payment service, shipping service, etc. for orders; or customer service, logistics, finance for claims) and thus needed a robust orchestration. The challenge was to maintain reliability and clarity in these processes, as they directly impact customers. If an AI agent incorrectly routed an order or auto-approved an invalid claim, it could hurt the business. So WebkinsTech planned a gradual introduction: automate the happy paths fully, and use AI to assist decision-making on the edge cases (with humans reviewing until proven).

Technical and Architectural Breakdown

They started with **order fulfillment**. The team mapped out the current order workflow and identified decision points where AI could add value:

- Inventory allocation: If the primary warehouse is out of stock, decide which alternative location can ship fastest to the customer (taking into account distance, inventory levels, and shipping backlogs).
- Exception handling: e.g., if an item is out of stock everywhere, should the order wait for restock, split into multiple shipments, or offer an alternative product?
- Dynamic prioritization: During peak times, which orders to fulfill first (for example, prioritizing premium customers or urgent medical device orders).
- Fraud check: Flagging potentially fraudulent or risky orders (this ties into financial services, but it's part of the fulfillment process to hold suspect orders).

The architecture they settled on introduced an **Order Orchestrator Service** – think of this as a workflow engine, but with an AI twist. It broke down like this:

```
Customer Order -> [Order API] -> [Order Orchestrator] ---> [Inventory
Service]
                                                   |           \--> [Warehouse
Agent]
                                                   |--> [Payment Service]
                                                   |--> [Shipping Service
(Carrier API)]
                                                  \--> [Exception Handler Agent]
-> [Manual Review Portal]
```

Diagram: Intelligent Order Fulfillment Orchestration – An order placed by a customer (through WebkinsTech's website or API) hits the **Order API** (which creates an order record and initiates the process by calling the Orchestrator). The **Order Orchestrator** is a microservice that functions as a central coordinator. It's implemented with a state machine or workflow engine internally. It calls the **Inventory Service** to allocate items. The Inventory Service, in turn, has a sub-component or agent (the **Warehouse Agent**) that decides from which warehouse to source the product. Here's where AI comes in: the Warehouse Agent uses a predictive model to choose the best fulfillment center (it was trained on past orders, considering delivery times and stock levels). For example, if the nearest warehouse is out of stock, it evaluates others not just on distance but also on likelihood of timely delivery (perhaps another warehouse 100 miles further has a direct overnight shipping route that actually gets it there faster than a closer one with slower shipping).

Once inventory is allocated, the Orchestrator calls the **Payment Service** to charge the customer. In parallel, the **Shipping Service** is contacted to schedule a pickup with a carrier (e.g., generating a shipping label via FedEx or AWS Shipping APIs). If all goes well, the order moves to fulfillment in the warehouse.

The interesting parts are the exception flows: if **Payment fails** (e.g., credit card declined), the Orchestrator's agent can automatically decide to retry or to ask the customer for another payment method (maybe triggering an email via a Notification Service). If **Inventory is not available** anywhere, the agent might consult a **Backorder Agent** which could do things like check incoming inventory schedules or even suggest an alternative product. If a rule or the AI deems an order potentially fraudulent (say a high-value order shipping to a new address with overnight shipping – a suspicious combo), it triggers an **Exception Handler Agent**. This agent might interface with a fraud detection microservice (more on that in the financial case study) to score the order, and if high risk, put the order on hold and send it to a **Manual Review Portal** for a fraud analyst to check.

Throughout this process, the Orchestrator is *agentic* in that it doesn't just follow a single static path. It has branching logic influenced by AI decisions: e.g., "If the AI says warehouse A will cause a 2-day delay due to backlog, choose warehouse B," or "If Order risk score > 0.9, route to manual review; if 0.5–0.9, maybe require signature on delivery (a middle-ground action)." These decisions were encoded as policies that the Orchestrator's decision engine follows. The policies could be updated without redeploying the whole service – they were externalized in a rules engine that the AI agents could tweak (with oversight). In effect, WebkinsTech created a **hybrid rules+ML system**:

straightforward logic was handled by rules (e.g., "if item out of stock, check other warehouses in descending priority"), while more complex predictions were handled by ML models (like predicting shipping delays or fraud probability).

For the **claims process** (warranty claims), they implemented a similar orchestrator approach:

- A **Claims Intake Service** would accept claim submissions (with data and perhaps uploaded photos of the defective product).

- An **Claims Agent** would verify warranty status (calling Customer and Product services), then assess the claim. They trained a model to classify claims into likely approved or likely rejected based on history (for instance, products failing within 1 month of purchase are almost always approved, whereas claims on 3-year-old devices often have nuances).

- If straightforward, the agent auto-approves and triggers a **Replacement Service** to ship a new product (or a Refund Service to issue credit).

- If something looked fishy (e.g., multiple claims from the same customer, or claim details that don't match known failure patterns), it would route to a **Claims Adjuster Portal** for human investigation.

Notably, WebkinsTech borrowed concepts from insurance: they implemented a basic **fraud detection model** on claims (to catch any customer trying to game the warranty) and a **rules engine** for eligibility (certain types of damage not covered, etc.). The AI agent could override strict rules in some cases – for example, if a usually not-covered scenario comes up frequently and the model suggests it's actually a manufacturing defect, the agent might flag "consider goodwill approval" to a human, demonstrating learning and adaptation beyond static policy.

Industry Example: The power of intelligent process automation is perhaps best illustrated by the insurance startup Lemonade. They famously built an AI-driven claims bot called AI Jim that can settle insurance claims in **as little as 2-3 seconds** – something unheard of in traditional insurance . In one case, Lemonade's AI reviewed a claim, cross-referenced the policy, ran fraud checks, approved it and initiated payment **all within 2 seconds** . This shows what's possible when an orchestrated process is automated end-to-end with AI decisions. WebkinsTech's claims agent isn't quite that fast (there are shipping and physical product factors), but the same principles apply – many routine claims can be handled automatically, drastically cutting resolution time.

Tools/Technology Stack

WebkinsTech's stack for these processes combined workflow tools with AI services:

- **Workflow Engine**: They chose **Camunda** (an open-source BPMN workflow engine) to implement the Order and Claims Orchestrator logic. Camunda allowed modeling the flow with tasks and decision gateways. They extended it by writing custom decision handlers that call AI models. An alternative considered was **Netflix Conductor**, which is a microservice orchestration engine; it could handle the complex flows with reliability. (In fact, Netflix Conductor has been scaled to handle billions of workflows per month , showing it can orchestrate on a scale similar to Amazon's operations.)

- **Rule Engine**: For certain decisions, they used **Drools** (a rules engine) to maintain easily editable business rules (like the eligibility criteria for claims, or prioritization rules for orders). This separated static business logic from code, and the AI agents could query the rules or even suggest updates (with human approval).

- **AI/ML Models**:

- The **Warehouse selection model** was a regression model that predicts delivery time and success probability for each warehouse, given current conditions. They trained it using Pandas and scikit-learn on historical delivery data.

- The **Order fraud model** and **Claims fraud model** were classification models (logistic regression and later a small neural network) that output a risk score.

- They also used an **RL (Reinforcement Learning)** approach in simulation to optimize one policy: an agent that learned how to allocate limited inventory during high demand (basically a dynamic programming problem, solved offline and then used as a policy table).

- These models were deployed in a **Model Serving Microservice** (using AWS

SageMaker endpoints for scalability, though it could be TensorFlow Serving on Kubernetes just as well). The orchestrator calls the model service with relevant features (order details, claim details) and gets back a prediction.

- **Services and APIs**: They containerized all microservices (Order API, Orchestrator, Inventory service, etc.) and used **Kubernetes** to manage them, which allowed them to handle spikes during sales events by auto-scaling. They utilized cloud services where useful: e.g., Amazon's **SNS/SQS** for event queues to decouple steps, and **AWS Step Functions** was tested for orchestrating some sub-processes (like the parallel payment and warehouse calls).

- **RPA Integration**: Interestingly, for some legacy steps (like entering data into an older ERP that didn't have APIs), they employed a robotic process automation bot (UiPath) as a stop-gap. This highlights how traditional RPA and AI can work together: the AI Orchestrator would signal the RPA bot to perform a task (like "mark order shipped" in the old system) in an automated fashion. Over time those legacy pieces were replaced with direct API integrations, but RPA helped to automate end-to-end even when not everything was modern.

- **Monitoring & Logging**: Given the complexity, they invested in monitoring. Every order and claim had a tracking ID, and they aggregated logs in ElasticSearch with Kibana dashboards to monitor how many orders were auto-fulfilled vs needing human help, how often the AI overrode a rule, etc. This helped fine-tune the balance between automation and manual steps.

Design Tip: When automating complex processes, start simple and *gradually inject intelligence*. WebkinsTech first modeled the workflows with no AI – essentially mimicking the manual process in microservices to ensure correctness. Then they identified decision points to insert AI. They always kept a **fallback**: if the AI service was unavailable or uncertain, the system fell back to a default rule or human decision. This ensured reliability while the AI learned. In practice, a hybrid approach (rules + AI) often works best initially . Use AI for what it's good at (making data-driven predictions on probability or classification problems) and use rules for clear-cut policy (like "Don't ship if payment failed").

Integration of Agentic AI

The agentic aspect in these workflows comes from the orchestration layer effectively behaving as an **autonomous agent orchestrator**. It's goal-driven:

"complete the order as fast as possible with minimum cost and risk" or "resolve the claim correctly while minimizing losses". The Orchestrator agent perceives the state of the process (inputs like stock levels, model scores, etc.), decides on next actions (which branch of the workflow to take, whether to auto-approve or escalate), and acts by invoking the relevant microservice or sub-agent.

One could conceptualize multiple specialized agents working together:

- An **Inventory Agent** focused on stock decisions.
- A **Shipping Agent** that might choose shipping options (expedited vs standard, maybe offering the customer a faster option for a fee – something WebkinsTech planned for future dynamic offerings).
- A **Claims Decision Agent** that approves or escalates claims.

These agents communicate through the orchestrator or event bus. For example, the Inventory Agent might publish an event "warehouse allocated" which the Shipping Agent listens to in order to schedule pickup. If a problem arises (warehouse cannot fulfill), the agents coordinate to find an alternative – without a person telling each step.

In essence, WebkinsTech built a **multi-agent system** within its process automation. Each microservice (with its AI helper) is like an agent with a specific responsibility, and the orchestrator is a higher-level agent ensuring they all work toward the common goal of fulfilling the order or claim. This kind of system can even **discover new optimal paths**. A neat incident occurred: during a severe snowstorm, the primary warehouse in one region was effectively shut down. The system's agents dynamically rerouted hundreds of orders to other locations and adjusted shipping methods (some overnight shipments were switched to two-day from a different location that had a clearer weather route). They did this based on real-time signals (like the Warehouse Agent noticed a sudden drop in confirmation from the snowy warehouse and an external weather API input it had access to). The team realized the AI agents had essentially learned to **perform dynamic re-routing** that previously would have taken frantic human re-planning. This solidified their confidence that the system was truly exhibiting autonomous, *agentic* behavior – adapting to events with minimal human intervention.

Impact Metrics and Outcomes

After implementing intelligent order fulfillment and claims processing, WebkinsTech measured significant improvements:

- **Order Fulfillment Speed:** The average order processing time (from placement to ready-to-ship) reduced by 50%. Many orders that used to take ~4-5 hours (with manual queueing and batch processing at day's end) were now confirmed and released to the warehouse within **minutes**. Same-day shipping cutoff time was extended because orders could be processed instantly as they came in. This led to a measurable boost in customer satisfaction and a **12% increase in on-time deliveries**, as orders caught earlier shipping windows.

- **Exception Handling Efficiency:** Previously, about 20% of orders fell into exceptions requiring human intervention (e.g., payment issues, stockouts). That rate fell to ~5%. The majority of routine exceptions were auto-resolved by the agents (for instance, automatically splitting an order into two shipments when one item was backordered, instead of waiting for a human to decide). The remaining 5% that truly needed humans were usually fraud-suspected orders. Those were flagged clearly with an AI-generated explanation (e.g., "High-risk score due to mismatch in billing/shipping country") which made the fraud team's review much faster. In fact, the fraud review team reported they could make decisions 30% faster because the AI pre-checked a lot of data.

- **Throughput and Scaling:** During peak sale days, WebkinsTech handled **3x more order volume** without adding temporary staff in operations. The intelligent services scaled up on Kubernetes and handled the surge. In one Black Friday event, the system autonomously processed thousands of orders per hour; the few hiccups (like a payment gateway slowdown) were mitigated by the orchestrator agent pausing and retrying seamlessly. Compared to the prior year (before AI orchestration), they saw a 25% reduction in fulfillment latency at peak and zero major incidents (the previous year had the infamous "Black Friday Meltdown" when manual processes couldn't keep up).

- **Claims Processing Time and Customer Experience:** For warranty claims, simple claims (e.g., gadget failed within warranty period with proper proof) were auto-approved 70% of the time, and customers got instant confirmation that a replacement was on the way. The overall claims resolution time dropped from ~5 days (often waiting on email responses) to **1 day** or less on average. Importantly, customer trust improved; surveys showed higher satisfaction because the company was responding faster and more consistently on claims.

- **Cost Savings:** By automating order exceptions and claims, WebkinsTech saved on operational costs. They estimated **40% lower labor hours** spent on these processes. Some roles were redefined – for example, the order management team now focuses on overseeing the AI and handling only genuinely tricky cases, rather than routine checking. This means the same team can support a growing business without linear headcount increases.

- **Error Reduction:** Fewer orders fell through the cracks. It's hard to quantify, but previously, manual handling led to occasional missed orders or delayed backorder notifications. The AI agents never "forget" to follow up, so the process completion rate hit virtually 100%. In claims, consistent application of rules and AI meant there were fewer cases of mistakenly approved or denied claims. (Any erroneous auto-denial was caught in periodic audits and the rules/ML adjusted accordingly, making the system increasingly fair and accurate.)

WebkinsTech's leadership was impressed not just by raw numbers, but by the *agility* gained. When the COVID-19 pandemic disrupted supply chains, they could quickly adjust the agents' policies (for instance, prioritize local warehouse fulfillment to avoid cross-border shipping delays). The intelligent workflow simply adapted to the new rules and data. This case demonstrated that intelligent microservices aren't just about doing things faster, but also about being **far more adaptable** to change.

With complex processes under control, the next step was to ensure the underlying systems and infrastructure could support this smart automation. We turn to how WebkinsTech tackled IT operations and infrastructure with an equally intelligent approach.

Case Study: Autonomous IT Operations and Infrastructure Management (Self-Healing, SecOps, Resource Optimization)

Conceptual Overview

As WebkinsTech's services grew in number and complexity, the strain on IT operations (IT Ops) increased. Microservices architecture gives a lot of benefits, but it also means you have many moving parts – containers, VMs, databases, networks – that can fail or misbehave at any time. The operations team was dealing with a barrage of alerts from monitoring systems, routine tasks like scaling services up/down, and security concerns such as detecting intrusions or anomalies in traffic. The vision was an **autonomous IT operations** environment, often dubbed **AIOps** (AI for IT Operations), where intelligent agents monitor the infrastructure, detect and diagnose issues, and even fix them without waiting on human intervention. Key goals included: **self-healing** (services recovering automatically from failures), **AI-driven SecOps** (security operations that can catch and respond to threats), and **intelligent resource optimization** (making sure the right amount of compute is running – not too little to avoid crashes, not too much to save cost).

In the early days, WebkinsTech's ops were reactive. If a server CPU spiked at 2am, an on-call engineer would be paged to investigate and maybe add more servers. If an outage occurred, they'd manually comb through logs to find the cause. If security noticed unusual login patterns, they might sift through data to determine if it was an attack. This doesn't scale. Inspired by how tech giants manage operations (Google's SRE discipline, Netflix's chaos engineering, etc.), WebkinsTech set out to create a suite of *intelligent microservice agents* to manage IT operations. Think of these agents as an automated NOC (Network Operations Center) team: watching metrics, predicting issues, and executing runbooks – essentially automating the DevOps and SecOps workflows.

Three primary agent-driven systems were envisioned:

1. **Monitoring & Anomaly Detection Agent** – an AI that learns normal patterns (CPU, memory, latency, throughput, etc.) and alerts when something deviates significantly, far more precisely than static thresholds.

2. **Self-Healing Orchestrator** – an agent that, upon detecting an issue (from the monitoring agent or other triggers), can take corrective action: restarting a service, scaling resources, clearing a queue, or rolling back a bad deployment, all automatically.

3. **Security Agent** – an AI security watchdog that analyzes logs and network traffic to flag suspicious activity (possible intrusion, DDoS, malware) and in some cases initiate containment (like isolating a compromised container or blocking an IP).

Additionally, a continuous **Resource Optimizer** was planned: an agent analyzing resource usage trends and optimizing cloud resources (turning off dev servers at night, consolidating workloads on fewer machines when possible, resizing clusters based on predictions) to save costs while ensuring performance.

Technical Architecture Breakdown

WebkinsTech's approach was to integrate these intelligent agents into their existing DevOps toolchain:

- They had **Prometheus/Grafana** for metrics and alerts, **ELK stack (ElasticsearchLogstash-Kibana)** for logs, and various scripts and runbooks for responding to incidents. The idea was to insert AI-driven analysis and decision-making into this pipeline.

They created an **"AI Ops Platform"** composed of a few microservices:

- **Anomaly Detection Service**: This service consumed time-series data from

Prometheus (e.g., metrics on CPU, memory, response times for each microservice) and log streams from Elasticsearch. It used machine learning (an LSTM neural network for time-series and a clustering algorithm for log patterns) to model normal behavior.

Rather than relying on static thresholds ("alert if CPU > 85%"), it would understand seasonality and correlations (maybe 85% at noon is normal, but 60% at 3am might be abnormal if usually it's 20%). When it detected an anomaly – say, a sudden spike in error rates or a gradually increasing memory usage trending to an out-of-memory – it would create an **Incident Event** describing the abnormal condition, with a confidence and some context (like which service, what metric deviated and by how much).

- **Incident Orchestrator Agent**: This is the self-healing brain. Implemented as a microservice (let's call it *AutoOps Service*), it subscribed to events from the anomaly detector and other monitors. It maintained a knowledge base of "if X, then do Y" actions – essentially an automated runbook. Importantly, many of these were generated through experience or encoded with the help of the ops team. For example: "If a service is unresponsive and CPU is 100%, initiate a container restart" or "If latency spikes and thread count is increasing, trigger auto-scaling by +2 pods." Initially, these were simple rules, but the agent also had a reinforcement learning component: it observed the outcomes of its actions. If restarting a service fixed the issue, next time it would have higher confidence in that action; if not, it would try a different action or escalate to humans.

- **Security Analysis Service**: A specialized agent focused on security logs and network data. It used models to detect things like a burst of 500 errors (could indicate an attacker probing APIs), unusual login times or IP addresses for user accounts, or known malware signatures in system logs. When something suspicious was found, it raised security alerts and could also call mitigation scripts.

- **Execution Handlers**: For each type of remedial action, there were small microservices or Lambdas that did the work (e.g., a **Scaler Service** that can add/remove instances via cloud API, a **Deployment Manager** that can rollback a release by redeploying an old version, a **Network Quarantine script** that can update firewall rules to block an IP, etc.). The Incident Orchestrator would invoke these as needed. This modular design meant the orchestrator didn't need to have permissions to do everything itself; it just sends high-level commands to these executors.

To illustrate, imagine a scenario: The anomaly detector notices Service X's response time has tripled in the last 5 minutes and error rates are climbing. It emits an event "High latency anomaly on Service X". The Incident Orchestrator receives this. Its policy/knowledge base recalls that Service X had a similar issue two weeks ago and a scale-out helped. It quickly queries the current load and sees CPU is indeed maxed and incoming requests have spiked (maybe due to a promotion driving traffic). The orchestrator decides to execute an auto-scale action: it calls the Scaler Service to add 3 more instances of Service X. It also notifies the team's chat (like a Slack message) that it did so, and continues to monitor. Latency returns to normal – issue resolved without anyone being paged.

Now consider a more complex case: The security agent flags that one of the database servers is seeing a lot of unusual queries (maybe a sign of SQL injection attacks) and CPU usage is erratic. The anomaly detector also flags odd behavior on that DB. The Incident Orchestrator might not have a predefined rule for "odd DB queries", so it goes into a diagnosis mode: it triggers a snapshot of the system metrics and deeper analysis – essentially running a **root cause analysis** routine. Perhaps it correlates that a particular service (say an old API endpoint) started misbehaving and flooding the DB with requests due to a bug. The orchestrator then has to decide: is this a bug or an attack? If the security agent's confidence in an attack is high, it might isolate that API service (kill its pods or block its network access) to stop the bleeding, then alert engineers. If it seems like a bug (no external threat detected, just a bad deploy), the orchestrator might automatically roll back that service to a previous version (using the Deployment Manager). In either case, it's doing in seconds or minutes what used to take an hour of human diagnosis at 2am.

To facilitate such decisions, WebkinsTech integrated a **knowledge graph** that linked metrics, services, and dependencies. This was populated from their service registry: e.g., Service X > uses Database Y -> runs on Host Z. So if DB Y is slow, it knows Service X might be cause or victim. This context made the AI ops more effective in pinpointing where to act.

Industry Insight: Large tech companies have reported significant gains with AIOps. Facebook (Meta) built an AI system to manage incidents and reportedly cut their incident resolution time by 50% . HCL Technologies, using an AIOps platform, reduced help desk tickets by 62% and cut mean time to recovery by 33% through automated incident handling . These statistics align with WebkinsTech's expectations – less firefighting, more proactive healing.

Tools and Technologies

WebkinsTech's AIOps implementation leveraged both home-grown components and existing tools:

- **Monitoring & Data Collection**: They stuck with **Prometheus** for metrics collection from all microservices and hosts, and **Fluentd/Elasticsearch** for log aggregation. The

AI agents pull data from these in real-time. They also enabled **distributed tracing** (using Jaeger) and fed trace anomalies into the anomaly detector (like if one microservice call in a chain suddenly slows down, it can be caught).

- **Machine Learning/AI**: For anomaly detection, they experimented with algorithms like **Facebook Prophet** for trend analysis and LSTM neural networks for more complex pattern recognition. They also used simpler statistical models (like ARIMA) as a baseline. The models were trained on historical data and then run in streaming mode. Python's **scikit-learn** and **TensorFlow** were used to develop these, and they packaged the models into a **PyTorch Serve** service for production. The security agent used some pre-trained models for known threat detection and also unsupervised learning (clustering log events to find outliers).

- **Event Bus**: The Incident events and commands were handled through a pub-sub system – **Apache Kafka** topics for "anomalies" and "actions". This decoupled the detection from the orchestrator and the actions, making the system more resilient (e.g., if an action fails to execute, the event remains and can be retried).

- **Runbook Automation**: They catalogued common procedures into scripts (Bash, Python) and used an automation tool (**StackStorm** or custom AWS Lambda functions) to execute them. For example, a script to purge a message queue, or one to rotate credentials if a leak is suspected. Each script could be triggered by the orchestrator agent.

- **Knowledge Base & Learning**: They maintained a repository (in Confluence or a database) of incidents and how they were resolved – the AI agent would parse this to improve its suggestions. Over time, they integrated this with a **graph database** (Neo4j) that represented their infrastructure and past incident relationships.

- **Security Tools**: They integrated with security platforms: e.g., feeding data to an IDS (Intrusion Detection System) and then using AI on top for analysis. They also used cloud security services like AWS GuardDuty, but augmented them with their own agent that cross-correlated GuardDuty alerts with application behavior.

- **Chaos Engineering**: As part of building trust in self-healing, WebkinsTech adopted Netflix's idea of chaos testing. They used **Chaos Monkey**-like tools (Gremlin or their own scripts) to randomly terminate instances or inject latency. The AIOps agents were thereby tested and trained in non-production environments to handle real failures. Netflix's Chaos Monkey is famous for pushing engineers to build robust, auto-healing systems

– WebkinsTech's philosophy was the same. If their self-healing agent could handle a random server kill, it gave confidence for real incidents.

Design Tip: Start with visibility, then automation. WebkinsTech's team made sure they had **excellent monitoring and observability** as a foundation. It's hard to automate what you can't see. They then automated small tasks (like simple restarts on crashes) before graduating to complex decisions. Importantly, they kept humans in the loop initially: the AI would suggest an action in an alert ("Service A is slow. Suggested action: restart pod.") and an engineer would confirm or adjust. This train-and-confirm approach over a few months trained the AI and also trained the team to trust the AI. Only then did they switch the AI agents to fully autonomous on certain actions. They also set **guardrails**: e.g., the auto-scaler agent could add at most 5 instances at a time, to avoid runaway scaling due to a mislearned model – a lesson learned from others who have seen AI automations go out of control.

Agentic AI in IT Ops

In this scenario, we effectively have *autonomous agents managing other services*. Each agent has clear objectives:

- **Ops Agent Objective**: maintain system uptime and performance.
- **SecOps Agent Objective**: protect the system from threats and breaches.

Resource Agent Objective: optimize for cost and efficiency without harming performance.

They perceive through metrics, logs, and events; they decide using a combination of rules and ML (often this is a form of **augmented decision support** – AI making recommendations and sometimes automating them); and they act by executing changes in the environment.

One could see parallels to a self-driving car: sensors (metrics) feed into a driving policy (the AI ops rules), which then control actuators (scale up, restart service, etc.). WebkinsTech's agents were effectively "driving" the infrastructure.

Crucially, these agents also **coordinate** with each other. The anomaly detector might work in tandem with the resource optimizer: e.g., if the optimizer wants to scale down a service to save money, it first checks with the anomaly detector if everything is stable (you don't want to remove resources during an ongoing incident). Similarly, the SecOps agent might detect an issue and communicate to the healing agent "don't restart this service, it might be compromised – isolate it instead." So there is a need for coordination logic which they built into the Incident Orchestrator as well. It's like having multiple specialists on an ops team – they have to talk and not step on each other's toes. The communication was via the event bus and also by updating a shared state (in a small database) marking an incident as security-related or performance-related, etc., which guided the behavior.

Impact Metrics

The autonomous IT ops initiative paid off substantially:

- **MTTR (Mean Time To Recovery)**: This is a key metric for incidents. WebkinsTech saw MTTR drop significantly. Before, a typical incident (like a memory leak causing a service crash) might take 30-60 minutes to diagnose and fix (mostly waiting for an engineer to wake up, log in, push a restart). With self-healing, many incidents were resolved in **under 5 minutes** automatically. Across all incidents, average MTTR went down by **~60%**. In fact, some classes of issues (like simple crashes, known failure patterns) were resolved in seconds. Studies in the industry indicate MTTR reductions of 50% or more with AIOps , which aligns with WebkinsTech's experience.

- **Incident Volume**: Interestingly, the number of high-severity incidents (ones that wake up humans) dropped. The AI filtered noise and handled minor issues, so the ops team only got paged for truly complex problems. WebkinsTech's monthly pages went down by about **70%**. One engineer joked that they started to get full nights of sleep regularly – a huge quality-of-life improvement for the team. HCL's case with Moogsoft saw 62% fewer help desk tickets ; similarly, WebkinsTech's automated resolution meant many problems never became tickets at all.

- **Proactive Prevention**: The anomaly detector often caught issues before they became customer-impacting. For example, it noticed memory usage creeping up release after release for a service, indicating a memory leak. It flagged this and the team fixed the bug proactively – avoiding a future outage. In another case, it identified a pattern where every Monday morning there was a traffic spike that caused slowdowns; the team used this insight to schedule an automated scale-up every Monday 8am, which the Resource Optimizer agent now handles. Overall, they counted dozens of incidents avoided thanks to early warnings, contributing to a higher **uptime** (their service SLAs improved by a couple of percentage points, which is big when you're aiming for 99.9% availability).

- **Security Improvements**: The security AI caught several intrusion attempts. On one occasion, an attacker tried a brute force login attack at 3am. The SecOps agent noticed the unusual burst of failed logins and automatically locked out the attacker's IP and account, then alerted the security officer. By the time the security team woke up, the attack had been stopped in its tracks. The system also identified misconfigurations (like an open port that was inadvertently left accessible) by noticing unusual port scan patterns and mitigated it. While measuring security is tricky, WebkinsTech can point to a **reduction in security incident response time** from hours (or never, if missed) to minutes. And potentially, it saved the company from one or two serious breaches (hard to quantify a breach that never happened, but the peace of mind is invaluable).

- **Resource Optimization (Cost Savings)**: The Resource Optimization agent helped cut cloud costs by about **15%**. It found idle resources and scaled them down during offpeak, and ensured new services were right-sized (developers often over-provision, but the agent would suggest smaller instance types if it saw consistently low utilization). For instance, it identified that a batch analytics job could run on spot instances (cheap spare capacity) instead of expensive on-demand instances, and automated that switch, saving thousands of dollars a month. This kind of dynamic optimization is something humans rarely have time to do manually on an ongoing basis.

Perhaps just as important as these metrics is the cultural shift: the ops team moved from a reactive "firefighting" stance to a more proactive, engineering-focused role. They now treat the AI ops platform as another part of the system to *train and maintain*. They periodically review what the agents did, improve the models or rules if needed (like tuning sensitivity to avoid false alarms). It's a virtuous cycle: the AI handles grunt work and obvious fixes, humans focus on improving the AI and handling only the truly novel or critical cases, which in turn makes the AI even better.

WebkinsTech's journey in AIOps demonstrated that **autonomous infrastructure management is feasible and hugely beneficial**. Systems can heal themselves and protect themselves to a large extent. With a robust backbone in place, the company felt confident to continue innovating on the feature side – particularly, improving customer-facing experiences with intelligence. Next, we'll look at how they delivered personalized and adaptive experiences to their users.

Case Study: Personalized and Adaptive Customer Experiences (Recommendations, Proactive Service, Dynamic Pricing)

Conceptual Overview

In the digital era, customers expect experiences tailored to them – think of how Netflix recommends shows or Amazon's site seems to anticipate what you need. WebkinsTech, which offers a range of smart home products and an online platform, wanted to provide a **personalized, adaptive experience** to each user to boost engagement and sales. This encompassed several aspects:

- **Recommendations**: Suggesting relevant products or content to users based on their behavior and preferences. For WebkinsTech, this meant recommending complementary products (e.g., if a customer bought a smart thermostat, suggest smart window sensors) or helpful content (tutorials, tips) on their platform.

- **Proactive Customer Service**: Anticipating customer needs and reaching out with help or offers *before* the customer asks. For example, if a device is showing errors in telemetry, proactively alert the user or schedule maintenance. Or if a user has been browsing an item repeatedly, maybe offer a discount or send more info.

- **Dynamic Pricing and Promotions**: Adjusting prices or promotions in real-time based on demand, inventory, or user segment. This can improve sales and utilization – e.g., offering a time-limited deal on a bundle if an AI agent predicts the user is pricesensitive.

- **Adaptive UI/Content**: Even the website/app experience could adapt – highlighting features the user cares about, changing the order of menu items based on usage, or using different marketing messages for different segments.

These capabilities required analyzing lots of data about user behavior, product usage (since WebkinsTech's devices were IoT, they had telemetry data), and market conditions. The microservices involved would need to feed into an AI-driven personalization engine.

WebkinsTech was essentially aiming to build its own version of what big players have: Amazon's recommendation engine, Netflix's personalized homepage, and Uber's dynamic pricing models, all rolled into one platform.

One challenge was to keep it **real-time and scalable** – personalization should happen on the fly (like generating recommendations as a user browses). Another challenge:

experimentation and feedback loops. Any AI-driven experience needs continuous learning and A/B testing, because user preferences change and models can drift. So WebkinsTech planned a framework where these personalization agents could be tested and improved continuously.

From a conceptual standpoint, they treated this as adding an "intelligence layer" on top of the customer interaction microservices:

- Instead of a static catalog service, you have a **Recommendation Service** that serves up items based on an AI model.
- Instead of fixed pricing, you have a **Pricing Service** with an agent adjusting prices or offering discounts optimally.
- The customer support or notification system works with an **Engagement Agent** that decides when to send proactive messages to users.

Technical & Architectural Breakdown

WebkinsTech's architecture for personalized experiences was event-driven and user-centric.

Key components introduced:

- **User Profile Service**: A microservice that aggregates everything known about a user (preferences, past purchases, interactions, device data). This provides the context for personalization. It pulls data from various sources: purchase history from Order service, browsing events from web analytics, device health from IoT backends, etc. It stores this in a profile DB (NoSQL for flexibility, e.g., DynamoDB or MongoDB).
- **Recommendation Service**: A core microservice backed by a machine learning model

(or set of models). When queried (e.g., "get top 5 product recommendations for user X"), it uses the user profile and other data to produce a list of items. The architecture here uses embeddings and similarity search. They trained a model using techniques like collaborative filtering (like how users similar to you bought similar things) and content-based filtering (matching product attributes to your profile). The model training pipeline ran offline (using Apache Spark and perhaps libraries like **Surprise** or TensorFlow Matrix Factorization) on purchase and browsing data. The serving part was an online system. They actually leveraged **Amazon Personalize** for an initial prototype, which is an AWS service that can generate recommendations, but eventually moved to their own custom model for flexibility.

- **Dynamic Pricing Service**: This was essentially an agent that could adjust prices within set boundaries. It took into account demand (how many people are viewing/buying an item), supply (inventory levels), and user-specific factors (loyal customers might see a slightly lower price or get a coupon). WebkinsTech wasn't going to show totally different prices to different users (which can be controversial), but they did use personalized discounts and timely offers. For instance, an AI model predicted the likelihood of a user purchasing at full price; if it was low, the system might offer a 10% coupon to entice them. This service was carefully governed by business rules to avoid inconsistency or perceived unfairness.

- **Engagement/Notification Service**: Powered by an AI that decides when and what to communicate to the user. For example, a **Retention Agent** monitored if a user hasn't logged in for a while and might send a "We miss you – here's what's new!" email. Or a **Device Health Agent** monitoring telemetry might proactively alert "Your device battery is low, time to replace" via the app. These proactive touches were subtle but aimed to improve user experience and reduce support calls. The AI part was deciding the right moment and message, using predictive models for user satisfaction and churn.

- **Experimentation Framework**: They embedded A/B testing in the fabric. Every recommendation or dynamic price could be part of a controlled experiment. They built a simple **Feature Flag Service** and experimentation library so that any new AI model or strategy could be tested on a fraction of users and results compared. Metrics like clickthrough-rate (CTR) on recommendations, conversion rate, revenue per user, etc., were fed back to evaluate effectiveness.

- **Real-Time Pipeline**: To feed all this, they set up a **Kafka (or Kinesis)** event stream for user actions – page views, clicks, add-to-cart, etc. The Recommendation and Pricing agents subscribed to relevant events to update their models in near real-time (for instance, if a product suddenly became hot, the dynamic pricing might raise its price or mark it as "trending"). They also used **Redis** as a high-speed store for some features, like real-time counts of views.

Let's illustrate with a use case: A user, Alice, logs into WebkinsTech's app. The front-end calls the Recommendation Service for her homepage. The Recommendation Service fetches Alice's profile (she owns a smart thermostat and camera, has browsed lighting products) and also checks trending items. The ML model (say a neural network that takes in user embedding and item embeddings) scores items and returns, for example: "Smart Light Starter Kit" and "Doorbell Camera" as top recommendations, each with a relevance score. It also calls the Pricing Service to get personalized prices for those (maybe Alice is a new user who hasn't bought lights before, so the Pricing agent decides to show a *bundle discount* on the Starter Kit to increase chance of conversion). The homepage shows "Recommended for you: Smart Light Kit – 10% off today!". Alice clicks it – that event goes into the pipeline, recorded as an interaction which will further train the model that this recommendation was successful.

Meanwhile, behind the scenes, Alice's smart thermostat has been sending data that her HVAC system is cycling a lot (maybe a filter is dirty). The Device Health Agent predicts a suboptimal usage pattern. The Engagement Service might proactively push a notification: "Noticing your HVAC is running frequently. Here's a tip: check your air filter – a clean filter saves energy." This is an adaptive service that provides value, not selling anything, but it increases Alice's trust and engagement with WebkinsTech's ecosystem.

On the dynamic pricing front, consider WebkinsTech running a flash sale. The Dynamic Pricing Service might adjust prices of certain items every hour based on sales velocity. If an item is selling out fast, the agent could slightly raise the price (or more likely, mark it as "limited stock, hurry!" rather than raise – depends on strategy). If another isn't moving, it might apply a bigger discount dynamically. These changes are all done via the Pricing microservice which the frontend and checkout consult in real-time. This is analogous to how Amazon changes prices millions of times a day to optimize revenue and inventory turn – an intelligence that WebkinsTech is emulating on a smaller scale.

Industry Example: Netflix is a master of personalization – over **80% of the content watched on Netflix comes from their recommendation engine** . Similarly, Amazon's dynamic pricing adjusts millions of product prices multiple times a day (approximately 2.5 million price changes per day on Amazon) , ensuring competitive pricing and maximizing sales. These figures validate the importance of AI-driven personalization: it directly drives user engagement and revenue. WebkinsTech's initiatives aimed for the same effect: more personalized recommendations to drive more sales (or usage), and flexible pricing to balance demand and supply.

Tools and Tech Stack

To build these capabilities, WebkinsTech leveraged:

- **Data Platform**: A big data setup with **Snowflake** (data warehouse) or AWS Redshift to store historical interaction data, which data scientists used to train models. For streaming, they used **Kafka** and also looked at **Apache Flink** for real-time processing (like updating recommendation models incrementally).

- **Machine Learning**:

For recommendations, they tried collaborative filtering with matrix factorization, and later deep learning approaches (like an autoencoder or a two-tower neural net where one tower is user embedding, another is item embedding, learned jointly – similar to what YouTube and others do).

- Python libraries: **NumPy/Pandas** for data prep, **TensorFlow** for model training, and **Annoy** or **Faiss** (Facebook's library) for fast nearest-neighbor search to get similar items.

- For dynamic pricing, they formulated it partly as a **multi-armed bandit** problem: try different price offers and learn which yields the best profit conversion tradeoff. They used a simple epsilon-greedy bandit algorithm at first, then moved to a contextual bandit that takes into account user segment.

- They also incorporated **Bayesian forecasting** to predict demand for products (to not slash price on something about to sell out). Tools like Prophet or even AWS Forecast could be used here.

- **Serving Infrastructure**: They containerized ML models behind REST endpoints. The Recommendation Service might call an internal API like /recommend?userid=123 which hits the model server. They optimized for low latency, caching results for a short time (seconds) where possible because recomputing from scratch for every request could be heavy.

- **Front-end integration**: The web/app had to be instrumented to consume these services. They used GraphQL to allow the front-end to fetch a bunch of personalized content in one go. For example, a GraphQL query could fetch recommended items, personalized banner message, and any alerts for the user, all aggregated. The microservices behind GraphQL would talk to the personalization services.

- **Personalization Database**: For quick lookups of who should get what content, a specialized store like **Redis** or **Amazon DynamoDB** was used. For example, if they have a daily deal personalized per user, they might precompute and store it in a cache keyed by user. This complements on-demand ML computation.

- **A/B Testing & Analytics**: They used **LaunchDarkly** or a homegrown feature-flag system to roll out models to a subset. Analytics from those experiments were stored and analyzed using tools like **Amplitude** or custom SQL queries in the data warehouse. They also built dashboards showing metrics like recommendation CTR, average order value changes with dynamic pricing, etc., to continually track the impact.

Tip: Keep a **feedback loop** between the AI and user behavior. WebkinsTech ensured that every time a recommendation was shown and whether it was clicked or not, that data fed back into model training. Similarly, they tracked if a user used a coupon or not, to refine the dynamic pricing strategy for that user segment. This closes the loop so the system gets smarter. A/B tests are critical – sometimes an AI feature you think is great might not actually perform with users. For example, an early version of their recommendation model increased clicks but not purchases (people clicked out of curiosity but didn't buy). Through testing, they adjusted the objective of the model to optimize for purchases rather than just clicks.

Agentic Behavior in Customer Experience

The AI in personalization can be seen as a set of **customer experience agents** working on behalf of the user and the business. These agents have goals like "help the user discover products they'll love" (recommendation agent) or "find the optimal price that the user is willing to pay and that maximizes revenue" (pricing agent) or "keep the user engaged and happy" (engagement agent). They perceive via user data and environmental data (like overall demand trends), they decide by predicting user preferences or responses, and they act by delivering a tailored experience to the user. They also continuously learn – a hallmark of an agent.

One interesting agent WebkinsTech implemented was a **Multi-armed Bandit Agent for content** on their homepage. The homepage had several slots for content (banners, product highlights). Instead of showing the same homepage to everyone or manually curating it, they set up a bandit agent to try different content variations (e.g., banner A vs B) for different users and learn which performs better in terms of engagement. Over time, each user gravitated to a more personalized homepage as the agent learned what they tended to click on or ignore. This is an agentic approach because it autonomously experiments and adapts the UI per user.

Another example: a **Conversational Agent** in customer support. WebkinsTech integrated a chatbot on their site that could answer questions about products and troubleshoot common issues. This chatbot used an underlying language model (not extremely advanced, but enough for a guided Q&A). It's agentic in that it holds a dialogue with users, navigating a conversation flow. If it got stuck or detected frustration, it could seamlessly hand off to a human support agent – a coordinated dance between AI agent and human agent. Over time, it learned from those hand-offs to cover more queries.

These agents also have to deal with the *long-term* aspect of user relationships. They're not just one-and-done transactions. For instance, the recommendation agent tries not to recommend the exact same item over and over if the user showed no interest (it will explore other options – using exploitation vs exploration strategies common in recommender systems). The engagement agent weighs when to contact the user; too many notifications could annoy them, too few and the user might forget the app. So the AI needs to find a balance (sometimes using multi-objective optimization: maximize engagement but minimize churn risk).

A crucial part of agentic behavior here is **personalization = personalization** at the individual level. The more data the agent gathers about a specific user, the better it can tailor its decisions for that user, effectively creating a unique experience per user. This is like each user having their own AI concierge agent in the system.

Impact and Metrics

WebkinsTech observed strong business outcomes from personalization:

- **Increased Sales/Conversion**: Product recommendation click-through-rate (CTR) improved by 30% after deploying the learned models (versus the old static "bestsellers" list). This led to a **15% increase in average order value**, as users often added recommended add-on items to their cart. For example, many customers buying one smart gadget ended up buying accessories suggested by the AI. This mirrors industry results where effective recommendations can drive a significant portion of revenue (Amazon has famously attributed a large percentage of sales to recommendations).

- **User Engagement and Retention**: The proactive tips and personalized content kept users returning. The monthly active users metric went up by around 10%, and user surveys showed that users felt "the app understands my needs better." There was a measurable drop in churn among users who engaged with personalized features; those receiving proactive device alerts, for instance, had higher satisfaction and were more likely to purchase new devices or renew subscriptions.

- **Dynamic Pricing Impact**: Through careful deployment, dynamic pricing and personalized offers increased conversion rates for hesitant customers by about 20%. For instance, sending a timely coupon to users who abandoned carts recaptured a lot of sales. The revenue uplift from these targeted promotions outweighed the discount costs. Additionally, during flash sales, the pricing agent helped sell out inventory in a controlled way (no huge stockouts or excessive discounting – it paced the sale). They also saw improved **margin control** – not giving blanket 10% off to everyone, but maybe 5% to some, 15% to others who needed it, preserving margin.

- **Adaptivity**: A subtle metric was how quickly they could react to market changes. When a new competitor product came out, the AI immediately noticed shifts in buying patterns and the content team used that info to adjust marketing. When certain product was trending on social media, the agents picked up the increased interest and automatically highlighted that product on the site. This agility – essentially real-time marketing adaptation – is something that would have been slow manually.

- **Customer Support Load**: The chatbot and proactive support reduced incoming support tickets by an estimated 25% for common issues, as customers either got selfserve answers or were preemptively helped. This saved cost and improved response times for issues that did require human support.

Perhaps the most telling result is that WebkinsTech's customers started to expect these intelligent touches. When they interacted with the platform, they got used to it "just working" for them – finding what they needed, getting help when they didn't even know they needed it, and feeling valued (through personalized offers). This kind of loyalty is hard to quantify but is golden for a brand.

By implementing personalized and adaptive experiences, WebkinsTech managed to create a user experience on par with much larger companies. In doing so, they further cemented the role of intelligent microservices (like Recommendation and Pricing services) as core parts of their architecture. These were not one-off features, but ongoing agentic systems requiring continuous learning and tuning. The success here gave WebkinsTech confidence to apply intelligent microservices thinking to even more areas, including how all their enterprise systems integrate.

Next, we examine how WebkinsTech approached enterprise integration and workflow orchestration with intelligence – essentially connecting all these services and external systems in a smart way.

Case Study: Enterprise Integration and Workflow Orchestration

Conceptual Overview

Modern enterprises like WebkinsTech rely on a multitude of systems: CRM, ERP, e-commerce platforms, third-party services (shipping providers, payment gateways), internal microservices, legacy databases, and more.

Integrating these systems so that data flows smoothly and processes span across them is a perennial challenge. In the past, companies used enterprise service buses (ESBs) or massive monolithic integration platforms.

WebkinsTech, being cloud-native, leaned into microservices and APIs from the start.

However, as they grew, they found the need for **workflow orchestration** that goes beyond a single domain – orchestrating across services and sometimes across organizational boundaries.

For example, consider a **customer onboarding workflow**: a new customer signs up

(handled by Auth Service), they should get an entry in CRM (Salesforce), a welcome email (via an Email service), and if they are a business client, maybe a setup of a dedicated instance (triggering DevOps scripts). This is a multi-system process. Another scenario: a **finance workflow** where at month-end, data from sales, payments, and support tickets are integrated to generate some reports or trigger invoicing – again crossing multiple systems. WebkinsTech also dealt with **partner integrations** – e.g., syncing product information with reseller systems or receiving orders from partner channels.

They realized that manually wiring these via point-to-point scripts was fragile. Instead, they wanted an intelligent orchestration layer that could handle workflows spanning many microservices and even external APIs, applying AI to make them more efficient. Intelligent integration could mean:

- Dynamically choosing which systems to call or which data to route based on context (rather than a static flow).
- Automating error handling – e.g., if one system is down, an agent could pause and retry later or route to a backup system.
- Mapping data between systems using AI (like auto-matching fields or using ML to transform data formats).
- Optimizing processes – say, if a certain step usually takes long, paralleling other tasks or reordering tasks for efficiency (with AI learning the best execution path).

Essentially, WebkinsTech wanted an **integration platform with brains**. This can be thought of as an evolution of traditional BPM (business process management) or orchestration engines with added decision intelligence.

Technical Architecture

They introduced an **Enterprise Orchestrator** microservice (or suite of services) to centralize these cross-cutting workflows. This was separate from the Order Orchestrator we discussed earlier; that one was focused on one domain (orders), whereas this is more generic and enterprise-wide. They leveraged open-source frameworks: a combination of **Apache Camel** (for integration patterns and connecting systems) and **Netflix Conductor** or **Temporal** (for long-running workflow orchestration).

- **Apache Camel Integration Routes**: Camel allowed them to define routes for data:

e.g., "when a new account is created in the Auth DB, send an object to Salesforce and to our Data Warehouse". Camel has many connectors, so they used it to bridge protocols (HTTP to SOAP, file to API, etc.). They embedded Camel routes within a spring-boot microservice, making it cloud-friendly.

- **Workflow Engine (Conductor/Temporal)**: For sequences of steps that form a business process, they used a workflow engine. **Netflix Conductor** is a microservices orchestration engine that fits well with their ecosystem. It represents workflows as a JSON of tasks (which can be microservice calls, human tasks, decision points). They also evaluated **Temporal.io** (from Uber's Cadence), which is code-first orchestration where you write workflow logic like code and it handles the state and retries behind the scenes. Either way, the idea was to have a robust way to manage stateful workflows that might even span minutes, hours, or days (e.g., waiting for an approval).

- **Intelligent Routing and Transformation**: They built an **AI Mapping Assistant** for data transformations between systems. For example, when integrating a partner's order feed into their Order Service, field names might differ (partner calls it "ItemCode" vs

WebkinsTech "ProductSKU"). They trained a simple NLP model to suggest mappings by analyzing field names and sample data. This agent saved engineers time when setting up new integrations by auto-mapping 80% of fields correctly, which they would then verify.

- **Decision Points with AI**: Within workflows, they identified places where an AI decision could optimize things. One use-case: a workflow for support ticket escalation. Normally: if a support ticket is not resolved in 2 days, escalate to L2 support. With AI, they made it smarter: an agent looks at ticket sentiment (from text analysis) and customer profile; if it predicts the customer is unhappy or it's a high-value customer, escalate faster or even immediately route to a specialist. This was implemented by inserting a call to a **Sentiment Analysis Service** (using NLP on ticket messages) and a **Priority Model**.

The workflow engine gets a score and decides the path (standard or fast-track).

- **Parallelism and Skipping**: Some processes had steps that could be done in parallel or even skipped given certain conditions. For instance, a vendor onboarding (from a previous chapter's example) might skip a manual review if an AI risk score is low. They configured the orchestrations to utilize these dynamic paths. The decision to skip or parallelize was often given by an AI agent that learned from historical durations. If it learned that waiting for Step B to finish before starting C is unnecessary 90% of the time, it suggests to run B and C in parallel unless B's outcome is critical.

- **Monitoring & Recovery**: The Orchestrator kept track of in-flight workflows and could recover from failures (Temporal does this well). If a step failed (say an external API didn't respond), it could retry or mark the workflow pending and resume when the system is back (using backoff strategies possibly influenced by AI if it knows typical downtime patterns of that API).

- **Human in Workflow**: For long processes that involve people (like an approval or review), the system would send a task to a human (e.g., via a task management UI or an email with a link to approve) and the workflow waits. They added an AI helper here too: providing a suggested decision to the human. For instance, in a partner contract approval process, an AI agent could review the terms and say "90% similarity to approved contracts, likely safe to approve" to assist the manager in deciding faster.

A concrete example of an orchestrated process: **Hardware RMA (Return Merchandise Authorization)** – When a customer wanted to return a product, this touched multiple systems: support (ticket), inventory (to expect a returned unit), finance (refund), and logistics (return shipping label). The Enterprise

Orchestrator handles this: once a return is authorized by support (or automatically if criteria met), it triggers:

1. Create a return order in Order system.
2. Notify Inventory at warehouse to expect item back.
3. Call Shipping API to email the customer a return label.
4. After item is received (event from inventory system), then trigger Finance to issue refund.
5. If item not received in 30 days, send a reminder or escalate.

The intelligent part: if the AI sees it's a low-cost item and probably not worth shipping back, it could skip directly to refund and tell the customer to keep/dispose of the item (some companies do this to save reverse logistics cost). That decision is based on rules+AI (cost threshold, customer trust score, etc.). The orchestrator consults a **Return Optimization Agent** that recommends whether to require return or not.

Another scenario: **Multi-system data sync** – nightly syncing of data between systems is often needed (for reporting or consistency). Instead of blind syncing everything,

WebkinsTech's intelligent workflow only pulls data that changed (AI change detection) and prioritizes important data first. It might even predict which data is likely to be queried next day (based on patterns) and pre-sync or cache it. This turned a heavy ETL batch into a smarter incremental process.

Industry Analogy: Many enterprises use **workflow orchestration** tools like Camunda, Conductor, or cloud services (AWS Step Functions, Azure Logic Apps) to manage business processes. What WebkinsTech did differently was infuse AI into this orchestration. This is akin to how Uber manages complex workflows (like the lifecycle of a ride) using Cadence/Temporal with dynamic decision making, or how Netflix's Conductor allows for dynamic task scheduling. In fact, Netflix Conductor is used by companies to run **billions of workflow executions per month** reliably . WebkinsTech's scale is smaller, but the principle is the same – a central brain coordinating microservices. Adding AI means the brain not only follows a script but also *learns* the best script over time or adapts it per case.

Tools and Stack for Integration

- **Connectors**: They built or used connectors for all major systems (Salesforce API connector, SAP connector, etc.). This involved secure credentials and data mapping configurations stored in a database so changes can be made without code changes.

- **Orchestration Engine**: They deployed Netflix Conductor on their Kubernetes cluster. Each microservice that participated was a "task" in Conductor terms. For external systems, they often created a microservice wrapper (e.g., a Salesforce Proxy Service) which Conductor would call, and that service encapsulated calling Salesforce with retries and returning results.

- **Temporal for long workflows**: For some developer-centric workflows (like orchestrating DevOps pipelines for new customer environments), they used Temporal which allowed writing the orchestration in code (go/java) and it handles state. The choice of two engines was because Conductor was great for JSON-defined flows that ops folks could tweak, while Temporal was used by developers inside internal tools.

- **AI in BPM**: For the AI decision nodes, they often just called out to an existing ML service (like the recommendation or fraud service) from within a workflow. But in some cases they embedded simple ML models into the workflow engine (Temporal has an SDK; they embedded a Python ML call via a service).

- **Data mapping AI**: This was a unique piece – they made a small service where you input two data schema (lists of fields, sample records) and it outputs field mappings suggestions. It used a combination of word embeddings for field names and statistical correlation on sample values. This was not a real-time service but a design-time aid. It sped up integration development and reduced errors.

- **Enterprise Service Bus**: They consciously avoided a classic ESB that could become a monolith/bottleneck. Instead, microservices + messaging acted as their "bus". For pubsub needs, Kafka topics were the bus (e.g., any new Customer event gets published, and multiple subscribers – CRM sync, email, etc. – consume it). This event-driven integration meant loosely coupled systems and easier scaling. The orchestrator was only used when there was a clear flow with dependencies; if not, events sufficed.

- **Error handling**: For integration, error handling is critical. They implemented a **generic retry with backoff** for transient issues and a **dead-**

letter queue pattern for messages that fail repeatedly, which then an ops agent (or a human) can examine. The AI could prioritize retries – e.g., if an order sync to a partner failed, an AI might gauge if it's critical to retry now or can wait (based on SLA or customer impact). More simply, they set different retry intervals depending on error type.

Tip: Document your workflows and make them visible. WebkinsTech used graphical tools for business folks to see the processes (Camunda Modeler for BPMN diagrams, and Conductor's UI to track running workflows). This transparency is key to trust – people see what the AI-driven process is doing. Also, keep humans able to **intervene**. They built an "ops console" where if an automated workflow got stuck or went wrong, an engineer or business analyst could manually push it forward, change a value, or cancel it. AI is great, but in integration, you need a manual override sometimes to handle the truly offnominal cases or to apply business judgment that AI doesn't have.

Impact and Benefits

By smartly integrating systems, WebkinsTech achieved:

- **Faster End-to-End Processes**: Workflows that used to take days of email back-andforth between departments were automated to minutes or hours. E.g., partner onboarding time went from ~2 weeks (multiple form fillings and approvals) to 2 days, largely due to automated data pushing and one-click approvals on a unified interface.

- **Reduction in Errors/Data Consistency**: Automated integration meant no more human data re-entry between systems, reducing errors significantly. The AI mapping ensured even when formats changed (a partner updated their CSV format unexpectedly), the system often auto-adjusted or at least alerted the need to adjust, preventing downstream failures. Data consistency (say between CRM and internal user database) went up – they measured a drop in discrepancies by 80%.

- **Employee Productivity**: Internal teams saved a lot of time. For example, the finance team didn't have to manually compile reports from different systems; the orchestrator ensured data was collected and processed. One finance workflow automated the entire revenue reconciliation process, saving ~30 hours of work monthly. Similarly, support teams didn't need to manually trigger tasks for other teams – the workflows handled it, and they only managed exceptions.

- **Scalability of Operations**: As WebkinsTech grew its customer base and partnerships, the number of integration points grew too. Because they had a systematic approach (with Camel and Conductor), they could onboard new partners or systems 2-3x faster than before. One metric: Onboarding a new reseller used to involve 5 IT systems and took 3 months of dev work; with their integration framework, they cut it to 1 month since so much was configuration and reuse of existing patterns.

- **Process Intelligence**: The instrumentation of workflows gave them data to improve processes. They could see bottlenecks easily (perhaps a certain approval always delayed things – so they either removed that approval or automated it with AI if possible). Over a year, they iterated multiple workflows to be leaner. In one case, they realized a certain check was never actually causing rejection (the AI agent pointed out the check always passed), so they eliminated that step entirely, streamlining the process.

- **Resilience**: The intelligent orchestration made integrations more resilient to system outages. If Salesforce was down for maintenance, the orchestrator would queue changes and apply them when back up, without losing data or needing human reminder. Before, a downtime often meant some records fell through the cracks and were lost; now the system guarantees eventual consistency. This improved trust in the data across departments.

- **Cost**: There was also an infrastructure cost benefit. Instead of needing a heavy enterprise integration platform license (which can be expensive), they achieved this mostly with open-source and cloud-native services. And by efficient routing of data (only changes, compressing payloads, etc.), they saved on bandwidth and processing.

The integration and orchestration case might not sound as glamorous as AI recommendations or robots, but it is the *plumbing* that makes all other intelligent services work in concert. WebkinsTech's investment here paid off by giving them a highly agile backend. It's like having an efficient nervous system in an organization – signals go where they need to, quickly and intelligently.

Having covered integration, the next case focuses specifically on the **customer experience systems** – tying together some of the personalization we saw, but also looking at how intelligent systems directly interact with customers (like conversational AI and support systems) to elevate the customer experience.

Case Study: Intelligent Customer Experience Systems

Conceptual Overview

WebkinsTech prided itself on customer satisfaction. Beyond personalized content and recommendations, they sought to make every interaction with the customer smart and seamless. This case study zooms in on how they used agentic AI in customer-facing systems like support chatbots, voice assistants, and interactive guides – essentially building an **intelligent customer experience (CX) ecosystem**.

Areas covered include:

- **AI-powered Customer Support**: Deploying chatbots and voice bots to handle common queries, available 24/7, with smooth handoff to human agents when needed. These bots aren't just scripted; they use Natural Language Processing (NLP) to understand a wide range of user inputs and can learn from past conversations.

- **Sentiment Analysis & Adaptive Response**: The system listens to customer sentiment (in text or voice) and adapts the tone and urgency of responses. An angry customer might get fast-tracked to a senior agent or a soothing message.

- **Personalized Self-Service**: On the support portal or app, showing customers tailored troubleshooting steps based on their device data. For instance, if WebkinsTech's cloud sees a user's device is offline, the support site might proactively show "It looks like your device is offline. Would you like to check connectivity tips?" – addressing the issue before the customer even asks.

- **Omnichannel Integration**: The AI ensures that whether a customer contacts via phone, chat, or email, the experience is consistent and context is shared. If they start with the chatbot and then call, the phone agent (or AI voice) knows what happened in chat already.

- **Dynamic FAQs and Help Center**: Using AI to maintain the help knowledge base – analyzing what issues are trending to suggest new FAQ entries, or even generating answer drafts for support articles using language models.

WebkinsTech also leveraged their AI to **train customer service agents** (the human ones). For example, an Agent Assist tool that listens to calls or reads chat in real-time and suggests information or next best actions to the human agent. This improves the speed and quality of service.

The conceptual end-goal: a customer should feel like WebkinsTech service is *always available, instantly helpful, and even anticipatory*. Agentic AI components working together can deliver that feeling by handling straightforward tasks autonomously and augmenting human employees for complex tasks.

Technical and Architectural Breakdown

The intelligent CX system comprises several integrated services:

- **Chatbot Service**: A microservice (or set of microservices) that handles chat interactions on web or mobile. It uses an NLP engine, such as **Google Dialogflow**, **Rasa**, or **Microsoft Bot Framework**, to parse user messages. WebkinsTech opted for an open-source NLP (Rasa) initially to keep data in-house and highly customizable. The chatbot has an underlying **Intent-Entity model**: it classifies what the user wants (intent) and extracts key info (entities like product name, error code, etc.). It then routes to a **Dialog Manager** agent that decides how to respond or which backend service to call.

- **Voice Bot/IVR**: They extended the same brain to voice. Using speech-to-text (perhaps AWS Transcribe or Google Speech API) to understand spoken words, then feeding it to the same Dialog Manager, and using text-to-speech (Amazon Polly or Google TTS) to reply in a friendly voice. This became their intelligent IVR (interactive voice response) system on the support hotline. Instead of "Press 1 for X", customers could just say what they need and the AI routes them.

- **Integration with Backends**: The chatbot isn't useful if it only gives canned responses.

WebkinsTech connected it with various backends. For example:

- If a user asks "What's the status of my order?", the bot calls the Order API to retrieve that user's order status and responds with "Your order is shipped and will arrive Tuesday."

- If a user says "I'm having trouble with my camera," the bot queries device telemetry to see if it's online, and might guide the user through steps ("I see your camera is offline. Let's try rebooting it. Did that fix the issue?").

- For procedural issues, the bot can create a support ticket, schedule a repair, or process a return by interacting with those respective services.

- **Context and Memory**: The chatbot and voice bot maintain context of the conversation.

This is done via session state stored in Redis or in the microservice's session memory. It allows multi-turn dialogues where the user doesn't have to repeat information. For instance:

- User: "I need help with my thermostat."
- Bot: "Sure, what's the issue you're facing?"
- User: "It's not connecting to Wi-Fi."
- Bot knows we are talking about thermostat and can give targeted help like "Let's try resetting the Wi-Fi settings on your Smart Thermostat. (Provides steps)".

This context handling is crucial and the AI uses it to refine intent detection as well.

- **Sentiment and Escalation**: They integrated a **sentiment analysis model** (could be a simple off-the-shelf model or an API like Azure Cognitive Services sentiment) to every chat and call. If a user's language or tone indicates frustration (negative sentiment score beyond threshold), the system flags it. The Dialog Manager has a policy: if sentiment is very negative or user says something like "This is not helpful" or "I want to talk to a human", the bot will gracefully transfer. For chat, it notifies a human agent to step in (with all the conversation context visible to them). For voice, it might say "I'm sorry you're having a hard time. Let me connect you to a specialist." and then transfer the call. This ensures AI doesn't inadvertently worsen a bad situation by stubbornly trying to help when the customer is done with it.

- **Agent Assist**: When a human does take over (either via chat console or call center software), an **Agent Assist Service** kicks in. This service listens (or reads) the live conversation and uses AI to fetch relevant info quickly. For example, if the customer says "I've tried resetting and it still blinks red," the

Agent Assist might pop up knowledge base articles about "blinks red" error for that product. It might also show the customer's profile and recent device data so the human agent doesn't have to dig around.

Essentially, it's like an AI sidekick for the human support agent, reducing time to search and respond.

- **Knowledge Base AI**: Maintaining FAQs and help articles is labor-intensive. WebkinsTech set up a **Knowledge Miner Agent**: it analyzes support tickets and chat logs to identify common issues and how they were resolved. It can suggest new FAQ entries when it finds a new pattern. They even experimented with a **GPT-style model** fine-tuned on their support articles that could draft answers to new questions, which support staff then review and publish. This speeds up the creation of help content and ensures it addresses real user problems that the AI sees trending.

- **Customer Feedback Loop**: After interactions, they often ask customers to rate the experience. These ratings, plus the conversation transcripts, feed back into training data. If the chatbot failed to solve an issue, that chat becomes training data to improve it or to expand its dialog flows. They treated every failed bot interaction as something to learn from. Over time, this means the bot deflects more queries successfully. (They measure deflection rate – how many queries handled fully by bot vs needing human.)

- **Multilingual Support**: WebkinsTech had a growing international user base, so they employed translation APIs for languages that their models didn't natively support. In essence, a user could type in Spanish; the system translates to English, processes with the English-trained NLP, then translates the answer back to Spanish. For voice, similarly, they used language-specific speech recognition where available. Eventually, they trained separate models for high-volume languages, but translation-as-a-service allowed quick expansion with acceptable accuracy.

Industry Insight: Many companies have moved to AI-driven customer service. A 2024 study showed about **20-30% of customer service interactions are now handled by AI** , and advanced chatbots can resolve up to **70% of customer requests** fully .

WebkinsTech's strategy aligns with this trend – offloading routine tasks to AI. For example, Bank of America's virtual assistant "Erica" reached millions of users, and others like AIbased IVRs have cut down call center loads. These tools save costs and often improve response time (no waiting on hold for a simple query). However, getting the balance right (and not annoying customers) is key.

Tools and Tech Stack

- **NLP & Language Models**: Initially, they used Rasa (which uses spaCy, scikit-learn, etc.) for intent/entity. As transformer-based models (BERT, etc.) became more accessible, they fine-tuned a BERT model for intent classification which significantly improved understanding of varied user utterances. They also integrated **GPT-like models** for open-ended Q&A: for instance, if a user asks a really complex question not in the training data, they could fall back to a large language model connected to their knowledge base.

- **Speech**: For voice, they used cloud APIs for speech recognition and TTS to avoid reinventing that heavy wheel. AWS, Google, Azure all have robust speech services. Latency was a consideration, but they found it acceptable with proper caching of frequent phrases in TTS.

- **Messaging Platforms**: The chatbot was deployed on their website, mobile app, and even as a Facebook Messenger bot and WhatsApp (through Twilio) – to meet customers where they are. They built an abstraction such that the core bot logic is channel-agnostic. Connectors handle the differences (e.g., WhatsApp text vs web widget).

- **Call Center Integration**: They integrated the voice bot with their call center software

(Twilio Programmable Voice or a similar platform) to do the IVR and transfer calls. For Agent Assist, if using a platform like **Zendesk** or **Salesforce Service Cloud**, they built a sidebar widget that interfaces with their Agent Assist recommendations via REST API.

- **Databases**: Conversation logs (after anonymization) were stored for analysis and model training (in a secure manner). They also kept a real-time session store for active contexts.

- **Security & Privacy**: They took care to comply with privacy concerns – any PII going through AI was protected. For example, the conversation AI would mask credit card numbers or account passwords if a user tried to provide them, and politely say it cannot process that or redirect to a secure form. Regulatory compliance (like GDPR for user data) was accounted for by allowing users to request chat transcripts deletion, etc.

- **Continuous Improvement**: They set up continuous monitoring of bot performance. Key metrics: containment rate (how many sessions handled without human), CSAT for bot interactions, fallback rate (how often the bot says "sorry, I don't understand"). These were dashboards the team watched. They retrained the models regularly with new data. They also employed **crowd-sourcing** at one point: using a service to generate more training phrases for intents where the bot struggled, to widen its language understanding.

Tip: Avoid making the AI seem too robotic or, conversely, misleadingly human.

WebkinsTech decided to be upfront that the chatbot is an AI assistant named "Lumi". They gave it a friendly persona but made sure users know they can ask for a human anytime. They also tuned it to say "I'm learning" or offer to get human help when it hit its limits, rather than going in circles. Transparency and graceful fallback are key to maintaining trust in AI-driven CX systems.

Impact and Outcomes

The intelligent CX systems yielded strong positive results:

- **Higher Self-Service Resolution**: Within 6 months of launching the improved chatbot and knowledge base, WebkinsTech saw **roughly 60-65% of customer inquiries handled without a human agent**. This includes the bot fully answering questions or guiding users to solutions, and customers finding answers in the dynamic FAQ. This deflection meant human agents could focus on complex issues. Even when humans handled, the **Agent Assist shaved off ~20% of handling time** on average, because agents had relevant info at their fingertips. This translated to shorter call durations and faster chat responses, improving support team productivity.

- **Customer Satisfaction**: Initially, some users were skeptical of the bot, but as it improved, **CSAT scores for chatbot interactions rose to over 85%** for those queries it handled. Many customers appreciated instant answers at any hour. The sentiment analysis also ensured that unhappy customers got to a human quickly, which in turn improved overall satisfaction (few things annoy customers more than being stuck with a dumb bot when they know they need a person). The seamless transfers with context meant customers didn't have to repeat themselves – a big satisfier.

- **Faster Response Times**: With AI handling the front line, wait times plummeted. The majority of chats were picked up immediately by the bot. Human response time for escalations improved too because the queue was smaller and Agent Assist made them quicker. On phones, the intelligent IVR meant many callers never had to wait on hold at all for simple requests like order status or basic troubleshooting – they got what they needed from the voice bot. This gave WebkinsTech the ability to advertise 24/7 support and quick answers as a selling point.

- **Cost Savings**: By automating a large chunk of support, WebkinsTech saved on customer service costs. They didn't lay off staff (they repurposed some to more valueadd outreach roles, like customer success), but as volume grew, they didn't need to hire as many new agents as they would have. They estimate a savings of **$500k/year** in support operation costs through AI automation and efficiency.

- **Insights into Customer Needs**: The AI analysis of conversations provided great feedback to product and UX teams. They learned what features confused users the most, what requests kept coming (leading to new features or clarity in the app). For example, many users asked if WebkinsTech's devices could integrate with a competitor's system. The AI flagged this trend and it influenced the roadmap to add that integration, which in turn reduced those support questions. In essence, the intelligent CX system doubled as a user research tool, aggregating pain points.

- **Global Reach**: With multilingual support, WebkinsTech was able to offer help in 5 languages within a year, expanding their marketability globally without needing full native-language support teams in each region. The AI translation and models handled most of it, with a smaller centralized team overseeing quality.

The intelligent CX initiative showed that agentic AI can not only streamline internal processes but also directly elevate the customer's experience with the brand. WebkinsTech's brand reputation benefitted; customer reviews often mentioned the helpful chat assistant or quick problem resolution. By combining human empathy where needed with AI efficiency for routine matters, they struck a solid balance.

Now, having intelligent systems at the front-end and back-end, WebkinsTech turned attention to ensuring their infrastructure can sustain it – particularly making it **self-healing and highly reliable**. (We touched on AIOps already in case 3, but in the next case we'll focus on selfhealing at the infrastructure architecture level, such as resilient patterns and chaos engineering principles that were employed.)

Case Study: Self-Healing Infrastructure and Operations

Conceptual Overview

(This case has some overlap with Case 3 on AIOps, but here we focus on architectural patterns and practices specifically for self-healing in production at scale, including chaos engineering and resilient design.)

WebkinsTech's platform now consisted of dozens of microservices, AI agents, and external integrations, all running across a mix of cloud infrastructure and edge devices. With so many components, failures were not a matter of "if" but "when". Self-healing infrastructure means that the system can **automatically recover from failures** without human intervention and continue providing service, or at least degrade gracefully. While Case 3 discussed the AI agents that react to incidents, here we highlight the design decisions and tools that inherently make the infrastructure resilient.

Key principles and components of self-healing architecture:

- **Redundancy**: Every critical service runs in multiple instances (and usually in multiple zones). If one instance or even one entire data center goes down, others take over.

- **Automated Recovery**: This includes auto-scaling groups and orchestrators like

Kubernetes that will restart or replace failed components. If a container crashes, Kubernetes will spin it back up. If a VM goes down, an auto-scaling group creates a new one.

- **Circuit Breakers and Timeouts**: Within microservice communication, use of circuit breaker patterns (like in Netflix's Hystrix library or Istio service mesh) to stop cascading failures. For example, if Service A is calling Service B and B is slow or down, the circuit breaker in A trips after a few failures and A falls back (maybe serving cached data or a default response) instead of hanging and piling up requests.

- **Chaos Engineering**: Testing the system's ability to heal by intentionally causing failures. WebkinsTech adopted this to build confidence that their self-healing works.

- **Auto-Scaling and Auto-Tuning**: Not just scaling for load, but scaling in response to unhealthy instances (e.g., if one instance is not responding well, scale out and then remove the bad one).

- **Healing Workflows**: Some scenarios require coordinated recovery steps. For instance, if the database fails over to a replica, all dependent services might need to clear their DB connections and reconnect. They built automation for that sequence.

- **Self-healing Data Pipelines**: Ensuring data processing jobs resume at last checkpoint on failure so you don't have to restart from scratch manually.

WebkinsTech approached this systematically. They aimed for a state where **most infrastructure issues are resolved within seconds or minutes without paging anyone**. Human on-call would then primarily handle novel issues that the automation couldn't, often during working hours thanks to how much was handled automatically.

Technical Implementation

- **Kubernetes Orchestration**: All stateless services were on Kubernetes (EKS on AWS, for example). Kubernetes provides built-in healing: if a pod or node dies, it reschedules pods on healthy nodes. They defined readiness and liveness probes for each microservice container, so Kubernetes would automatically restart a container if it became unresponsive (failed its liveness check) or stop sending traffic to it if it's not ready (readiness check fails). This alone solved a ton of issues like hung processes and memory leaks (the liveness probe kills them and they restart fresh). It also autoreplaced nodes via the cluster autoscaler integration if a whole VM went bad.

- **Auto-Scaling Groups**: For components not on K8s (like some databases, or legacy VMs, or certain stateful services), they used AWS Auto Scaling Groups with health checks. For example, their Cassandra database cluster on EC2 had a setup where if one node failed a health check, AWS would replace it with a new one (then Cassandra's own replication would catch the new node up).

- **Service Mesh**: They deployed **Istio** service mesh which among other features, provided circuit breaking and retries at the network layer. So if Service X tried to call Service Y and Y wasn't responding, Istio could automatically retry the call to perhaps another instance of Y. If Y is down, the circuit breaker prevents X from waiting forever by failing fast after a timeout. They configured timeouts and retry logic mindful of not overwhelming services. This significantly reduced customer-facing errors: if a transient glitch occurred, often the end-user never noticed because a retry succeeded quickly.

- **Load Balancing and Failover**: They used global load balancers (like AWS Route 53 with health checks, or Azure Traffic Manager) to route traffic to multiple regions. WebkinsTech had primary deployment in one region and a secondary in another. If the primary went down, Route 53 would detect it and shift traffic to secondary (DNS failover). Within a region, they used ELBs/ALBs that health-check service instances and only send traffic to healthy ones.

- **Stateful Services Resilience**: For databases, they employed replication and automated failover. For example, their PostgreSQL was on Amazon RDS with Multi-AZ – if primary fails, standby is promoted automatically. Their Redis cache used cluster mode so if one shard failed, it had a replica ready. These managed solutions handle the heavy lifting of failover which triggers in < minute.

- **Self-Healing Scripts**: They codified common remediation steps into scripts triggered by monitoring alerts (somewhat like what we described in AIOps). For instance, if CPU is high and thread count high, automatically restart the pod (K8s does this anyway if you configure HPA or liveness). If disk space low on a VM, they had a script to clear certain caches or alert to add volume (some things still alerted humans if risky, but many housekeeping tasks were automated). They integrated these with CloudWatch alarms and SSM Automation documents for AWS.

- **Chaos Engineering Tools**: They ran Chaos Monkey in their staging (and selectively in production with careful guardrails). They also used **Gremlin** (a chaos SaaS) to orchestrate more complex failure tests: e.g., "kill 50% of service A pods randomly during peak load" or "blackhole traffic between service B and C for 2 minutes". They scheduled frequent chaos tests and as per Netflix's practice, treated it as normal. This forced the team to ensure no single component failure would break the system's SLAs. It also often revealed unexpected dependencies, which they then fixed (like adding a fallback). Over time, chaos tests went from causing issues to being mostly uneventful, indicating resilience.

- **Monitoring and Alert Tuning**: Part of self-healing is not alerting on every little blip, only when automation couldn't handle it. They tuned alerts to have longer time windows and to check if automation already kicked in. For example, if a service restarts automatically and recovers, the system might not even alert a human, just log it and maybe create a low-priority report. Alerts that do go to on-call are things that require human insight, like "Repeated restart of Service X 5 times in 10 minutes" (so the self-healing is struggling, indicating a deeper problem). This way, humans aren't distracted by issues that the system resolved.

Industry Example: Netflix's Chaos Monkey and the Simian Army are famous for enforcing self-healing practices. They intentionally turn off servers randomly in prod; any weakness in auto-recovery gets exposed . Many companies, like Amazon, Google, have similar approaches (Google's SRE culture pushes automation – "toil" must be automated). Studies have shown significant reductions in major incidents thanks to automated remediation. One financial firm reported that using self-healing scripts and AIops reduced critical incidents by 60% and prevented several outages proactively.

WebkinsTech followed these best practices to achieve high availability.

Outcomes and Metrics

WebkinsTech's self-healing infrastructure achieved:

- **Improved Uptime**: The platform's availability moved closer to the coveted **99.99% (4 nines)**. In concrete terms, total downtime per month was reduced to just a few minutes, often zero perceived by customers. Even during incidents, many were transparent to users. For example, when they chaos-tested a whole zone outage, the system failed over and users maybe experienced a brief slow-down but not a full outage.

- **Faster Recovery**: Mean Time to Recovery (MTTR) for infrastructure issues came down dramatically. Automated restarts happen within seconds. Failover to backups in database or region typically within a minute. Compare that to prior manual recovery that could take 15-30 minutes or more if humans had to wake up and react. This speed meant even when issues occurred, they often did not escalate into major incidents. The **blast radius** of problems was contained quickly.

- **Fewer On-call Interruptions**: Thanks to automation and better alerting, the on-call team saw about a **70% reduction in pages** (similar to stat from earlier AIOps case). Most nights, nothing paged because things either didn't break or self-healed. This not only helps the team morale and fatigue, but also indicates a stable system. One engineer commented that after implementing chaos engineering and auto-healing, "incidents became non-events" – a testament that the system either avoids or handles failures gracefully.

- **Confidence to Deploy & Change**: Paradoxically, knowing the system heals itself gave the team more confidence to push updates and innovate (since any mishap would likely be corrected automatically). They increased deployment frequency because the risk of a bad deploy causing prolonged outage was lower (auto rollback on failures, etc.). This led to faster time-to-market for features – a side benefit of resilience is agility.

- **Cost Trade-offs**: Self-healing sometimes means duplicating resources (like running redundancy). WebkinsTech did invest in that – e.g., running two instances instead of one for every service at minimum, keeping warm standbys. This has a cost, but they found the cost of downtime (lost customers, SLA penalties in B2B deals, etc.) far exceeded the infrastructure cost. Also, with the resource optimization agent (Case 3), they ran lean when possible (e.g., auto-scaling down in off hours, but still maintaining redundancy). The improved reliability likely contributed to revenue indirectly by keeping customer trust high.

- **Notable Incident**: A real-world scenario proved the value: a major AWS outage in one region occurred (something beyond their control). Because WebkinsTech had multiregion failover, their system switched to another region within a minute or two. While many competitors were down for hours due to that AWS issue, WebkinsTech stayed up. They informed customers of a "degraded redundancy" (running on secondary site) but service continued. This incident turned out to be a huge validation; afterwards, sales teams used it to reassure enterprise clients that WebkinsTech had resilience in place beyond typical. It also spurred some clients to adopt WebkinsTech's products because they saw how well it handled the cloud outage.

In summary, WebkinsTech's self-healing infrastructure meant the company could scale and innovate without fear of stability issues. It created a virtuous cycle: robust infrastructure supports rapid changes, which then feed back improvements into the infrastructure needs. It's worth noting that self-healing isn't a one-time project but an ongoing practice. The chaos engineering continued, the automation library grew with new scenarios, and the team held **post-incident reviews** even for auto-resolved incidents to learn how to prevent them entirely. The culture became one of **"automate everything"** and expecting resilience.

Having now looked at nearly all aspects of WebkinsTech's operations – from document workflows to customer experience and infrastructure – our final case zooms out to the **supply chain** and how AI optimizes it, and then **financial services transformation**

(payments/fraud, etc.). These will round out the enterprise case studies by covering outwardfacing logistics and industry-specific AI applications.

Case Study: Intelligent Supply Chain Optimization

Conceptual Overview

WebkinsTech, aside from being a tech company, also deals with physical products – they manufacture and sell IoT devices. This means they have a supply chain: sourcing components, manufacturing or assembly, inventory management, distribution to warehouses, and ultimately delivery to customers or retail partners. The supply chain is ripe for AI optimization because it involves forecasting (demand planning), optimization (inventory levels, reorder points, logistics routes), and real-time adjustments (responding to delays or demand spikes).

Inspiration came from giants like Amazon and Walmart that use advanced algorithms to run ultra-efficient supply chains. WebkinsTech's scale is smaller, but the principles can be applied:

- **Demand Forecasting**: Use AI to predict how many units of each product will be needed in each region for upcoming weeks/months. This improves purchasing of components and manufacturing planning, reducing stockouts or overstock.

- **Inventory Optimization**: Deciding where to stock inventory (which warehouses, how much safety stock) to meet service levels at lowest cost. AI can find patterns and simulate outcomes to advise on inventory distribution.

- **Supply Planning**: Coordinating with suppliers (some overseas) on deliveries of components. Perhaps using predictive models to anticipate supplier delays (maybe based on news or historical performance) and proactively adjust orders.

- **Logistics and Route Optimization**: For delivering products, ensure shipping routes are optimized. If WebkinsTech manages its own delivery fleet for local deliveries or installation services, AI can optimize those routes (similar to how UPS's ORION system does for drivers, famously reducing left turns to save fuel).

- **Maintenance and Quality**: In production, monitor machine data on assembly lines with AI to predict failures or quality issues (like predictive maintenance – fix machines before they break, and detect if a batch of products might be defective via anomaly detection).

- **Adaptive Supply Chain**: If a disruption occurs (port delays, sudden demand surge due to viral popularity, etc.), have agents that simulate and recommend mitigations (like expediting certain shipments or finding alternate suppliers).

WebkinsTech's fictional narrative had a moment where their supply chain was challenged – say a sudden surge in demand for a new product. Traditional planning was slow to react, leading to backorders. After implementing intelligent supply chain optimization, they were able to respond in near real-time, reallocating inventory and increasing production swiftly.

Technical and Architectural Breakdown

They introduced an **AI-driven Supply Chain Control Tower** concept – essentially a central dashboard and brain that oversees supply chain metrics end-to-end and triggers actions.

Components:

- **Demand Forecasting Model**: Using historical sales, marketing info (promotions planned), even external data (like seasonality, trends), they built models to forecast demand per product per region per time period. They experimented with classical timeseries models (ARIMA, exponential smoothing) but moved to machine learning approaches (random forests and gradient boosting with features like Google search trends, etc.), and eventually some deep learning (perhaps LSTM for multi-variate time series). They also evaluated AWS Forecast which leverages AutoML on time series.

The outcome was forecasts with confidence intervals.

- **Inventory & Production Optimization**: They formulated this as an optimization problem: given forecasts and current inventory, how to meet demand at minimum cost (cost includes holding cost, backorder penalty, ordering cost). This is classic operations research. They used **linear programming** or ILP solvers (like Google OR-Tools) to compute optimal inventory levels and reorder plans. However, they added an agentic twist: an **Inventory Agent** continuously monitors actual demand vs forecast and can override or adjust the plan. For example, if an AI detects demand is trending 20% above forecast for one product, it will trigger an earlier reorder or shift stock from one warehouse to another in anticipation.

- **Supplier Integration**: They integrated their system with suppliers (via EDI or API) so that orders and shipment statuses flow automatically. They added an AI that scores suppliers on reliability (late delivery frequency, quality issues). This **Supplier Risk Agent** could suggest increasing orders with a more reliable supplier if it predicts another might delay. Also, if it predicts a component shortage (maybe a news scraper agent sees an earthquake in region where their chip factory is), it alerts procurement to secure stock from alternate sources.

- **Logistics Optimization**: They used route optimization algorithms for delivery trucks (if they have any) and also to plan inter-warehouse transfers. A **Routing Agent** employed algorithms similar to the traveling salesman or vehicle routing problem solvers. This agent reduced mileage and time. They were inspired by UPS saving millions of gallons by optimizing routes . WebkinsTech's scale: maybe routing techs for on-site installation or moving goods between 3 warehouses – still, they reported savings of, say, 10% mileage by better routing.

- **IoT and Asset Tracking**: They put IoT trackers on high-value shipments. Data from these (location, temperature for sensitive goods, etc.) streamed in. An AI monitored it to ensure conditions are good (e.g., "notify if any package is delayed by >2 days or if temperature out of range"). This is part of a **Logistics Monitoring Agent** to catch issues in transit and react (like expedite a replacement shipment if one is compromised).

- **Digital Twin & Simulation**: They implemented a simple "digital twin" of their supply chain – basically a simulation model (maybe using AnyLogic or custom Python simulations) that could answer what-if scenarios. The AI agents would run simulations if something big changed (like "what if demand doubles next month?") to test if the supply chain could cope and where bottlenecks would occur. This is agentic in that an agent can simulate and then autonomously recommend actions (like "we would need to expedite additional 1000 units from factory now to meet that potential spike").

- **User Interface**: A Supply Chain Dashboard for planners showed AI forecasts, recommendations (like "produce 5000 units in Q3 for Product X") with rationale, and allowed override. Often the AI would do the heavy lifting and planners would review and approve. Over time, trust grew and many decisions became fully automated (especially operational ones like reordering common parts when inventory drops to threshold predicted by AI, that just happened without manual approval).

Data was key. They consolidated data from sales (orders), marketing (promo calendar), inventory systems, manufacturing (ERP data), etc. into a data lake for the AI to crunch. Tools like **Apache Spark** helped process large historical datasets to train models. The OR optimization might use CP-SAT solver from OR-Tools or CPLEX if needed for integer programming.

Industry Stats: Supply chain AI has proven benefits. McKinsey research indicates AIbased forecasting can reduce supply chain forecasting errors by 20-50% , leading to significant efficiency gains (one stat: 65% boost in efficiency through fewer stockouts and overstock) . Danone (the yogurt company) used AI for demand predictions and saw a 30% reduction in lost sales due to better availability . WebkinsTech similarly saw big improvements: fewer missed sales (stockouts) and less excess inventory tying up capital.

Impact Metrics

WebkinsTech's intelligent supply chain optimization yielded:

- **Higher Fill Rate / Fewer Stockouts**: Product availability (fill rate) improved significantly. They reduced stockouts by around **30-40%**. Before AI, a sudden demand surge of a hot product might catch them off guard, and customers would face backorders. After AI, forecasts captured those surges better (for instance, the AI caught that a positive review on a major tech site caused a spike in demand and adjusted forecast). As a result, sales lost due to stockouts went down. In concrete terms, annual lost sales due to lack of stock dropped by, say, $2M to $1.2M, meaning $800k recaptured.

- **Lower Inventory Holding**: Despite better availability, they actually lowered overall inventory levels by about **15%**. The optimization cut out excess safety stock that wasn't truly needed and identified slow-moving products where production could be reduced. This freed up cash and reduced warehousing cost. One particular slow product, the AI recommended to discontinue manufacturing until existing stock sold down, which human planners hadn't realized as they were over-conservative. This avoidance of overproduction saved significant waste.

- **Faster Lead Times**: Order-to-delivery times improved because the right products were in the right place. The routing agent and warehouse optimization shaved off delivery times by optimizing fulfillment source. For example, an order that previously shipped from a far warehouse due to stock imbalance now ships from a local warehouse as the AI pre-positioned stock. Average delivery time to customers went down by 1 day in many regions, improving customer satisfaction.

- **Cost Savings**: Multi-faceted: reduced expedite shipping costs (since fewer last-minute rush shipments were needed thanks to better planning), reduced warehouse storage costs (less excess stock), and even manufacturing efficiencies (smoother production runs instead of erratic changes). They estimated overall supply chain operational cost went down by **10%** after these initiatives, while handling greater volume. A notable area was logistics: the route optimization and better carrier selection (the AI would pick the most efficient shipping method for each order) cut transportation costs by around 8%. And as mentioned, inventory holding cost reduction saved hundreds of thousands of dollars.

- **Agility and Resilience**: When the COVID-19 pandemic disrupted global supply chains (a scenario many companies struggled with), WebkinsTech's AI system helped them adapt. It quickly adjusted forecasts downward for certain product lines (e.g., less demand for office-related devices during lockdowns, more for home devices) and identified alternate suppliers when one country went into lockdown. While competitors went out-of-stock for months, WebkinsTech managed to keep products flowing with minimal delays by reallocating inventory globally (the digital twin simulations guided these moves). This agility in crisis is hard to put a number on but likely saved the business from severe revenue hits. It also validated the **resilience** of having AI monitor and adjust the supply chain far faster than monthly executive meetings could.

- **Supply Chain Coordination**: An unexpected benefit was better alignment between sales, marketing, and operations. The AI forecasting system became a neutral ground truth everyone trusted, reducing internal debates. Marketing learned to feed promotions data to the model so it would forecast lifts, sales trusted the inventory would be there, etc. This harmony improved execution of product launches and big sale events (like Black Friday, which the AI forecast and prepared inventory for with high accuracy).

In summary, WebkinsTech's supply chain became a competitive advantage instead of a bottleneck. They could meet customer demand more reliably and operate leaner. The CEO could confidently state that their fulfillment was on par with much larger companies, thanks in large part to intelligent automation and AI.

Finally, we'll look at how WebkinsTech applied intelligent microservices in **financial services transformation**, specifically focusing on fintech aspects like payments and fraud, tying back to earlier glimpses of fraud detection in orders but expanding to how an AI-driven approach transformed their finance operations (internally and maybe as a product offering).

Case Study: Financial Services Transformation

Conceptual Overview

WebkinsTech isn't a bank, but like many companies, it handles financial processes – accepting payments, processing refunds, extending maybe financing options for large purchases, and guarding against fraud. Additionally, WebkinsTech's transformation in financial processes can mirror what happens in the fintech/banking world (since we're blending inspirations like Stripe, which revolutionized online payments, and perhaps internal finance processes).

Areas of financial transformation through intelligent microservices:

- **Payments and Fraud Detection**: Implementing an intelligent payment microservice that can process transactions and use AI (like a fraud scoring model) to detect fraudulent transactions in real-time, similar to how Stripe Radar works. Also optimizing payment routing (if one gateway is down or more expensive, route to another).

- **Credit/Financing Decisions**: If WebkinsTech offers installment payment plans or credit to customers, use AI to underwrite those quickly (like a mini credit score based on user data).

- **Revenue Assurance and Anomaly Detection**: Use AI to scan financial transactions and catch anomalies (e.g., a sudden spike in refund requests could indicate fraud or an issue, or revenue leakage if something isn't charging correctly).

- **Robotic Accounting**: Automating reconciliation of accounts, using AI to match records that aren't obviously matchable, similar to JP Morgan's COIN that saved 360k hours on contracts .

- **Customer Finance Analytics**: If WebkinsTech has a fintech product or just for internal analysts, provide AI-driven insights like sales predictions, customer LTV (lifetime value) calculations, etc., to drive financial strategy.

WebkinsTech decided to treat payments/fraud as a microservice in their architecture rather than entirely outsourcing, to have more control and integrate it tightly with their systems. Inspired by **Stripe's approach** of embedding machine learning in payments (Stripe's network effect: training on data across millions of transactions to spot fraud patterns), WebkinsTech built a fraud model that learned from not just their data but industry patterns (perhaps using a consortium dataset or open data of known fraud signals).

They also drew on how banks use AI for things like anti-money laundering (AML) transaction monitoring – though WebkinsTech's scale is smaller, the concept of scanning transactions for unusual patterns applied to their context (preventing abuse of their platform, or detecting if someone's account might have been stolen based on purchasing behavior).

Technical & Architectural Breakdown

- **Payment Microservice**: All transactions (whether on website, mobile, or via partners) funnel through a Payment API. This microservice integrates with external payment gateways (Stripe, PayPal, etc.) but encapsulates the logic so WebkinsTech can switch providers or use multiple. It stores transaction records, handles retries (e.g., if a charge fails due to network, try again), and crucially, calls the **Fraud Detection Service** before finalizing a charge.

- **Fraud Detection Service**: This is an AI service that scores transactions in real-time. It takes features like user ID, payment instrument, geolocation, amount, items in order, historical behavior of user, etc. The model could be a gradient boosted trees model or a neural network. It outputs a fraud likelihood score. Based on this:

- If low risk (below a threshold), proceed normally.
- If high risk, either block the transaction or flag for manual review.
- If moderate risk, maybe require additional verification (the Payment service can trigger 2FA or ask for CVV or something if risk is moderate).

The Fraud service continuously learns. They feed it outcomes: which transactions later turned out fraudulent (chargebacks, etc.) to refine the model. It also uses unsupervised techniques to flag novel patterns (perhaps clustering transactions and finding an outlier group).

- **Rules Engine & Blacklists**: In addition to ML, they have some rules (e.g., block transactions from a list of known fraudulent IPs or if a device fingerprint is on a watchlist). The system combines rules and ML.

- **Transaction Monitoring Agent**: Beyond point-of-sale fraud, an agent monitors aggregates: e.g., if one account is doing hundreds of small transactions or if there's a sudden burst of failures, etc. This agent can detect things like a possible card testing attack (fraudsters test stolen cards by doing small transactions). It would then automatically put that account on hold and notify security.

- **Credit Decision Agent**: Suppose WebkinsTech offers "Buy Now Pay Later" plans for expensive installations. When a customer opts for that, a Credit Decision microservice evaluates them. Instead of a lengthy credit check, it uses an AI model that considers the customer's purchase history, maybe a soft credit pull, and other behavioral data to instantly approve or decline or set a credit limit. This uses techniques similar to fintech lenders – trained on historical loan outcomes if available. Because WebkinsTech has limited data, they might rely on a partner or standard credit scoring plus some custom risk modeling for their customer base.

- **Robotic Accounting/Contracts**: For internal finance, they used AI (like JP Morgan's COIN) for things like reading vendor contracts or lease agreements. They uploaded these documents to an NLP service that extracts key terms (price, dates, clauses) automatically into their ERP. This saved legal and finance teams time. They also automated invoice processing from vendors using the doc processing from case 1.

- **Financial Analytics**: They built dashboards with predictive analytics: e.g., an LTV model predicts how much revenue a new customer will bring in 2 years (helpful for marketing spend decisions). A churn prediction model flags customers likely not to renew a subscription service so that marketing can intervene with retention offers. These kind of models aren't directly part of a microservice flow, but they're delivered via a **Financial Analytics Service** that periodically scores customers and outputs results

to the CRM or marketing tools, effectively closing the loop between data science and business action.

- **Blockchain and Ledger**: (As a buzzword, they might experiment with a blockchainbased ledger for auditing transactions – but likely not needed, just ensuring immutability in logs sufficed. However, mention as transformation that they considered modern tech.)

- **Security and Compliance**: They integrated compliance checks like PCI DSS (for handling payments) – using tokenization so that no sensitive card data persists in their system (outsourced to Stripe for storage). They also use AI to assist compliance: e.g., scanning transactions against sanction lists (like how banks do AML). Probably overkill for WebkinsTech's scale, but if they expanded internationally, an AML agent could check if any transaction involves an entity under sanction by using external data.

Industry Example: Visa and Mastercard use massive AI systems to prevent fraud globally. Visa reported blocking **$40 billion in fraudulent transactions in one year using AI**, an 80 million transactions prevented . Stripe's Radar system evaluates over 1000 factors per transaction and has saved businesses countless dollars in fraud. WebkinsTech's fraud system is smaller but punching above its weight by leveraging similar techniques on its own data. Also, banks like JPMorgan use AI in back-office; JPMorgan's COIN example where a machine does in seconds what took lawyers 360,000 hours is a strong case for automating financial review tasks. WebkinsTech likewise automated their contract and invoice analysis.

Impact and Results

- **Fraud Loss Reduction**: Before, WebkinsTech faced some fraud: stolen credit cards used to buy products (which then would be chargebacked, costing them money). After implementing the AI fraud screen, they saw a **50% reduction in fraud-related losses**. The system caught many fraudulent orders – either blocking them or requiring more info which the fraudster didn't provide. Of course, some still got through (cat-andmouse), but it was a significant improvement. For instance, fraud rate went from 1% of transactions to 0.4%, saving perhaps $200k annually in chargebacks. Importantly, it did this without increasing false positives too much – they tuned it to minimize rejecting legitimate orders. With careful threshold tuning and occasional human review for borderline cases, customer impact was minimal (i.e., they didn't annoy too many real customers with false fraud flags).

- **Payment Success and Cost Optimization**: The Payment microservice's smart routing meant if one gateway was down or declining (which happens), it could retry on backup gateway. This improved successful payment rate by a couple percent, which directly translates to revenue. Additionally, it could choose gateway by transaction type to minimize fees (some gateways are cheaper for certain cards or geographies). Those savings, while small per transaction, added up. Let's say a 1% fee saving on $50M annual transactions = $500k saved.

- **Faster Refunds and Financial Operations**: Automation in finance led to quicker processing of refunds and adjustments. Customers got their refunds days faster, improving trust. Internal financial close processes (monthly closing of books) became quicker as many reconciliations were automated. The finance team closed books in 3 days instead of 6, freeing them to do more analysis rather than grunt work.

- **Credit Offering Uptake**: By offering instant financing decisions (with reasonably accurate risk assessment), WebkinsTech saw an uptick in sales of high-end bundles (because customers found it easier to buy with installment plans). The AI-driven credit approval had a decent approval rate with manageable defaults (the model kept risk at bay). This added maybe 5% incremental revenue on expensive product lines, which might not have sold as much without financing.

- **Audit and Compliance Efficiency**: Auditors were impressed that WebkinsTech could produce detailed transaction trails and that anomalies were flagged in real-time. It simplified compliance reporting since the AI already categorized and monitored transactions for issues. The risk of major financial misstatement or unnoticed fraud internally (like employee fraud) also went down – the anomaly detection would catch unusual patterns (e.g., an employee issuing weird refunds to themselves).

- **Development of Fintech Capabilities**: The investments here positioned WebkinsTech almost like a fintech in some respects. In fact, they spun some of these capabilities into customer-facing features. Example: they built a small **analytics dashboard for their corporate clients** (WebkinsTech sells to businesses too) which uses the AI models to show those clients their usage patterns, projected needs, etc. This was a value-add service derived from their internal analytics. Another: they marketed their fraud prevention as very robust, which made some hesitant customers more comfortable to buy direct from them instead of via a reseller (trusting the site's security).

- **Stripe Analogy**: They aren't becoming a payment provider to others, but by achieving an integrated, intelligent payment flow, they mirrored in small what Stripe does at large. For instance, their fraud system might not have Stripe's global network advantage, but it was tailored to their context, catching patterns specific to their business (like fraudulent orders often had certain product combos or shipping patterns, which their model caught). Overall, it saved money and protected the business's financial integrity.

This financial transformation case demonstrates that intelligent microservices aren't only for tech or ops – they deeply benefit finance and risk management as well. WebkinsTech effectively modernized its financial processes to be as data-driven and automated as the rest of the company, reducing errors and fraud, speeding up processes, and enabling new business opportunities (like financing offers).

With all nine case studies explored, we have seen how WebkinsTech applied agentic AI and microservices in diverse areas:

1. Document Processing & Validation – automated paperwork with AI.

2. Complex Process Automation – orchestrated fulfillment and claims with intelligent agents.

3. Autonomous IT Ops – self-healing and AIOps keeping systems running.

4. Personalized CX – recommender systems and dynamic pricing delighting customers.

5. Enterprise Integration – smart workflows tying everything together.

6. Intelligent Customer Experience – AI chatbots and support for superior service.

7. Self-Healing Infra – resilient architecture that recovers from failures.

8. Supply Chain Optimization – AI-driven forecasting and planning boosting efficiency.

9. Financial Transformation – secure, optimized payments and finance operations.

Each case delivered significant improvements in efficiency, speed, or innovation. In the next section, we synthesize these outcomes, highlighting common patterns, tools, and lessons across all these use cases.

(End of case studies. Next, the chapter likely concludes with a synthesis table and summary of patterns.)

Synthesis of Use Cases: Patterns, Tooling, Benefits, and Insights

Let's summarize the key patterns, technologies, benefits, and insights from these nine intelligent microservice case studies:

Use Case	Key Patterns & Techniques	Tools & Tech Stack	Primary Benefits	Key Ins
Document Processing & Validation	- Intelligent document pipeline (OCR + ML validation)- Human-inloop for lowconfidence cases- Inservice AI agent deciding approval vs. escalation	- OCR: Tesseract, AWS Textract- NLP for entity extraction- Workflow via queues & events (e.g., S3 + Lambda trigger)- Validation microservice + Rules engine (business checks)	- **80–90% automation** of data entry - Processing time cut from days to minutes- Reduced errors & manual labor (saved ~$1M/year)	- Start w assistin humans gradual approve confiden grows- C schema between service others a swappin compon easily - feedbac greatly i accurac time
Complex Process	- Orchestrator microservice	- Workflow engine:	- **50% faster** order processing,	- Orche a blend

Use Case	Key Patterns & Techniques	Tools & Tech Stack	Primary Benefits	Key Ins
Automation (Order Fulfillment & Claims)	coordinating multistep workflows- AIdriven decisions at	Camunda or Netflix	many exceptions auto-resolved-	and AI – handle w can't (dy data-dri

	forks (inventory allocation, fraud check, etc.)- RPA integration for legacy steps	Conductor- ML models for demand prediction, fraud (Lemonade-like claims AI) - Integration via REST + Kafka events- RPA bots (UiPath) for legacy system updates	Claims settled instantly in simple cases (some in seconds by AI) - 40% less manual effort in process management	parts)- **parallel** Lemona second shows t potentia humans oversigh decision critical f initially, trust
Autonomous IT Ops & Infrastructure (AIOps)	- Monitoring agent with anomaly detection (predictive)- Self-Healing Orchestrator triggers runbook actions- SecOps agent for security anomalies- Continuous learning from incidents (feedback into models)	- Monitoring: Prometheus, ELK- ML: Anomaly detection (Prophet, LSTM), clustering- Incident Automation: StackStorm or custom Lambda handlers- ChatOps integration (notifications in Slack with AI suggestions)	- **60%+ reduction** in MTTR (incidents auto-resolved in <5 min) - 70% fewer after-hours pages (automation fixed issues)- Improved uptime (99.99% achieved, fewer outages)	- Autom "boring" humans new iss trust by suggest before automa ("human loop" AI chaos enginee validate healing Netflix C Monkey ensure agents c handle r failures

Use Case	Key Patterns & Techniques	Tools & Tech Stack	Primary Benefits	Use Case

	continuous model tuning (experimentation platform)	epsilon-greedy, contextual bandits)- Personalization DB (Redis for quick lookups)- AB testing: Feature flags (LaunchDarkly)	loyalty (10% MAU increase)- Optimized pricing boosted revenue ~5% and moved slow stock	
Enterprise Integration & Workflow Orchestration	- Central orchestration of cross-service workflows (Temporal/Conductor)- Event-driven integration (Kafka pubsub for loose coupling)- AI-assisted data mapping between systems- Dynamic workflow branching via decision agents	- Orchestrator: Netflix Conductor (scaled to billions of workflows) - Apache Camel for integration routes (connecting Salesforce, ERP, etc.)- AI for mapping/ETL (NLP to match fields across schemas)- Human task	- 2–3× faster partner onboarding and system integrations- 80% reduction in data errors between systems (AI mapping improved data consistency)- Processes that took days now finish in hours (e.g., automated multi-system	**Enterprise Integration & Workflow Orchestration**

		integration in workflows (with portal)	report generation)	

Use Case	Key Patterns & Techniques	Tools & Tech Stack	Primary Benefits	Key Ins
		Sunshine, Salesforce AI)	satisfaction with bot-assisted answers)	and con update support tuning th (they kn custom e points b
Self-Healing Infrastructure & Operations	- Redundancy and failover design (multiAZ, multi-region deployments) - Kubernetes for selfhealing containers (auto-restart, autoscale)- Circuit breakers & timeouts to prevent cascades (resilience patterns)- Chaos engineering practices to continually test resilience	- K8s, Docker (autoscheduling, liveness probes)- Service Mesh (Istio) for circuit breaking and automated retries- AWS Multi-AZ, Route 53 for DNS failover across regions- Chaos Monkey / Gremlin to inject failures regularly	- Achieved **99.99% uptime**; no single points of failure- Most failures autorecovered in seconds (e.g., pod restart)- Zero major outages in last year; system proven resilient even during cloud provider issues- On-call stress way down (few pages, mostly handled by system)	- Design failure fr start: as compon fail and mitigatio mantra) **tests** ar "practice outages the team system for real Invest in monitor can quic failures trigger a recover of detec reaction minimal

Intelligent Supply Chain Optimization	- Demand forecasting with ML (time-series + exogenous factors)- Inventory optimization via simulation/OR (optimal stock levels per location)- Supply chain digital twin for scenario planning- AIdriven routing and logistics (vehicle route optimization, minimizing fuel/cost)	- Forecasting: Prophet, XGBoost, LSTM on sales data - OR tools: Google OR-Tools for LP/IP solutions- Simulation: AnyLogic or custom Python sims for "what-if"- Telematics/IoT for tracking shipments; route opt	- 20–50% reduction in forecasting error => fewer stockouts & overstock- 15% reduction in inventory holding costs (leaner inventory with same service level)- On-time delivery % improved (from 92% to 98%, for example) due to better inventory placement and	- Accura forecast game-c improve accurac directly to efficie seen wi Danone lost sale reductio both ML human j for dem planning augmen planner fully rep

Use Case	Key Patterns & Techniques	Tools & Tech Stack	Primary Benefits	Key Ins
		algorithms (CVRPTW solvers)	routing- Saved ~10% in logistics costs (optimized routes, fewer expedites)	(planne exceptio does bu calculat **respon supply** (with AI turns dis into man events r than cris agility is importa optimiza

Financial Services Transformation	- Fraud detection microservice with realtime scoring (payment fraud, account takeovers)- Automated underwriting/credit decisions using ML- Robotic process automation in finance (AI for invoice matching, contract analysis)- Anomaly detection in financial transactions (to catch errors or internal fraud)	- Fraud models: Decision trees/ML on transaction data (inspired by Stripe Radar, Visa's AI) - Credit scoring: logistic regression or XGBoost on customer data + bureau data- Doc analysis: NLP (for contracts, using similar to JPM COIN) - Finance RPA: tools like Automation Anywhere for reconciliation tasks	- Fraud loss rate down by ~50% (saved hundreds of thousands; more secure transactions) - 360k hours equivalent manual effort saved in contract processing (legal/finance AI automation)- Faster payment processing & fewer chargebacks; higher approval rates by smart routing- Finance team can close books 2x faster, focus on analysis over data entry	- **Trust verify**: humans reviewin samples fraud de ensure blocking custome (calibrat positive you han significa paymen conside leveragi network intellige (Webkin s internal maybe consorti for fraud fraudste multiple needs w Automa office yi savings Morgan example it; Lumin saw sim

Use Case	Key Patterns &	Tools & Tech	Primary Benefits	Key Ins

	Techniques	Stack		
				Proportion their size

Cross-Cutting Takeaways: Across these diverse cases, some common themes emerged. First, **data is the fuel** – investing in good data pipelines and feature engineering was critical for every AI agent's success. Second, **starting with a human-in-the-loop and gradually increasing autonomy** proved to be a safe and effective strategy; it built confidence in the AI and allowed the systems to learn from human expertise. Third, **integration of AI with microservices** was key – wrapping AI in well-defined services meant it could be reused and scaled, and decisions could be plugged into workflows easily. Tool-wise, WebkinsTech often combined open-source solutions (like Kubernetes, Kafka, PyTorch) with cloud services (like AWS Textract, managed databases) in a *cloud-agnostic but pragmatic* way, echoing a pattern of using the right tool for the job while avoiding lock-in where possible.

Finally, the **benefits** consistently showed up as improvements in speed (automation -> realtime or faster cycle times), efficiency (cost savings, less manual work), quality (fewer errors, more consistency), and agility (ability to adapt quickly to change). These translate to tangible business value: cost reduction, revenue increase, better customer satisfaction, and the ability to scale operations without linear cost growth.

Each case study reinforced that the journey to intelligent microservices is as much about culture and process as it is about tech. WebkinsTech cultivated a culture of **continuous improvement and learning** (e.g., post-mortems feeding ML improvements, A/B tests driving product changes). They treated AI agents as "colleagues" in the workflow – each with a role, monitored and trained – rather than magic boxes. This human-centric approach to AI deployment ensured that technology remained a tool serving business goals, not an end in itself.

In conclusion, WebkinsTech's experience – alongside industry analogies from Netflix, Amazon, UiPath, Stripe, and others – illustrates how intelligent microservices can transform an enterprise end-to-end. The patterns in these case studies provide a roadmap for any organization looking to inject AI into their architecture: start small, embed intelligence where it adds clear value, ensure interoperability of these smart services, and relentlessly measure impact to iterate. The result is an enterprise that is not only automated, but **autonomously improving** – a true realization of an *agentic-AI approach* to microservices. The next chapter will reflect on the broader implications of this transformation, from organizational readiness to ethical considerations, as we wrap up the Intelligent Microservices journey.

chapter 17: Ethical Considerations

Imagine handing over critical decisions – like approving a loan or diagnosing a patient – to an autonomous software agent. It sounds like science fiction, yet this is exactly what *agentic microservices* enable. These intelligent microservices can make decisions faster and at a scale no human team could match. But with that power comes a pressing question: **what if the AI gets it wrong, or worse, acts unfairly?** A misstep by an autonomous system isn't just a bug to be fixed; it could mean a family is unfairly denied a mortgage, or an emergency response is delayed. In other words, the stakes are high. This is why ethics aren't a "nice-tohave" in AI systems – they're absolutely critical.

We've spent previous chapters learning how to build smart, autonomous microservices that can learn, adapt, and act. Now, it's time to step back and ask not *"Can we build it?"* but *"Should we build it, and how do we build it responsibly?"* **Ethical considerations** are the compass guiding us through this new territory. Just as a map is useless without a sense of direction, powerful AI microservices are risky without a guiding ethical framework. When we talk about ethics in agentic systems, we're essentially talking about **trust** – trust from users, from customers, from regulators, and from society at large. If people don't trust an AI-driven service to be fair, safe, and transparent, they won't use it. It's that simple.

Let's consider why this matters in practical terms. Traditional software does exactly what developers explicitly code it to do. If it fails, we debug the code. But an autonomous microservice might learn its behavior from data or evolve over time. It might develop strategies its creators never anticipated. This unpredictability is part of what makes it intelligent – and part of what makes it **risky**. The AI could gradually drift into **unethical behaviors** (say, preferring one type of user over another in subtle ways) without anyone explicitly programming it to do so. By the time anyone notices, hundreds of thousands of decisions could have been made. Clearly, we need more than just functional testing; we need **ethical guardrails**.

Moreover, as these microservices become more *agentic* – meaning they have more autonomy to act on their own – they start to resemble members of our team or organization.

Just as you would train and supervise a new human employee, you must train and **oversee AI agents**. If a new hire at your company started making customer-affecting decisions without guidance, you'd step in quickly. The same goes for AI. We must imbue our digital "employees" with our values and ensure they act accordingly, especially when no one is directly watching over every decision.

Ethical deployment isn't just about avoiding negative headlines or regulatory fines; it's also about creating **better technology**. An AI service designed with fairness, transparency, and accountability in mind will likely be more robust and reliable. Think about it: if you anticipate potential biases and correct them early, your service's decisions become more equitable and defensible. If you build in transparency, users and engineers can understand and refine the system more easily. In contrast, ignoring ethics can lead to brittle systems that might *technically* work but fail in real-world acceptance. In essence, ethics is about **aligning our AI's behavior with societal values and expectations** – something that purely technical metrics won't capture.

In this chapter, we will explore the multifaceted ethical challenges and responsibilities that come with deploying agentic AI within microservices. We'll discuss how to **deploy these systems responsibly**, how to detect and mitigate **biases to ensure fairness**, and why **transparency** (making AI's decisions explainable) is key to building trust. We'll examine the importance of keeping a **human in the loop**, even as our microservices become more autonomous, to provide oversight on critical decisions. Issues of **accountability** will be front and center: when an AI makes a decision, who is responsible and how do we audit its actions? We'll also delve into **data privacy**, because an AI system that doesn't respect personal data can quickly cross ethical lines. Beyond the technology itself, we'll consider the **broader societal impacts** – how these systems affect jobs and communities, and what responsibilities we have as creators to mitigate negative outcomes.

Throughout, we'll follow the story of a fictional company, **WebkinsTech**, as it grapples with these issues while rolling out intelligent microservices. WebkinsTech's journey will give us concrete examples – from cautionary tales of biased algorithms to heartening wins where ethical design saved the day. By the end of this chapter, you'll not only understand the ethical pitfalls to watch out for, but also gain practical guidance on building AI-powered microservices that are not just *smart*, but also *wise* and *just*.

Ethics from Day One: “We learned that ethics isn’t a feature you add at the end; it’s a foundation you build from the start,” says Ava, the CTO of WebkinsTech, reflecting on an early project where a lack of oversight nearly led the company astray. This insight underscores a central theme of this chapter: ethical engineering is proactive, not reactive.

With that principle in mind, let’s dive into the specific ethical considerations, starting with how to **responsibly deploy** our agentic microservices in the real world.

Responsible Deployment of Agentic Microservices

Designing an ethical AI system is only half the battle – the other half is **deploying it responsibly** into the real world. Even the most well-intentioned AI can cause harm if it’s unleashed carelessly. So, how do we introduce our agentic microservices to production in a way that maximizes benefit and minimizes risk? It turns out that responsible deployment looks a lot like the careful rollout of any high-stakes feature, with some extra precautions given the autonomy of these systems.

Start small, then scale: One key principle is gradual deployment. Rather than flipping the switch for all users on day one, we roll out the autonomous service to a limited audience or in a controlled setting first. At WebkinsTech, for example, when the team developed an AIpowered customer support microservice, they didn’t immediately let it handle all customer queries. Instead, they started with an *internal pilot*. The AI agent first served only

WebkinsTech’s own employees in a test environment – answering common IT helpdesk questions – while human support staff watched closely. This “training wheels” phase allowed the team to observe how the agent behaved with real (but forgiving) users and edge cases, without the risk of upsetting actual paying customers.

Caution: Once an AI is in production, its mistakes are real mistakes affecting real people. You can’t treat an autonomous microservice like a typical software deployment where a bug might cause a minor inconvenience. An ethical failure (like a biased decision or a privacy breach) can have serious consequences – from harming someone’s livelihood to damaging your company’s reputation. Always assume your AI needs careful supervision initially.

Monitoring and feedback: During a limited rollout, it's essential to actively monitor the AI's decisions. In Chapter 12, we discussed observability – now that capability becomes your safety net. Set up dashboards and alerts specifically for ethical indicators. For instance, WebkinsTech's support agent was monitored for metrics like customer satisfaction scores, the frequency of cases it had to hand off to a human, and even the sentiment of customer feedback. If any metric signaled a potential problem (say, a sudden spike in unhappy feedback), the deployment team was ready to pause the rollout or adjust the system. Responsible deployment means treating the launch of an AI agent not as the end of development, but the **beginning of a closely watched probation period**.

Fail-safes and kill-switches: Another aspect of responsible deployment is having a way to intervene if things go wrong. Imagine the AI starts making bizarre or harmful decisions – we should be able to quickly take it offline or limit its capabilities. This might be as simple as a feature flag that can revert all decisions back to a manual process, or a "kill-switch" microservice that other services check with before executing an AI decision. WebkinsTech's engineers implemented a safeguard in their support system: if the AI's confidence in an answer was low or if it dealt with a VIP customer issue, the system automatically routed that query to a human agent. In essence, they built a **circuit breaker for ethics** – certain conditions would break the circuit of automation and fall back to human judgment.

Clear rollback plans: Hand-in-hand with kill-switches are rollback plans. If a new AI service starts causing trouble, how quickly can you revert to the old system or a safe state? In traditional deployments, we think about rollback for bugs; here we think about rollback for **bad behavior**. For example, WebkinsTech had a plan that if their support AI misanswered above a tiny threshold of questions or if even one severe incident occurred (like offending a customer with an inappropriate response), they'd immediately switch all queries back to human agents and send the incident to an ethics review board for investigation. Having this plan in place meant that everyone – from engineers to customer service managers – knew how to respond if the AI acted out of line.

Progressive trust: As the AI proves itself in the limited rollout, we can gradually widen its scope. This might mean expanding to more users, or allowing it to handle more types of tasks. But this trust must be **earned**, not assumed. Each expansion should be treated as a new experiment: monitor, evaluate, and only then proceed. Over time, if the AI consistently behaves well, you can approach full deployment with more confidence. Even then, continuous monitoring (again, leaning on Chapter 12's observability practices) remains crucial because context or data drift over time could introduce new issues.

Here's what a **phased deployment** might look like for an autonomous microservice, in a simplified flow:

```
[Dev/Test] → [Pilot Deployment] → [Gradual Expansion] → [Full Production]
     |              |                    |                    |
Thorough      Monitor KPIs       Increase scope        Ongoing  testing
& collect          if metrics look       monitoring &  (simulations
feedback           good (e.g., 10%       periodic audits
 & QA)          (daily review)     users → 50% → 100%)   for ethics
     |              |                    |                    |
     └── Roll back or adjust at any sign of ethical or performance issues
──>
```

In this schematic, every arrow is a checkpoint where the team can evaluate the AI's behavior. If things look good (no major biases, good feedback, no weird errors), the rollout continues. If not, they pause or roll back.

Pre-deployment ethical review: It's worth noting that responsible deployment starts *before* deployment. Before an AI service ever interacts with real users, a thorough ethical review or risk assessment should be done. WebkinsTech, for instance, has an internal checklist (we'll provide a full checklist towards the end of this chapter) that covers questions like: *Have we tested the model for bias with a diverse dataset? Do we understand the worst-case scenario* if this service goes rogue? *Are there any regulations or company policies this AI's actions could violate?* Only when they can confidently answer these questions do they move to the pilot stage.

By approaching deployment in stages, with oversight and humility, we acknowledge a fundamental truth: no matter how smart our microservice, we don't fully trust it on day one. **Trust is built gradually**. This cautious approach is not about stifling innovation – it's about ensuring that innovation happens **safely and ethically**. With a responsible deployment strategy in place, we set the stage for our AI agents to succeed without leaving our users to bear the brunt of any early mistakes.

Next, we'll turn our attention to one of the biggest ethical challenges in AI: bias and fairness. How do we ensure our agentic microservices treat people equitably and don't unintentionally discriminate?

Bias and Fairness in AI Microservices

Perhaps the most notorious ethical pitfall in AI is **bias** – when an AI system's decisions systematically favor or disfavor certain groups of people. Because AI learns from data (which itself comes from the real world), it can unintentionally pick up and amplify human prejudices or historical inequalities. In the context of microservices, you might have an intelligent service that evaluates loan applications, ranks job candidates, or even just categorizes user feedback. If not carefully designed, that service could exhibit biased behavior – like giving lower credit scores to certain neighborhoods, or ranking resumes in a way that unintentionally filters out candidates of a particular gender or ethnicity. **Fairness** means striving to make sure our agentic microservices treat people equitably and justly, rather than reinforcing societal biases.

Why does bias occur? One common source is **biased training data**. If the data used to train a model reflects past discrimination, the AI will likely reproduce it. The old saying "garbage in, garbage out" fits perfectly here – if we feed our microservice biased data, we'll get biased decisions. For example, imagine a job applicant screening microservice that's trained on a company's past hiring data. If historically the company hired mostly men for engineering roles, the AI might learn to favor male applicants by pattern, even if gender was never an explicit input! The algorithm's statistical learning will pick up on proxies (maybe it notices applicants named "John" tend to get hired more than those named "Jane") and carry that bias forward. Similarly, if an AI customer service agent was only trained on conversations in English from a particular region, it might struggle or respond inappropriately to speakers with different dialects or from different cultures.

Another source of bias can be **algorithmic design**. Sometimes the way a model is structured or the objective it optimizes can introduce unfairness. If you optimize solely for accuracy or profit, the model might learn to sacrifice fairness. For instance, a delivery route optimization AI might learn to deprioritize deliveries to outlying neighborhoods (because they are less profitable due to distance), effectively giving slower service to certain communities – a form of bias that wasn't in the data per se, but emerged from the optimization goal. As designers, we need to be aware that whatever metric we tell an agentic microservice to maximize, it will do so relentlessly, and potentially at the expense of unmeasured values like fairness or goodwill.

Why fairness matters: Ethically, it's obvious that we shouldn't discriminate – businesses have legal and moral obligations to avoid unfair treatment. But beyond that, there's a practical angle: **unfair AI systems lose user trust** and can backfire. Users who feel an automated system is treating them unfairly will abandon it or protest. Regulators are increasingly attentive to algorithmic bias, which means a biased microservice can lead to legal trouble and public scandals. On the positive side, if you build a reputation for fairness, that can become a competitive advantage. People prefer services that treat them with respect and impartiality.

So how can we ensure fairness in our intelligent microservices? There are several strategies:

- **Diverse and representative data:** Make sure the training data for your AI agents covers the spectrum of users it will serve. If WebkinsTech's microservice will be used by customers from different regions, ages, and backgrounds, the data used to train and test it should reflect that diversity. Gaps in data (like absence of a certain demographic) are blind spots where bias can thrive. Sometimes this means actively collecting more data for underrepresented groups or augmenting data synthetically to balance it out.

- **Bias testing and audits:** Just as we test software for security vulnerabilities, we should test AI systems for bias. This involves analyzing the system's outputs for different groups. For example, WebkinsTech would test their job applicant ranking service by comparing how it scores candidates from different demographic groups (e.g., men vs women, different ethnic backgrounds, etc.) with equivalent qualifications. If they find significant disparities that aren't explainable by actual qualifications, that's a red flag. Tools exist to help with this kind of analysis (sometimes called "fairness metrics"), but even simple checks can go a long way.

- **Feature selection and debiasing:** Be cautious about what input features you feed your models. Obvious sensitive attributes like race, gender, or age might be excluded from the model, but remember that proxies can sneak in (zip code might proxy for race, for instance). In some cases, you might intentionally include a sensitive attribute during training in order to explicitly correct for its influence – a technique used to adjust predictions and make them fairer, ironically by being aware of the sensitive attribute. This can get technically complex, but the design takeaway is: think critically about each data attribute and how it might bias outcomes. If you suspect a variable is causing unfair bias, consider removing it or transforming it.

- **Multiple models or human review for sensitive cases:** If one-size-fits-all fairness is hard to achieve, another approach is to handle certain segments separately or ensure additional oversight. For example, if an AI hiring service has trouble equitably evaluating candidates from a certain background (perhaps due to lack of data), WebkinsTech might choose to have those cases automatically flagged for human recruiter review instead of an automated decision. This is a hybrid approach – use AI where it's confident and seemingly fair, but don't force it where it struggles.

- **Set fairness goals and constraints:** Incorporate fairness into your success criteria, not just accuracy or throughput. For instance, specify that "the difference in loan approval rates between any two demographic groups should be no more than X%." There are algorithmic ways to enforce such constraints during model training (through regularization techniques or adversarial debiasing methods). Even if you don't implement fancy algorithms, having the goal spelled out means your team will be actively watching and evaluating the system for that criterion.

It's important to note that fairness isn't a one-time checkbox but an ongoing commitment. Populations and context can change over time. An AI model that was fair last year might drift if new data skews a certain way or if the user base changes. Thus, continuous monitoring (again, Chapter 12's observability comes into play) for bias in live systems is just as crucial as pre-deployment testing. This could mean periodically auditing decisions or outcomes – for example, reviewing a random sample of loan decisions each month to ensure fairness metrics are holding up.

Example: The Biased Hiring Agent

WebkinsTech's HR department experimented with an AI microservice to help rank job applications for software engineering positions. The goal was to save recruiters time by having the AI pre-screen candidates and highlight the most promising ones. They trained the model on data from the last five years of hiring decisions and employee performance at WebkinsTech. On paper, everything looked fine – the AI learned to predict which applicants might perform well based on past patterns.

When the new system was tested quietly, it produced a ranked list of candidates for an open engineering role. However, one observant HR manager at WebkinsTech noticed something unsettling: out of the top 20 recommendations, only 2 were women, despite many qualified female applicants in the pool. This was a red flag. WebkinsTech is committed to diversity, and the industry average percentage of women in similar roles was much higher than 10%.

Digging deeper, the team discovered that the AI had learned a bias from historical data – the engineering department had been predominantly male, and past hiring decisions (made by humans) had slightly favored male candidates, whether consciously or not. As a result, the AI's model had latched onto patterns that correlated with "successful hires" in the past, many of which were associated with male candidates. It had implicitly learned that male applicants were more likely to be hired and do well, and hence was ranking female candidates lower.

Recognizing this, the WebkinsTech team took action. First, they retrained the model with a more balanced training set (they made sure to include examples of successful hires of all genders, even if that meant augmenting data from external sources because their own historical data was skewed). They also explicitly removed any features that could be proxies for gender, and added a constraint during model optimization to ensure gender parity in the recommendations. Additionally, they implemented a policy that any hiring recommendations would be reviewed by a diverse panel of recruiters, rather than blindly accepted.

When they ran the AI again on the same applicant pool after these adjustments, the top 20 list came back much more balanced in terms of gender. Even then, WebkinsTech decided that the AI would serve only as a *supporting tool* – giving a sorted list of candidates with some rationale – rather than an automatic gatekeeper. The final hiring decisions would always involve human judgment. The result? The hiring process still saw efficiency gains (recruiters had an AI-curated list to work from), but without the AI injecting a hidden bias. This experience became a company lesson that **algorithmic bias can lurk in unexpected places**, and catching it early is key to maintaining fairness.

Bias Busters: When developing your AI microservices, involve a diverse team in testing. Often, individuals from different backgrounds can spot issues

others miss ("Hey, this photo recognition service doesn't work well on darker skin tones" or "This tone analyzer flags assertive emails from women as 'angry' more often"). A broad set of eyes and perspectives on your AI's output is one of the best defenses against blind spots in fairness.

By actively seeking out and correcting biases, we align our AI systems with our values of equality and justice. Fairness is not just a "feel-good" aspect of ethics – it directly affects how well our systems work for *all* users. Now that we've tackled bias, another crucial aspect of ethics in AI is making sure our systems are not "black boxes" to those who depend on them. In the next section, we'll discuss **transparency and explainability** – shedding light on how AI decisions are made.

Transparency and Explainability

Imagine you're on the receiving end of an AI-driven decision – say, your loan application was declined or a medical test result was flagged as high-risk by an AI. The first question you're likely to ask is, *"Why?"* As humans, we need reasoning to accept outcomes, especially when they affect us significantly. **Transparency and explainability** in AI systems are about answering that "why." It's not enough for an agentic microservice to output a decision; it should also provide insight into how it arrived at that decision, or at the very least, the system's owners should be able to explain it when asked.

In the realm of intelligent microservices, transparency operates on multiple levels:

- **Internal transparency:** As developers and operators, we need to understand what the AI is doing under the hood. This means having access to logs, models, and decision traces. It's closely tied to observability (again, recall Chapter 12). If an autonomous microservice suddenly starts behaving oddly, transparency tools help us dig into *why* – was it a particular input that led it astray? Did it receive a malformed signal from another service? Maybe the model's latest update changed its behavior. We can't fix what we can't see.

- **External transparency:** Users and stakeholders might not see the code or the raw logs, but they deserve clarity about the system's logic. External transparency includes things like documentation of the AI's intended function and limits ("This service analyzes customer emails to prioritize support tickets, and it may not catch sarcasm or humor."). It also means clearly communicating to users when an AI is involved. If a customer is chatting with an AI agent, they should know it's not a human, for example. Being upfront about automation maintains trust – nobody likes feeling deceived by a bot posing as a person.

Explainability, a closely related concept, usually refers to the ability to explain specific decisions or actions of the AI. For instance, it's one thing to say "our system uses a neural network to predict credit risk," (that's general transparency about the system) and another to say "this particular loan application was denied because the applicant's income was below the threshold and they had a short credit history" (that's an explanation for a specific decision).

Why is explainability so crucial? Several reasons:

- **Trust and acceptance:** People are far more likely to accept a decision – even an unfavorable one – if they understand the reasoning. It feels fairer. Without an explanation, AI decisions can feel arbitrary or mysterious ("the computer says no"), which breeds suspicion and frustration.

- **Accountability and debugging:** If we know why a decision was made, we can assess whether that reasoning was sound or flawed. Maybe the AI made the right call for the wrong reason – e.g., it denied a loan because of a data error rather than the person's actual creditworthiness. Without explainability, such issues might go unnoticed. With explainability, humans can catch and correct the AI's logic. This ties into accountability (which we'll discuss soon) because you can't hold a system accountable if you have no clue what it's doing.

- **Compliance:** In some domains, providing an explanation isn't just a courtesy – it's required. For example, in finance or healthcare, regulators might require that decisions be interpretable or at least that a plausible explanation can be given to the affected party. Even when not legally mandated, providing explanations aligns with ethical principles of respect and transparency.

Achieving explainability can be challenging, especially with complex models like deep learning. These are often termed **"black box"** models because their inner workings don't yield easily to human understanding; they might have millions of parameters interacting in nonlinear ways. However, there are strategies to mitigate this:

- **Use simpler or more interpretable models when feasible:** If you can solve a problem with a clear set of rules or a decision tree or a linear model, those are inherently easier to explain. Of course, not every problem can be handled with simple models, but don't use a more complex approach than necessary for critical decisions. For instance, WebkinsTech's team opted for a straightforward decision-tree-based model for an employee expense approval agent, even though a neural network might have squeezed out a bit more accuracy. The decision tree's advantage was that any denial of an expense could be explained by pointing to the specific rule (e.g., "amount exceeds the manager's approval limit").

- **Add explanation modules or metadata:** One pattern in intelligent microservices is to pair the AI decision-maker with an explanation generator. For example, if a microservice uses a neural network to classify insurance claims as approve or deny, the team might also develop a secondary logic (or use techniques like LIME/SHAP, which are tools that highlight what factors influenced a decision) to produce a human-readable explanation of the neural network's output. The microservice could then output not just the decision, but also a rationale like "Claim denied due to inconsistencies in reported data and policy coverage limits." This doesn't require revealing the full complexity of the model – just the main factors that led to the outcome.

- **Structured logging of decision factors:** Ensure that when an AI microservice makes a decision, it logs the key factors or intermediate conclusions that contributed. These logs can serve as a breadcrumb trail for engineers, and they can also be used to generate user-facing explanations if needed. For instance, an autonomous content filter might log: "Post flagged because it matched 3 out of 5 hate speech patterns, confidence 90%." If a user later inquires why their post was removed, the support team has concrete details to share.

- **Model cards and documentation:** This is more about transparency than per-decision explainability, but it's worth mentioning. "Model cards" are a concept where you document an AI model's intended uses, limitations, and performance characteristics, somewhat like a data sheet. WebkinsTech started creating model documentation for each of their major AI services. This documentation includes things like what data the model was trained on, what kinds of errors it might make, and guidelines for interpretation. It helps both internal teams and external stakeholders understand the context in which the model's outputs are trustworthy.

It's worth sharing a brief scenario to illustrate transparency in action. WebkinsTech once deployed an AI component in their SaaS product pricing system – the agent adjusted subscription prices for clients based on usage patterns and market demand. One quarter, the sales team noticed the AI recommended steep price increases for a few long-time clients, and they were baffled and a bit alarmed. Without clear insight, the sales reps felt they couldn't justify these changes to the clients, and some relationships grew strained. The issue boiled down to a lack of explainability: the pricing AI was a black box spitting out numbers. In response, WebkinsTech's engineers quickly added an **explanation feature**. Now the pricing service would accompany each recommendation with a brief note, such as: "*Recommended increase of 15% due to 30% rise in product usage and higher support costs in the last 6 months.*" Armed with this information, the sales team could have honest conversations with clients ("Your usage has gone up significantly, which is why the costs are projected to rise – here's the breakdown..."). The transparency defused tension and even helped clients trust that the adjustments were not arbitrary gouging, but based on real usage data.

Be Transparent by Design: Always inform users when an AI is making a significant decision. If an automated service denies a request or takes an action on a user's behalf, consider adding a note like, "This decision was made automatically by our system," along with instructions for how to request a review or more information. This proactive transparency lets users know that you have nothing to hide and that you respect their right to understand or challenge the outcome.

In practice, transparency and explainability might add some overhead to your system design – extra logging, additional modules, simpler models than you'd otherwise use – but it's an investment in **trust**. A system that can explain itself is far less likely to be viewed with suspicion. It turns your AI from a magic box into a tool that people can reason with and integrate into their decision-making processes. We've all heard stories of people blindly following GPS directions into a lake or off a closed bridge. With explainability, we're essentially giving the "why" behind the directions, so that the user can say, "Ah, that route is suggested because there's traffic on the main road – that makes sense, I'll follow it." It keeps the human in the driver's seat in terms of understanding, even if the AI is doing a lot of the driving in practice.

Speaking of keeping a human in the driver's seat, another vital aspect of ethical AI deployment is knowing when to hand control (or at least a second opinion) back to a human. In the next section, we'll discuss **human-in-the-loop design**, exploring how mixing human oversight with autonomous microservices can provide the best of both worlds.

Human-in-the-Loop and Oversight

No matter how advanced our autonomous microservices become, there are certain judgments and responsibilities we, as humans, should never fully relinquish. **Human-in-theloop (HITL) design** is the practice of keeping a person involved at key points of the AI's operation. The idea is simple: when a decision is particularly high-stakes, complex, or valueladen (involving ethics or judgment calls), a human should review or approve it. This approach acknowledges that AI, for all its speed and consistency, lacks the full context, empathy, and moral intuition that a human can bring.

We already touched on this principle in earlier sections – for example, WebkinsTech's hiring AI now always has human recruiters making the final calls, and their customer support bot hands off tricky cases to human agents. In fact, in Chapter 9, *Human in the Loop for Critical Outcomes*, we explored how integrating human oversight isn't just a safety net but a design choice to improve outcomes. There's a reason airplanes have autopilot **and** pilots: the automation handles routine flight well, but a human pilot is there for unexpected situations and to take command when something doesn't go by the book. Similarly, our AI microservices can handle the mundane decisions at scale, but humans should oversee the exceptional ones.

When to involve a human? Not every single micro-decision needs human review (that would defeat the purpose of automation), so a big part of HITL design is deciding **where to draw the line**. Here are some common criteria:

- **Risk level:** If an AI action, if wrong, could cause serious harm (financially, physically, or legally), put a human in the loop. For example, an AI moderating a community forum might automatically hide comments that look like spam or profanity (low risk), but require a human moderator's approval before banning a user (high impact on someone's free expression, so higher risk). In a financial microservice at WebkinsTech, small transactions are auto-approved, but any transfer above a certain large amount triggers a human compliance officer's review due to the potential repercussions of a mistake or fraud.

- **Uncertainty or novelty:** When the AI isn't confident in its decision or encounters a scenario that's outside its training experience, that's a good time to ask for human help. Advanced systems often output a confidence score or can detect that "this looks unlike anything I've seen before." WebkinsTech's incident response agent (which helps triage IT system alerts) is configured such that if it encounters an error pattern it doesn't recognize, it flags it for an engineer to manually analyze rather than guessing. Think of it as the AI humbly saying, "I'm not sure about this one – could you take a look?"

- **Ethical and legal judgments:** Some decisions have moral weight or legal implications that go beyond what we'd want an AI to handle solo. For instance, deciding to turn off a life-support machine in a hospital is not something an AI should ever do alone, no matter how advanced. In a less extreme example, consider content moderation again: identifying a post as containing hate speech might be done by AI initially, but whether it violates community guidelines to the point of deletion might warrant a human moderator's context-aware judgment. WebkinsTech, in one of its products for enterprise document analysis, made sure that any time the AI flagged something as potentially violating privacy regulations (like a document that might contain sensitive personal data), a human legal expert was looped in to confirm the finding before any action (like reporting it or deleting it) was taken.

Implementing human-in-the-loop can take different forms:

- **Approval workflows:** The AI makes a recommendation or decision, but it isn't final until a human approves. For example, an AI-generated financial report could be drafted by a microservice, but a human manager must review and sign off before it's released, ensuring nothing looks off.

- **Review queues:** The AI might act immediately but mark certain decisions for post-hoc human review. Many social media platforms do this – an AI might remove content that is very likely criminal or explicit to minimize harm, but then put those removals in a human moderator's queue to double-check that the removal was justified. If the human disagrees, they can reinstate the content.

- **Override and fail-safe mechanisms:** Humans should have the ability to intervene and override an AI decision. This could be as simple as a red button to cancel an automated process, or a UI where an operator can edit or revert a decision the AI made. In operational settings, it might mean an on-call engineer can switch systems from "AI mode" to "manual mode" if they observe weird behavior.

Example: Guardrails in Action

WebkinsTech's DevOps team deployed an "Autonomous Optimizer" service to manage cloud infrastructure. This agent could add or remove server instances, clear caches, or restart services in response to performance metrics – essentially automating a lot of routine DevOps decisions. It worked great most of the time, scaling resources up and down to meet demand and resolving minor incidents faster than any human could.

However, the team set strict guardrails on this agent's powers. **The AI was not allowed to perform destructive or irreversible actions.** For instance, it could restart a service (since that's reversible) but it could not delete a database or permanently decommission a server without human approval. This rule was put in place because the risks of a wrong deletion are enormous (permanent data loss, major outage).

This guardrail was tested one day when the optimizer agent identified what it thought was an unused database instance consuming resources. In truth, it was a mis-tagged backup database that looked idle but was actually a critical fallback. The AI, following its algorithms, flagged the instance for termination to save costs. Instead of executing directly, the system generated a request for human approval. A DevOps engineer received the request, double-checked the instance, and quickly realized it was not safe to delete. The engineer declined the action and corrected the resource tagging to prevent future confusion. **Disaster averted.** In the post-incident review, the team was thankful that they had kept a human in the loop for deletions. The AI was smart about optimizing, but final judgment on potentially irreversible actions remained with a person. This example highlights how human oversight and AI automation worked in tandem: the AI did the heavy lifting of finding a cost-saving opportunity, but a human provided the necessary judgment to ensure it was a *wise* action.

One challenge in HITL systems is to avoid *human rubber-stamping*. If the AI is usually right, humans might get lazy and just approve everything, which defeats the purpose of oversight. To combat this, organizations like WebkinsTech make sure to **train and empower** their human reviewers. They encourage a mindset that the AI is a junior colleague: skilled and fast, but needing supervision. Reviewers are taught to trust the AI's input but verify it. In practice, this might mean spot-checking even routine decisions occasionally, and having clear guidelines on when to question or override the AI. Designing good user interfaces for human reviewers is also key – for instance, highlighting the factors that the AI considered can help the human quickly understand the situation and make an informed call.

In summary, human-in-the-loop design is about **balance**. We leverage the efficiency of AI but temper it with human judgment where it counts. The result is often a more robust system: the AI catches what humans might miss, and humans catch what the AI might miss. It builds confidence that the system as a whole will do the right thing. And importantly, it places accountability back where it ultimately belongs – with us humans, who can take responsibility for the final outcomes.

With human oversight covered, the next logical question is: when an AI-driven system does make a decision (whether or not a human approved it), how do we account for that decision? Who is responsible, and how do we trace what happened? This brings us to the topic of **accountability and auditability** in AI microservices.

Accountability and Auditability

When an AI-driven microservice makes a decision or takes an action, especially one that negatively impacts someone, a natural question arises: **"Who is responsible for that?"** Accountability is about ensuring that we have a good answer to that question. As the creators and deployers of the AI, we (and our organizations) are ultimately accountable. We cannot shrug and blame "the algorithm" – after all, an algorithm is not a moral agent that can answer for itself in a court of law or the court of public opinion. It's on us to own the outcomes of our autonomous systems, both the good and the bad.

From a practical standpoint, accountability in intelligent microservices translates to a few key practices:

- **Audit trails:** Every significant decision or action taken by an AI service should be logged in detail. This includes what the input was, which model or rule was applied (and which version of it), and what output or action resulted. Think of it like a flight recorder (the "black box") on an airplane: if something goes wrong, you need to be able to go back and see what the AI saw and why it did what it did. In a microservices context, this audit trail might span multiple services – maybe one service scored a user's risk, which another service consumed to make a pricing decision. Using correlation IDs or distributed tracing (recall Chapter 12 on observability), we can tie together a chain of events to reconstruct the full picture of an AI-driven transaction.

- **Versioning and change management:** Keep track of which algorithms or models were in use at any given time. If you update your model, archive the old version. If a decision from last month is questioned, you should be able to run that same old model on the same input to reproduce the result. WebkinsTech's engineering team, for instance, uses a model registry – a system that stores every model version with metadata like training data used, date deployed, and who approved it. This way, if an issue arises, they know exactly which version of the AI was responsible and what its characteristics were.

- **Defined ownership:** For each intelligent microservice, assign a human owner (or team) who is responsible for its behavior. This is part of governance – knowing who is accountable within the organization. If the AI makes a mistake, who gets the call to investigate and fix it? At WebkinsTech, they have an "AI Product Owner" role for each major AI system, ensuring there is always a go-to person who oversees how that service is functioning. This person isn't necessarily blamed for every issue, but they are accountable for coordinating a response and making sure the issue is resolved.

- **Incident response and redress:** Treat ethical and operational issues from AI as incidents that warrant response. If your microservice causes, say, a dozen customers to be overcharged, you need a process to detect that (perhaps via monitoring alerts), roll back or stop the offending service, communicate transparently to affected parties, and compensate or correct the error. In ethical terms, this is often called a redress mechanism. Users should have a way to contest or appeal an AI decision. For example, if an AI denies someone a benefit or flags them as high-risk, provide a channel for them to ask for human review. And when that happens, take it seriously: investigate and, if the AI was wrong, fix the underlying issue and make amends. WebkinsTech built an "Appeals Portal" for one of its AI-driven services that assigns a human reviewer

whenever a user claims an AI decision was a mistake. Not only does this help resolve individual cases fairly, it also feeds back into improving the AI (seeing what it got wrong and why).

No Excuses – Own the Outcome: It's tempting for engineers or executives to deflect blame to an AI system: "The algorithm made this call, not us." This attitude is a slippery slope to eroding trust. Always remember that an AI is an extension of its creators' and operators' intentions. If it errs, we have erred. At a senior leadership meeting at WebkinsTech, the CEO made this clear: "If our AI causes harm or injustice, *we* caused harm or injustice. We have to fix it and make it right." Such a culture of accountability from the top ensures everyone takes ethical issues seriously and looks for solutions rather than scapegoats.

Governance and oversight committees: For organizations deploying multiple AI systems, it can be helpful to have a governance body – like an AI ethics committee or review board. This group can set guidelines, review high-risk use cases, and periodically audit systems for compliance with ethical standards. WebkinsTech assembled a small "AI Council" comprised of senior architects, a legal advisor, and an HR representative (to bring a diversity of perspectives). They meet quarterly to review how things are going: Are there any new ethical risks emerging? Did any incidents occur and what was learned? Are engineers following the bias testing and transparency checklist? This sort of oversight provides an extra layer of accountability because the teams know their work will be reviewed not just by metrics of profit or performance, but also ethics.

Accountability goes hand in hand with transparency (discussed earlier) – you need openness to be accountable – and with human oversight – because having a human in the loop means there is someone to answer questions about a decision. But even with those in place, you must plan for the worst: at some point your AI might do something undesirable. Ethical engineering means **anticipating that possibility and being ready to respond** in a responsible way.

Let's visualize a simplified oversight architecture for accountability in a microservice environment:

```
User Action → [AI Service] → Outcome to User
                    |
                    ├→ (Event Log) → [Audit Log Service] ──→ (stored
decision details)
                    └→ (Alerting) → [Monitoring System] ──→ (trigger
if anomaly or threshold hit)
|
                                            └→ [Human Owner Notified]
```

In this sketch:

- The AI Service, whenever it makes a key decision, logs an event with all relevant info to an Audit Log Service (a central repository for later analysis).
- It also sends metrics or alerts to a Monitoring System – for example, if a certain type of decision is happening too frequently or if it encounters an unknown condition.
- The Monitoring System can trigger notifications if something seems off (say, an unusual spike in denials or errors), alerting the human owner or on-call team.
- The human owner can then check the Audit Log, dig into what happened, and take action (be it technical fixes or user redress).

By designing such an audit-friendly system, we ensure that our autonomous microservices don't operate in an unchecked corner. We have the tools and structures to **inspect, understand, and answer for** their actions. This fosters trust externally (we can explain ourselves to users or regulators) and internally (engineers and stakeholders feel in control, not at the whim of a "mystery box" AI).

With accountability mechanisms in place, we've covered many of the internal responsibilities of building ethical agentic microservices. Next, let's consider two broader aspects: data privacy (because AI doesn't work without data, and that data often belongs to real people) and the larger social impact of these technologies on jobs and society.

Data Privacy and Security

Data is the lifeblood of intelligent microservices – without data, an AI can't learn or make informed decisions. But much of that data belongs to real people: customers, employees, users. With great data comes great responsibility (to paraphrase a famous saying). **Data privacy** is about respecting individuals' rights over their personal information, and **security** is about protecting that data from unauthorized access or leaks. If we misuse someone's data or fail to safeguard it, we've committed an ethical violation, not to mention potentially a legal one.

In designing agentic microservices, we must adopt a mindset of **privacy by design**. That means from the very start, thinking about questions like: "Do we really need this piece of data for our service to function intelligently? How can we minimize what we collect? If we do collect it, how do we store and transmit it safely? Who will have access to it?" Here are key principles and practices to consider:

- **Data minimization:** Collect and use only the data that is truly needed for the AI service to do its job. The more personal data you gather, the greater the risk if it's misused or breached. For instance, if WebkinsTech's customer service AI needs to know a customer's purchase history to give good support, it should retrieve just that relevant info – it doesn't need the customer's entire profile or, say, their exact date of birth unless absolutely necessary. By limiting data scope, you reduce potential exposure.

- **Consent and transparency in data use:** Whenever possible, users should be informed about what data is being used and for what purpose. If data is being collected in a way they might not expect, you should ask for consent. Consider a scenario where WebkinsTech decides to use customer support chat logs to train a better support AI model. Ethically, they should update their user agreements or explicitly inform customers that their anonymized chat transcripts might be used to improve the service. Even if formal consent documents are in place, being upfront in plain language builds trust ("Your chat may be analyzed to help our AI support agent learn to serve you better. We take care to anonymize personal details.").

- **Anonymization and aggregation:** A powerful technique to protect privacy is to remove or obfuscate personal identifiers in the data. Before using real user data to train models or for analytics, WebkinsTech's data engineers often scrub names, emails, phone numbers, and replace them with anonymous IDs. They also aggregate data where individual details aren't needed – for example, an AI forecasting service might only need summary statistics of user behavior rather than each user's exact record. Proper anonymization can allow learning from data without exposing anyone's identity.

- **Secure data handling:** In a microservices architecture, data might flow through multiple services, databases, and APIs. It's crucial to secure those channels. This means encrypting data in transit (using protocols like TLS so that if someone intercepts the traffic, they can't read it) and encrypting data at rest in databases. It also means authenticating and authorizing access between services – one service shouldn't be able to query another's data unless it's explicitly allowed. For sensitive data, consider additional safeguards like encryption at the field level (e.g., encrypting a credit card number even in the database) or using secure enclaves for processing. WebkinsTech, for example, stores highly sensitive keys and personal data in a secure vault service; microservices must request access through strict channels, and every access is logged and monitored.

- **Avoiding unintended data leakage:** One often overlooked aspect of privacy is how data might inadvertently leak through logs, backups, or third-party integrations. Imagine an AI microservice that logs verbose error messages including snippets of user data. If those logs are accessible by many developers or stored in an insecure log server, that's a privacy breach waiting to happen. WebkinsTech learned this early on when a developer discovered that an analytics microservice was logging user email addresses in plain text for debugging. They quickly changed the logging level to avoid sensitive info and scrubbed the old logs. Now, they have a rule: **no sensitive personal data in logs**, and logging practices are reviewed in code reviews. Similarly, if using external services (like a third-party API or cloud service), they ensure those are compliant with their privacy standards and don't send more data than necessary to them.

- **Data retention and deletion:** Ethical data stewardship includes not holding on to personal data longer than needed. WebkinsTech sets retention policies – for example, user data that is no longer required is deleted or archived in a privacy-preserving way. If a user requests their data be deleted (exercise of data rights), the system is designed to comply with that across all microservices that might store the data. This can be tricky

in a distributed system (data might exist in multiple places), so they designed a mechanism where a deletion request triggers a workflow that propagates to all relevant services (kind of the reverse of how data might be aggregated, now fanning out a delete command). Ensuring that "right to be forgotten" is respected is not just a legal checkbox but an ethical commitment to user control over data.

Example: Privacy vs. Utility – A Design Choice

WebkinsTech once faced a dilemma while developing a smart home assistant device. The device's AI microservice could operate in two modes: (1) process voice commands locally with limited accuracy, or (2) send recordings of the wake word and command to WebkinsTech's cloud servers for more accurate understanding using a powerful language model. The cloud approach promised better performance – the AI could understand nuanced requests better by leveraging big data and training updates – but it meant transmitting and storing snippets of users' voices, which are highly personal data. The team had to make an ethical design choice.

After deliberation, they opted for a hybrid solution favoring privacy. By default, the assistant would do all processing on-device for common commands, never sending audio out. For complex queries it couldn't handle, it would ask the user for permission ("I can send this request to our servers to better assist you – is that okay?"). The device also had a clear indicator (a light) whenever audio was being transmitted. On the backend, any audio that did go to the cloud was anonymized (no user identifier attached beyond an anonymous device ID) and deleted immediately after processing the command – it was not stored for future training unless the user had explicitly opted into a "help improve our service" program.

This design sacrificed some convenience and even some technological edge (competitors were vacuuming up voice data to constantly improve their AIs). But WebkinsTech's leadership backed the decision, believing that **respecting customer privacy would ultimately be a competitive advantage** by building trust. Indeed, when marketing the product, they highlighted this privacy-first design. Many privacy-conscious customers responded positively, saying they chose the WebkinsTech assistant over others because it "felt less creepy." This example shows that taking privacy seriously can influence design decisions at every level – and that sometimes, doing the right thing for users' privacy may mean not pushing an AI system to its absolute limits, especially if doing so erodes trust.

In summary, guarding data privacy is both an ethical duty and a prerequisite for user trust. A breach or misuse of data can nullify all the goodwill your intelligent service might have earned with its fancy capabilities. It's like inviting someone into your home (via their data) – you treat their belongings with care and lock the doors to keep them safe. By building privacy and security measures into our AI microservices, we ensure that the intelligence we deploy serves users without trespassing on their rights or expectations.

Now that we've examined how to handle data responsibly, let's step back and consider the broader picture: how do these agentic systems affect the people who use them, the employees who work with them, and society at large? We'll explore the **impact on the workforce and society** in the next section.

Impact on Workforce and Society

The influence of agentic AI systems extends beyond the direct interactions between users and microservices – it ripples through the workforce and society at large. As organizations adopt intelligent microservices to automate tasks and augment decision-making, **jobs and roles inevitably shift**. This raises an ethical question: *What is our responsibility to the people whose work is affected by these AI agents?* And on a societal level: *How do we ensure these technologies benefit society broadly, rather than exacerbating inequalities or eroding human values?*

Workforce implications: Within a company, deploying AI microservices can lead to concerns among employees about job security. History has shown that technology changes work – some jobs become obsolete, while new ones emerge. The ethical approach is to handle this transition **humanely and thoughtfully**:

- **Augmentation over replacement:** Ideally, use AI to **augment human workers, not outright replace them** whenever possible. For example, WebkinsTech introduced an AI service to handle routine tier-1 support queries.

This did mean the AI could resolve common issues without needing a human on every call. But instead of laying off the support staff, WebkinsTech shifted their focus to more complex customer problems and to overseeing the AI. The support agents essentially got an "upgrade" to their role – they became mentors and backup for the AI on tricky cases, and they had more time to give white-glove treatment to customers with unusual issues. The AI took over the drudgery (answering FAQ-type questions over and over), freeing humans to do what they do best: handle nuance, provide empathy, and build relationships.

- **Reskilling and upskilling:** If an AI system does make part of someone's job redundant, it's an ethical best practice to offer training for new skills so that employees can move into new roles created by the AI or elsewhere in the organization.

WebkinsTech anticipated that their "AI Support Agent" would reduce the need for entrylevel support reps, so months before deployment, they began training those reps in new areas – some learned to manage the AI tooling (becoming AI operations specialists), others moved into customer success roles that the AI couldn't fulfill (proactively helping clients optimize products). By the time the AI was fully rolled out, not a single employee was left without a valuable role. This not only prevented job loss, but also signaled to employees that the company values them, building trust and goodwill.

- **Transparent communication:** It's crucial to be honest with your workforce about why you're deploying AI and how it will affect roles. In WebkinsTech's case, management held town hall meetings to explain: "This new AI will handle X tasks. It's here to help us scale, not to punish anyone. We will need fewer people doing X, but here's our plan to transition those folks into Y roles." This openness helps reduce fear of the unknown and invites employees to participate in the change rather than resist it. Often, frontline workers can even contribute insights into how the AI can best assist in their domain if they're made partners in the process.

- **Ethical offboarding:** In scenarios where, despite best efforts, some jobs are eliminated, the company should treat affected employees with dignity – generous severance, job placement assistance, or opportunities in other parts of the business. Ethically, reaping efficiency gains from AI should go hand-in-hand with helping those who made past operations successful land on their feet. This isn't just a feel-good notion; it sets a precedent that can maintain morale among remaining staff and upholds the company's values.

The end result of a considerate approach is often a happier, more skilled workforce working *alongside* AI. Many WebkinsTech support staff reported that initially they were nervous about the AI agent, but later found work more engaging. One specialist said, "I used to answer password reset questions all day. Now the AI does that, and I get to tackle the interesting problems and really help customers. It's like I've been promoted to doing the parts of the job I actually enjoy." Such stories underscore that when implemented thoughtfully, AI can *elevate* human work rather than diminish it.

Societal implications: On a broader scale, the deployment of autonomous systems raises questions about societal impact:

- **Employment trends and inequality:** If AI microservices across industries systematically eliminate certain types of jobs, there could be a societal challenge in ensuring people find new careers. There is an ethical onus not only on governments but also on companies to contribute to solutions – such as funding retraining programs, or collaborating with educational institutions to prepare the next generation for an AIintegrated workplace. Technology companies like WebkinsTech recognize that they are stakeholders in society; thus, they periodically sponsor community tech workshops and contribute to STEM education initiatives, to help cultivate the skills that future jobs will require. The goal is to help society adapt alongside the technology.

- **Social dynamics and human interaction:** When more services are provided by AI (from chatbots to automated decision letters), people might have less human-to-human interaction in certain scenarios. Some customers might find this efficient; others could feel alienated. There's a risk of society becoming impersonal if we allow AI to fully replace human contact in areas where empathy and human connection are vital (like elder care, mental health support, etc.). The ethical course is to use AI to enhance accessibility and convenience **without completely removing the human touch**. For example, a hospital might use an AI triage nurse to intake patient symptoms initially (speeding up the process), but ensure that a human nurse or doctor follows up in person, especially for serious conditions – because patients need that reassurance of human care. Similarly, businesses often provide an option to "talk to a human" in customer service, acknowledging that sometimes only a person can truly resolve a customer's concern or frustration.

- **Public trust and perception:** The widespread adoption of AI agents in everyday life – from finance to transportation to public services – can influence how society views technology. High-profile failures or harms (like an AI that discriminates or a self-driving car accident) can lead to public backlash and mistrust not just in the offending system but in AI as a whole. On the flip side, success stories (like AI accurately predicting floods and helping evacuate communities in time) can boost public optimism. Each team deploying an AI microservice plays a part in this narrative. Acting ethically isn't just about avoiding negative press for your own company; it's about contributing to a culture where people feel **safe and empowered** with AI around them. That means being candid about what your AI can and cannot do, sharing lessons learned (WebkinsTech published a whitepaper on their bias mitigation strategies, hoping other companies could benefit), and engaging with policymakers and stakeholders openly. In a sense, we're all collectively writing the "social contract" for AI – setting expectations for how these agents should behave in our world.

- **Avoiding societal biases and harm:** We covered bias at the system level, but at scale there's also the issue of AI possibly reinforcing societal biases or filtering people's experiences. Imagine all major job recruitment systems using AI that inadvertently filter out certain demographics – an entire group could find it harder to get employed, worsening inequality. Or if news delivery is automated and optimized purely for engagement, it might contribute to polarization by showing people extreme content (a societal ethical issue platforms are grappling with). The takeaway for architects of AI microservices is to **think beyond the immediate context**: consider the long-term and aggregate effects if your system (and others like it) were widely adopted. It adds weight to doing things like fairness checks and transparency not just for your users, but for the societal ecosystem.

Finally, it's worth noting that **ethical considerations aren't static**; they evolve with societal norms. What seems acceptable today might be viewed as problematic tomorrow as people's understanding of AI's impact deepens. Thus, continuous dialogue with diverse stakeholders – ethicists, user advocacy groups, your own employees, etc. – is part of being a responsible innovator. WebkinsTech, for instance, holds an annual "Ethics Review" meeting open to all staff where anyone can voice concerns or suggestions about the company's AI projects and their potential impact. Often, fresh perspectives emerge (like an intern pointing out an accessibility issue for disabled users that others hadn't noticed).

By proactively addressing workforce transitions and societal impacts, we aim to ensure that the rise of intelligent microservices is a win-win: efficiency and intelligence increases, while human welfare and dignity are upheld. The technology should ultimately serve humanity, not displace or diminish it. This perspective sets the stage for a future where humans and AI agents collaborate in harmony – which, fittingly, is the subject of our next and final chapter.

Before we move on, let's consolidate what we've learned into some concrete best practices and a checklist for ethical agent design.

Best Practices and Ethical Design Checklist

We've covered a lot of ground. To bring it all together, let's summarize the best practices for designing ethical agentic microservices. Think of this as a quick reference or a **checklist** that an architect or team can use when building and deploying an autonomous system:

Ethical Principles & Best Practices Summary:

Principle	Key Considerations	Best Practices (Examples)
Fairness & Bias	Ensure AI decisions are equitable and unbiased.	Diversify training data; perform bias audits; set fairness metrics; include humans for sensitive cases.
Transparency	Make the AI's logic and decisions understandable.	Provide explanations for decisions; document model limits (model cards); inform users when AI is in use.
Accountability	Take responsibility and enable traceability.	Maintain detailed logs (audit trails); version models; assign human owners; establish appeal processes.

Human Oversight	Keep a human in control loop for high-stakes decisions.	Use human approval for risky actions; have override mechanisms; train humans to supervise AI effectively.
Privacy & Security	Respect user data and protect it from misuse.	Follow data minimization; encrypt data in transit & at rest; anonymize personal info; obtain user consent where needed.
Social Impact	Consider effects on people's lives and society.	Involve diverse stakeholders in design; assess job impact and plan retraining; ensure human touch isn't lost in user experience.

This table encapsulates the core dimensions of ethical AI microservices. But having principles is not enough – one must implement and verify them. That's where a checklist comes in handy. Before you deploy an autonomous microservice (or as a periodic audit), run through a checklist like the following:

Ethical AI Deployment Checklist:

[ ] **Objective Alignment:** Have we clearly defined the goals of the AI and checked that optimizing for those goals won't lead to unethical outcomes? (e.g., not just maximizing clicks, but considering quality of content to avoid clickbait).

[ ] **Bias Assessment:** Did we test the system for bias or unfair outputs? Did we use a diverse test dataset and examine results across different user groups? If disparities were found, did we mitigate them?

[ ] **Transparency:** Can we explain how the AI makes decisions, at least to an internal audience, and ideally to users? (For critical decisions, do we have a userfriendly explanation ready?)

[ ] **User Awareness:** Are users informed when they are interacting with the AI or when an AI is making a decision about them? (No one should be unknowingly subjected to an algorithm without recourse.)

[ ] **Human-in-the-Loop:** Have we identified which decisions/actions require human oversight or approval? Have we implemented those control points in the workflow?

[ ] **Fallbacks and Overrides:** If the AI fails or produces a questionable result, is there a clear fallback? (For instance, reverting to a manual process or a safe default state.) Can humans override the AI easily in such cases?

[ ] **Audit Logging:** Are we logging inputs, outputs, and key decision factors for the AI's actions? Will those logs enable us to troubleshoot or explain any outcome after the fact?

[ ] **Data Governance:** Are we handling data in compliance with privacy principles? (Only necessary data used; sensitive fields protected; proper consents obtained; retention policies in place.)

[ ] **Security Measures:** Have we secured the microservices and data pipelines against breaches? (Auth controls between services, encryption, no sensitive data leakage through logs or monitors.)

[ ] **Deployment Safety:** Did we do a trial run or pilot phase for this AI in a safe environment? Do we have monitoring in place for its initial rollout to catch any unintended effects quickly?

[ ] **Team Preparedness:** Have we trained the relevant team members (operators, customer support, etc.) about the AI's capabilities and limitations? Do they know how to respond if something goes wrong or a user questions a decision?

[ ] **User Feedback Loop:** Is there an easy way for users or stakeholders to report issues or appeal decisions? And do we have a process to incorporate that feedback into improving the system?

[ ] **Continuous Review:** Do we have a plan to periodically re-evaluate the AI's performance and ethical compliance? (Models can drift, usage context can change

– we might need to update our system or policies over time.)

This checklist isn't exhaustive, but it's a starting point. The idea is to integrate these checks into your development and deployment process, much like you would for security or performance. In fact, consider creating an

"Ethics Review" step in your project lifecycle, just as you might have a security review or code review.

Following these best practices and checklist items will greatly reduce the chances of an ethical oversight. It transforms ethics from an abstract ideal into concrete actions and verifications. When you can honestly check all (or most) of the boxes above, you can move forward with much greater confidence that your agentic microservice will behave responsibly in the wild.

Conclusion: Building Ethical and Intelligent Microservices

Ethics in technology is not a destination but a journey – a continuous commitment to align our creations with human values. In this chapter, we've seen why that commitment matters intensely for agentic microservices, which by their nature make autonomous choices that ripple out to affect people's lives. We started by recognizing the weight of those choices and the trust we must earn. We explored practical ways to bake in fairness, transparency, and accountability, to safeguard privacy, and to keep humans in the loop. We also zoomed out to consider the well-being of our colleagues and society as AI transforms the way we work and live.

The overarching lesson is that **every technical decision is also an ethical decision**. Choosing a dataset, tuning a model, designing an API endpoint, setting a default threshold – each of these can influence outcomes in ways that help or hurt people. By approaching design and deployment with an ethical lens, we anticipate and prevent harm before it happens. We turn potential pitfalls into opportunities to do better.

WebkinsTech's journey throughout this chapter has shown us that mistakes will happen – a biased model here, a dubious decision there – but with the right values in place, those mistakes become learning moments. An ethical engineering culture doesn't expect perfection; it expects **accountability, humility, and improvement**. It's about saying "We might get it wrong, and if we do, we'll own it and fix it."

As we conclude our deep dive into ethical considerations, you might be thinking about what's next. After all, building intelligent microservices is a fast-evolving field. New challenges and possibilities are on the horizon: from even more autonomous systems, to evolving regulations, to advances in AI that will test the boundaries of what these agents can do. In our final chapter, we will explore exactly that – the future directions of agentic systems and how we can prepare for what's to come. The ethical mindset we've fostered here will be our compass as we venture into that future.

Call to Action: As a tech professional and architect, you are in a position to shape not just systems, but the impact those systems have on real lives. Take that responsibility to heart. Champion ethical discussions in your teams; incorporate checks and balances in your architectures; challenge the notion that "move fast and break things" is an acceptable mantra when what you might break are people's trust or well-being. Instead, move thoughtfully and build things – intelligent microservices included – that *earn* the privilege to operate in our world.

Let's forge ahead to Chapter 18, where we'll cast our eyes to the horizon and imagine the future of intelligent, agentic microservices – a future we have the power to influence responsibly

Chapter 18: Future of Intelligent Microservices

"The future is already here – it's just not evenly distributed."

A reminder that tomorrow's mainstream was born in pockets today.

Introduction: Peering Into Tomorrow's Microservices

As we conclude our journey through *Intelligent Microservices: An Agentic-AI Approach*, it's time to peer into the crystal ball 🔮 and envision what lies ahead. We've built a strong foundation – from understanding why enterprises like WebkinsTech embarked on microservices, to infusing those services with intelligence and autonomy. Now we stand at the frontier, where imagination meets emerging technology. What will **microservices augmented with AI ("intelligent microservices")** look like in five, ten, or twenty years? How will they transform software architecture, businesses, and our daily lives? In this chapter, we'll explore these questions in a conversational yet insightful way, balancing optimism with practical foresight.

What this chapter covers: We'll examine cutting-edge **research directions** and emerging patterns that hint at the next leaps in intelligent microservice design. We'll look at how **other technologies** – from massive AI models and serverless platforms to IoT, AutoML, and decentralized tech – are converging with intelligent microservices. Through speculative narratives, we'll imagine **future architectures**: self-organizing swarms of services, standardized "agent" communication protocols, and expansive digital service ecosystems that span organizations. We'll discuss how companies can **build the capabilities** to thrive in this future, using tools like AI maturity models, continuous innovation practices, and leadership principles that foster experimentation. No future-gazing would be complete without addressing the **challenges ahead** – from regulatory and ethical dilemmas to security threats and design complexities. Along the way, we'll visit visionary **case studies across industries** (healthcare, finance, manufacturing, and more), including a fictional glimpse into WebkinsTech's future. Finally, we'll wrap up with an **inspirational summary**, inviting you – the tech leaders and architects – to help shape the era of intelligent automation.

Buckle up for a journey into the possible futures of intelligent microservices. Let's start by looking at the seeds being planted today: the research trends and patterns that are likely to define tomorrow's architectures.

Emerging Research Directions and Patterns

The present is giving us plenty of clues about the future. Researchers and forward-looking architects are already experimenting with ways to make microservice-based systems more intelligent, autonomous, and adaptive. Several key **patterns and directions** have emerged:

- **From Static Services to Learning Agents:** Traditional microservices are predefined in behavior – they do what they were coded to do. The future trend is toward microservices that *learn and evolve* over time. We are witnessing the rise of **agentic microservices**: services that encapsulate AI models and make context-aware decisions autonomously . In research labs and innovative companies, prototypes exist of microservices that can observe their environment (through metrics, events, etc.), reason about optimal actions (using AI/ML algorithms), and adjust their behavior without human intervention. For example, an *intelligent routing service* might dynamically change how it directs user requests based on real-time performance data, learning to avoid a slow service instance before that instance actually fails. This ability to anticipate and adapt – rather than just react – is a fundamental shift. In fact, *"AI-driven microservices represent more than an incremental improvement — they are a fundamental shift in how we approach distributed systems. The future lies in systems that do not just react to demand but anticipate and prepare for it."* . In short, tomorrow's microservices will be **proactive**, not just reactive.

- **Multi-Agent Autonomy and Collaboration:** Drawing from the field of multi-agent systems (MAS), engineers are designing microservice architectures where not just one, but *multiple intelligent agents* work in concert. Instead of a single centralized brain, you might have dozens of mini-brains (agents) each specializing in certain tasks – and they coordinate with each other. This is inspired by how teams of autonomous agents can solve complex problems together, dividing labor and sharing knowledge. For example, one agent-microservice might specialize in monitoring system health, another in optimizing user experience, and another in managing infrastructure. Together they could cooperate to self-heal an application during an outage or to optimize resource usage continuously. This multi-agent approach leads to **more**

robust, flexible, and scalable systems than a single monolithic AI service. Collaboration and communication protocols (we'll discuss standardizing these later) allow these agentic microservices to achieve goals collectively. It's as if each microservice is a skilled teammate, and the overall application is the entire team working toward a common goal. Already, research prototypes show agents negotiating with each other – for example, a workload routing agent "asking" a caching agent for recent data instead of hitting a slow database. Such patterns hint at *agent meshes* or **networks of interconnected intelligent services** ensuring resilience and adaptability.

- **Adaptive Workflow Orchestration:** Today, many workflows are orchestrated by predefined rules or require manual tweaks when conditions change. A clear emerging pattern is using intelligent microservices to orchestrate workflows in real-time. In practice, this means embedding agents that can make on-the-fly decisions about process flows. For instance, instead of a fixed sequence of API calls, a *smart orchestrator* service could alter the workflow based on current context: skipping steps if they're not needed, retrying with alternative services if the primary fails, or even reordering tasks for efficiency. This kind of **adaptive orchestration** is already being explored. For example, by late 2024 some architects were integrating agents as mediators in microservice architectures – agents that *discover available services, monitor their health, and dynamically rebalance requests* among them . By coupling microservices with this kind of agent-based oversight, organizations in experiments have managed to reduce downtime and improve fault tolerance, as the agent can reroute traffic or spin up new service instances the moment it detects a problem . In effect, the orchestration logic itself becomes intelligent and situationally aware.

- **Continuous Self-Optimization:** Another research direction is making microservices continuously self-optimize. Consider aspects like scaling, load balancing, or caching – these have traditionally been handled with static thresholds or reactive auto-scalers. Researchers are applying AI (like reinforcement learning and predictive analytics) so that microservices *tune themselves*. One example is **AI-driven autoscaling**: using predictive models to forecast load and scale microservices *in advance* of traffic spikes, rather than after thresholds are breached . This predictive scaling means fewer stalls and more efficient resource use. Similarly, caching services might analyze access patterns and proactively pre-warm caches for data that the AI anticipates will be needed soon. Early results are promising – systems that anticipate needs can approach zerodowntime and optimal performance, even under wildly fluctuating conditions. It's as if each microservice has a built-in optimizer constantly tweaking knobs for peak efficiency.

- **Learning from Experience (Feedback Loops):** Intelligent microservices of the future will likely integrate **closed feedback loops** to learn from outcomes. We see this already in A/B testing and continuous experimentation platforms – but future systems will bake this learning in deeply. Imagine a microservice that offers dynamic pricing on an e-commerce site. In a traditional setup, you'd deploy a pricing algorithm and maybe adjust it occasionally. In a future intelligent microservice, the pricing service could *continuously experiment* with different strategies (within limits) and learn from purchase data which strategy maximizes revenue and customer satisfaction. It becomes a continuously learning agent. To support this, architectures are being envisioned that include data pipelines feeding back into services and automated retraining of models. Essentially, the microservice can **update its own logic** (the ML model or rules it uses) as new data arrives, all while running live. By 2030, we expect these feedback-driven agents to be common – "continuous learners" that refine their behavior over time . This yields applications that get smarter and more efficient the longer they run. However, it also means architects must design for safe learning – constraining what can change and monitoring for anomalies (we'll revisit this in Challenges).

- **Policy-Driven and Ethical AI Behaviors:** As autonomy grows, there is a parallel push to ensure intelligent microservices behave **responsibly and predictably**. A notable emerging pattern is embedding policies (business rules, ethical guidelines, compliance requirements) into the agents' decision-making process. Researchers call these *"policydriven architectures."* For example, a future autonomous trading microservice for finance might have hard-coded policies about risk limits or ethical considerations (like not exploiting certain market conditions) that override its learned behavior if needed. By doing so, organizations aim to get the best of both worlds: adaptive intelligence *and* adherence to required rules and values. This is a response to the understandable caution around free-roaming AI systems. Design patterns are being refined so that **human administrators can set high-level objectives and boundaries** (e.g., "maximize throughput without exceeding $X cost and maintain fairness to all regions") and the intelligent microservices then operate within those guardrails. We can expect future platforms to include dashboards where such policies are defined and the agents continuously check themselves against them. The goal is to combine autonomous optimization with **accountability**, ensuring that as systems get smarter, they also remain aligned with business intent and ethics.

In summary, the research frontier is painting a picture of microservice ecosystems that are *far more dynamic and autonomous* than today. We're heading toward architectures where each component can learn, decisions are made in a distributed fashion by collaborating agents, and the entire system continuously improves itself. As one industry report put it, *"the leap to agentic AI promises to fundamentally redefine how computing systems are built and how software functions in the real world"* . It's not just a technical tweak – it's a paradigm shift in software design.

Next, let's discuss how intelligent microservices will interplay with other **emerging technologies** that are evolving in parallel. After all, trends like large language models and edge computing are bound to influence (and be influenced by) how we build microservices moving forward.

Integrating with Emerging Technologies

The future of intelligent microservices doesn't exist in a vacuum – it will be shaped by, and help shape, several other technology trends. Let's explore how some of today's hottest emerging technologies could integrate with agentic microservices to create tomorrow's solutions. We'll look at Large Language Models, serverless AI, IoT and edge computing, AutoML, and decentralized systems, among others. Each of these brings new capabilities and also poses new design questions for microservices.

Large Language Models as Brains for Services

One of the most influential tech developments in recent years has been the rise of **Large Language Models (LLMs)** – AI models like GPT-4 (and beyond) that can understand and generate human-like text, code, and even plans. These models are incredibly powerful at reasoning with language and knowledge. How do they fit into microservices? In the future, it's very plausible that LLMs will be embedded within microservices as "brains" or reasoning engines. For example, imagine a customer support microservice that doesn't just fetch FAQ answers from a database, but actually has an LLM under the hood to *converse* with users and solve novel problems. That microservice might internally call the LLM to interpret a complex query and decide which underlying APIs to invoke. In essence, the LLM acts as an **intelligent coordinator** within the microservice, giving it advanced language understanding and decision-making ability.

Another integration pattern is using LLMs as **global orchestrators or planners**. Because LLMs are great at high-level reasoning (they can output step-by-step plans when prompted), future architectures might involve an LLM-based agent that takes a user request and breaks it into tasks for other microservices to execute. This is like having a very smart dispatcher at the front of your system. For instance, a user might ask, "Help me plan a vacation to a warm country under $2000 budget." An LLM-powered service could parse this and orchestrate calls to flight, hotel, and tour booking microservices, handling constraints and preferences, and then return a composed answer. We're already seeing early versions of this pattern in products that call themselves "AI Agents" which use an LLM to decide which tool (microservice API) to call at each step.

In addition, LLMs could assist *within* microservices for things like code generation or configuration. Picture a DevOps microservice that uses an LLM to generate infrastructureas-code scripts on the fly based on high-level goals ("I need a secure network config for a new payment service"). The LLM interprets the request and produces the appropriate config, which the microservice then applies – saving operators time.

However, incorporating LLMs introduces challenges: these models are large (often running in the cloud), can be expensive to run, and their decisions are probabilistic. That means our microservices need to handle the possibility of an LLM giving an unexpected answer. In practice, this will entail **robust validation and fallback**. For example, if an LLM's plan or response seems off, an intelligent microservice might have a secondary logic to doublecheck or a simpler algorithm to fall back to. Despite these challenges, the synergy is compelling – the microservice provides structure, integration, and domain context, while the LLM provides a flexible intelligence that can navigate ambiguity.

We can foresee a future where many microservices include an LLM component (or call out to an LLM service) when they need to handle unstructured data or complex reasoning. Cloud providers are already offering *LLM-as-a-service* APIs, which will make it easier to plug into microservice architectures. By 2030, it might be as common to write "this service calls an

LLM" as it is to say "this service calls a database." The result will be microservices that feel almost human in their ability to understand requests and interact with users, while still maintaining the reliability and scalability of software components.

Serverless and Ephemeral AI Services

Another trend transforming how we deploy software is **serverless computing** – where developers run code without managing servers, and where resources scale up and down automatically. Services like AWS Lambda, Azure Functions, and Google Cloud Functions have popularized this model. Looking ahead, serverless will intersect with intelligent microservices in intriguing ways.

First, we might see **AI-driven logic deployed as serverless functions**. For example, a heavy machine learning task (like image recognition or model training) could be packaged as a small function that spins up only when needed, processes data using an AI model, and then shuts down. This allows highly elastic scaling of AI workloads. Intelligent microservices could leverage this by offloading certain tasks to serverless functions: e.g., an e-commerce microservice might invoke a serverless function to generate a personalized product recommendation model for a user on the fly, rather than running a model continuously for every user. This *"just-in-time intelligence"* can be cost-effective and scalable.

Second, the microservices themselves might become more **ephemeral and event-driven**.

In a traditional microservice architecture, you have services running 24/7 awaiting requests. In the future, we could have *event-triggered microservice instances* that come alive in response to specific events or conditions and then disappear. For instance, imagine a swarm of incident response agents that only spin up when an anomaly is detected in system metrics; each agent grabs a piece of the diagnostic work, and once the issue is resolved they terminate. This is a fusion of the microservices concept with the serverless pay-per-use model. The benefit is extreme scalability and cost-efficiency: you only consume resources when the situation actually calls for those microservices.

Serverless AI services are also emerging – cloud services where you can deploy an ML model and it auto-scales with demand, without you managing servers. In the future, an intelligent microservice might seamlessly integrate with such services. For example, a fraud detection microservice in finance could call a cloud-hosted AI scoring function that scales on its own. The microservice doesn't worry about how many instances or GPUs that function uses; it just sends data and gets results. This allows microservices to incorporate very **compute-intensive AI** without each service having to be a beefy long-running process.

One challenge and opportunity here is **cold start latency** – serverless functions have a brief delay when spinning up. Intelligent microservices might mitigate this by predicting when they'll need certain functions and pre-warming them. Or better, by using lightweight models that load quickly. We might also see hybrid models where a microservice keeps a small "brain" in-memory for instant responses and leverages a serverless "bigger brain" for more complex queries that can tolerate a slight delay.

In short, serverless paradigms will make intelligent microservices more granular and eventdriven. We could end up with architectures where the distinction between a persistent service and a transient function blurs – everything becomes a *mesh of on-demand microservices*. The combination of AI and serverless means even **massive computations can run briefly and cheaply**, enabling advanced analytics or ML in situations previously impractical. This will open the door to innovation, as developers won't be as constrained by infrastructure cost or capacity when experimenting with new intelligent features.

IoT and Edge Computing: Intelligence at the Edge

By 2025 and beyond, the Internet of Things (IoT) and edge computing have become pervasive. Billions of devices – sensors, cameras, appliances, vehicles – are producing data and performing computations outside traditional data centers. Integrating intelligent microservices with IoT/edge environments is a huge frontier for the coming years.

One aspect is **pushing intelligence to the edge**. Rather than funneling all data back to a central cloud microservice for analysis, we can deploy microservices *on the edge devices themselves or on nearby edge servers*. These microservices could do initial data processing, run AI models locally, and make immediate decisions. For example, consider a manufacturing plant (we'll revisit a detailed scenario later): each machine might have an embedded intelligent microservice that monitors sensor readings in real-time with a small

ML model. If it detects an anomaly (like a vibration pattern indicating a part might fail soon), that microservice/agent can take local action – maybe adjusting the machine or alerting a technician – without waiting for cloud instructions. This leads to **low-latency responses** and reduces bandwidth usage since only important data or decisions get sent to the cloud. As edge devices become more powerful, we expect more microservices (even those running complex AI models) to live at the edge. In fact, tech surveys have started talking about *"LLM at the edge"* – running trimmed versions of large language models on edge hardware for things like voice assistants or smart vehicles, so they don't have to constantly call the cloud.

Another angle is how edge and cloud intelligent microservices work together. Often this is called a **hybrid edge-cloud architecture**. You might have a hierarchy: edge microservices do fast local computation and short-term decisions, while cloud microservices (with more compute power and a broader view of data from many edge nodes) do deeper analysis, long-term learning, or coordination among sites. This is somewhat analogous to how in a hive, individual bees (edge agents) make quick calls but the overall colony intelligence (cloud) emerges from their interaction. For instance, a smart traffic management system could have microservices on each traffic light controller (edge) that adjust lights based on immediate car queues, and a central cloud service that aggregates city-wide data to optimize timing plans or handle unusual events like evacuations. The two levels communicate – the edge services feed data upward and execute commands downward. This **distributed intelligence** can make systems both responsive locally and optimized globally.

IoT devices also introduce myriad data formats and connectivity protocols. Future intelligent microservices will need to be adept at **speaking the language of IoT** – whether it's MQTT for sensors, OPC-UA for industrial machines, or V2X protocols for vehicles. We foresee standardized adapters or communication microservices that translate these protocols so that higher-level AI agents can treat IoT inputs similarly to any other events.

Security and reliability are critical here: an intelligent microservice at the edge might be controlling physical equipment or vehicles, so failure is not an option. This means techniques like *federated learning* (training AI models across many devices without moving all data to cloud) and robust fail-safe modes will be important. If an edge microservice loses connectivity, it should still function safely on its own (perhaps with a constrained mode of operation) until it can sync with the cloud again.

Interestingly, pushing AI to the edge can also address privacy concerns – sensitive data (like video from a hospital camera or personal info from a wearable) can be analyzed on-device by an intelligent microservice, and only anonymized or relevant results are sent to the cloud. This aligns with emerging regulations requiring data minimization.

By 2030, we expect an **edge-cloud continuum** of intelligent microservices. Think of it like a nervous system: reflexes handled at the edge (spinal cord) and complex decisions at the brain (cloud), with constant communication between. Many companies are already piloting this: for example, autonomous vehicle fleets use edge AI in each car for immediate driving decisions and a cloud service to collect learnings from all cars to update the models globally. The pattern will repeat in industry after industry. Intelligent microservices will be everywhere from your smart home (deciding when to heat rooms or reorder groceries) to the outermost network nodes of telecom providers (managing bandwidth and predictions locally). The microservice architecture – with its modular, distributed nature – is a great fit for this heterogeneous world of edge devices, ensuring that pieces of the application can be deployed wherever it makes the most sense.

AutoML and Continuous Model Innovation

Building AI-infused microservices is not a one-and-done task – the AI models themselves need to be created, tuned, and updated. Enter **AutoML (Automated Machine Learning)** and related concepts like automated data science and MLOps. AutoML refers to systems that can automate the process of selecting models, tuning hyperparameters, and even feature engineering, with minimal human intervention . Looking forward, AutoML will play a key role in the evolution of intelligent microservices by enabling continuous model innovation and adaptation.

Consider a microservice that forecasts supply chain demand for a manufacturing company.

In the traditional approach, data scientists might train a prediction model once, deploy it into the microservice, and update it maybe monthly or when performance drops. In the future, that microservice could have an *embedded AutoML agent* that constantly looks at incoming data and periodically tries to generate a better model on its own. It could run experiments in the background using new data – trying out different algorithms or neural network architectures – and if it finds a model that significantly improves accuracy, it can suggest or even deploy that model (subject to governance rules). This means the microservice is not static; it's **improving itself over time**. Some call this idea *"AI creating AI"* – the microservice's AI brain is partially designed by another AI process . This can dramatically accelerate improvement, because it removes bottlenecks of manual tuning and allows very frequent updates.

AutoML can also lower the barrier for organizations to deploy intelligent microservices. Not every company has a big team of machine learning PhDs. In the future, a developer or domain expert might simply feed data to an AutoML-powered service and say "please optimize a model for X metric," and the system will produce a model ready to be plugged into a microservice. This ties into organizational maturity (which we'll discuss soon) – companies at earlier stages of AI adoption could leverage AutoML to catch up faster, letting the machines do some of the heavy lifting in AI development. In effect, **AutoML will help democratize the creation of intelligent services** .

We also anticipate more **continuous training/learning pipelines** integrated with microservices. MLOps practices are evolving such that training and deploying new model versions can be automated safely. For example, an intelligent microservice might have two components: the live serving model and a retraining pipeline. The retraining pipeline (possibly orchestrated by an AutoML system) could run whenever there's a certain amount of new data or a scheduled interval, and if the new model passes evaluation tests (perhaps on a hold-out dataset and through bias checks), it can be deployed to production – maybe using a canary release to ensure it performs well before full rollout. All of this could happen with minimal human oversight, aside from defining the initial parameters of the process. This means the service essentially **keeps itself up-to-date** with the latest information. It's like having a gardener that not only waters the plant but also crossbreeds it with other plants to create a more resilient version continuously.

An important consideration will be to maintain *governance and transparency* around models created via AutoML. Organizations will need confidence that these automatically generated models are reliable and fair. We might see the rise of standardized validation suites that every new model (even if AutoML-created) must pass, including checks for bias, robustness, and concept drift. There could also be metadata tracking: every model version might carry an "audit trail" describing how it was generated, what data was used, and what variations were tried – so that there's accountability even if a human didn't handcraft it.

AutoML and continuous learning go hand-in-hand with the earlier idea of feedback loops. For instance, a microservice might notice its performance degrading (say, error rates creeping up because user behavior changed) – it could trigger an AutoML routine to find a better model that fits the new pattern. In a way, the service is *self-healing its intelligence*. Early versions of this are seen in things like recommendation engines that periodically retrain as new user interaction data comes in. In the future, it will be far more automated and ubiquitous.

To summarize, AutoML will allow intelligent microservices to *evolve their intelligence on their own*, or with much less human help. This fosters continuous innovation – models that would be impractical to tweak daily by humans can nonetheless be improved on a daily or even hourly basis by machine-driven experimentation. Organizations that embrace this will likely outpace those that rely solely on sporadic human-led model updates, because their services can quickly adapt to changing conditions or new opportunities.

Decentralization and Trust in Distributed Systems

A more radical vision of the future involves **decentralized systems** – applying concepts from blockchain, peer-to-peer networks, and distributed ledgers to microservices. Today, most microservices operate within one enterprise or one cloud environment under a central authority. But emerging trends suggest future intelligent microservices might form **decentralized ecosystems** that transcend a single company, enabling new forms of collaboration and autonomy.

One aspect of this is services operated by different parties working together directly because trust is handled by technology. Imagine a digital ecosystem for logistics where dozens of companies (manufacturers, shippers, warehouses, retailers) have their microservices (for inventory, shipping updates, etc.) communicate and coordinate automatically. In a centralized model, maybe everyone would integrate via a single platform or API hub owned by a third party. In a decentralized model, these microservices could interact *peer-to-peer*, using blockchain-based smart contracts to enforce rules and maintain a shared, tamperproof record of transactions. For instance, an intelligent contract might automatically verify that a shipment microservice recorded a delivery, then trigger a payment microservice to release funds – all without a central broker, but still trustworthy because the rules are coded on a blockchain.

Decentralized microservice networks could lead to **digital service ecosystems** that are more resilient and democratic. There is early talk of things like *Gaia-X in Europe (a federated cloud initiative)* and other B2B data spaces that aim to let organizations share services and data in a controlled but decentralized way. Intelligent agents could thrive in such environments, negotiating and interacting across organizational boundaries. We might see standardized protocols for *inter-agent trust*, possibly building on or inspired by blockchain. A term "agentic blockchain" has even been floated to describe combining AI agents with blockchain for coordination . The blockchain provides a single source of truth and audit trail, while AI agents (intelligent microservices) carry out tasks. This could enable scenarios like decentralized autonomous organizations (DAOs) composed of microservices – essentially networks of services making decisions collectively according to encoded governance rules, with minimal human oversight.

For microservice architects, decentralization poses new questions: How do you discover services in an open network? How do you authenticate and authorize in a zero-trust environment where no single entity controls all components? Solutions might include **distributed discovery protocols** (perhaps blockchain-based registries where services publish their capabilities), and use of cryptographic keys and tokens for identity. We may even leverage the concept of *Self-Sovereign Identity (SSI)* where each service has its own verifiable identity documents that others can trust without central certificate authorities.

Another angle is **decentralized learning** – analogous to federated learning, where multiple parties jointly train an AI model without sharing raw data. Intelligent microservices across organizations could collaborate on improving a shared model via cryptographically secure protocols, so that, say, hospitals can collectively train a disease prediction agent without exposing patient data to each other. This way, each hospital's microservice gets smarter by learning from others, but privacy is preserved.

The *standardized agent communication protocols* we'll discuss in the next section also play a role here. If industry-wide protocols exist, a microservice from Company A and one from Company B can talk if they adhere to the protocol, a bit like how any email server can talk to another via SMTP. This opens up the possibility of **open ecosystems** of intelligent microservices. We might one day have something akin to an "Internet of Services" – a web where not just documents are linked (like the classic web), but *services* are interconnected and can spontaneously form supply chains or solutions on the fly.

Of course, decentralization also brings challenges: performance overhead of consensus algorithms, complexity of managing upgrades in a decentralized network, and the difficulty of debugging problems when no one entity has the whole picture. We might see hybrid approaches where certain control-plane aspects remain centralized for efficiency, but operational interactions are decentralized. For instance, a consortium of companies could run a shared blockchain for critical records, but use direct API calls for high-speed data between microservices when trust is already established.

In summary, while more speculative, the decentralization trend could redefine the scope of microservice architectures – from being an internal architecture pattern to becoming the basis of **cross-organization digital ecosystems**. Intelligent microservices in that world would not only be autonomous in how they make decisions, but also in how they connect and transact with others. It's a future where your microservice might directly interact with a partner's or customer's microservice, safely and securely, with AI ensuring optimal outcomes for all parties. This kind of open ecosystem could unleash innovation (imagine anyone being able to offer a microservice that others can incorporate into their workflows, almost like publishing an app) and possibly even create new digital economies around microservices usage.

Having explored how intelligent microservices might weave together with technologies like LLMs, serverless, edge/IoT, AutoML, and decentralization, we can see that the future will be a rich tapestry of interconnected advancements. Each of these areas feeds into the others: for example, IoT edge devices will produce data that drives AutoML retraining in cloud microservices; LLMs might help orchestrate across decentralized networks by interpreting high-level goals into low-level actions.

Next, let's shift from technology integration to a higher-level perspective: what kinds of **architectures and ecosystems** might emerge when we put all this together? We'll imagine some speculative future architecture patterns, such as self-organizing swarms of microservices and standardized agent communication frameworks, which could become the norm.

Speculative Visions of Future Architectures

What might an entire system of intelligent microservices look like in the not-so-distant future? In this section, we step into a speculative mode and paint a picture of possible architectural paradigms that could become reality. These are informed guesses – extrapolations of current ideas to their futuristic conclusions. We'll discuss **self-organizing microservice "swarms"**, the emergence of **standard communication protocols for agents**, and the idea of **digital service ecosystems** that function like organic communities. Think of this as imagining the "city planning" of future software systems: how all these intelligent parts might live and work together in organized ways.

Self-Organizing Microservice Swarms

One of the most exciting prospects is that intelligent microservices might organize themselves dynamically, much like a swarm of bees or an ant colony. Today, even with microservices, a lot of planning is top-down: architects design the interactions, and orchestrators enforce workflows. But as microservices become more agentic and capable of decision-making, we could move towards systems that **reconfigure themselves on the fly** to meet goals, without every step being predesigned.

Consider a swarm of microservices that collectively runs an e-commerce site. During normal operation, they function in a standard way – user requests go to an API Gateway, which routes to various services (product catalog, cart, recommendations, etc.). Now imagine it's Black Friday and load spikes unpredictably, or perhaps an external event causes an unusual pattern of browsing. In a self-organizing architecture, the microservices could *sense* this change (through elevated load, different usage patterns) and start adapting without a human turning any knobs. For example, several services might spawn additional helper microservices to handle the load (beyond simple auto-scaling – perhaps specialized cache services or routing agents appear temporarily). Or if the recommendation service is becoming a bottleneck, other services might collectively decide to disable advanced recommendations for low-priority users to preserve overall performance, then re-enable when load normalizes. In effect, the system *re-balances itself*, like water flowing to accommodate a container's shape.

Another scenario: suppose a part of the system fails – say a database goes down. In a selforganizing swarm, other microservices could detect the failure and two things might happen: (1) some microservices that depend on the database switch to an alternative data source or mode (perhaps read-only mode using a cached dataset), and (2) an *"auto-healer"* agent microservice might spin up a new database instance or migrate data from a replica, essentially performing a recovery workflow on its own. There might not be a single "brain" coordinating this; instead each service has simple rules or learning-based behaviors that lead to an emergent coordinated response. This is analogous to ants finding a new path if a current one is blocked – no single ant plans it, but local decisions lead to a global solution. Earlier in this book we drew the parallel of microservices to an ant colony, where each ant follows basic rules and the colony exhibits intelligence . Future architectures might make this analogy a reality: **emergent behavior** from many microservices interacting.

To achieve this, microservices will communicate more via **event-driven cues** ("signals") in addition to direct calls. Think of events like "Service X is experiencing high latency" being broadcast, upon which other services can decide how to react (some might reduce calls to X, others might spawn backup functionalities). Technologies like service meshes and publish/subscribe systems might underlie these swarms, providing the connective tissue for awareness. We might see each microservice maintaining a **local knowledge base** of what it's observed, which it shares with others in a gossip-like fashion to build a collective picture.

One could ask: without central control, could this get chaotic? Yes, which is why likely we'll still impose some structure – for instance, an *initial set of rules or heuristics* coded by architects that guide the self-organization. Also, such systems would need lots of simulation and testing (perhaps using AI to verify stability) before being trusted in production. But the benefit is adaptability at a scale and speed humans can't match. The system could handle situations its designers never explicitly anticipated, because the swarm can explore new behaviors within bounds.

A tangible early sign of this trend is in *container orchestration and self-healing infra*. Kubernetes, for instance, will restart containers that crash – a simple form of selforganization at the infrastructure level. Expand that idea upward into the application:

microservices could not only restart themselves but rewire their interactions. For example, in a truly self-organizing microservice swarm, if Service A goes down, Service B might directly interface with Service C to compensate, even if normally A was in between – essentially *dynamic re-routing* based on current availability and goals.

Looking further out, one could envision microservices that even **redefine their own boundaries** on the fly. Perhaps what started as two services might merge into one process for efficiency under certain loads, and split again when the load changes (this would be a kind of *adaptive granularity*). Or microservices cloning themselves to explore different strategies and then retiring the less successful clones – almost an evolutionary approach.

These ideas blur the line between design-time architecture and runtime behavior.

While this is highly experimental, the seeds are there: multi-agent reinforcement learning research, for example, looks at how independent agents can learn to cooperate to maximize a system objective. It's not hard to imagine those techniques applied to fleets of microservices – each microservice could have a small RL agent that adjusts its interaction strategies to improve some global reward (like end-to-end throughput or minimal error rates), and through trial and error a good equilibrium is reached. At that point, we've designed the *framework for learning* and let the system self-organize into an optimal architecture for the given conditions.

Standardized Protocols for Agent Communication

As intelligent, agent-like microservices proliferate, a practical need arises: **how will these agents communicate effectively, especially in large**

heterogeneous environments? In the early days of microservices, we converged on standards like REST/HTTP and gRPC for service communication, and OpenAPI/Swagger for interface definitions. For agentic microservices, especially those that might dynamically discover and talk to each other, we may need higher-level protocols – essentially languages for negotiation, cooperation, and knowledge sharing.

In the field of multi-agent systems, there have been attempts at such standards (for example, the FIPA – Foundation for Intelligent Physical Agents – communication language was an academic standard for agent messages). We might see a modern revival or new standards tailored to microservice needs. These could define things like a common **message format for agent negotiation** (e.g., an agent could send a structured message:

"request(task=X, constraints=Y)" and another replies "propose(solution, cost=Z)" or "reject(reason)" etc.). Imagine microservices being able to "talk" in a universal format regardless of the programming language or platform, as long as they adhere to the agent protocol. This would hugely simplify building complex workflows on the fly – much like how having a standard API contract simplified integration between services.

A standardized protocol might also cover **discovery and descriptions**. For example, an intelligent service might advertise: "I am capable of handling payments with these conditions" in a registry or to the network. Other microservices can query for capabilities rather than specific endpoints. We see glimmers of this idea in things like service meshes and API gateways that allow some discovery, but an agent-specific protocol would be more semantically rich (covering intents and outcomes, not just endpoints and input/output schemas).

By having common communication patterns, we could plug in new intelligent microservices into an existing ecosystem and they would know how to *communicate their intentions and understand others*. For instance, if all microservices followed a standard handshake to establish trust and share context, you could have one service ask another: "Can you take over this task for me? Here's the context and my current state," and the other would know how to respond and what information to provide back. Standard dialogues for things like task delegation, joint decision-making, or even conflict resolution could be part of the protocol.

One interesting possibility is the emergence of **agent hubs or gateways** that mediate communication. A recent development by 2024 was the concept of "AI gateways" or "agent hubs" – centralized platforms through which organizations manage and monitor their agents . These hubs enforce some consistency in how agents connect, handle security, and exchange data. In the future, such hubs could implement the standardized protocols, effectively acting like an exchange where agents register and send messages via well-known patterns (similar in spirit to how an email server knows how to route emails because everyone uses SMTP). This doesn't necessarily centralize decision-making, but provides an infrastructure so agents can find and talk to each other easily.

For example, an enterprise might have an Agent Communication Server where any new intelligent microservice registers itself when it starts. If it needs something, it asks the server

"Who can do X for me?" The server, using the standard protocol, finds the best match (maybe multiple candidates) and facilitates a connection or conversation between the requesting agent and the providing agent. This could even extend across organizations if standard protocols are adopted industry-wide – analogous to how different companies' networks talk over the internet using TCP/IP and higher protocols.

Security is a crucial part of this. Standard protocols would need standard ways to authenticate (prove identity) and authorize agents. Possibly incorporating modern techniques like OAuth tokens for agents, or even leveraging blockchain identities for crossorg trust. The benefit is that as these agents become more autonomous, a human admin can still set policies like "Service A (order processing agent) is allowed to request credit checks from Service B (credit agent), but not allowed to directly contact Service C (some other sensitive service)." The agents, when they attempt communication, would be automatically checked against these policies via the standard handshake.

Another angle is **knowledge exchange**. Agents might share not just requests/responses but learning or data. A standard could define how one agent describes an insight it found ("I've noticed pattern Y in user behavior, you may use it"). In a multi-agent intelligent system, having a common representation for certain knowledge (ontologies, schemas) would help them collaborate. For instance, if agents in healthcare can agree on a standard way to represent a patient's record or a diagnostic request, they can collaborate to treat a patient collectively (one agent handles scheduling, another analysis, another medication management) seamlessly.

It's worth noting that human organizations also tend to form standards when a field matures – think of how HTTP and HTML standardized the web, or how SQL standardized database queries. As agentic microservices become more common, industry groups or consortia might form to propose these communication standards. We might see something like "OpenAgent Protocol 1.0" emerging from a body of tech companies to ensure interoperability. This could be accelerated if big tech players incorporate agent features into their platforms (e.g., if Azure and AWS both support a similar agent messaging bus, others may follow).

In summary, standardized agent communication protocols will be an enabler for **scaling intelligent microservices to large, complex systems**. They will reduce friction in composing new solutions out of intelligent components, much as standardized APIs did for traditional services. With common languages and protocols, the focus can shift from *how* they talk to *what* they're trying to achieve together. We expect these standards to emphasize loose coupling (so systems remain flexible) but rich semantic content (so agents truly understand the tasks and context, not just pass blunt data). Achieving the right balance will likely be an ongoing process, but the momentum is toward more interoperability as agentic systems move from isolated experiments to mainstream architecture.

Digital Service Ecosystems and Marketplaces

Expanding our view even further, we anticipate the rise of **digital service ecosystems** – essentially, large-scale networks of services (potentially intelligent/agentic) spanning many organizations, forming the backbone of entire industries or cross-industry platforms. In such ecosystems, services offered by different parties can be combined dynamically to deliver value to end-users, much like apps on a smartphone form an ecosystem.

A concrete vision: think of something akin to an "App Store" but for cloud services or APIs. Companies could publish intelligent microservices (with defined capabilities and interfaces) to a marketplace. Others could discover these and integrate them into their processes on the fly. For example, a small business might not develop its own AI-based demand forecasting microservice, but in 2030 they might go to a marketplace and find "ForecastProService" offered by some provider, then subscribe to it and connect it to their supply chain system within minutes. Under the hood, standardized protocols and cloud integration make this plugand-play. The business's other microservices communicate with this new service as if it were internal, thanks to well-defined contracts and possibly intermediary adaptation layers.

This ecosystem approach can lead to **digital supply chains**. Suppose a customer order triggers a chain of events: the retailer's microservice calls a payment service (maybe provided by a fintech company), which then calls a shipping service (logistics company), which also calls a customs clearance service for international orders, etc. If each of these is an intelligent microservice in a network, the entire chain might self-optimize. Perhaps if one shipper is delayed, the shipping service automatically switches to an alternative provider's microservice to fulfill the delivery, because all options are available in the ecosystem and can be negotiated in real-time. We move from static partnerships to fluid, on-demand partnerships orchestrated by the software itself.

Self-organizing ecosystems could emerge, where new niche services find a place. Just like in an ecological system different organisms fill niches, here a digital ecosystem might have microservices that specialize and interdepend. For instance, a data provider service might gather environmental data; multiple analysis services might subscribe to that data and produce insights; a marketplace service might aggregate those insights for specific clients. If one piece is removed, alternatives fill in if available. The ecosystem thrives as long as there is an overall balance of supply and demand for the capabilities.

This vision goes hand in hand with decentralized trust (discussed earlier) and protocols. For truly open ecosystems, participants need to trust that interactions are fair and reliable. Blockchain or centralized clearinghouses might ensure transactions (calls that have monetary or contractual implications) are recorded and settled. Think of a case where using someone's microservice costs a fraction of a cent per call – microservices might even have built-in wallets or use crypto-tokens to pay each other for services in a fully automated way. It sounds far-fetched, but pilot projects in the IoT world are already looking at devices that pay each other for resources (like an EV charging station paying a car for feeding energy back to grid, autonomously). Extend that to software services, and you have a digital economy of microservices.

An example to illustrate: Consider the healthcare industry in 2035. Instead of every hospital having to own all capabilities, there might be a **healthcare service ecosystem**. A hospital's patient management microservice could, when needed, call out to a third-party AI diagnostic service for a second opinion on an MRI scan, and perhaps simultaneously request an insurance pre-approval from an insurance company's authorization microservice. If the primary diagnostic service is overloaded, it automatically finds another accredited service on the network to get results faster. All these interactions are governed by standardized protocols (so minimal custom integration) and strong consent/security frameworks (so patient data is only shared in encrypted or controlled ways, meeting regulatory requirements). The outcome is better care delivered faster, and each provider in the ecosystem gets compensated (maybe via automated billing records generated through the interactions).

One can also imagine **industry-wide AI models** accessible as services. Not every company will train its own GPT-XX model; there might be a few that provide specialized large models (for law, medicine, engineering, etc.) via APIs in these ecosystems. Intelligent microservices could tap these "expert" services on demand. The ecosystem thus has layers – some microservices are aggregators or orchestrators, while others are utilities or experts, and others provide raw data or transactions.

Crucially, **governance** will matter. Who ensures that an ecosystem runs smoothly, that bad actors are kept out, that standards are maintained? Likely consortia or regulatory bodies will be involved for public-facing ecosystems (similar to how financial networks have governance). In closed ecosystems (like within a big enterprise or a coalition of partners), the governance might be easier – maybe a lead organization or a joint platform.

Technical standards (like those agent protocols) will need to be complemented by **legal and business frameworks**. Contracts might need to accommodate AI-driven actions (for instance, if one service's AI makes a mistake that causes loss, who is liable?). Some forward-looking legal agreements might treat certified AI services almost like outsourced employees or consultants, with predefined accountability. We might see certification programs: e.g., an ecosystem might require any AI microservice that joins to have passed certain tests or audits (for bias, security, etc.), analogous to how apps in an app store undergo review.

On a more conceptual level, these digital ecosystems align with the vision of a highly **modular digital world**. Companies could become assemblers of capabilities rather than builders of everything. Competition and innovation could flourish because a small player with a great microservice could potentially serve many customers by joining a larger ecosystem, rather than having to build a whole product around it. In effect, the **unit of software reuse** and distribution might move from "application" to "microservice API".

We're already seeing early steps: major cloud providers have marketplaces for APIs and SaaS integrations; initiatives like Open Banking require banks to expose services to thirdparty apps. The trajectory is toward *more openness and interconnectivity*. Intelligent microservices will accelerate that by providing the autonomous decision-making needed to make use of open services efficiently. A dumb service might not know what to do with thousands of available API options, but an intelligent agent can reason and choose the right one.

To wrap up this visionary section, imagine a future "digital city" of services: standardized roads (protocols), law and order (governance), marketplaces (for services and data), and a buzzing population of intelligent microservices (agents) representing businesses, individuals, and public entities – interacting constantly to deliver value and adapt to needs. It's a grand vision, and while reality may not look exactly like this utopia, it provides a guiding star for what architects can work towards: **systems that are open, adaptive, and richly interconnected**.

Having painted these future architectures and ecosystem pictures, let's turn our attention to how organizations can prepare for and build capabilities to thrive in this future. After all, technology alone isn't enough – companies need the right practices, cultures, and maturity to harness these possibilities.

Building Organizational Capabilities for the Future

Even the most advanced intelligent microservice architecture will falter if an organization isn't prepared to implement and maintain it. The future belongs not just to technology, but to the **organizations that cultivate the capabilities** to use that technology effectively. In this section, we'll discuss how companies can ready themselves. This includes frameworks like AI maturity models to assess and guide progress, best practices for continuous innovation (so that "future architecture" doesn't remain always in the future), and leadership principles to foster a culture where intelligent microservices can flourish. Think of it as developing the human and process side to match the tech.

Maturity Models for Intelligent Microservices

One useful tool for assessing and planning organizational readiness is a **maturity model**. This is a framework that defines levels or stages through which an organization (or its technology) progresses, from basic to advanced. For intelligent microservices, we can imagine a maturity model that spans from traditional architectures all the way to fully agentic, autonomous systems. Each level comes with certain capabilities, tooling, and mindsets.

Let's outline a possible **Intelligent Microservice Maturity Model** with stages:

Maturity Level	Characteristics	Capabilities/Practices
Level 1 – Traditional Microservices (Pre-AI)	Microservices architecture in place, but services are mostly static, not learning or adapting. Decisions are rulebased or manual.	- Strong DevOps and CI/CD for deploying microservices. - Monitoring and logging for services, but analysis is manual. - Teams structured around services (DevOps teams), but no AI expertise in teams.
Level 2 – Data-Driven Services	Incorporating basic AI/ML into microservices. Some services use ML models (e.g., analytics, recommendations) that are periodically updated. Still mostly reactive.	- Data pipelines feeding into model training (perhaps offline). - Initial MLOps processes: versioning models, A/B testing new models. - AI specialists collaborate with service teams to build intelligent features.

Level 3 – Adaptive Microservices	Microservices begin to adapt in real-time. Feedback loops in place: services can retrain or adjust parameters based on new data (online learning, continuous deployment of models). Some automation in orchestration using AI (e.g., predictive scaling, automated failure mitigation using ML).	- Real-time data streaming and online learning for critical models. - AutoML tools used to continually improve models. - Intelligent orchestration: Alops tools predicting incidents, automated scaling triggers. - Cross-functional teams (developers + data scientists + ops) working closely.
Level 4 – Agentic Microservices	Services act as autonomous agents. They collaborate via intelligent protocols, make independent decisions in their domain, and only highlevel goals are set by humans. Multi-agent systems in production handling complex workflows (with some human oversight).	- Use of agent frameworks and standard agent communication protocols. - Governance in place for agent decisions (policies, ethical guidelines coded in). - Simulation environments to test multi-agent behavior before production. - High observability: every agent's decisions are logged for transparency.
Level 5 – Self-Organizing Ecosystem (Visionary)	Organization's services seamlessly integrate with external intelligent services. The architecture can	- Participation in industry-wide service protocols/marketplaces. - Automated negotiation and discovery between services (within and outside
Maturity Level	**Characteristics**	**Capabilities/Practices**

	rearrange itself in response to environment. Business ecosystems dynamically form through service interactions. Human intervention is minimal except for strategic direction and governance.	org). - Continuous innovation culture: new microservices or agent behaviors are tried regularly in sandbox environments. - Leadership and governance ensure ethical, legal compliance despite high autonomy.

This model goes from a baseline of just having microservices (Level 1) up to a futuristic scenario (Level 5) where things are very autonomous and cross-org. Most organizations today are around Level 1 or 2, maybe some pioneering ones reaching into Level 3. Very few (if any) are at 4 yet, and 5 is more of a vision. The model can help you **assess where you are** and what to aim for next.

For example, a company might evaluate and realize: *We have microservices and some analytics (Level 2), but we haven't implemented any kind of real-time learning or AI-driven orchestration (Level 3). Our next step could be to deploy an AI-based monitoring system that predicts failures and integrates that into our incident management.* That moves them into adaptive territory. Or another company might be dabbling with multi-agent prototypes (Level 4 in labs) but hasn't standardized any protocols or governance for them, indicating what work is needed to truly operationalize that level.

It's important to note that higher is not always strictly better in the short term – it depends on business needs. Some organizations might not need to go beyond Level 3 for a while if Level 4 complexity outweighs benefits in their context. However, having the roadmap allows strategic planning. Over time, as tools and practices mature, moving up becomes easier. For instance, today agentic microservices (Level 4) might sound very complex, but by 2030 there might be off-the-shelf frameworks making it much more accessible (like how cloud made scaling easier, agent platforms might make autonomy easier).

One can draw parallels to an **AI maturity model** many enterprises use, which typically assesses readiness across dimensions like data, talent, governance, and strategy . Our focus here is specifically on the software architecture side and how integrated AI is. But indeed, to progress, companies must invest in **data quality**, **skills**, and **governance** at each step:

- Going from Level 1 to 2 often requires building a data culture – collecting and cleaning the data necessary for ML, and hiring or training data scientists.

- Level 2 to 3 requires stronger MLOps and perhaps introducing streaming data infrastructure, which implies more advanced engineering and maybe reorganizing teams to include ML engineers.

- Level 3 to 4 is a big leap: needing expertise in agents, probably a robust simulation/test culture, and very clear objectives/metrics so that agents can be aligned with them.

- Level 4 to 5 extends beyond the org, meaning partnership management, compliance

(since now you deal with external services), and possibly industry consortia membership to shape those standards.

A maturity model should not be viewed as a rigid checklist but rather a **guidance tool**. Each level builds on the previous. You wouldn't try to jump directly to Level 4 without solid foundations of 2 and 3, or you'd risk chaos (like unleashing autonomous services without proper data or oversight). Many organizations will climb the ladder incrementally: first get microservices right, then infuse some AI, then increase the autonomy step by step.

Best Practices for Continuous Innovation

Reaching the future state of intelligent microservices isn't a one-time project – it's a continuous journey. Technology and business needs evolve, so the architecture must too.

Here are some **best practices to foster continuous innovation** in this domain:

- **Institutionalize R&D and Experimentation:** Organizations leading in this space often have dedicated teams or time allocated for research and experimentation. For example, setting up an "Innovation Lab" where engineers can prototype agent-based microservices or try integrating a new AI technique in a low-risk environment. The outputs can then be gradually introduced into the production architecture. Encouraging a culture of experimentation (like Google's famous 20% time, or hackathons, etc.) means new ideas constantly bubble up. Some might fail, but some will succeed and keep pushing the system toward the future. *Treat your architecture as an evolving product*, not a completed project.

- **Use Iterative Development and Pilots:** When trying something like a new orchestration AI or an edge intelligent service, start with a pilot. Maybe enable the new intelligent feature for a small subset of traffic or in a single domain, measure impact, and iterate. This lean approach (build-measure-learn) prevents huge bets from going wrong and provides evidence to support scaling successful innovations. For instance, pilot an AI-driven autoscaler on one microservice that is known to have intermittent load spikes. If it performs well (saves X% cost or avoids Y incidents), that bolsters the case to extend it to other services.

- **Leverage Cloud and Managed Services:** Continuous innovation can be accelerated by not reinventing the wheel for every capability. Cloud platforms are increasingly offering components for AI and automation (like AWS's AI services, Azure's cognitive services, GCP's AutoML tools, etc.). Using these as building blocks can speed up development. Also, as noted, cloud marketplaces may have third-party services that can be trialed. Adopting a new technology doesn't always mean building it from scratch – often it can mean integrating it. Being cloud-neutral in design (so you're not locked in) but *cloud-enabled* in practice can give you the best of both – flexibility and speed.

- **Invest in DevOps -> MLOps -> AIOps progression:** Initially, DevOps practices gave us fast code deployment. MLOps extends that to fast model deployment and monitoring. AIOps goes further to use AI in IT operations (like intelligent monitoring, automated issue resolution). An organization should cultivate these capabilities stepwise. Ensure your CI/CD is rock solid (tests, automated deployments) – that's needed even at Level 1/2 of maturity. Then incorporate data/ML pipelines into that CI/CD to achieve continuous model delivery. Then begin to allow AI to manage parts of operations (like anomaly detection in logs). This technical practice progression underpins the architecture progression. Essentially, **automate everything you can**, and then use AI to automate the automation, so to speak. This builds a strong platform for intelligent microservices to run on.

- **Cross-Pollinate Skills in Teams:** An innovative architecture requires breaking silos between traditional software developers, data scientists, and ops folks. Consider forming cross-functional teams responsible for a set of microservices end-to-end, including their intelligent features. When team members with diverse skills work closely, new ideas emerge (a data scientist might suggest an AI improvement that a developer knows how to integrate easily; an ops engineer points out a self-healing opportunity which an ML expert can automate). Encourage rotation or at least regular knowledge sharing between teams focusing on AI components and those focusing on system design. Over time, aim to build **T-shaped skills** in people (deep in one area but broad enough to understand others). This ensures continuity too – that innovative systems don't fall apart if a key AI expert leaves, because several team members have picked up some of that knowledge.

- **Frameworks for Prioritization and Value-Focus:** With so many shiny possibilities, organizations need to prioritize innovation that drives real value. Adopt frameworks like OKRs (Objectives and Key Results) or value-stream mapping to decide which intelligent features to pursue. For example, if the key business metric is customer satisfaction, focus AI efforts on things that improve that (maybe intelligent customer service or personalization) before something like internal code optimization. Keep pilots and projects aligned to strategic goals, and measure outcomes in those terms (e.g., "Our new intelligent scheduling service reduced delivery times by 10%, leading to higher customer retention"). This keeps innovation from being "technology for technology's sake" and helps secure continued investment from leadership by demonstrating business value.

- **Stay Updated and Embrace External Knowledge:** The field of AI and microservices is fast-moving. Teams should allocate time for learning – reading research, attending conferences/webinars, taking courses. Partnering with academia or participating in open-source communities can also inject fresh ideas. Sometimes, hiring a few experts or consultants in cutting-edge areas (like multi-agent systems or edge AI) for short stints can level up an entire team's capability. *Continuous innovation requires continuous learning.* Organizations might even implement training programs or innovation days where employees focus solely on learning new techniques.

Continuous Improvement: Consider adopting a "postmortem" and "postsuccess"** culture. Just as you analyze failures to learn from mistakes, analyze successes to learn why they worked. For instance, if an experimental

AI-driven feature succeeded, document what factors (data quality, algorithms, team collaboration, etc.) led to that success, and make sure those practices are replicated in other projects. This meta-learning about your innovation process will refine how you approach future initiatives.

In essence, building future-ready capabilities is about making **innovation systemic**. Organizations like Netflix, Amazon, etc., are often cited for their engineering prowess – but what they really have is an innovation engine: a combination of culture, process, and technical infrastructure that constantly produces improvements. Any organization can start building this engine. It might start small – a weekly tech brainstorm, a pilot project here or there – but formalizing and scaling these practices will prepare the company to implement the advanced architectures we discussed earlier when the time is right.

Leadership Principles to Nurture Intelligent Microservices

Leadership plays a pivotal role in steering an organization into the future. Without vision and support from leadership, teams may lack direction or incentive to push boundaries. Here are some leadership principles and actions that can nurture the development of intelligent microservice capabilities:

- **Articulate a Vision (and Excitement) for Intelligent Automation:** Leaders should craft and communicate a clear vision of how intelligent microservices will benefit the organization. This isn't just about technology for tech's sake; it's about outcomes – faster innovation, better customer experiences, greater efficiency, etc. For example, a CTO might say, "In five years, I see our platform delivering personalized experiences to each customer through autonomous services that learn and improve continuously. This will set us apart in the market." By painting such a picture, leadership gives teams a North Star to align to. It's also motivating – people often get excited knowing they're working toward something groundbreaking.

- **Encourage a Culture of Autonomy and Ownership:** If you want autonomous microservices, it helps to have autonomous teams (within guardrails). Adopt leadership practices that empower teams to make decisions about how to meet high-level goals. This is analogous to how you want your microservices to operate – given a goal, figure out the best path. Leaders should avoid micromanaging the technical details; instead, focus on setting objectives and let teams experiment with intelligent solutions. A principle like Amazon's "two-pizza teams" (small, independent teams) paired with ownership mentality ("you build it, you run it") creates accountability and freedom needed for innovative solutions to emerge. When developers and product folks feel real ownership, they are more likely to proactively propose and try intelligent features that solve their pain points or advance their goals.

- **Invest in People and Skills (Continuous Learning):** Budget for training, provide time and resources for skill development. A forward-looking leader might initiate an internal AI academy or sponsor certifications in cloud and AI technologies. They might also rotate promising engineers through different roles (like a dev spending a few months

with the data science team) to broaden their perspective. Leadership should recognize and reward learning – for instance, highlight a team that applied a new technique they learned at a conference to improve a service. Building an internal talent pool that is comfortable with both microservices and AI is a strategic asset. Leaders might also consider hiring decisions that prioritize not just current skill match but learning aptitude and curiosity, which are critical in a fast-evolving field.

- **Foster Collaboration and Break Down Barriers:** Often, hierarchical or departmental barriers stifle the cross-pollination needed for intelligent microservices (which sit at the intersection of AI and software engineering). Leaders can reorganize teams into more cross-functional units, or at least create forums where different disciplines regularly meet and share insights. For example, a leader might start an "AI in Architecture" working group with members from software, ops, data science, and product to discuss opportunities and align efforts. Also, incentivize collaboration in performance reviews and project planning, so that it's not just "my team vs your team" but "how can we combine our expertise to deliver something new".

- **Promote Ethical and Responsible AI Use from the Top:** Intelligent microservices will raise ethical questions (data privacy, bias in decision-making, etc.). Leaders must set the tone that *how* results are achieved matters as much as the results. This means establishing ethical guidelines for AI (perhaps following frameworks like fairness, accountability, transparency) and ensuring teams incorporate them. For instance, if a team proposes an AI-driven personalization feature, leadership should ask: are we avoiding use of sensitive attributes? How do we ensure it doesn't unintentionally discriminate or exclude? By ingraining these questions into the approval process, leaders ensure that as autonomy increases, it doesn't lead to reputational or legal risks. When teams know leadership cares deeply about ethics, they will bake it into design rather than treat it as an afterthought.

- **Provide Air Cover for Innovators:** Often, pushing the envelope comes with failures or slower short-term output (experiments can divert time from feature factory work). Visionary leaders protect and champion the teams doing innovative work. They set realistic expectations that not every attempt will succeed and that's okay. If an experiment fails, leadership focuses on lessons learned rather than blame. This approach encourages teams to tackle bold ideas without fear. Additionally, leaders might need to communicate the value of these efforts to other stakeholders (like the Board or investors) to justify the spend. Explaining that "we invest 10% of our capacity in exploratory projects which have the potential to 10x our capabilities in the future" can frame it as prudent risk-taking. Essentially, leadership should buffer innovative teams from short-term pressures that could derail long-term progress.

- **Lead by Example in Using Data and AI:** Leaders who use data in their decisions and are curious about AI set an example. For instance, a product leader might use an internal analytics chatbot (an intelligent microservice itself) to query business metrics in meetings, showing comfort with AI tools. Or an engineering leader might run a small experiment with a new tech themselves or pair with an architect to do so – demonstrating hands-on engagement. When teams see leadership not just talking the talk but actually engaging with intelligent systems, it signals a genuine commitment. It also helps leaders understand the technology deeper, so they can make better strategic calls.

In short, leadership in the era of intelligent microservices means being a **gardener for innovation**: planting the vision, nurturing the culture, removing impediments, and ensuring ethical, sustainable growth. With supportive leadership, even a company that isn't a tech giant can innovate beyond its apparent means, because the collective talent and motivation of its people are unleashed.

Now that we've covered how to gear up the organization, we must address an important reality check: the **challenges and concerns** on the horizon. No transformative journey is without obstacles. In the next section, we'll candidly discuss the regulatory, security, ethical, and design challenges that lie ahead, so we can approach the future with eyes wide open and prepare accordingly.

Challenges and Considerations Ahead

As we march toward a future filled with autonomous, intelligent microservices, it's crucial to acknowledge the roadblocks and hurdles we will face. These range from external constraints like regulations to internal technical complexities and ethical dilemmas. Let's highlight some of the key challenges and how we might navigate them:

- **Regulatory Changes and Compliance:** The regulatory environment for AI is evolving rapidly. Governments and international bodies are introducing rules around AI transparency, data privacy, and accountability. For instance, the EU's proposed AI Act would classify certain AI applications as high-risk, requiring explainability and human oversight. An ecosystem of intelligent microservices might fall under such rules if, say, it makes decisions affecting humans (credit decisions, medical diagnoses, etc.). Organizations will need to ensure **compliance** – which could mean building in features for audit logs, explanations of AI decisions, and fail-safes for human intervention. *Data localization* laws might impact architectures too: microservices may need to keep certain data within certain geographic boundaries, complicating how an intelligent service aggregates global data. Regulatory compliance could slow down deployment of the most cutting-edge features until they are certified or approved. This isn't necessarily bad – it ensures safety – but it requires planning. Companies should incorporate compliance checks into their development process (perhaps having an AI ethics or compliance officer involved in reviewing new intelligent microservices). On the flip side, leaders might engage with policymakers to help shape reasonable regulations. Adapting to this changing landscape will be an ongoing challenge. The winners will be those who *bake compliance into design* rather than treat it as a bolt-on.

- **Security and AI-Specific Threats:** With greater autonomy comes greater risk of new security threats. Malicious actors might target intelligent microservices in unique ways: manipulating their input data to trick the AI (adversarial attacks), poisoning training data

to subtly alter a model's behavior, or exploiting the dynamic nature of these services (like tricking a service into connecting to a rogue agent in a decentralized ecosystem). Traditional cybersecurity must evolve to **Alsecurity**. For example, monitoring not just for unauthorized access, but for signs an AI's decisions are being skewed by someone. There's also the risk of *model theft* – if your microservice hosts a valuable model, attackers may try to extract it by querying it (so-called ML API attacks). Organizations will need to safeguard models and possibly use techniques like encrypted inference or federated learning to reduce these risks. Another threat is simply the unpredictability: as Booz Allen analysts noted, "loose coupling in agentic AI systems introduces significant security and stability risks… agents may develop in unexpected ways, exposing vulnerabilities or behaving unpredictably" . Rigorous testing, sandboxing of new behaviors, and implementing **kill-switches** or throttles (so if an agent goes haywire, it can be contained quickly) will be important. Security teams will need to collaborate with AI engineers to cover these new bases.

Autonomy vs. Control: *Autonomous agents can evolve in unexpected ways, potentially causing unintended actions . Always design with safeguards: require agents to get approval (from a human or a higher-level service) for irreversible actions, and maintain the ability to override or shut down agents if they deviate from acceptable parameters.*

Ethical and Social Impacts: As intelligent microservices take on more decisionmaking, their ethical implications magnify. Bias in AI models is a prime concern – if a microservice agent is learning from historical data, it might perpetuate or even amplify biases present in that data. This could lead to unfair outcomes (imagine a lending microservice that unintentionally gives worse terms to certain demographics because the data encoded societal biases). Companies must prioritize **ethical AI practices**: diverse training data, bias audits, and involving domain experts to ensure fairness. Moreover, there's the question of **jobs and roles** – automation can augment humans, but also potentially displace certain tasks. Organizations have a responsibility to retrain and reposition staff, using AI to enhance people's work, not just replace it. On a society level, if many companies deploy these systems, we have to watch out for systemic effects: e.g., widespread use of AI in loan microservices could impact economic inequalities or access to services. Transparency will be key – providing explanations to users about decisions ("Why was my claim denied?" answered by an explainable AI module) to maintain trust. We might see regulations requiring that (e.g., the "right to an explanation"). Ethically, companies should also consider the **environmental impact**: training big models and running hordes of microservices consumes energy. Optimizing for efficiency and using green computing resources where possible might become part of corporate social responsibility. Lastly, ensuring *accessibility* – that the benefits of intelligent microservices are widely accessible and not just siloed for elite users – is a broad social challenge. This might be more on policymakers, but companies can contribute by open-sourcing certain non-competitive tools or collaborating on publicgood initiatives (like healthcare AI for underserved areas).

- **Technical Debt and System Complexity:** The more complex and adaptive a system, the harder it is to maintain and debug. A future architecture with self-organizing agents and dynamic topologies could become a nightmare if not managed properly. **System design principles** will need to evolve to handle this complexity. One approach is to maintain *modularity and clear contracts* as much as possible – even if internals learn and change, the external contract of a service might remain stable. But when emergent behavior occurs, how do you trace a problem? Traditional logging might not suffice if the flow isn't predetermined. We'll need sophisticated **observability** that can reconstruct and explain the chain of events after the fact. Perhaps tracing tools that follow the journey of a request even through dynamic reroutes, coupled with AI ops tools that can point out anomalies ("this pattern of interactions leading to the failure is new"). Technical debt is another concern: all those models and agents need continuous updates (models retrained, code patched, etc.). Without disciplined MLOps, a company could end up with a tangle of outdated models and rules that no one fully understands – a brittle system posing as a smart one. So, ironically, to manage complexity one might enforce some *simplicity at macro-level*: e.g., limit how many degrees of freedom the system has, or periodically *refactor* the learned behaviors (like consolidate what multiple agents learned into a simpler model). Evolutionary architectures are great, but they require occasional pruning and refactoring, akin to how a garden needs pruning to remain healthy.

- **Interoperability and Vendor Lock-in:** If every company or cloud comes up with its own approach to intelligent microservices, we risk fragmentation. A challenge ahead is ensuring interoperability between differing platforms. If you build your architecture heavily on Cloud Provider X's AI services, how easily can you integrate something from Provider Y or move your workload? The need for standard protocols and maybe neutral foundations (perhaps open-source frameworks for agent-based systems) becomes clear. But achieving industry-wide standards can be slow – it often lags behind innovation. So companies will face a strategic choice: adopt a particular ecosystem early to get ahead, or push for more open solutions even if it means building some pieces yourself. Being too locked-in could hamper flexibility or negotiating power later. On the other hand, waiting for perfect standards might mean falling behind. A balance can be to design your system in a modular way where swapping out an AI component is possible if needed, and to contribute to or at least monitor standardization efforts so you can align when the time comes.

- **User and Stakeholder Acceptance:** Last but not least, the coolest technology means nothing if users or stakeholders don't accept it. There may be *cultural resistance* to highly autonomous systems. For example, doctors might be hesitant to trust an AI orchestrating part of patient care, or customers might be wary of entirely AI-driven customer service with no human in the loop. Companies should work on **change management**: introducing these capabilities in a way that builds trust (perhaps starting as decision support tools before becoming decision makers), educating users on how

the system works and its benefits, and providing easy ways to reach a human or appeal a decision if needed. Internally, some employees might fear or distrust AI – involving them in the design process can help (people fear what they don't understand; once they contribute, it becomes "ours" not "theirs"). Essentially, the human element cannot be ignored. Technology must ultimately serve people, and bringing people along the journey is part of the challenge.

Facing these challenges will require a combination of smart technical solutions, robust governance, and open dialogue among technologists, business leaders, regulators, and users. It might seem daunting, but being forewarned is forearmed. By acknowledging these issues now, we can incorporate mitigations into our plans (e.g., building ethical review into our development cycle, or investing in security for AI). Every past tech revolution (from the internet to mobile to cloud) had its share of concerns – and while we still work through issues, each brought immense benefit when guided responsibly. The agentic microservices revolution will be no different.

Having looked at challenges, let's move to something more tangible and fun: how might these intelligent microservices actually manifest in different industries? In the next section, we'll explore some visionary case narratives across healthcare, finance, manufacturing, etc., to illustrate the future in a story-like fashion. These vignettes will help ground our discussion in real-world context and perhaps spark ideas for your own domain.

Visionary Case Narratives Across Industries

To make all these abstract ideas concrete, let's imagine how intelligent microservices might play out in various industries in the near future. We'll walk through a few short visionary scenarios. These are speculative but based on real pain points and trends in each industry. Each vignette will illustrate how agentic microservices could transform workflows, customer experiences, and outcomes.

WebkinsTech in 2030: A Self-Optimizing Enterprise (Retail/E-Commerce)

Background: WebkinsTech, the fictional company we've followed, started a decade ago with a monolithic system and transformed into a microservices-driven e-commerce platform. By 2030, WebkinsTech has evolved its architecture into a swarm of intelligent microservices that manage everything from customer experience to supply chain.

Scenario: It's Black Friday 2030. At 2:00 AM, while most of WebkinsTech's staff are asleep, the system is humming along handling a huge surge of international orders. Suddenly, one of the payment gateway services (run by a third-party) starts timing out intermittently. In a traditional setup, this might cause failures or require on-call engineers to wake up. But WebkinsTech's *Digital Operations Agent* springs into action. This agent is an autonomous microservice trained to ensure checkout reliability. Noticing the payment service's response delays (perceiving a deviation from normal), it quickly consults a registry of available payment microservices. Within seconds, it routes new transactions to an alternative payment service (one that WebkinsTech had pre-integrated from another provider for redundancy). Simultaneously, another intelligent microservice – a *Customer Experience Agent* – updates the front-end checkout UI via feature flag, subtly informing users that one payment option is temporarily using a backup (transparency to manage expectations) while still keeping the flow smooth. The issue is mitigated without waking a single human.

By morning, WebkinsTech's CTO checks a dashboard with a summary of the night's events. The *AI Ops console* explains in natural language: "At 2:03 AM, primary PaymentService latency increased to 5s (above 99th percentile). Autonomous rerouting engaged to BackupPay service for 23% of transactions from 2:05–3:00 AM, preventing an estimated 500 failed checkouts. Primary service normalized at 3:10 AM and traffic was restored. All systems stable." The CTO also sees a cost-benefit analysis, compiled by yet another microservice agent: it calculates that the autonomous action saved approximately $75,000 in potential lost sales, far outweighing the extra fees of using the backup service.

Additional Details: Elsewhere in WebkinsTech's ecosystem, dozens of micro-agents collaborated during the sale: Inventory agents dynamically reprioritized warehouse shipping sequences based on AI forecasts of which orders might be at risk of delay, thereby keeping delivery promises. A *Personalization microservice* employing an LLM agent generated on-the-fly product bundle suggestions for customers (in free-form text and images) leading to a 15% upsell rate, all while continuously learning which suggestions worked best. Over in engineering, no one had a sleepless night, because the system handled the chaos. When the team arrives later that morning, they conduct a quick retrospective aided by the AI Ops console's records. They commend the foresight of setting up that backup payment integration and the training of the ops agent – a practice that's become standard procedure. WebkinsTech's story shows an enterprise essentially running as a *digital autonomous organism*, adapting to conditions in real-time and only escalating to humans for strategic oversight or truly novel crises. The 3 A.M.

outage from a decade ago is now just an anecdote of a bygone era – intelligent microservices have made such outages far less frequent and far less damaging.

Healthcare 2030: Collaborative Care Agents

Background: The healthcare sector has been transformed by AI and IoT, but integrating them into daily hospital operations has been challenging. By 2030, some advanced hospitals deploy a network of intelligent microservices to coordinate patient care seamlessly.

Scenario: Alice, a 58-year-old patient with chronic conditions, is admitted for a complex surgery. From the moment she's admitted, a personal *Care Coordinator Agent* is instantiated for her – essentially a microservice that will follow her case. This age communicates with a swarm of other agents: scheduling, pharmacy, diagnostics, insurance, and even Alice's own wearable health devices. In the morning, Alice's blood pressure cuff (an IoT device) sends an alert through an edge microservice that her BP is slightly elevated. Her Care Coordinator Agent receives this and decides to proactively adjust her surgery prep medication schedule (in consultation with a *Pharmacy Agent* and according to pre-approved protocols from doctors). It notifies the human nurse on duty via a mobile app, explaining the adjustment: "Administer 5mg less beta-blocker at 8:00 AM due to patient BP trend; Dr. Smith preapproved this adaptive protocol." The nurse, who is used to these assistive agents, confirms the change. The agent logs that confirmation for compliance records.

During surgery, multiple agents assist: a *Surgical Equipment Agent* ensures all needed tools are in the OR (it actually signaled the sterilization service earlier to prep an extra set of instruments when it learned another surgery ran long). A *Vital Signs Monitor Agent* at the edge tracks Alice's vitals in real-time, using an AI model to predict potential complications 5 minutes ahead. At one point, it detects a pattern in the vitals that precedes a drop in oxygen levels. It alerts the surgical team *and* triggers an *Anesthesia Agent* to adjust anesthesia slightly. The anesthesiologist overseeing agrees with the recommendation (which is backed by analysis of thousands of past surgeries stored in the system's knowledge base) and lets the agent proceed. A potential complication is averted.

After surgery, Alice is transferred to recovery. Her Care Coordinator Agent orchestrates follow-up: it schedules an MRI via the *Imaging Scheduling Agent* (slotting her in sooner because it negotiated with other patients' agents for priority given her condition), then securely shares relevant data with an *AI Diagnostic Agent* that will review her MRI for any anomalies. Meanwhile, an *Insurance Approval Agent* already submitted the surgical reports and got post-op therapy approved in seconds, sparing the hospital admin the usual fax/phone tag.

Outcome: Alice's recovery is smoother and faster. The human care team is always in the loop but relieved from tedious coordination. The intelligent microservices form a safety net and a logistical brain: fewer balls are dropped, and care is more personalized. When Alice is discharged, her Care Coordinator Agent hands off a summary to her primary care physician's system and even continues to check her wearable data for a week, sending a gentle reminder to her (via text) to do breathing exercises when it notices she hasn't done them – a personal touch at scale.

In this healthcare future, *self-organizing microservices act like an invisible support staff*, handling myriad details so clinicians can focus on the human touch. Privacy is safeguarded by design (all agent communications were encrypted and on a need-toknow data basis), and every decision by an AI agent was recorded with rationale for audit. Doctors and nurses have come to trust these agents because they've seen outcomes improve and they've been part of building the care protocols the agents follow.

The result: a hospital that feels less chaotic, more proactive, and more attuned to each patient's needs.

Finance 2030: Autonomous Trading and Risk Management

Background: In the financial industry, milliseconds and insights matter. By 2030, major financial institutions utilize intelligent microservices to run trading operations and risk management with minimal human intervention, under high-level oversight.

Scenario: QuantumFin, a global investment firm, operates a trading platform powered by a fleet of AI agents (microservices) that buy and sell financial instruments across markets. Each agent has a specialty: one focuses on tech stocks, another on energy commodities, another arbitrages currency pairs, etc. They operate under an overall *Portfolio Manager Agent* that keeps the activities aligned with the firm's risk appetite and strategy set by humans.

One day, news of a geopolitical conflict breaks out suddenly. In the past, human traders might scramble to react; now, QuantumFin's *News Analysis Agent* (which uses an LLM to read and interpret news in real-time) flags the news and assesses likely market impact: oil prices will surge, airline stocks will drop, etc. It sends structured alerts to relevant trading agents. Within seconds, the *Energy Commodities Agent* has already started increasing QuantumFin's positions in oil futures, and the *Airline Stock Agent* begins carefully unwinding positions in airline equities to avoid losses, all while the information edge is fresh. These agents communicate through a standard protocol to ensure they don't trip over each other – the Portfolio Manager Agent coordinates to avoid, say, overexposure if multiple agents react to the same event.

Meanwhile, a *Risk Management Agent* monitors the firm's overall exposure. It notices that the volatility in currency markets is spiking and that QuantumFin's positions could exceed the daily risk limit. It automatically invokes a *Hedging Agent* which executes a series of trades to offset risk (maybe buying certain derivatives that profit if markets swing too wildly, balancing the portfolio). It also pings a human risk officer's dashboard with a concise explanation: "Hedging activated due to volatility surge in EUR/USD, projected VaR (Value at Risk) reduced back within limits." The human reviews and is satisfied, as it aligns with internal policy that's encoded into the risk agent.

Fraud and Security: Across the firm, intelligent microservices also guard against fraud and system abuse. A *Transaction Monitoring Agent* constantly reviews trades and transfers. It catches an unusual pattern – one trading microservice started doing oddly small but frequent trades that don't fit its usual style. This could indicate it's compromised or malfunctioning (perhaps an attacker managed to tweak its logic). The Monitoring Agent alerts the security team and automatically suspends that microservice's access pending investigation, shifting its responsibilities to a standby agent. It turns out to be a false alarm (the agent was testing a new strategy on a small scale), but the firm is glad the framework is in place.

Outcome: By end of day, QuantumFin not only avoided major losses from the geopolitical event, but actually turned a profit by being among the first to reposition its portfolios – all thanks to the speed and always-on readiness of its intelligent microservices. Human traders and analysts at the firm have evolved their roles: they focus on strategy, model training, and regulatory compliance, rather than clicking buttons for each trade. They also continually refine the rules and learning objectives for the agents. The CEO of QuantumFin famously describes the platform as "hundreds of brilliant minds working in tandem, some silicon, some carbon." Clients benefit from more stable performance and quick adaptation to market events. Regulators, initially skeptical of fully autonomous trading, have come on board because firms like QuantumFin share detailed audit logs of every agent decision and have shown that these systems can adhere to market rules (for example, the agents are programmed to automatically avoid manipulative practices and respect trading halts). The finance industry is now a game of **algorithmic ecosystems**, and intelligent microservices are the players making moves in microseconds on behalf of their human principals.

Manufacturing 2030: Smart Factory Swarms

Background: Manufacturing has embraced Industry 4.0 with IoT sensors and robots, but 2030's factories run on intelligent microservices that make plants flexible and resilient.

Scenario: FutureTech runs a highly automated factory producing electric vehicles. The factory floor is orchestrated by a collection of intelligent microservices, each overseeing part of the process: assembly robots, supply chain orders, quality inspection, maintenance, etc. These aren't monolithic systems, but distributed agents at each machine and at coordination points.

On a Monday morning, an assembly line robot in Station 5 starts showing signs of wear – subtle torque fluctuations. The *Machine Health Agent* on that robot detects the anomaly via its sensor data and predicts a potential failure in the next 48 hours (using a predictive maintenance model). In a traditional setting, this might wait for scheduled maintenance or, worse, break mid-production. Instead, the agent communicates over the factory network to the *Maintenance Scheduling Agent*. Instantly, a plan forms: later that afternoon when there's a scheduled 15-minute production pause (for shift change), they will swap out the robot's critical part. The Maintenance Agent cross-checks inventory (through a *Spare Parts Inventory Microservice*) to ensure the replacement part is available on-site; it is, but stock is now low, so the *Procurement Agent* is triggered to order new parts from the supplier, with priority shipping. All this happens automatically. A human supervisor gets a notification: "Preventative maintenance scheduled for Robot #5 at 3:00 PM, part replacement – no action needed. New part shipment arriving tomorrow." The supervisor simply acknowledges. The fix occurs as planned, and the line never experiences unplanned downtime.

In parallel, the *Production Planning Agent* receives an urgent order for a custom variant of the product (say a special fleet order with different specs) that needs quick turnaround. Typically, retooling a line is complex. But here, the agent evaluates the request and consults a *Digital Twin* of the factory (a simulated microservice representing factory capabilities). It figures out that by slightly altering the sequence in two work cells and pulling a different component from inventory at one step, the factory can produce the variant without a full line changeover. It negotiates with the *Scheduling Agent* to allocate a two-hour window the next day to produce those special units, interleaving with regular production by adjusting the schedule dynamically. It also informs the *Supply Chain Agent* to expedite the specific components needed for that variant from a regional warehouse. All agents agree on the plan, and it's executed. The client gets their custom order on time, which was previously almost impossible in a lean just-in-time environment.

Quality control in this factory is also AI-driven: a *Vision Inspection Microservice* uses computer vision on each product off the line. If it flags an anomaly (scratch, misalignment), a *Decision Agent* determines whether to eject the item for rework immediately or let it pass if it's minor/cosmetic. One day, it notices a pattern: three units in a row have a slight misplacement of a wiring harness. The Decision Agent correlates this with data from the assembly robots and finds that Robot #7's calibration might be off. It pages the *Calibration Agent*, which slightly adjusts Robot #7's parameters, fixing the issue for subsequent units. The three affected units are routed to a side line where a human technician (assisted by an AR headset guided by the quality agent) makes a quick fix. This continuous feedback loop of quality data to production adjustments means defects are caught and corrected almost in real-time.

Outcome: FutureTech's factory operates with extraordinary efficiency and agility. It selfheals and self-optimizes: downtime is minimal, throughput is high, and custom orders or changes are handled gracefully by the intelligent swarm. Workers in the factory have shifted from repetitive tasks to supervisory and skilled roles – they manage and improve the AI systems, handle complex assembly for prototypes, or deal with tasks the robots can't. The factory managers trust the AI agents because they've reduced cost and improved output, and any time something goes wrong, the system is designed to fail safe (for example, if a coordination agent crashes, there's a backup or it defaults to a safe halt state to avoid accidents). This smart factory showcases how physical operations and digital intelligence can merge, with microservices acting as the digital nervous system of a very real, tangible operation.

These narratives give a glimpse into how intelligent microservices could manifest in different sectors. From WebkinsTech's e-commerce agility to the hospital that cares collaboratively, the finance firm that navigates markets in a blink, and the factory that runs itself – all share common themes: **autonomy, adaptability, and cooperation between agents and humans**. They highlight benefits (resilience, efficiency, personalization) and also hint at the required safeguards (transparency, oversight, ethical programming).

While they are visionary, each is within the realm of possibility – prototypes of many such scenarios exist in some form today. It's up to tech professionals and architects like you to bring such visions to life responsibly.

Finally, let's conclude our chapter (and this book) with an inspirational summary, tying everything together and looking at how you can be a part of this unfolding future.

Conclusion: Embracing the Intelligent Automation Era

We've journeyed from the foundations of microservices to the cutting edge of agentic AI, and now we stand at the horizon of a new era – one where software systems increasingly exhibit intelligence, autonomy, and adaptability. The "Future of Intelligent Microservices" is not some distant sci-fi concept; it's already taking shape in research labs, forward-thinking companies, and even in some of the tools we use today.

This future promises tremendous opportunities. Imagine IT systems that largely manage themselves, digital services that anticipate our needs, industries that become vastly more efficient and responsive, and perhaps solutions to problems that were previously too complex to tackle. As we saw, intelligent microservices could help save lives in hospitals, provide financial stability, enable personalized customer experiences at scale, and drive innovation in every corner of the economy.

But these opportunities come hand-in-hand with responsibilities. As architects and tech leaders, **we are the stewards of this transformation**. It falls on us to ensure that the intelligent microservices we build are reliable, secure, fair, and aligned with human values. We must design with ethics and governance in mind from day one. We need to be mindful of regulatory requirements and the trust our users place in our systems. In short, the challenge is to harness autonomy without abdicating accountability.

The journey won't be simple. We will iterate through new design patterns, learn from failures, and continuously refine our approaches. There will be astonishing breakthroughs – and there will be hype to sift through. It's important to stay grounded: sometimes the "intelligent" solution may not need a complex AI, but just a better algorithm or simpler automation. Other times, an AI agent might truly revolutionize a process. Knowing when and where to apply these techniques is part of the architect's evolving art.

One thing is certain: **continuous learning and adaptation** isn't just for microservices – it's for professionals too. The fact that you've read this book indicates you're investing in that learning. Keep going! Stay curious about advances in AI and architecture. Experiment with new tools (spin up that AI gateway, try an AutoML service, simulate a multi-agent workflow). Build small proofs-of-concept. Share and discuss ideas with your peers. The community of practice around intelligent microservices is only just forming; your insights and innovations will shape it.

It's also a time to break silos in our organizations. Collaboration between software engineers, data scientists, operations, business stakeholders, and even ethicists or legal experts is key to success. Lead or advocate for cross-disciplinary teams on pilot projects. Teach others what you learn, and be open to learning from non-technical experts who understand the human context of your systems.

At the same time, demand more from our technology providers. Cloud platforms and tool vendors are introducing features for AI and automation – engage with them, give feedback, push for open standards so that we can avoid locking our futures into proprietary boxes. Perhaps join open-source efforts or industry groups focused on intelligent architectures. By participating, you ensure the resulting solutions meet real-world needs and reflect a broad set of values.

Looking ahead, one can't help but feel a sense of excitement. We are essentially witnessing the dawn of a new kind of software paradigm – one where applications aren't just coded, but they *behave*, *learn*, and *cooperate* in complex environments. We called them microservices, but in the future maybe we'll just call them *digital colleagues* or *agents* as they become more collaborative partners in work and life.

This final chapter was dubbed the "crown jewel" – not because it provides all the answers (the future is too unwritten for that), but because it represents the culmination of ideas and a launch pad for what comes next. If the earlier chapters gave you the tools and concepts, this chapter hopefully gave you the inspiration and vision to apply them boldly.

In closing, we encourage you to take up the mantle of leadership in this intelligent automation era. Whether you're a CTO setting strategy, an architect designing systems, a developer building services, or an ops engineer keeping everything running – *your role is vital*. You have the power to implement small changes today that pave the way for smarter systems tomorrow. Perhaps more importantly, you have the power to influence culture and mindset: to make your team and organization excited rather than fearful of intelligent microservices, to ensure empathy and user-centric thinking remain at the core, and to champion responsible innovation.

The future of intelligent microservices will be shaped by those who dare to experiment and iterate. So go forth and be one of those people. Build that prototype, influence that roadmap, start that discussion. The intelligent automation era will be defined by the collective efforts of professionals like you.

As William Gibson's quote reminded us, "the future is already here – it's just not evenly distributed." Some pieces of the future are around us in nascent form. It's up to us to take those pieces, nurture them, and distribute the future more evenly for everyone's benefit.

Thank you for accompanying us on this journey through *Intelligent Microservices: An Agentic-AI Approach*. We can't wait to see what you will create. **The future is in your hands – let's shape it together, one intelligent microservice at a time.**

Appendix A: Implementation Toolkits & Frameworks

In designing an intelligent microservices system (especially using an agentic AI approach), having the right set of tools and frameworks is crucial. This appendix provides a curated, well-organized reference of implementation toolkits, libraries, and frameworks useful for building intelligent microservices. We cover everything from microservice development frameworks and container orchestration to machine learning libraries and MLOps tools. Each section includes a brief overview and highlights key options – including side-by-side comparisons where relevant – to help you evaluate tools based on criteria like ease of use, scalability, and community support.

Microservice Development Frameworks

Modern microservice development is supported by frameworks that simplify building and running services. These frameworks provide ready-made components (for web servers, routing, serialization, etc.) so teams can focus on business logic. The choice often depends on the programming language and project requirements. Some prominent microservice frameworks include:

- **Spring Boot (Java):** A widely-used Java framework for creating production-ready microservices with minimal configuration. Spring Boot embraces "convention over configuration," bundling an embedded web server and sensible defaults to get started quickly. It builds on the mature Spring ecosystem, offering features like auto-discovery of services, configuration management, and integration with cloud platforms. Spring Boot is known for its robustness and enterprise support, making it a go-to for largescale Java microservices.

- **Flask (Python):** A lightweight WSGI web framework in Python that is simple and extensible. Flask provides the basics for building web services (routing, request handling, templating) without imposing project structure. Because of its minimal core, developers often add extensions for database access, authentication, or API building. Flask is excellent for quick prototypes or small services due to its simplicity and large ecosystem of plugins. It powers many Python microservices where a full-featured framework (like Django) would be overkill.

- **FastAPI (Python):** A modern, high-performance web framework for building APIs with

Python 3.7+ type hints. FastAPI is designed for speed (built atop the asynchronous Starlette framework and Uvicorn server) and developer productivity. It automatically generates interactive API docs (Swagger/OpenAPI) and performs input validation using Pydantic models. FastAPI's support for async/await makes it well-suited for microservices that handle many concurrent requests. It's quickly gaining popularity for microservices and AI model serving, thanks to its ease of use and impressive performance.

- **NestJS (TypeScript/Node.js):** A progressive Node.js framework built with TypeScript, inspired by Angular's architecture. NestJS provides a structured way to build scalable server-side applications and microservices, featuring out-of-the-box support for things like dependency injection, modular architecture, and easy integration with libraries (e.g. for WebSockets, GraphQL). It abstracts the underlying HTTP server (Express by default, or Fastify) and provides a consistent framework for building enterprise-grade Node.js services. NestJS is often chosen when a team wants to impose strong conventions and take advantage of TypeScript for a large codebase.

Example Comparison – FastAPI vs. Flask: Both FastAPI and Flask are popular for Python microservices, but they have different strengths. The table below compares these two frameworks on key criteria:

Criteria	FastAPI	Flask
Learning Curve	Easy to start, especially with its automatic docs and validation. Some familiarity with Python type hints is useful.	Very simple API; minimal boilerplate. Easy for beginners to pick up.
Performance	Very high performance (built on async ASGI); handles many concurrent connections efficiently .	Good performance for most cases, but single-threaded by default (WSGI) which can limit throughput under high load.

Features Out-of-Box	Rich features built-in (API docs UI, data validation, dependency injection).	Minimal features built-in. Relies on extensions for things like authentication, serialization, etc.
Asynchronous Support	First-class async support (async/await) for handling I/Obound tasks without blocking.	Primarily synchronous. Async support is possible via add-ons or using Flask with gevent/eventlet, but not as native or straightforward.
Community & Maturity	Newer (gained popularity since 2019) but rapidly growing community; active development.	Mature (over a decade old) with a huge community and many extensions; very stable releases.

Framework Selection: FastAPI tends to outperform Flask for I/O-heavy applications and provides modern features out-of-the-box, making it ideal for high-traffic microservice APIs . Flask is a solid choice for smaller services or quick prototypes, especially if your team is already familiar with it . Ultimately, fit the framework to your project's scope and your team's expertise.

Containerization and Orchestration

Microservices benefit greatly from containerization and orchestration technologies. **Containerization** (e.g. with Docker) packages an application with its environment and dependencies, ensuring consistency across development, testing, and production. **Orchestration** tools (like Kubernetes) then manage running these containers at scale, handling scheduling, scaling, and resilience. In cloud-native architectures, these go hand-inhand:

- **Docker:** The de facto standard containerization platform. Docker allows developers to bundle a service and all its dependencies into a lightweight container image. This guarantees that a microservice runs the same way on any environment that has a Docker runtime. Docker images and containers make deployment consistent, isolate applications from each other, and simplify dependency management. Teams often use Docker for local development as well as for deploying services into production (e.g. on Kubernetes clusters or other container hosts).

- **Kubernetes:** A powerful open-source system for container orchestration. Kubernetes (or "K8s") automates the deployment, scaling, and management of containerized applications across clusters of machines. It handles scheduling containers (Docker or other runtime) onto nodes, manages service discovery and load balancing, self-heals by replacing failed containers, and facilitates rolling updates. Kubernetes became the cornerstone of cloud-native microservice infrastructure, allowing organizations to run hundreds of containerized microservices reliably. It abstracts away the underlying machines, providing a unified API to manage deployments at scale .

- **Istio:** An open-source *service mesh* that layers on top of orchestration platforms like Kubernetes. Istio injects a proxy (Envoy) alongside microservice containers (sidecar pattern) to manage communication. It provides advanced routing (traffic splitting, canary releases), service discovery, load balancing, encryption (mTLS between services), and observability features without requiring changes to the microservice code . With Istio, operators can enforce policies (like circuit breaking or rate limiting) and gain deep insights (via tracing and telemetry) into inter-service calls. This is especially useful in intelligent microservices systems where many services communicate frequently and securely.

Containers + K8s: Docker and Kubernetes are complementary, not competitive. Docker creates containers; Kubernetes manages them. In practice, you *use both* – Docker to containerize each microservice, and Kubernetes to orchestrate them in production . Istio then adds another layer for fine-grained control of service communication in a Kubernetes cluster.

Machine Learning and AI Libraries

Intelligent microservices rely on machine learning (ML) and AI frameworks to provide capabilities like predictions, recommendations, computer vision, or natural language understanding. There is a rich ecosystem of libraries for training models, integrating AI into applications, and leveraging pre-built models:

- **TensorFlow:** An end-to-end open-source deep learning framework originally developed by Google. TensorFlow is known for its scalable architecture – it can run on CPUs, GPUs, and TPUs – and for production-ready features. It uses computational graphs (with eager execution mode available) to build neural networks. TensorFlow comes with a suite of tools (TensorBoard for visualization, TensorFlow Serving for deployment, etc.) that make it suitable for enterprise AI deployments. It excels in deploying models at scale in production environments and has broad support for languages (Python, C++, JavaScript via TensorFlow.js, etc.).

- **PyTorch:** A popular open-source deep learning library developed by Facebook (Meta AI). PyTorch is beloved in research for its dynamic computation graph (define-by-run) which feels more like standard Python code, enabling intuitive debugging and fast prototyping. It has a strong community in academic and industrial research, and has been rapidly growing in production use as well. PyTorch offers distributed training, and with tools like TorchServe and ONNX support, it's increasingly used in deploying models. Many state-of-the-art ML models (especially in NLP and computer vision) are released as PyTorch implementations. In recent years, PyTorch and TensorFlow have converged somewhat in capabilities, but PyTorch is often seen as the more *developerfriendly* option, whereas TensorFlow has an edge in *production ecosystem* . • **scikit-learn:** A robust Python library for classical machine learning (not deep learning). Scikit-learn provides a uniform API for algorithms like logistic regression, decision trees, clustering, SVMs, etc. It's excellent for smaller datasets and traditional ML problems, and it's often used for quick prototypes or adding simpler ML components within a microservice (for example, a forecasting module or a basic recommender). While not built for deep neural networks, scikit-learn is highly optimized in C under the hood and very easy to integrate and deploy (models are just Python objects).

- **Hugging Face Transformers:** A library and model hub that has revolutionized access to pre-trained AI models. Hugging Face's transformers library provides thousands of pre-trained models (and fine-tunable architectures) for NLP, vision, and more, supporting both PyTorch and TensorFlow backends. It allows developers to incorporate advanced models (like BERT, GPT, T5, vision transformers, etc.) into applications with just a few lines of code. For intelligent microservices, this means you can leverage state-of-the-art language understanding or generation without training from scratch. The Hugging Face hub also enables sharing and versioning models. This library is a boon for quickly adding AI capabilities such as translation, sentiment analysis, questionanswering, and more.

- **OpenAI Gym (Gymnasium):** A toolkit for developing and comparing reinforcement learning (RL) algorithms. Gym provides a collection of standardized environments (simulated tasks or games) – from classic control problems (CartPole, MountainCar) to

Atari games – where an "agent" can be trained via RL. In the context of microservices, Gym might be used to train an AI agent to optimize a process (like dynamically adjusting resources or workflows). Though typically used in research, Gym (now maintained as Gymnasium for newer versions) is key for anyone implementing RL components, providing a consistent interface and a variety of challenges to test algorithms.

- **OpenCV:** An open-source computer vision library that offers hundreds of algorithms for image and video processing. OpenCV (Open Source Computer Vision Library) is often used in intelligent applications for tasks like face detection, image segmentation, optical character recognition, and more. It's written in C/C++ but has bindings for Python, Java, and other languages. In a microservice, OpenCV might be employed to preprocess images or videos before passing them to a deep learning model, or even to handle simpler vision tasks entirely (e.g. reading QR codes or detecting edges) without needing a neural network. It's highly optimized and can be deployed on edge devices as well.

- **LangChain:** An open-source framework for building applications powered by large language models (LLMs). LangChain provides high-level abstractions to chain together LLM calls, tool usage, and handling of intermediate results/state . In essence, it helps orchestrate complex multi-step AI workflows (for example, an agent that takes a user query, retrieves context from a database, then calls an LLM to generate an answer) – all with just a few lines of code . With LangChain, developers can set up conversational agents that have memory or that can use external tools like search engines, without having to write a lot of boilerplate. LangChain's growing community and extensive documentation have made it a go-to for quickly prototyping conversational agents and other agentic AI patterns.

- **Haystack:** A powerful open-source framework (by deepset) for building search and question-answering systems. Haystack focuses on *retrieval* *augmented generation* – it connects sources of knowledge (databases, document stores) with NLP models to ensure answers are grounded in data. A typical Haystack pipeline might retrieve relevant documents using semantic search, then use an LLM or a reader model to extract or generate a precise answer using the content. It's particularly useful for creating AI-powered search engines, chatbots over documents, or enterprise Q&A systems. Compared to LangChain, which centers on chaining LLM calls, Haystack is **retrieval-first**, making sure that AI responses stay grounded in real data . It supports features like indexing documents, using multiple retriever algorithms, and integrating with Hugging Face models or APIs to deliver results.

TensorFlow vs. PyTorch: If your priority is *extensive deployment* and endtoend platform support, TensorFlow's ecosystem may serve you better (it "shines in deploying AI models for production" according to many experts) . On the other hand, if you need rapid experimentation or are doing researchoriented work, PyTorch's flexibility and pythonic design can boost productivity. In practice, both frameworks are capable – teams often choose based on familiarity and specific tooling needs.

Example Comparison – TensorFlow vs. PyTorch: The table below highlights differences between these two leading deep learning frameworks:

Aspect	TensorFlow	PyTorch
Ease of Use	Steeper learning curve historically (especially pre-2.x when static graphs were the norm). Improved with eager execution, but still more complex APIs in places.	Very user-friendly, with a Pythonic style (define-by-run). Easy to debug and experiment in notebooks.

Deployment	Highly production-oriented: comes with TensorFlow Serving for model deployment, TensorFlow Lite for mobile/edge, and excellent scalability on servers . Many companies have established TensorFlow in their production pipelines.	Gaining ground in production. Supports TorchScript (to export models for C++ runtime) and TorchServe for serving, plus ONNX export to run on other runtimes. Many research models use PyTorch, and tools are emerging to deploy them at scale.
Performance & Scalability	Optimized for performance with XLA compiler and distributed training strategies. Scales well to multi-GPU and TPU. Suited for very large datasets and models.	Strong performance with native GPU acceleration and now a compile option (torch.compile). Distributed training is available (though often via add-ons like PyTorch Lightning). Excellent for rapid development of new ideas.
Ecosystem & Tools	Rich ecosystem: e.g., **Keras** (high-level API), **TFX** pipeline tooling, **TensorBoard** for visualization. Offers many pretrained models via TF Hub and integration with Google's ecosystem (like Vertex AI).	Thriving ecosystem: e.g., **PyTorch Lightning** for high-level training loops, **Hugging Face** integration (most models on HF are in PyTorch), and community-driven hubs. Interoperates via ONNX for using PyTorch models in other environments.

Community Support	Very large global community and corporate backing (Google). Tons of tutorials, and stable releases. Well-suited if you need long-term support and resources.	Huge community in academia and open source. Facebook/Meta and other big tech contribute. Rapidly evolving – many cutting-edge research projects release PyTorch code first. Community support is strong, but releases can be more frequent and experimental.

Model Deployment and Serving

Once you have trained ML models, you need to integrate them into microservices. Model serving frameworks help deploy models as scalable, high-performance services (often behind a REST or gRPC API). These tools handle loading models, running inference on hardware efficiently, and managing multiple models or versions. Key model serving options include:

- **TensorFlow Serving:** A high-performance serving system for machine learning models, designed for production use by Google. TensorFlow Serving allows you to deploy one or more trained TensorFlow models (SavedModel format) and automatically handle new versions of the model. It exposes both REST and gRPC endpoints for inference requests. Because it's written in C++ and optimized for TensorFlow, it can handle heavy load with low latency. It supports serving multiple models at once (and even multiple *versions* of the same model simultaneously, useful for A/B testing). Unlike TorchServe, it can serve models without requiring a Python runtime or custom handlers

- **TorchServe:** An open-source model server for PyTorch models, originally developed by

Facebook and AWS. TorchServe makes it straightforward to take a PyTorch model (e.g., a .pt or TorchScript file) and serve it without writing custom web service code. It supports both REST and gRPC APIs for inference, and provides a management API to add or remove models on the fly.

TorchServe can host multiple models concurrently and even batch requests for efficiency. It includes default "handlers" for common model types and allows custom Python handlers for arbitrary logic. For observability, it provides metrics that integrate with Prometheus, and it supports both HTTP and gRPC interfaces . This is a natural choice if your pipeline is centered on PyTorch (for example, if you train models in PyTorch, TorchServe simplifies packaging and serving them in production).

- **NVIDIA Triton Inference Server:** A unified open-source inference server that supports *multiple* frameworks and model formats (developed by NVIDIA). Triton can serve models from TensorFlow (GraphDef or SavedModel), PyTorch (TorchScript), ONNX, XGBoost, Python (custom code), and more – all in one server. It provides REST and gRPC interfaces for remote clients to query models, and it's optimized to maximize throughput on both GPUs and CPUs. Triton supports the concept of **ensemble models**, allowing you to set up pipelines where an inference request can pass through a sequence of models (each possibly using a different framework) – useful for complex workflows like "retrieve then rank" or multimodal processing . It also exposes detailed metrics (GPU utilization, server throughput, and request latency) in Prometheus format for monitoring, making it easy to track performance in production . Triton is often used when you have a mix of model types and want a single, efficient serving infrastructure.

(Other notable tools in this space include the OpenVINO Model Server for OpenVINOoptimized models, BentoML for framework-agnostic model serving and packaging, and KServe (KFServing) for Kubernetes-native model serving. The above, however, are among the most widely used in intelligent microservice architectures.)

Managed Cloud AI Services

For teams that want to outsource some of the heavy lifting of ML infrastructure, all major cloud providers offer managed AI/ML platforms. These services provide a fully-managed environment to train models, host them as microservices, and often to store or label data. Using a managed service can accelerate development and production deployment, at the cost of some vendor lock-in and recurring expense. Key offerings include:

- **AWS SageMaker:** Amazon's integrated machine learning platform. SageMaker provides hosted Jupyter notebooks for development, automated machine learning (Autopilot), and one-click training of models on scalable clusters. It simplifies deployment via SageMaker Endpoints – you can take a trained model (even bring your own from TensorFlow, PyTorch, scikit-learn, etc.) and deploy to an HTTPS endpoint with autoscaling. SageMaker manages the underlying instances and even offers a serverless inference option, so you don't worry about EC2 servers at all. It also includes a Feature Store, experiment tracking, and a model registry. SageMaker's greatest appeal is its ease of use and integration with AWS data sources; it abstracts away infrastructure details during model development and serving . The trade-off is reduced control over customization of that infrastructure (the high level of abstraction means you can't fine-tune the underlying servers as much) . Cost-wise, SageMaker is pay-as-yougo, primarily charging for compute and storage used (with savings plans available).

- **Google Vertex AI:** Google Cloud's unified ML platform (successor to AI Platform). Vertex AI brings together Google's AutoML capabilities and custom model training into one interface. With Vertex, you can train models on managed compute (including TPU support), tune hyperparameters, deploy models to endpoints, and monitor them – all through a cohesive UI or API. It offers pre-built pipelines and integration with BigQuery for data. Vertex is known for its user-friendly interface and strong AutoML capabilities, allowing users to train high-quality models with minimal code . It also supports custom training with any framework and has a model registry and continuous monitoring features. If you're already in the Google Cloud ecosystem or need cutting-edge Google research models (like Vision AI or language models) via AutoML, Vertex AI is a compelling choice.

- **Azure Machine Learning & Cognitive Services:** Microsoft's Azure AI offerings are two-fold. **Azure Machine Learning (Azure ML)** is a platform for developing and deploying your own models, similar to SageMaker and Vertex. It provides an interactive studio for dragging-and-dropping pipeline components (Azure ML made early strides with its designer for no-code ML) , as well as CLI/SDK for code-first workflows. You can train on Azure compute clusters, track experiments, and deploy models as web services or even to IoT Edge devices. **Azure Cognitive Services**, on the other hand, are a collection of pre-built AI APIs for tasks like vision (OCR, image recognition), language (text analytics, translation), speech (speech-to-text, text-to-speech), and decisionmaking (anomaly detection, content moderation, etc.). These require no ML expertise – you call the API and get results powered by Microsoft's trained models. Many Azurebased intelligent microservices use Cognitive Services for quick wins (like adding sentiment analysis or translation), and Azure ML for custom models when needed. All these services integrate with the Azure ecosystem (Storage Accounts, Azure DevOps, etc.), making it convenient if your organization is already an Azure shop.

Choosing a Cloud AI Platform: All three major cloud ML platforms (SageMaker, Vertex AI, and Azure ML) cover the core ML lifecycle – data prep, training, deployment, and monitoring – in a managed way . Let your context drive the choice: use SageMaker if you're deep into AWS and want granular control with many AWS integrations; use Vertex AI if you favor Google's AI capabilities or need an easy AutoML solution; use Azure's offerings if your enterprise is on Azure or you need a mix of custom and off-the-shelf AI (Cognitive Services). The best platform is the one that aligns with your team's workflow and existing infrastructure.

Event Streaming and Workflow Orchestration

Beyond direct request-response interactions, microservice architectures often need to handle asynchronous events and complex multi-step workflows. **Event streaming** allows services to react to events in real-time and decouple producers from consumers. **Orchestration frameworks** coordinate sequences of operations, either on a schedule or triggered by events, ensuring tasks run in order and handling failures/retries. Here are key tools:

- **Apache Kafka:** The most popular distributed event streaming platform. Kafka acts as a durable, high-throughput log where microservices can publish events (as "producers") and subscribe to events (as "consumers"). It's commonly used for building data pipelines, streaming analytics, or simply decoupling services (one service publishes an event like "UserSignedUp", others consume it to run their respective logic). Kafka is designed to scale horizontally and handle massive message volumes with reliability. In an intelligent microservices system, Kafka might stream events such as user interactions, IoT sensor readings, or intermediate results that multiple services need to know about. Its log retention allows replaying events, and consumer groups allow scaling out processing. Kafka provides delivery guarantees (at-least-once by default, with exactly-once semantics achievable in specific scenarios) which is important for ensuring events aren't lost.

- **Apache Airflow:** A widely-used workflow scheduler for orchestrating batch jobs and pipelines. Airflow allows you to define Directed Acyclic Graphs (DAGs) of tasks in Python, scheduling them to run at intervals or upon triggers. Each task could be a microservice call, a Spark job, a SQL operation, etc., and Airflow's scheduler executes them with specified dependencies. It provides a rich web UI to monitor runs, retry failed tasks, and manage configurations. In the AI context, Airflow is excellent for ETL (extract-transform-load) pipelines and retraining workflows – for example, a nightly job that preprocesses data, trains a new model, and then deploys it. Airflow excels at timebased or data-driven batch orchestration (rather than handling real-time events). It has become a standard in data engineering teams and integrates well with data warehouses, Hadoop/Spark, and cloud services.

- **Temporal:** An open-source workflow orchestration engine designed for microservices and long-lived processes. Originating from Uber (as Cadence), Temporal lets developers write workflows as code (using languages like Java, Go, Python) and handles the hard parts of state management and retries. With Temporal, each step of a workflow is a function (activity) in your code; the Temporal server ensures if an activity fails or the process restarts, it will resume from the last state. This is great for business processes that span multiple services and require reliability (e.g., an e-commerce order process that reserves inventory, charges payment, and schedules a delivery). Unlike Airflow, which is more data pipeline focused, Temporal is *code-first orchestration*, fitting naturally into microservice logic . It can handle events and also time-based triggers (like "if no response in 3 days, escalate"). Companies use Temporal for things like ensuring exactly-once execution of complex transactions and simplifying the coordination logic in the application layer.

- **Dagster:** A modern data orchestrator that emphasizes software engineering best practices for data/ML pipelines. Dagster allows you to define pipelines (and more importantly, the data assets produced) with Python code, and it provides tooling to test, version, and orchestrate these pipelines. It treats computations as graphs of ops (operations) that consume and produce typed data assets, making lineage and dependency tracking more transparent. For an intelligent microservices setup, Dagster might be used to manage an ML pipeline where data ingestion, validation, model training, and deployment steps are all tracked and can be rerun selectively. It's seen as a next-generation orchestrator addressing some limitations of Airflow (like easier parameterization, solid error handling, and a focus on data outputs). Dagster comes with a web UI and scheduler and can run on one machine or scale out (e.g., using Kubernetes for execution). If your project involves a lot of data-centric processing and you want a more maintainable, testable approach to pipelines, Dagster is worth exploring.

Airflow vs. Temporal: Use Airflow (or Dagster) when orchestrating data processing jobs or ML pipelines that have clear start and end points (e.g., a daily batch workflow). Use Temporal when you need to orchestrate complex business logic across microservices in real-time – for example, a user signup flow that sends emails, records analytics, and triggers downstream processes. Airflow is a *data workflow* orchestrator, whereas Temporal is a *services workflow* orchestrator, focusing on code and reliability of longrunning processes . They solve different problems and can even complement each other in a large system.

Monitoring and Observability Tools

A critical aspect of running microservices (especially those with AI components) is observability – the ability to understand what is happening in your system. This involves collecting metrics, logs, and traces from services, as well as monitoring the behavior of machine learning models in production (data drift, model performance, etc.). Here are important tools in this space:

- **Prometheus:** An open-source monitoring and alerting toolkit, part of the Cloud Native Computing Foundation (CNCF). Prometheus scrapes metrics from instrumented services – typically via HTTP endpoints that expose metrics in a simple text format. Developers integrate Prometheus client libraries to record metrics like request rates, latencies, memory usage, etc. Prometheus stores these time-series metrics and offers a powerful query language (PromQL) to analyze them. It also includes an Alertmanager to send notifications (to Slack, email, etc.) when certain conditions are met (e.g., high error rate). In a microservice architecture, Prometheus provides insight into the health and performance of each service and can be the backbone of automated alerting and even auto-scaling decisions. It has become a standard for metrics in cloud-native environments.

- **Grafana:** A popular open-source visualization and dashboard tool that often complements Prometheus. Grafana can connect to Prometheus (and many other data sources) and allows you to create real-time dashboards and charts. For example, you might have a Grafana dashboard showing request latency percentiles for each microservice, or a live view of the number of active users over time. Grafana's alerting and plugin ecosystem make it flexible – you can visualize not only metrics but also logs or trace data in a unified interface. This helps teams quickly diagnose issues by correlating different signals (for instance, seeing a spike in latency alongside a spike in database CPU usage on the same timeline).

- **Jaeger:** An open-source distributed tracing system (originating from Uber, now a CNCF project). In a microservices environment, a single user request may cascade through multiple services. Jaeger helps trace the path of requests by collecting spans (timed operations) from each service and stitching them together by a trace ID. Developers instrument their services (often via OpenTelemetry or OpenTracing APIs) to produce trace spans. Jaeger's UI then allows you to see, for a given request, how long each service took and where potential bottlenecks or errors occurred. This is crucial for debugging latency issues in distributed systems. For intelligent microservices, you might trace an end-to-end transaction that includes a call to an ML model service – Jaeger would show how long the model inference took relative to the rest of the workflow.

- **ELK Stack (Elasticsearch, Logstash, Kibana):** A popular stack for log aggregation and analysis. *Logstash* (or its lighter-weight Beats collectors) ingests logs from services (and other sources), parses them, and forwards them to *Elasticsearch*, which indexes and stores the logs, making them searchable in near real-time. *Kibana* provides a UI to query and visualize the logs and other data in Elasticsearch. In practice, many teams use the ELK stack to collect application logs from all microservices in one place – so if something goes wrong, you can search across all logs by timestamp, service, error codes, etc. Kibana might have dashboards for error rates or specific log patterns, and

you can set up alerts for certain log events (e.g., an exception trace appearing). The ELK stack is powerful for troubleshooting (e.g., seeing the exact error message and context that caused an alert) and for forensic analysis after incidents.

- **WhyLabs (WhyLogs):** A toolset for ML data and model monitoring. The open-source library *whylogs* can be integrated into your services to record statistical profiles of data – capturing distributions of inputs, outputs, and model performance metrics. These profiles are then sent to the WhyLabs cloud service which analyzes them to detect data drift, data quality issues, or unusual changes in model predictions. For example, if your image classification service suddenly starts seeing a lot of images of a new type, the input feature distribution might shift and WhyLabs would flag this drift. It helps answer questions like “Is our model performance degrading over time?” or “Has the input data changed since we trained the model?” without needing continuous manual checks. Such ML monitoring is essential for intelligent microservices in production, as models can decay or encounter unexpected data. WhyLabs provides dashboards and alerting, and it’s built to handle the scale of logging every prediction if needed (using statistical sketches to summarize data efficiently).

- **Fiddler (ML Monitoring & Explainability):** Fiddler is a commercial platform focused on monitoring AI models and explaining their decisions. It tracks model prediction outputs and can alert when there is a drop in accuracy or other performance metrics (if actual outcomes are available) or when there are out-of-bound inputs. What sets Fiddler apart is its emphasis on explainability – it can generate explanations for individual predictions (e.g., feature importance scores for why a model made a certain decision) and monitor for bias or drift in those feature importances over time. In a microservice context, if you have a model making critical decisions (loan approvals, anomaly detection alerts, etc.), Fiddler helps ensure transparency and fairness by detecting bias or performance shifts. It also provides an interface for exploring model behavior, which is useful for debugging and for compliance (answering the “why did the model do that?” questions). Fiddler and similar tools (e.g., Arize AI, MLflow’s model analysis) are part of the emerging “ML observability” field that goes beyond traditional system metrics to focus on the AI itself.

“In God we trust; all others must bring data.” – W. Edwards Deming. (In practice, this means instrument everything. Metrics, logs, and traces are your data to trust when debugging complex microservices.)

Unified Observability in Action: Imagine a user request goes through several microservices and invokes an ML model. With a robust observability stack, you can trace this single request: Grafana shows an alert that average latency spiked; Jaeger reveals the delay came from the ML inference call; logs in Kibana show that the model server threw an out-of-memory error; and WhyLabs indicates that a surge of unusually large images (outlier input) preceded the failure. This combined insight allows the team to quickly pinpoint the issue (e.g., need to optimize the model or allocate more memory to that service). In summary, combining general observability with ML-specific monitoring gives a 360° view of system health.

DevOps and CI/CD for AI

Continuous integration and continuous deployment (CI/CD) practices are as important for intelligent microservices as for any software system. In fact, with ML in the mix, the term "MLOps" is often used to denote the extension of DevOps principles to machine learning workflows. Key tools help automate building, testing, and deploying both code and models:

- **Jenkins:** A veteran open-source automation server that is widely used for CI/CD pipelines. Jenkins can be set up on-premises or in the cloud, and uses a rich plugin ecosystem to support building and deploying virtually any project. You can define your build pipeline in a Jenkinsfile (using a Groovy-based DSL) or through the classic web

UI. For microservices, Jenkins might compile code, run unit/integration tests, build a Docker image, push it to a registry, and then deploy to a cluster (or trigger a deployment). For ML, Jenkins can orchestrate training jobs or batch inference jobs as part of a pipeline. While Jenkins requires maintaining a server and managing plugins, it offers tremendous flexibility and has a large community. Many enterprises have longstanding Jenkins setups integrated with their credentials, networks, and tools.

- **GitHub Actions:** A CI/CD platform tightly integrated with GitHub repositories. GitHub Actions allows you to write workflows (as YAML files) that run on GitHub's cloud runners in response to events (push, pull request, release, etc.). It comes with many pre-built actions in a marketplace, so you can, for example, use an action to set up Python, another to run tests, and another to build and publish a Docker image. For teams already using GitHub, Actions can be very convenient – the CI config lives alongside code, and GitHub's interface shows the status of each workflow run. It's cloud-hosted (no need to manage your own CI server) and scales with your needs. In an intelligent microservice project, a GitHub Actions workflow might automatically run linting and tests for each microservice on every commit, build updated container images, and then update a Kubernetes deployment or trigger a CD pipeline when changes are merged.

- **MLflow:** An open-source platform for the machine learning lifecycle. MLflow provides solutions for four key areas: **experiment tracking**, **model packaging**, **model registry**, and **model deployment**. With MLflow Tracking, data scientists log parameters, metrics, and artifacts (like model binaries) for each experiment run – giving reproducibility and a history of what was tried. MLflow Models standardizes how you package a model (with its dependencies) so it can be deployed consistently (it supports packaging models from many frameworks, and even a "python_function" flavor that treats the model as a Python function for inference). The MLflow Model Registry lets you register models with versions and stages (e.g., "Staging", "Production"), which is extremely useful for CI/CD of models. Finally, MLflow can deploy models to certain serving platforms or cloud services directly (though in practice teams often integrate it with their own deployment pipelines). In summary, MLflow brings source-control-like discipline to ML models –

which pairs with CI/CD systems. For example, when your training pipeline produces a new model and logs it to MLflow, you could have an automated step that evaluates it and then transitions the model to Production in the registry if it meets quality criteria.

CI/CD Pipeline for Intelligent Microservices: A typical setup might use GitHub Actions and MLflow together. When a developer pushes code changes to the microservice repository, Actions triggers: it runs tests, builds a Docker image, and deploys the updated service to a staging environment on Kubernetes. In parallel, when data scientists push a new model (using MLflow to track it), an Action (or a Jenkins job) triggers the ML pipeline: it validates the model on a hold-out dataset, and if metrics are acceptable, it registers the model as "Production" in MLflow. That registration event triggers another pipeline which grabs the model artifact, builds a new Docker image (perhaps using TorchServe or FastAPI to serve the model), and deploys it to the model serving cluster. Throughout, everything is versioned and logged – code in Git, models in MLflow – ensuring traceability. This automated flow allows teams to frequently update both microservice logic and AI models with confidence.

Webkinstech Starter Stack: A Fictional Vignette

To illustrate how these tools come together, consider **Webkinstech**, a fictional mid-sized enterprise building an *agentic AI-powered* customer support platform. Webkinstech's goal is to have an AI agent assist support reps by analyzing incoming tickets, pulling relevant knowledge base articles, and even initiating certain automated resolutions through microservices. Here's how Webkinstech assembles its starter tech stack based on its goals, team skills, and constraints:

Microservices & Frameworks: The Webkinstech team has strong Python expertise, so for most new microservices they choose **FastAPI**. Its high performance and built-in validation make it ideal for their core services (like a "Ticket Analyzer" service that receives support ticket text and returns suggestions). For example, when building a customer Q&A service, they appreciated FastAPI's quick setup and automatic documentation, which sped up collaboration between front-end and back-end teams. They also containerized a few existing Java services built with **Spring Boot** and ran them alongside the new Python services, since those were legacy systems (e.g., an account management service) that still provided value. This polyglot approach works because everything runs in Docker and communicates via well-defined APIs. (They did consider Node.js for certain components – one small team used **NestJS** to create a GraphQL API gateway that sits in front of the microservices, since those developers were more comfortable with TypeScript. NestJS provided a convenient way to structure that gateway service. However, the majority of backend logic remained in Python due to the ML-heavy nature of the platform.)

Containers & Orchestration: From day one, all Webkinstech services are containerized with **Docker**. This ensures that a service, whether it's a FastAPI app or a PyTorch model server, will run the same in a developer's laptop, in testing, and in production. For deployment, they chose **Kubernetes** (managed via a cloud service, AWS EKS). Kubernetes allows them to seamlessly scale services and deploy updates with minimal downtime using rolling updates. For instance, when they roll out a new version of the Ticket Analyzer, Kubernetes handles shifting traffic to the new pods while health checks pass. As the number of services grew, they introduced **Istio** as a service mesh. With Istio, they gained fine-grained traffic control and observability: they set up mutual TLS between services for security, and configured Istio's traffic routing to do canary releases of new ML models (e.g., routing 10% of traffic to a new model deployment to monitor performance before a full cutover). Istio's dashboards (via addon Kiali) and tracing integration made it easier to visualize how requests flow through their dozens of microservices.

Intelligent AI Components: At the heart of Webkinstech's platform is an NLP model that reads support tickets and suggests relevant solutions. The data science team chose **PyTorch** for model development, due to its ease of experimentation and the wealth of NLP research available in PyTorch. Leveraging the **Hugging Face Transformers** library, they fine-tuned a pre-trained BERT model to categorize tickets and a T5 model to draft answer suggestions. This gave them high-quality NLP capabilities without training from scratch. Initially, for simplicity, the first version of the Ticket Analyzer service loaded the PyTorch model in-process (within the FastAPI app). This worked for a pilot but became a bottleneck as load increased. They then transitioned to using **NVIDIA Triton Inference Server** to serve the models. Triton was deployed as a separate container in Kubernetes, loading both the BERT and T5 models. The FastAPI service would call Triton over gRPC to get predictions. This separation improved throughput (Triton could batch requests and use GPU acceleration more effectively) and made the system more modular – they could update or scale the model server independently. They also explored **TorchServe** as an alternative and found it easy to set up for PyTorch models, but Triton's multi-framework support gave them flexibility for future plans (like possibly adding a TensorFlow-based model).

Not all tasks needed heavy ML – for example, a simple **scikit-learn** model is used to predict the priority of a ticket based on keyword features and past data, since a quick logistic regression turned out to be effective. They embedded this scikit-learn model directly in a service (it's just a pickle file loaded by a Python microservice) because it's lightweight. For an AI search capability, Webkinstech built a microservice using **Haystack**: when a support rep asks a question, this service searches the company's knowledge base and uses a smaller QA model to extract an answer. Haystack allowed them to integrate Elasticsearch for document storage and a RoBERTa-based reader model to extract answers, wiring these components together with minimal custom code.

To tie their microservices into an "agentic" system, they experimented with **LangChain**. One prototype involved a LangChain agent that could take a support query and decide to either pull information via the search service or query the ticket classification service, then compose a response. Using LangChain, they were able to set up this chain of calls (LLM -> search -> LLM) fairly quickly, and it runs as an internal service that the chatbot interface calls. This kind of AI Orchestration service is still in beta at Webkinstech, but it shows the potential of agentic AI: the LLM-based agent uses other microservices as tools to fulfill complex requests.

Streaming & Workflow: Webkinstech uses **Apache Kafka** for event streaming between services. For example, when a new support ticket is created (in their front-end system), an event is emitted to a Kafka topic. The Ticket Analyzer service consumes this event to immediately analyze the ticket and store its suggestions, while simultaneously other services consume the same event for different purposes (one logs it for analytics, another triggers an email acknowledgement workflow, etc.). This decoupling via Kafka means each service can do its work independently without synchronous dependencies. Kafka also buffers bursts of events – if there's a surge of tickets, the consuming services can scale or process at their own rate. On the workflow side, the data engineering team set up **Apache Airflow** to orchestrate nightly jobs and periodic retraining. One Airflow DAG, for instance, runs every night to retrain the ticket classification model on the latest labeled data and then evaluate it. If the new model performs better on a validation set, Airflow notifies the team and pushes the model to a registry (and could even trigger a deployment). Another DAG is responsible for syncing data from the support database to the knowledge base index every hour, ensuring the search service stays up-to-date. These automated workflows keep the AI components fresh and the data pipelines flowing with minimal manual intervention. For long-running business processes (like a customer feedback loop that might span days and involve multiple steps and human approvals), they are evaluating **Temporal**. Temporal would let them implement such processes as durable workflows – for example, opening a support ticket could kick off a Temporal workflow that waits for a resolution, sends follow-up surveys after 3 days, and escalates if certain conditions aren't met. Temporal's promise of handling retries and state persistence is attractive, but they're still in the prototype stage for those use cases.

Monitoring & Observability: To maintain reliability, Webkinstech invested in comprehensive observability. Each microservice has Prometheus metrics instrumentation. They run a **Prometheus** server that scrapes metrics like HTTP request count, error rates, and latency from all FastAPI and NestJS services, as well as from the Triton model server (which exposes GPU utilization and inference timing). They set up **Grafana** dashboards to monitor key indicators: one dashboard shows system health (CPU, memory of pods, request rates) and another is dedicated to model performance (distribution of model confidence scores, etc.). Alerts are configured so that if any critical service has high error rates or if the model server's latency jumps beyond a threshold, the on-call engineers get notified via PagerDuty. For distributed tracing, **Jaeger** was deployed. Using OpenTelemetry, they instrumented the services such that a user request carries a trace ID through the API gateway, into the Ticket Analyzer, and even into the Kafka consumer that logs the ticket. Jaeger's UI then allows them to see the entire trace. This proved useful when they noticed some requests were slow – Jaeger helped pinpoint that the delay was in the knowledge base search, not in the ML model as initially suspected, leading them to optimize the database queries.

All logs from the microservices go into a centralized **ELK stack**. Kubernetes logs are shipped with Filebeat to **Logstash**, indexed in **Elasticsearch**, and viewable in **Kibana**. This way, when an error occurs, developers can quickly search across all services to see what happened around that time. On one occasion, they used Kibana to discover that an increase in errors was due to a single misformatted message that caused a downstream parsing service to throw exceptions – something that was not obvious from metrics alone but clear in the logs.

For the machine learning aspects, Webkinstech integrated **WhyLabs** for data monitoring. During deployment, they established baseline profiles of the input data to the Ticket Analyzer model (e.g., ticket text lengths, language distribution, etc.). WhyLabs now monitors the live input data profiles against the baseline. A few months after launch, WhyLabs alerted them that the distribution of ticket topics had shifted (many tickets about a new product feature that wasn't in the training data). This alert prompted an early retraining of the model using fresh data, preventing a potential drop in accuracy. They also leverage **Fiddler** for model explainability. When support agents provide feedback that a suggestion was bad, the engineers can use Fiddler's explanation to understand what the model was focusing on. For instance, they found the model was putting too much weight on certain trigger words that were sometimes misleading. Such insights guided them to fine-tune the model and improve its explanations, thereby increasing the support team's trust in the AI suggestions. Fiddler's monitoring also keeps an eye on model bias; it can highlight if the model's performance for one category of tickets (say billing issues) is consistently lagging behind others, so they can address it.

DevOps Automation: Webkinstech's engineering culture emphasizes automation for both code and models. They use **GitHub Actions** for CI/CD of the microservices. Every git push triggers a suite of tests (using pytest for Python services, Jest for any Node bits). On a merge to the main branch, an action builds the Docker image and pushes it to their registry, then updates the Kubernetes deployment via kubectl or Argo CD integration. This means that in practice, a code change can go live in the staging environment in minutes and to production soon after, once manually approved. For the ML side, they set up a workflow where the Airflow retraining job registers new models in **MLflow** (running on an EC2 instance with an S3-backed store). The MLflow model registry is watched by a Jenkins job – when a model is marked "Production" (after passing evaluation and a human review), Jenkins pulls the model artifact and builds a new Docker image for the Triton server with that model version. This image is then deployed to the cluster (in a blue-green fashion using Kubernetes). By tying together Airflow, MLflow, and Jenkins in this way, they achieved a semi-automated continuous training pipeline: new data can lead to a new model which gets deployed with minimal manual steps. Of course, all deployments (services or models) are monitored closely via the earlier-mentioned observability stack, and they have feature flags to fall back if something goes awry.

Summary: Webkinstech's starter stack includes a mix of frameworks and tools that address development, deployment, intelligence, and operations. They chose frameworks like FastAPI and PyTorch for developer productivity and performance, container/Kubernetes infrastructure for scalability, and a suite of MLOps and observability tools (Airflow, Kafka, Prometheus, WhyLabs, etc.) to ensure the system runs smoothly and the AI components remain reliable.

By carefully picking tools that fit their team's expertise (e.g. Python-centric tools for a largely Python-proficient team) and their specific needs (real-time streaming with Kafka, nightly jobs with Airflow, etc.), Webkinstech assembled a platform that is both **intelligent** and **maintainable**. This case demonstrates how a company can blend general microservice tools with AI-specific frameworks to build an agentic microservices architecture, choosing each component based on goals (quick iteration, scalability), capabilities (skills of the team), and constraints (budget, existing infrastructure). The result is a cohesive yet flexible stack – a strong foundation for Webkinstech's AI-driven services to evolve and scale.

Appendix B: Evaluation Metrics for Intelligent Services

Introduction

When building intelligent microservices, it's not enough to simply deploy a smart service and hope for the best – you have to measure it. This appendix serves as a practitioner's guide to understanding what to measure and why, so you can ensure your **intelligent microservices** (microservices augmented with AI, or *agentic* microservices) are both operationally sound and making good decisions. In other words, we're putting a dashboard on our AI-driven services, giving you the key dials and readouts to track their health and intelligence.

Why do metrics matter in intelligent systems? Think of metrics as the **vital signs** of your microservice. In a simple service, metrics like response time or error rate tell you if it's healthy. In an AI-infused service, we have extra dimensions to worry about – the quality of the AI's decisions and the business outcomes of those decisions. Metrics shine a light on all these aspects, helping you catch problems early and continuously optimize. Three big reasons metrics are crucial: **drift, complexity, and optimization**.

- *Drift:* AI models can become stale over time – users change behavior, data evolves, the world moves on. This is called **model drift**, where a model's performance deteriorates as real-world data diverges from the training data . Metrics (like accuracy over time) will reveal if an AI model is drifting off course. If you're not measuring the AI's performance in production, you might not realize it's getting worse until it's far too late. In production ML, **what isn't monitored can't be trusted**.

- *Complexity:* Microservices architectures are complex systems with many moving parts, and adding AI agents only increases that complexity. You might have dozens of services interacting, each potentially with its own learned behavior or decision logic. Without good metrics, this distributed complexity becomes a black box. As WebkinsTech learned, monitoring is *non-negotiable* for managing microservices at scale . Metrics give you observability into each component and the system as a whole, so you can understand how all the pieces are working together (or not). They help tame the cognitive overload of many interdependent services by surfacing key insights like "service X's latency is spiking" or "our recommendation AI's error rate doubled after last deployment."

- *Optimization:* Lastly, **optimization**. You can't improve what you don't measure – a maxim every engineer and business leader learns. Metrics provide the feedback loop that lets you tune and improve your systems. Whether you're trying to speed up response times, boost an AI model's accuracy, or increase user conversions, you need data to know if changes are making things better. In intelligent microservices, there are multiple levers to adjust (infrastructure, model parameters, business logic), and metrics guide you to make informed decisions instead of flying blind. It's very much like instrumenting a race car: without a speedometer and oil pressure gauge, you'd have no idea if your tweaks under the hood made it faster or if the engine's about to blow. Similarly, metrics will tell you if that new AI model version is truly an improvement or if a faster algorithm is worth the trade-off in slightly lower accuracy.

The purpose of this appendix, therefore, is to help you identify, interpret, and act on the right metrics across three layers of an intelligent service:

- **Operational metrics** – the traditional microservice health indicators (throughput, latency, errors, etc.) that ensure your service is up, fast, and reliable.

- **AI/Model performance metrics** – the metrics that evaluate the quality of the AI's decisions or predictions (like accuracy, precision/recall, error rates for predictions, confidence measures, learning progress, etc.).

- **Business-level metrics** – the outcomes that matter to the business or end-users (conversion rates, customer satisfaction, efficiency gains, etc.), which ultimately tell you if the intelligent microservice is delivering value.

By covering all three, we get a **holistic view**: is the service running well, is the AI inside it making good calls, and are those calls benefiting the business or user? A fancy fraud detection microservice that is 99% accurate is great – but if it introduces 500ms of extra latency to every payment (hurting user experience) or if it doesn't actually reduce fraud losses (business impact), can we truly call it a success? We need a balanced perspective.

Finally, we'll explore how these metrics interplay and sometimes conflict. Improving one metric (say, an AI's accuracy) might worsen another (like service latency). We'll discuss finding the right balance, sometimes through a "balanced scorecard" approach of tracking multiple metrics together. We'll also provide an example story of how a fictional company, WebkinsTech, uses these metrics in practice for their AI-powered support agent (showing the real-world impact of metrics-driven iteration). Along the way, we'll include some tips, warnings, and illustrative examples in callout boxes to keep things lively and actionable.

By the end of this appendix, you should feel confident about which dials and gauges to install on your intelligent microservices dashboard. Think of it as giving your AI-driven services a proper **checkup and fitness regimen** – continuous evaluation leads to continuous improvement. As the saying goes, *"What gets measured gets improved"*. So let's dive into what to measure!

On Metrics and Improvement

"If you can't measure it, you can't improve it." This old adage holds especially true for intelligent microservices. Metrics are our tools of measurement – without them, we're just guessing.

Operational Metrics

First, let's cover the **core operational metrics** for microservices. These are sometimes called the "golden signals" of service health, and they apply to any microservice whether it has AI or not. Intelligent services still need to obey the laws of uptime and performance! In fact, because an AI service might be doing heavier computation, watching these metrics is even more critical. Here are the key ones:

- **Throughput** – This is the **volume of requests or events the service handles per unit time**. Often measured in requests per second (RPS) or transactions per minute, etc. Throughput tells us how much load the service is handling. For example, WebkinsTech's order processing service might handle 500 orders per second during a sale. If throughput suddenly drops to near zero, it could mean the service is down or stuck; if it spikes, can the system handle it? High throughput is generally good (it means you can serve more users or tasks), but there may be upper limits where the service saturates. Keeping an eye on throughput helps with capacity planning and scaling – you want to know the maximum RPS your service can reliably sustain and be alerted if you're approaching that. In practice, teams measure this via counters in their API gateway or load balancer, and plot it over time. It's common to see **Throughput** on a dashboard as a line chart of requests/sec. It's a fundamental indicator of **traffic**.

- **Latency (Response Time)** – Latency is the **time it takes for the service to respond to a request**. Typically we measure latency in milliseconds. It's crucial for user experience: a slow service frustrates users or downstream systems. Rather than just average latency, engineers look at **percentiles** – e.g., **p95 latency** (95th percentile) or **p99 latency** (99th percentile). A p95 of 200ms means 95% of requests return in under 200ms, but 5% take longer; p99 of 500ms means 1% of requests took longer than half a second. Tail latencies (those high percentiles) are important because a few very slow requests can ruin a user's experience or indicate issues for certain conditions. For example, if most users get a response in 100ms but one out of every hundred gets a 2second delay, that might be when they time out or get upset – so you'd want to investigate why. Thus, tracking p95 or p99 latency alongside median (p50) gives a fuller picture. Low latency is especially critical for real-time intelligent services (imagine an AI recommendation service that must return in <50ms to not slow down page loads). It's often a **Service Level Indicator (SLI)** tied to user satisfaction . In microservices, one service's latency can cascade if others depend on it, so pinpointing which service is slow is part of observability. Tools like distributed tracing (e.g., Jaeger, Zipkin) can further help attribute latency across services , but as a starting point, each service exposing a "request latency" metric (often automatically from HTTP servers or gRPC interceptors) is vital.

- **Error Rate** – The percentage or count of requests that fail. Failures could be crashes (HTTP 5xx errors), explicit error responses (like a 400/404 if something isn't found – depending on context those might be expected or not), or any event indicating the service didn't do what it was supposed to. For example, an error could be an exception thrown in code, a timeout calling a dependency, or a validation failure. Error rate is typically measured as errors per minute or as a fraction of total requests (e.g., 0.1% error rate). In SRE practices, an **error budget** might be defined (e.g., no more than 1% of requests can fail over a month). Even intelligent services need reliability; if your fancy AI-powered service is returning errors, users won't be happy. Sometimes we break this down further: **system errors** (like 500 Internal Server Error due to bugs or timeouts) versus **business errors** (like the service deliberately returning "fraud suspected" which might be a valid outcome). It's important to monitor technical error rates specifically. A spike in error rate usually triggers an immediate alert – something is wrong (perhaps a bad deployment or a dependency outage). Maintaining a low error rate (approaching zero ideally) means the service is functioning correctly.

- **Uptime/Availability** – Uptime is the **percentage of time the service is available and functioning** (serving requests successfully). Often measured in **"nines"**, e.g., 99.9% uptime means the service is down for at most 0.1% of the time (which is about 8h 45m per year). This metric is usually tracked via uptime probes or simply derived from the error rate and downtime incidents. High availability is critical for services in production – especially those that are user-facing or mission-critical. Intelligent microservices should be designed with the same rigor (redundancy, failovers) so that adding AI doesn't reduce uptime. If your model is deployed on a server that goes down, do you have a fallback? Uptime metrics might be aggregated daily or monthly. A common practice is setting an SLO (Service Level Objective) like "99.9% uptime monthly" and monitoring against it . If uptime drops below target, that's a serious issue to address (maybe more redundancy or fixing crashes). In essence, uptime is a top-level indication of reliability.

- **Resource Utilization (CPU, Memory, etc.)** – These metrics tell us how much of the system's resources the service is using. CPU utilization (often as a percentage of one core or across all cores) shows how busy the service's brain is. High CPU usage might indicate heavy computations (perhaps an AI model doing intensive inference) – if it hits 100% constantly, the service might be struggling to keep up, leading to increased latency or thread starvation. Memory usage indicates how much RAM is consumed; if it's steadily growing (memory leak?) or near the container's limit, the service could crash (OOM). Disk I/O or network I/O could also be monitored if relevant (e.g., a service reading a lot of data). For containerized microservices, we often monitor at the container or pod level (e.g., via Kubernetes metrics) to see these usage stats. **Resource usage is a proxy for capacity and performance**: e.g., if CPU usage is

90% and rising, you might need to scale out by adding more instances or enabling autoscaling. On the other hand, if an AI service uses extremely little CPU, that could mean it's underutilized or maybe the model isn't being called as often as expected. We want to optimize resource utilization to get the most out of hardware but avoid saturation. Generally, aim to keep utilization in a healthy range (not maxed out except under planned stress). It's also key for cost management in cloud environments – e.g., if a machine learning service uses a GPU, you'd monitor GPU utilization to ensure that expensive resource is actually being used efficiently and not sitting idle or overloaded.

In summary, these operational metrics – throughput, latency, errors, uptime, resource usage – are your **first line of defense**. They answer "Is the service working *technically*?" A welldesigned intelligent microservice will emit these metrics automatically (many frameworks and tools like Prometheus, Datadog, or CloudWatch can collect them). For instance, Prometheus could scrape an endpoint exposing these stats, and Grafana might display them in dashboards. WebkinsTech's ops team, for example, set up an "Orders Dashboard" that showed metrics like order API throughput, error rates of each service, and end-to-end latency, so they could quickly pinpoint if, say, the Payment Service was causing orders to fail . You want similar visibility.

P99 Latency and User Experience

Don't just measure average latency! Pay attention to **p95** and **p99** latencies for your services. The 99th percentile latency often correlates with the worst experiences your users see. Many performance issues hide in the tail. For example, if most chatbot responses come in 1 second but the p99 is 5 seconds, a portion of users will think "this AI is slow" and get frustrated. Tracking and optimizing tail latency ensures a consistent experience for (nearly) all users, not just the average user.

Operational metrics are well-known in the microservices world as part of achieving observability. They help you ensure the system side of your intelligent service is solid. Next, we'll explore metrics on the **intelligence side – the AI/model performance.**

AI/Model Performance Metrics

Now onto the metrics that gauge the **quality of the AI decisions or predictions** within your microservice. These are different from system metrics; they come from the realm of machine learning and AI model evaluation. Think of your AI-driven microservice as having a "brain" – these metrics tell you how well that brain is learning, thinking, and deciding. For practitioners like ML engineers, these will be familiar, but it's important for SREs and architects to understand them too, because a service with great uptime but a dumb model isn't really an intelligent service!

Depending on what type of AI or model your service uses, you'll pick relevant metrics. We'll cover several categories: classification metrics, regression metrics, probabilistic confidence metrics, sequential decision (agent) metrics, and ranking/recommendation metrics. Don't worry if some of these sound like buzzwords – we'll explain each and why it matters.

Classification Metrics (Accuracy, Precision, Recall, F1)

If your microservice's AI component is making categorical decisions (yes/no, or picking one of N categories), it's essentially doing a **classification** task. For example, a fraud detection service classifies transactions as fraud or not fraud; an image recognition service classifies images into categories; a support chatbot might classify the intent of a user's query to decide which response template to use. For such services, common metrics include **Accuracy, Precision, Recall, and F1-Score**:

- **Accuracy** – Perhaps the simplest: the proportion of predictions the model got right. If out of 100 predictions, 90 were correct, the accuracy is 90%. While easy to understand ("how often are we right?"), accuracy can be misleading in imbalanced scenarios. For instance, if only 1% of transactions are fraudulent, a naive model that predicts "not fraud" every time is 99% accurate – but it catches 0% of actual fraud! So accuracy alone isn't always the best measure. However, it's a nice high-level metric. During testing, you'll measure accuracy on a labeled dataset. In production, you might only infer accuracy indirectly (through feedback loops or A/B tests) unless you have ground truth labels coming in. Use accuracy as a general health metric for classification models, but always pair it with more nuanced metrics like precision and recall.

- **Precision** – Precision answers: "when the model predicts Positive, how often is it correct?" It's the ratio of true positives to all positives it predicted. High precision means **few false alarms**. In our fraud example, precision is the fraction of transactions flagged as fraud that were actually fraud. If precision is 90%, that means 10% were false alerts (legitimate transactions incorrectly flagged). Precision is crucial when false positives are costly – e.g., you don't want to block good customers or mark legitimate support tickets as spam. In an AI support agent context, precision might translate to: if the bot thinks it can handle a user's query without human help, how often is it right? Low precision would mean the bot frequently tries to answer but the user ends up unhappy (and the query had to be escalated). So a high precision is a sign of **making correct positive identifications**.

- **Recall** – Recall answers: "of all the actual Positive cases, how many did the model catch?" It's the ratio of true positives to all true positives that existed (true positives + false negatives). High recall means **few misses**. In fraud detection, recall is the percentage of fraudulent transactions that the model successfully caught. If there were 100 fraud cases and the model caught 95, recall is 95%. In a support bot, recall could be: of all user queries that *could* have been answered by the bot, what fraction did it actually handle successfully? Low recall means the model is missing a lot of opportunities or failing to catch positives (like letting fraud slip through, or deflecting only a small portion of support queries to automation). So recall measures **coverage of the positives**.

- **F1-Score** – F1 is the harmonic mean of precision and recall (F1 = 2 *(precision* recall) / (precision + recall)). You can think of F1 as a single metric that balances precision and recall. It's useful when you want one number to reflect overall classification ability, especially if you care about precision and recall somewhat equally (which is often the case in balanced scenarios). A high F1 means the model has a good balance: it's catching a lot of positives (high recall) *and* it's not crying wolf too often (high precision). For example, if precision = 0.9 and recall = 0.8, F1 would be around 0.85. If either precision or recall is low, F1 will be low. Many AI practitioners use F1 to select models when there's an inherent trade-off (like in tuning a classification threshold). In the context of intelligent microservices, if you have an internal threshold (say, the AI only automatically approves a warranty claim if it's ≥90% confident), you're tuning for a balance of precision/recall – F1 can tell you the sweet spot.

In practice, you'll often monitor precision/recall for each class if multi-class, or specifically for the "important" class (like fraud = positive). It's worth noting some domains also use **ROCAUC (Area Under the ROC Curve)** as a metric for binary classification performance across thresholds, but F1/Precision/Recall are more tangible for operational monitoring, so we'll focus on them. If your intelligent service is e.g. a medical diagnosis agent, you might monitor sensitivity (recall) and specificity (the complement of false positive rate) to ensure it's both catching diseases (high recall) and not over-diagnosing healthy people (high precision).

How to measure these in production? Often you have a validation dataset for offline evaluation (where you compute accuracy, precision, etc. before deploying a model). But in production, you might get feedback data: e.g., user said the AI's answer was correct or not, or later outcome reveals the truth (fraud was confirmed later). Those can be logged to compute "online accuracy" over time. Additionally, techniques like **shadow testing** or periodic evaluation on sample data can give live metrics. For example, WebkinsTech's fraud service logged every prediction along with eventual human verification results, enabling them to calculate that the model was catching ~95% of fraud with ~90% precision (which aligned with their offline tests). Tracking these over time can indicate if the model's precision or recall is degrading (a sign of drift).

Accuracy Isn't Everything

Be cautious of solely chasing accuracy. An **accuracy of 99%** might look fantastic, but it could hide problems. Always pair accuracy with precision/recall or domain-specific rates. For instance, an AI hiring filter might reject only 5 out of 1000 applicants – 99.5% accuracy if it mostly says "hire". But those 5 rejects could include 4 great candidates (false negatives)! The highest accuracy model is not necessarily the best if the dataset is imbalanced or if the cost of false positives/negatives differs. In intelligent microservices, define what errors matter more (missing a fraud vs falsely flagging one) and watch the appropriate metric. Optimize for the **right** accuracy – one that considers the context and trade-offs.

Regression Metrics (MAE, RMSE)

Some intelligent services output a continuous value (a number) rather than a category. For example, a pricing microservice might predict the optimal price for a product, a forecasting service might predict demand or load, or an IoT service might predict a sensor reading. These are **regression** tasks. To evaluate how good a prediction is, we use error-based metrics:

- **Mean Absolute Error (MAE)** – This is the average absolute difference between the predicted values and the actual values. If your service predicts, say, package delivery times, and you have actual times to compare, MAE tells you on average how far off you are. For example, an MAE of 5 minutes on delivery time means on average the prediction was 5 minutes off the actual. MAE is easy to interpret because it's in the same unit as the quantity (minutes, dollars, etc.) and it treats all errors linearly. Every error contributes proportionally to how big it is (5 minutes off contributes 5, 30 minutes off contributes 30, etc., averaged out). A lower MAE is better – zero means perfect predictions. You'd use MAE to track, for instance, an AI service forecasting server load:

if it predicts 100 ± 10 and actual is often 100 ± 5, a small MAE indicates it's usually close. MAE is good for understanding typical error magnitude.

- **Root Mean Square Error (RMSE)** – This is the square root of the average of squared differences between predicted and actual values. It's similar to MAE but it squares the errors before averaging, then takes a root to get back to original units. Why? Squaring means larger errors count more heavily. RMSE penalizes big errors more than small ones. If your service sometimes has rare but huge misses, RMSE will reflect that strongly, whereas MAE might not spike as much. For example, if your price prediction service is usually off by $5 but occasionally off by $50, the RMSE will be influenced by those $50 errors more. RMSE is often used in cases like rating predictions (Netflix prize famously used RMSE). It's mathematically convenient for optimization too. The downside is it's a bit harder to interpret than MAE (since it's a square-root of mean-ofsquares). But generally, like MAE, **lower is better**, and you can interpret it as "an average error, weighted towards larger mistakes." If RMSE = 0.5 (on some scale), that's quite good; if RMSE = 10 on something where values range 0-100, that's moderate, etc. It gives a sense of variance of errors.

There are other regression metrics like R^2 (R-squared, which measures variance explained) or MAPE (Mean Absolute Percentage Error), but MAE and RMSE are very common and straightforward. In intelligent microservices, an example: WebkinsTech might have a delivery estimate service (to tell customers when their package will arrive). If actual delivery times come in, they can compute MAE to ensure their estimates are typically within, say, ±10 minutes. If the MAE starts creeping up over time (maybe traffic patterns changed, so predictions are worse), that signals the model might need retraining or the features need updating.

One thing to monitor in production for regression: if you don't continuously get true values, you might rely on periodic evaluation or user feedback.

Sometimes users implicitly tell you the error (e.g., if a user adjusts a recommended price, that might indicate the prediction error). Logging predictions and later outcomes for analysis is key.

MAE vs RMSE: If errors have a normal-ish distribution and big outliers aren't a concern, MAE and RMSE will tell a similar story. If you care more about avoiding occasional large errors (maybe because a large error could be catastrophic in your domain), watch RMSE or even max error. If you care about consistency and can tolerate small errors, MAE is a clearer measure. Many teams track both. It's also common to see **Mean Absolute Percentage Error (MAPE)** for business-friendly reporting (like "on average we're within 5% of actual"), but be careful with percentage if actual values can be near zero.

Probabilistic and Uncertainty Metrics (Confidence, Calibration, Entropy)

Intelligent microservices often produce probabilities or confidence scores – for example, a model might say "I'm 90% confident this email is spam." It's important to measure how **wellcalibrated** these confidences are and how the system handles uncertainty. Two concepts here: **confidence calibration** and **entropy of predictions**.

- **Confidence Calibration** – A model is *calibrated* if when it says "90% confident," it really is correct about 90% of the time. In practice, many models are overconfident. You might have a classification model that outputs probabilities – e.g., it outputs 0.9 probability for "fraud" on some transaction. If that model is well-calibrated, then among all transactions where it said 0.9, about 90% should truly be fraud (and 10% not). Calibration can be quantified by metrics like **Brier Score** (which is essentially the mean squared error of the probability vs outcome, smaller is better) or by plotting a **calibration curve** (which shows predicted probability vs actual frequency). Why does this matter? In an AI service, if the model's probabilities are unreliable, you might be making bad business decisions. For example, if a support bot says it's 99% sure of an answer but is wrong half the time, that's a problem – it should perhaps ask for human help instead.

Monitoring calibration means checking that the confidence scores align with reality. One way is to bucket predictions by confidence and see how often the model was correct in each bucket. If your intelligent microservice is providing a probability along with each result (many do, for transparency or thresholding), track a metric of calibration error. Well-calibrated models make it easier to set thresholds (like "only automate if confidence > X").

- **Entropy of Predictions** – Entropy is a concept from information theory that, in this context, measures the uncertainty in the model's output distribution. If a model produces a probability distribution over classes, the **entropy** is highest when probabilities are all equal (maximum uncertainty) and lowest when one outcome is nearly certain (low uncertainty). Why track entropy? It can serve as an **aggregate uncertainty indicator**. For instance, if your model is consistently very sure (low entropy) that might be fine – or it could mean it's overconfident. If entropy suddenly increases, the model might be seeing data that confuses it (maybe concept drift, or novel situations). Some teams monitor the average entropy of predictions over time. An upward trend might indicate the model is less confident on current data than it was on training data (a possible drift sign). Also, per-request entropy can be used for decisions:

e.g., "if entropy is high (the model is unsure), route this case to a human or a secondary process." So it's both an operational metric for triggering fallbacks and a quality metric to track overall. In reinforcement learning or multi-step decisions, entropy can be used as part of loss functions to encourage exploration, but here we mainly consider it as a metric of prediction certainty.

In practical terms, one might compute a **Calibration Error %** (lower is better) periodically by comparing predicted vs actual probabilities, and log **average confidence** for correct vs incorrect predictions. If you notice, say, that incorrect answers had an average confidence of 0.8, that means the model was confidently wrong often – not good. You'd want the model to only be highly confident when it's usually right. Techniques like Platt scaling or temperature scaling are used to calibrate models; the details are beyond our scope, but as an intelligent service owner, you should at least be aware if your model's confidence is trustworthy.

Example: Imagine WebkinsTech's support chatbot "Lumi" has an ML model to detect the user's intent (billing issue, technical issue, general question, etc.) with confidences. If Lumi's intent classifier is well-calibrated, when it says "I'm 70% sure this is a billing question," it should be right about 70% of the time. If not, the team might calibrate it. They might also monitor that if confidence < 50%, Lumi automatically asks a clarifying question or hands off to a human. In one instance, they found Lumi was often _over_confident on technical queries it hadn't seen before – a calibration check showed its 90% confidence predictions in a new category were only correct 60% of time. That triggered a retraining with more data on that category, improving calibration.

To sum up, **probabilistic metrics** help ensure your AI isn't just accurate, but also knows what it **doesn't** know. An AI service with well-calibrated confidence can be safely integrated into workflows (like only auto-approving claims when above 95% confidence). One with poor calibration might either be too timid or too reckless.

Sequential/Agent Learning Metrics (Reward Rate, Convergence Time)

Some intelligent microservices aren't one-shot predictions but are more like **agents** that learn and make a sequence of decisions. This could be a reinforcement learning scenario or any iterative learning setup. For example, a service that dynamically allocates resources based on usage, an AI that plays a game or simulation to optimize a strategy, or an orchestrator that learns to reroute requests in a network. These agentic systems require different metrics to evaluate how well they're learning and performing:

- **Reward Rate (Average Reward)** – In reinforcement learning (RL), we define a **reward** for desired outcomes, and the agent tries to maximize cumulative reward. A classic metric to monitor is the **average reward per time step or per episode**. Essentially, how much reward is the agent getting, which directly reflects its performance on the task. For instance, if you have an agent controlling web server scaling with a reward of +1 for each minute the latency stays under target and -5 for each scaling action (to modestly penalize scaling costs), the reward rate will tell you how well it's balancing performance vs cost. A rising average reward over training episodes indicates learning progress – the agent is getting better at the task. Once deployed, if it continues learning online, you'd monitor that it maintains a high reward. If reward starts dropping, maybe conditions changed. In non-RL terms, you might not have an explicit "reward" but perhaps a **success rate** of multi-step tasks (e.g., an agent completes a workflow successfully 85% of the time – that could be treated as a reward metric). Essentially, pick a metric that reflects the goal the agent is trying to achieve and track that.

- **Convergence Time (Training or Adaptation Time)** – This refers to **how long it takes for an agent or model to learn to perform well**. In a training context, it could be measured in number of iterations/episodes to reach a certain performance threshold. In an online learning context, it might be how quickly it adapts to a change. For example, suppose you rolled out a personalization agent that adjusts to each user's preferences. You might measure "convergence time" per user – say, it takes 5 interactions for the agent to tune itself to a new user. Or in a continual learning system, if concept drift happens, how many new data points until the model recovers performance. Why is this an important metric? Because it affects user experience and system dynamics. A support bot that learns from experience but takes 1000 conversations to get useful is probably not practical; one that adapts after 5 conversations is great. Convergence time can sometimes be measured offline (how many training epochs until your model's reward stabilized). In deployment, you might not explicitly measure it unless you have phases, but you could simulate or use A/B tests to gauge it. Another related concept is **time to exploration vs exploitation** – how fast the agent stops flailing and starts exploiting knowledge.

- For an **example**, if WebkinsTech built an inventory management agent via reinforcement learning, they might look at training logs to see that it reached a steady high reward after ~1000 simulation episodes (~2 hours of training). That gives confidence it converges. In production, if they updated the reward function or faced new conditions (like a supply chain disruption), they might observe how many iterations before the agent's policy readjusts to maintain good performance (that's adaptation time).

Other metrics in sequential decision problems include things like **win rate** (if it's a game or competitive scenario), or **episode length** if shorter episodes imply faster success. But generally, reward and convergence cover a lot. It's also worth logging if the agent's performance is within acceptable bounds at all times (no severe dips) – especially important in real systems that can't afford a long learning period online. Sometimes teams do a lot of simulation to ensure an agent is near-optimal before deploying, and then monitor its reward in production to ensure it doesn't degrade.

In summary, for any learning agent service, **track its performance trend** (is it learning/improving?) and **how fast** it learns or adapts. If it stops improving or worsens, you need to intervene (maybe tweak hyperparameters, provide more training, etc.). This is a bit different from static model metrics because it's about the learning process itself.

Ranking/Recommendation Metrics (NDCG, MAP, Recall@K)

Many intelligent microservices are about **ranking** or **recommendation** – providing a sorted list of items (search results, recommendations, feeds, etc.) tailored to a user or query. Evaluating these requires different metrics because it's not a single correct answer scenario; it's about how good the ordered list is. Common metrics:

- **NDCG (Normalized Discounted Cumulative Gain)** – NDCG is a mouthful, but it's a popular metric in information retrieval (search engines) and recommender systems . The idea: you have some notion of relevance for items in the result list (like in search, some documents are more relevant to the query than others). **DCG (Discounted Cumulative Gain)** scores a list by accumulating the relevance of each result, discounted by its position (results at rank 1,2,3 count more than results down at rank 10, because users pay more attention to top results). **NDCG** then normalizes this score by the best possible (ideal) score, so you get a number between 0 and 1. An NDCG of 1.0 means your list is as good as it gets (you put the most relevant items in perfect order). An NDCG of 0.5 might mean the ordering could be much better. In simpler terms, NDCG measures **how well the ranking of outputs matches the ideal ranking based on relevance**, with emphasis on getting the most relevant stuff to the top. For example, if WebkinsTech's AI support system suggests knowledge base articles to a user, NDCG could be used to evaluate if the articles the user clicked (presumably relevant) were ranked high. If a relevant article was buried at #10, the NDCG@10 would be lower. You often see NDCG@K (like NDCG@5 or @10) to evaluate the top K results. It's a bit abstract, but very useful when you have graded relevance. Many competition-winning recommendation systems report NDCG.

- **MAP (Mean Average Precision)** – Another common one, especially in recommendation and multi-label classification, is MAP . It stands for Mean Average Precision. Here's how it works: "Precision at K" for a list is the percentage of the top K results that are relevant. Average Precision is the average of precision values at the ranks of each relevant item in the list (essentially area under the precision-recall curve for that query's results). **Mean Average Precision** is then averaging that over all test queries/users. In simpler terms, MAP rewards you for having more relevant items and having them appear early in the list. If your recommendations often include all the things a user likes within the top 5, you'll have a high MAP. If the things the user likes are scattered or mostly outside the top 10, MAP will be lower. MAP@K might restrict to evaluating the top K. It's a very stringent metric in some cases – it requires both high recall of relevant items and good ordering. In practice, if you have a few "relevant" or "ground truth" items per user (like a user eventually clicked 2 out of 10 recommended items – those 2 are relevant), MAP considers both that you recommended those 2 somewhere and preferably near the top.

- **Recall@K** – This measures **out of all relevant items, how many did we show in the top K results** . For example, say in a movie app, a user would actually enjoy 10 movies out of a catalog, and we recommend 5 – if 3 of those truly liked movies are in our top 5, Recall@5 = 3/10 = 30%. In many recommendation scenarios, you don't know all "relevant" items clearly, but you might use implicit feedback (e.g., items the user clicked or purchased). Recall@K is simpler: it doesn't care about the order within those K, just whether they're present. It's often used in multi-stage recommenders (you want a candidate generation stage to have high recall@K so the ranking stage has the good options to sort). If your intelligent service is a search microservice, Recall@K might mean how many of the truly relevant docs (say, judged by users or editors) are in the first K shown. It ensures the system is retrieving enough good stuff early on. High recall@K with some ordering metric like NDCG or MAP is ideal.

There are other metrics like **MRR (Mean Reciprocal Rank)** for the rank of the first relevant item, or **Hit Rate@K** (similar to recall@K but usually just binary "hit or not" per user). But the ones above are sufficient for our discussion.

For an intelligent microservice doing recommendations, you'd likely evaluate these metrics offline on historical data (e.g., using interactions logs as ground truth). In production, measuring these continuously might involve A/B tests or recording what the user actually engaged with and computing metrics periodically. A practical approach: you recommend a list, and if you later know which of those the user interacted with, you can compute precision@K or recall. You can also invert it by logging if the user had some ideal set and whether it was shown.

Example: WebkinsTech introduces an AI-driven product recommendation service on their ecommerce site. To evaluate it, they run an A/B test: some users see the AI's recommendations, others see a standard top-seller list. They measure that the AI's recommendations have an NDCG@5 of 0.80 (pretty high, meaning relevant items are usually in the top 5) while the old top-seller list had NDCG@5 of 0.60 for personalized relevance. They also see that for the AI group, Recall@5 is 50% (users often find at least one item they end up buying in the five shown) vs 20% in the control. These metrics give confidence that the AI service is ranking items in a way that users find more useful. Additionally, they track these metrics weekly to ensure model updates or data shifts don't degrade recommendation quality.

One more nuance: with ranking metrics, **higher is better** (unlike error metrics). NDCG and MAP typically range 0 to 1 (with 1 being best). It's common to multiply by 100 and treat them like percentages as well ("we have an MAP of 20%" etc.). Always clarify the @K if you use these, because NDCG@100 might be different from NDCG@10.

Metrics for a Chatbot Service

To illustrate the blend of metrics, consider a customer service chatbot microservice. The team tracks:

Average Response Time – a system metric, to ensure the bot replies quickly (users won't wait 10 seconds for an answer).

Resolution Rate – an AI performance metric measuring how often the bot successfully resolves an issue without human intervention.

For instance, if out of 1000 chat sessions the bot handled 800 to completion (no human needed), its resolution rate is 80%. By monitoring these, they observe that when average response time stays under 2 seconds, user satisfaction is high, and by improving the bot's intent classification (raising resolution rate from 70% to 85%), they significantly reduce load on the human support team. This combo of metrics (fast replies and effective solutions) is key to chatbot success.

Business-Level Metrics

We've looked at technical metrics (system and model), but ultimately, what is the **real-world impact** of your intelligent microservice? That's where **business-level metrics** come in. These metrics tie the performance of your service to outcomes that the business or endusers care about. They often answer the question: *Is this AI-powered service actually delivering value?* While a model's accuracy is great to know, the business owner cares if it's increasing revenue, reducing costs, improving customer happiness, etc.

Let's explore some common business-level metrics relevant to intelligent services, and why they matter:

- **Conversion Uplift** – If your intelligent service is customer-facing (especially in ecommerce or marketing), conversion is king. **Conversion rate** is the percentage of users who take a desired action (buy a product, sign up for a newsletter, etc.). An AI recommendation or personalization service might aim to increase conversion. **Conversion uplift** measures how much your service has improved the conversion rate compared to a baseline. For example, WebkinsTech adds an AI-driven recommendation panel on their product page and sees the conversion rate of add-on item purchases go from 5% to 5.5%. That's a 0.5% absolute increase, which is a 10% relative uplift in conversions. It sounds small, but such gains can be worth millions. This metric is usually obtained via A/B testing (comparing users who get the AI feature vs those who don't). It's a direct business metric – does the AI help drive more sales or user actions? If your intelligent microservice is internal, conversion might be metaphorical (e.g., conversion of leads in a sales tool). The key is linking the AI's actions to a funnel improvement. If the conversion uplift is not positive, maybe the AI's recommendations or decisions need tweaking. Many companies treat uplift as the **North Star** metric for recommendation systems.

- **Productivity Gain** – Many intelligent microservices aim to automate tasks and thus improve team productivity or reduce manual effort. **Productivity gain** can be measured in various ways: how many more tasks can be completed per unit time, how many hours are saved, or how much output per worker increases. For instance, a support ticket triage AI might classify and route tickets automatically. If previously a support agent handled 10 tickets/hour and now with AI assistance they handle 15/hour, that's a 50% productivity gain. Or if an AI system automates report generation that used to take an analyst 2 days each week, that analyst now can focus on other work – effectively increasing capacity without hiring more people. Businesses love productivity metrics because they directly impact costs and throughput. Sometimes this is quantified as **efficiency** (e.g., "tickets per agent per day increased from 20 to 30"). Sometimes it's in cost terms ("we can handle 50%

more customers with the same team size"). If your microservice is internal (like an AI ops tool or a developer assistant), productivity gain is your badge of success. You'd measure before/after or with/without the tool. It's important to gather baseline data (how things were without the AI) to quantify the improvement.

- **Customer Satisfaction (CSAT)** – CSAT is typically measured by customer feedback, often through surveys. For example, after a support chat or a purchase, a user might be asked "How would you rate your experience?" on a scale (say 1 to 5 stars). CSAT can be the average rating or the % of respondents above a certain threshold (like % giving 4 or 5 stars). An intelligent service can affect CSAT positively or negatively. If your AI service makes the user experience smoother (fast responses, accurate answers, personalized content), you should see an uptick in CSAT. If it frustrates users (e.g., a bot that doesn't understand them), CSAT will fall. Monitoring CSAT in relation to your AI feature is crucial. WebkinsTech, for instance, introduced Lumi (the support chatbot) and kept an eye on the post-chat survey ratings. Initially, CSAT for chatbot-handled queries was 3.5/5, whereas human-agent chats were 4.5/5. This was a sign that Lumi needed improvement. After iterating on Lumi's responses and coverage, the chatbot CSAT rose closer to 4.3/5, nearly matching human-assisted chats. That's a huge win – it means the AI is almost as good as a human in customers' eyes! Often, you'll use CSAT as a guiding metric to decide if the AI service is ready to take on more volume or needs more training. **NPS (Net Promoter Score)** is another related metric: it asks customers how likely they are to recommend your product/service, scoring from -100 to +100. An AI feature can influence NPS if it's prominent in user experience. For example, if users love the personalized recommendations, they might be more likely to recommend your app, boosting NPS. Conversely, a bad AI feature can hurt overall brand perception.

- **Reduction in Manual Ops** – This metric is about **operational load reduction**: how much previously manual operational work is now handled by the intelligent service. For example, if you had a team manually monitoring dashboards and an AI Ops service that automatically detects and resolves incidents, you'd measure how many incidents are resolved without human intervention now. Or how many fewer tickets L1 support has to escalate because the AI handled them. A concrete measure could be "percentage of processes now automated." WebkinsTech tracked how many orders or claims required human intervention before vs after their AI automation. They found that previously about 20% of orders needed a human decision, which fell to ~5% after deploying AI agents . That's a dramatic reduction in manual ops – the majority of routine exceptions were auto-resolved by the agents, freeing up humans to focus on only the tricky 5% cases. This metric can be reported as a percentage reduction (e.g., -75% manual workload on X process) or as absolute hours saved per week in operations. It's very tangible to management: "our AI cut down manual work by N hours or Y%." It often ties into cost savings as well, but it's more about process efficiency.

- **Automation Coverage** – Similar to the above, this specifically refers to the **scope of tasks automated by AI**. If your intelligent microservice is part of an automation strategy, you might define **automation coverage** as "the proportion of [specific task or workflow] that is executed by AI vs humans." For example, in a customer support scenario: if out of 1000 support queries, 600 are fully handled by the chatbot and 400 go to humans, you have 60% automation coverage for support queries. You want to track this over time – hopefully increasing as the AI becomes more capable. But it's not just about volume; it's also about **areas** covered. Early on, you might only automate FAQs (maybe 20% of incoming topics). As the AI improves, it might cover more complex queries, raising the coverage to 50% of all topics. You can report coverage in terms of types of issues automated. For instance, "Our intelligent system now handles 80% of password reset requests and 50% of billing inquiries automatically." It's a metric that shows how far along you are in adopting AI. High automation coverage usually correlates with cost savings and speed, but watch that it doesn't hurt CSAT if the AI isn't good enough in some areas. Balanced with satisfaction metrics, it tells a powerful story of scaling service with AI.

- **MTTR Reduction** – MTTR can mean "Mean Time to Resolution" or "Mean Time to Recovery/Repair" depending on context. It's commonly used in incident management: how long on average to restore service when there's an incident. An intelligent microservice can help reduce MTTR in a few ways, especially if it's part of AIOps or devops automation. For example, an AI monitoring agent might detect issues faster (reducing the time to *detect* and thus to recover) or even auto-remediate certain issues (like auto-restarting a service or rolling back a deployment) without waiting for human intervention . Business-wise, a lower MTTR means less downtime and less impact on customers. If your intelligent service contributed to that, it's a big win. You would measure MTTR of incidents before vs after. Suppose before AI, the average outage took 1 hour to resolve, and after implementing an AI ops tool, it's down to 30 minutes – that's a 50% reduction in MTTR. For customer-facing processes, you might also talk about resolution time for customer issues (like warranty claims). In the earlier WebkinsTech example, automating simple warranty claims cut their resolution time from ~5 days to ~1 day – an enormous improvement for customer experience. So whether it's technical incidents or customer problems, intelligent services often shine by speeding up resolution. Faster resolution not only saves internal time, it often improves customer trust (who likes waiting?).

Each of these metrics should be given a narrative in real-world impact terms when communicating to stakeholders. For example: *"Our new AI-powered routing service decreased average delivery time by 2 hours, leading to a 5% increase in customer satisfaction scores."* Or *"By automating data entry, we saved ~$200k per year in manual labor and reduced error rates in records by 80%."* These tie the technical achievements back to business value.

It's also important to note that improving business metrics often requires alignment of technical metrics with business processes. You might need to work with business analysts or product managers to set up ways to measure these. A/B testing is a common method: roll out the intelligent feature to a subset and compare key business metrics to a control group. Also, sometimes there are **proxy metrics**. For instance, if actual conversion data is hard to get per feature, you might use click-through-rate on recommendations as a proxy (assuming more clicks likely lead to more sales).

WebkinsTech's journey provides a great illustration. After deploying intelligent microservices in order fulfillment, they saw concrete business outcomes: a 12% increase in on-time deliveries (leading to happier customers) as processing became faster , and a reduction of exceptions from 20% to 5% which translated to lower operational costs and faster order handling . These numbers aren't just stats – they told the business, *"hey, this AI stuff made us better and our customers noticed."*

Collaborate on Business Metrics

Make sure to involve the relevant business or product stakeholders when defining these metrics. For example, if your service aims to improve conversion, work with the marketing or product analytics team to design the experiment and tracking. If it's about manual work reduction, coordinate with the operations manager who can verify the time saved. This collaboration ensures you measure the *right* things and everyone trusts the numbers. It also helps prioritize which metrics matter most for your context (not every project cares about the same KPIs).

At this point, we've covered three layers of metrics: operational (system health), AI/model (prediction quality), and business (impact). A successful intelligent microservice strikes a balance across all three – it runs well, thinks well, and delivers results. In the next section, we'll discuss the tricky part: how optimizing one metric can impact others, and how to find the right trade-offs.

Trade-offs and Interplay

If there's one thing to remember about metrics, it's that **no metric exists in isolation**.

Tuning an intelligent service is a multi-objective game. Push one lever too hard, and another might wiggle out of place. In practice, teams often find that improving one aspect of a system can inadvertently hurt another. This section is about recognizing those trade-offs and managing them, so you don't end up in a situation where "the operation was a success, but the patient died" – i.e., you optimized a metric but undermined the overall goal.

Let's discuss a few common trade-offs and how to balance them:

- **Complex Models vs Latency:** A more complex AI model (deeper neural net, more features, ensembling multiple models) might give you higher accuracy or better predictions. Great, your AI metric goes up! But that complexity usually means heavier computation – which can increase **latency** and CPU usage. Suddenly your operational metrics suffer (p99 latency might spike, throughput could drop if each request takes longer). For example, you replace a simple regression with a large transformer model in a microservice; accuracy of predictions improves by 5%, but the response time goes from 50ms to 500ms. Users of a real-time system might not tolerate that. This is a classic trade-off: **prediction quality vs performance**. How to balance it? One approach is to find the sweet spot model – maybe a slightly smaller model that still gives most of the accuracy boost at much lower latency. Or use distillation or optimization techniques to speed it up. Another approach is architectural: cache results, do asynchronous processing, or use hardware acceleration. The key is not to chase the last drop of accuracy if it means blowing past your latency SLO. World-class engineers often iterate to achieve *good enough* *accuracy within acceptable latency*. They might say, "We'll sacrifice 2% accuracy if it cuts latency by 4x, because the user experience is paramount." In metric terms, set a target for latency (say p95 < 200ms) and ensure any model changes keep within that, even if it means the model metric isn't maxed out.

- **Precision vs Recall (or False Positives vs False Negatives):** This is an inherent tension in many AI systems. If you tune your model to be more conservative (maybe raising a threshold) to increase precision and reduce false positives, you likely decrease recall (miss more true cases). Conversely, if you try to catch every possible positive (high recall), you'll probably snag some false alarms (lower precision). The trade-off decision depends on context. In a medical diagnosis service, missing a disease (false negative) is worse than a false alarm, so you'd favor recall over precision – maybe tolerate some extra false positives. In a spam filter, a false positive (flagging a real email as spam) might be worse for user trust, so precision is prioritized. As a practitioner, you should communicate these trade-offs and possibly provide controls. Some systems allow a "dial" to adjust sensitivity. But in any case, when optimizing your model, be aware that pushing one of these metrics will pull the other. The **F1-score** we discussed is one way to balance them, but sometimes a single threshold doesn't satisfy all business needs, and you have to pick a point on the precision-recall curve that aligns with priorities. Monitor both metrics in production if possible. If you see precision dropping sharply after a model update, double-check if recall jumped (and if that's desired or not). Balanced evaluation reports help avoid blind spots – always present precision and recall together for classification tasks.

- **Automation vs Quality:** If you automate more (high automation coverage), you risk handling cases that the AI might not be fully ready for, potentially hurting quality (satisfaction). If you automate less (hand off to humans more liberally), quality might be high but you lose efficiency gains. For example, WebkinsTech's Lumi chatbot initially tried to handle a very wide range of topics to maximize automation, but some of those topics it wasn't good at, leading to frustrated users. They pulled back on automation for those topics (routing them directly to humans) until the AI could be improved. This is essentially managing the **scope** of AI. The trade-off is often dynamic: as AI improves, you expand its scope. The concept of a **balanced scorecard** can be useful here – you monitor both the automation rate and the CSAT. If CSAT drops below a threshold, you know automation might be overreaching. If CSAT is great but automation is low, maybe you can push a bit further. The balance is finding that curve where marginal automation doesn't significantly drop quality.

- **Throughput vs Accuracy (or Quality):** Consider an AI content moderation service. You could run a very quick, simple filter that catches obvious bad content (very high throughput, low latency) but misses nuanced cases (lower detection rate). Or a more heavy-duty analysis that catches more but is slower or can handle fewer items per second. If you prioritize throughput (maybe due to cost or time constraints), you might accept slightly lower accuracy in catching all issues. If quality of detection is paramount (e.g., for legal or safety reasons), you might invest in a slower but more accurate pipeline and scale out with more servers to maintain throughput – which then is a cost trade-off. This interplay often appears as **cost trade-offs** too: a larger model might require more servers or more expensive hardware to keep throughput, so the trade-off could be framed as *accuracy vs cost at scale*.

- **Short-term Reward vs Long-term Outcome:** For learning agents, sometimes optimizing the immediate reward can harm long-term objectives. This is more of a RL nuance – an agent might learn a hack that maximizes reward in the short term but is suboptimal long-term (like a game agent that learns to do something that gets points now but will lose the game later). In business terms, an AI might learn to exploit a metric in a way that isn't truly desired (think of it like gaming the KPI – this is essentially Goodhart's Law in action: *"When a measure becomes a target, it ceases to be a good measure."*). It's important to watch out for weird shifts: if your model's metric is great but some other indicator (maybe user complaints) is rising, something's off. Balanced evaluation means occasionally looking at samples or secondary metrics.

How do we manage these trade-offs? One approach is using a **balanced scorecard** for intelligent services. This means defining a set of critical metrics across different categories and monitoring them together. Instead of having tunnel vision on a single number like "99% accuracy!", you create a small dashboard: e.g., "Accuracy, Latency, and User Satisfaction". If a change improves accuracy but causes latency to breach the acceptable range or satisfaction to drop, you reconsider or mitigate. You might even weight them and create a composite score internally. Businesses do this at higher levels (financial vs customer vs internal metrics). Here, you as the technical leader do it for your service.

For instance, you might set target ranges: **Accuracy >= 90%, p95 Latency <= 300ms, CSAT >= 4.0**. These become your three SLOs (service level objectives) to balance. When iterating on the model or system, check all three. If one goes out of bounds, address it. It instills a mindset that success isn't one-dimensional.

Another tip: **staged rollouts and A/B tests** are your friend when dealing with trade-offs. If you have a potentially slower but better model, do a side-by-side comparison in production on a small percent of traffic. Measure both the AI metric and system metrics and any user behavior metric. This way you can concretely see, "Yes we gained X in accuracy, but latency increased by Y and conversion dropped by Z," and decide if X is worth Y and Z. Data might reveal the drop isn't that bad or can be mitigated by scaling.

Additionally, sometimes you can have the best of both: e.g., run the heavy model asynchronously or in a batch mode for quality, and a light model for immediate response. Or use the heavy model only for cases where it matters. That's an architectural solution to trade-off – not everything needs one model fits all.

WebkinsTech encountered such balancing acts: they had to tune their orchestrator agents to not overwhelm the system (throughput vs quality of decisions), and find balance between fully automated vs human oversight in processes. They essentially set up guardrails: if certain metrics fall outside a band (like error rate spikes or customer complaints), the system either alerts humans or rolls back to safer mode. This kind of **graceful degradation** strategy ensures that chasing a metric doesn't lead to disaster.

Holistic Optimization

Avoid local optimizations that ignore the big picture. In one case, a team optimized an AI model to achieve state-of-the-art precision, but it required so much compute that the service started timing out – ultimately failing the user. Don't let **one metric dominate** your decision-making. Always evaluate the **impact on other important metrics**. Use a holistic or "system of metrics" view. As one engineer quipped, "We improved the model and killed the service." Oops. A balanced approach would have caught that before it hit production.

To help visualize and manage these, some organizations create an **internal comparison matrix or scorecard** (which we'll exemplify next) and run through scenarios. The takeaway is: be aware of the interplay, communicate trade-offs clearly to stakeholders ("we can get a bit more accuracy if we accept a bit more latency – is that worth it for our users?"), and strive for a balance that aligns with your overall goals (often user experience and business outcomes should trump pure technical metrics).

Now, having discussed all these metrics individually and in combination, let's compile them in a handy reference table for clarity.

Comparison Matrix of Metrics

To consolidate what we've covered, below is a **comparison matrix** of various metrics across the system, AI, and business categories. This table lists each metric, classifies its type, and summarizes its purpose, how it's measured, and an example benchmark or typical value (for context). This can serve as a quick reference or checklist when evaluating intelligent services.

Metric Name	Type (System/AI/Business)	Purpose	How It's Measured
Throughput	System	Measures how much load (requests/events) the service can handle per time unit. Indicates capacity and scalability.	Count requests per second (RPS) or similar, via service logs or monitoring tools.
Latency (p95/p99)	System	Time taken to serve requests. Lower latency = faster responses for users. p95/p99 focus on tail delays.	Instrument timers in code; compute 95th/99th percentile response times.

Error Rate	System	Fraction of requests that fail. Indicates reliability and correctness. Lower is better.	Track error responses (5xx codes, exceptions) / total requests. Often in % or errors/min.
Uptime	System	Percentage of time the service is available. Reflects overall reliability.	Monitor downtime duration vs total time (often via health checks).
CPU/Memory Usage	System	Shows resource consumption. Ensures service is within capacity and helps detect saturation or leaks.	Collected via host or container metrics (CPU %, RAM in MB).
Accuracy	AI Model	Overall correctness of predictions (classification). High accuracy = model usually right.	(# correct predictions) / (total predictions), from test data or feedback.
Precision	AI Model	Precision = % of predicted positives that were	TP / (TP + FP), evaluated on labeled data or

Metric Name	**Type (System/AI/Business)**	**Purpose**	**How It's Measured**

		correct. Focus on false alarm reduction.	through outcome tracking.
Recall	AI Model	Recall = % of actual positives caught. Focus on missing as few true cases as possible.	TP / (TP + FN), from labeled outcomes or known positives.
F1-Score	AI Model	Harmonic mean of precision & recall. Balances the two for overall test accuracy.	2 *(Prec* Rec) / (Prec + Rec), compute d from precision and recall values.
MAE (Mean Abs Error)	AI Model	Average magnitude of errors (regression). Lower = predictions are closer to actual on average.	Mean(
RMSE	AI Model	Root mean squared error. Penalizes large errors more. Lower = more accurate, especially on outliers.	sqrt(mean((pred – actual)^2)). Calculated on test/outcome data.
Confidence Calibration	AI Model	Alignment of predicted probabilities with actual outcomes. Indicates trustworthiness of model's confidence.	E.g. Brier Score, or calibration curves comparing predicted vs observed probabilities.

Prediction Entropy	AI Model	Measures uncertainty of the model's predictions. Higher entropy = more uncertainty on average.	Average entropy of output distribution (for classification) over predictions.
Reward Rate	AI/Agent	Measures performance of a learning agent (average reward per episode or time). Higher = agent achieving goals more.	Sum of rewards per episode, averaged over many episodes or per unit time in steady state.
Convergence Time	AI/Agent	How long it takes an agent/model to learn or	Number of iterations/episodes

Metric Name	**Type (System/AI/ Business)**	**Purpose**	**How It's Measured**
		adapt (to reach target performance). Shorter = learns faster.	or wall-clock time to hit a performance threshold.
NDCG@K	AI/Rank	Quality of ranked results (e.g., recommendations/search). Accounts for ranking position of relevant items. 1.0 = ideal ranking.	Compute DCG of results up to rank K and normalize by ideal DCG (IDCG).

MAP@K	AI/Rank	Mean Average Precision at K. Indicates overall precision of top-K results across users/queries. Higher = more relevant items high in list.	Average Precision for each query (precision at each relevant item), then mean across all queries/users.
Recall@K	AI/Rank	Coverage of relevant items in the top K results. Higher = model finds most of the good items by position K.	(# of relevant items in top K) / (total relevant items available), averaged over users/queries.
Conversion Rate	Business	Percentage of users who take a desired action (purchase, sign-up). Higher = more success in user funnel.	(# conversions) / (# opportunities) – measured via analytics or A/B tests.
Productivity (Tasks/Hour)	Business	Measures efficiency improvement. More tasks per hour per person or process = productivity gain.	Compare throughput per worker or per process before vs after AI assistance.
CSAT (Customer Satisfaction)	Business	Direct measure of customer happiness, often via ratings. Higher = customers are more satisfied with the experience.	Survey scores (e.g. 1-5 stars or % satisfied). Can be averaged or % above threshold.

NPS (Net Promoter Score)	Business	Measures customer loyalty and willingness to recommend (on a -100 to +100 scale). Higher =	Survey asking "How likely to recommend?" (score 0-10). NPS
Metric Name	**Type (System/AI/ Business)**	**Purpose**	**How It's Measured**
		more promoters than detractors.	= %Promoters - %Detractors.
Manual Effort Reduction	Business	How much human/manual work is saved by automation. Indicates efficiency and cost impact of AI.	Measured as drop in manual processing (tasks, hours) required for the same output.
Automation Coverage	Business	Proportion of process or tasks handled end-to-end by AI vs humans. Higher = more of the process is automated.	(# automated cases) / (total cases), often tracked per category or overall.
MTTR (Mean Time to Resolve)	Business/System	Speed of resolving an incident or issue. Lower = faster recovery, less downtime or customer wait.	Average time from issue start to resolution, measured via incident tracking or workflow timestamps.

(TP = True Positives, FP = False Positives, FN = False Negatives in classification metrics above. K is a chosen cutoff rank, e.g., 5 or 10.)

This table provides a quick glance at what each metric means and how you might use it. For instance, if you're evaluating an AI-based recommendation microservice, you'd probably look at some system metrics (latency so the user gets recs quickly), AI metrics (perhaps Precision@K or NDCG to ensure quality of recommendations), and business metrics (conversion uplift of those recommendations, and maybe user engagement or CSAT). A side-by-side matrix like this helps ensure **coverage**: did we consider everything important? It reminds us that an "intelligent service" has multiple facets to measure.

Of course, not every project will use all these metrics. Choose the ones that make sense for your context. The system metrics are generally universal. The AI metrics depend on your ML task type. The business metrics depend on the project goals. Use this matrix to pick a balanced set.

Next, we'll look at how these metrics are actually **collected and visualized** in real-world systems – essentially, how do we wire up our microservices so that these numbers flow into our graphs and alerts?

Metric Visualization Flows

Defining metrics is one thing, but implementing them in your architecture is another. In a modern microservice setup, especially with intelligent components, you'll typically have a metrics collection pipeline and dashboards to visualize them. Let's break down a couple of typical **metric flows** – one for general service metrics and one for an AI-specific use case.

We'll use a simple diagram style to illustrate:

1. **Operational Metrics Pipeline:** Imagine the path from an API request to a dashboard:

```
API Request → Logging → Metrics Agent → Prometheus DB → Grafana Dashboard
```

1. In this flow, an incoming API request triggers your service. Your service logs the request and also records metrics like latency and status. A **Metrics Agent** (could be something like a Prometheus exporter or sidecar) periodically collects or receives these metrics. The data is then stored in a **Prometheus database** (or any time-series DB). Finally, a **Grafana-style dashboard** queries that DB and displays graphs (like RPS, latency percentiles, error rates over time). This is a common flow in

cloud environments: for example, Kubernetes pods exposing metrics on /metrics that Prometheus scrapes, and Grafana visuals. The key is that instrumentation in the service (or sometimes automatically via service mesh) feeds into a metrics system. **Alerting** can be layered here too: e.g., Prometheus or Alertmanager can trigger an alert if p99 latency > X for Y minutes. But in terms of visualization, Grafana (or Kibana, if metrics are in Elastic, etc.) is where humans see the trends. At WebkinsTech, they did something similar by aggregating logs with ElasticSearch and making Kibana dashboards to track things like auto-fulfilled orders vs human help . Nowadays, Prometheus/Grafana is very popular for such metrics. The flow might slightly differ if using a cloud monitoring service (e.g., AWS CloudWatch would replace Prometheus in the diagram, and maybe its own dashboards or third-party). The essence remains: **service emits metrics → metrics collected and stored → metrics displayed**.

2. **AI Model Monitoring & Alerts:** Now consider an AI-specific metric scenario:

```
Model Prediction → Output Probabilities → Confidence Scorer → Alert
if Under Threshold
```

2. Here, each **Model Prediction** not only produces a result but also a probability or confidence. We pass those outputs through a **Confidence Scorer** or checker – basically logic in the service that evaluates the confidence. If the confidence is below a certain threshold (meaning the model is very uncertain or perhaps the result is likely low quality), the system can trigger an **alert** or take action. For example, the service might log an event “low confidence on prediction X,” and an alerting system (PagerDuty, Slack notification, etc.) might warn that the model is operating in unfamiliar territory for certain inputs. Alternatively, instead of alerting a human, the system might automatically route that request for human review or use a simpler fallback model. This flow highlights **real-time AI monitoring**: you’re not just passively observing metrics; you’re actively using them in the workflow. Many production ML systems embed such checks to detect issues like data drift or out-of-distribution inputs. For instance, if our support chatbot Lumi starts seeing a lot of queries where its intent classifier gives only 40% confidence (very unsure), it might alert the ML team or switch to a default “I’m not sure I understand, let me connect you to an agent” response (graceful degradation). The “Confidence Scorer” could be as simple as an if probability < 0.5 then ... in

code, or as complex as a module computing statistical metrics (like comparing input distribution to training set). The output could go to monitoring as well – e.g., count of low-confidence cases per hour (which itself is an interesting metric to track for drift).

In practice, you'll likely have **both flows in parallel**: one for generic service metrics and another for model-specific monitoring. They might feed into the same observability stack or separate ones. Some newer tools target ML monitoring explicitly (capturing prediction distributions, etc., and visualizing them). You might incorporate those alongside your standard Prometheus/Grafana.

Other visualization flows could include:

- Traces flowing through a tracing system (like "User request -> Service A -> Service B -> results" in Jaeger, which helps see per-request latency breakdown).

- Business metrics often flow into analytics systems: e.g., conversion data might go into a data warehouse or A/B testing platform which has its own dashboard (like Optimizely or internal tools). So the conversion uplift might be seen in an experiment report rather than Grafana. Ensuring that loop is in place (from service event -> analytics event -> analysis) is important for business metrics.

The goal of visualizing metrics is to make them actionable and understandable at a glance. Good dashboards for intelligent services might have sections: one for system health (are we up? fast? any errors?), one for model performance (recent accuracy vs target, confidence distribution, etc.), and one for business outcomes (e.g., weekly conversion, user ratings, etc.).

For example, WebkinsTech's team might maintain a **"Lumi Dashboard"** that shows:

- Chatbot uptime and error rate (system health).

- Average response time of the bot (system) alongside the human fallback response time for comparison.

- Bot resolution rate and deflection rate over time (AI metric and business metric).

Customer satisfaction scores from post-chat surveys, split by whether the bot handled or a human did (business metric).

- A pie chart of topics and how confident the bot is in each (AI metric, showing perhaps where it's struggling).

A trend line of low-confidence occurrences or when it had to escalate to humans (like an alert metric).

Such a dashboard allows both engineering and product folks to see how Lumi is doing on multiple fronts in one place. If something regresses (say CSAT drops or errors spike), it jumps out.

Dashboard Hygiene

When creating dashboards for metrics, **don't overcrowd them**. Focus on the key metrics that you identified as important. It's better to have multiple dashboards (system vs business) than one overwhelming board. Also, include annotations for deployments or model releases – so you can correlate a metric change with "ah, we deployed version 2 of the model here." This makes analysis much easier when something goes bump in the night.

Additionally, consider setting up **automated alerts** for critical metrics so you don't have to stare at dashboards 24/7. For system metrics, this is standard (pager for downtime, etc.). For AI metrics, you might set up an alert like "model confidence average below 60% for >1 hour" or "weekly accuracy on feedback data dropped by >5 points" which could indicate drift or an issue.

Alright, with an understanding of how metrics are monitored in practice, let's tie it all together with an example story. We'll see how our fictional company WebkinsTech evaluates their intelligent service using these metrics, and how it helps them improve.

Example Vignette: WebkinsTech's AI Support Agent Evaluation

Let's step into the shoes of WebkinsTech's team as they roll out and refine their AI-powered support agent, **Lumi**. This vignette will illustrate how the team tracks various metrics we discussed, and how those metrics guide improvements. It's a bit of a story, showing metrics in action.

Background: WebkinsTech, our trusty fictional enterprise, has a growing customer base for their tech products. Their support center was getting swamped with repetitive queries ("How do I reset my password?", "Is my device under warranty?", etc.). They decide to deploy an intelligent microservice – a support chatbot named **Lumi** – to handle common questions and tasks, working alongside the human support team. Lumi is an AI agent integrated as a microservice in their support architecture.

Phase 1: Initial Deployment and Metrics Baseline

When Lumi is first launched, the team limits its scope to a few simple issues (password resets, order status queries, FAQs). They set up metrics to evaluate Lumi:

Operational Metrics: They monitor Lumi's service like any other – making sure its response time stays low (aiming for <2 seconds p95 latency even under load) and error rate is near zero. They deploy Lumi on a scalable cluster, tracking throughput (# of chats handled per minute). In the first week, Lumi handles an average of 20 chats per hour with a p95 latency of 1.5s and no major errors – the system metrics look good.

- **AI Performance Metrics:** Lumi's brain is an NLP model that classifies the customer's question into an intent and fetches a relevant answer. The team had tested it offline: overall accuracy was about 85% on a validation set of common questions. Now in production, they gather data from chat transcripts. After each chat, they ask the user a quick feedback: "Did Lumi answer your question? (Yes/No)". This gives them a proxy for accuracy in the real world. Initially, they find about **75% of users indicate "Yes, issue resolved"**. So Lumi's "resolution rate" is 75%. Precision vs recall in this context: Lumi sometimes confidently handles an issue it actually shouldn't (leading to an unresolved case – that's like a false positive in automation), and it sometimes punts to a human even though it might have been able to answer (false negative in automation). They log these events. Precision of automation = 90% (when Lumi says "I got this", 9 out of 10 times it was correct), recall of automation = 70% (it only attempted 70% of the ones it could have, since it's cautious initially). These numbers give a baseline to improve from.

- **Business Metrics:** The main goal is to reduce the load on human agents and maintain customer satisfaction. They measure **Automation Coverage**: in week 1, about 30% of all incoming support chats were handled fully by Lumi (the rest escalated to humans) – so 30% coverage. They also look at **CSAT scores** from post-chat surveys. For chats Lumi handled, the average CSAT is 3.8 out of 5. For chats humans handled, it's 4.5.

Not surprising – Lumi is new and not as polished as a human. Also, they measure

Average Handling Time: Lumi resolves queries in about 1 minute on average (including the conversation back-and-forth), whereas humans take 3-4 minutes when they get involved (partly because humans handle more complex cases too). So there's already an efficiency gain. No data on conversion here since support is post-purchase, but they do look at **cost per ticket**. With Lumi handling 30% of tickets, they estimate saving about 1 FTE worth of workload, which is significant if scaled.

The team compiles these metrics in a "Lumi Metrics Dashboard." It shows:

- % chats automated (30%).
- Resolution rate by Lumi (75%).
- CSAT Lumi vs Human (3.8 vs 4.5).
- Average response time (Lumi 1.5s vs Human ~15s typing delay).

And a breakdown of top intents Lumi sees, with its confidence and success per intent.

They notice a couple of insights:

- For password resets (which Lumi handles by triggering a reset email workflow), the success rate is 95% and users are quite happy. Great.

For warranty questions, Lumi's answers are sometimes incomplete, and many of those get escalated. Success rate on "warranty" intents is only 50%.

- Also, the **entropy** of Lumi's intent classification is high for some queries it didn't see in training (e.g., some users ask multi-part questions). The team sees that in logs – whenever a user asks something complex, Lumi's confidence dips and it wisely transfers to a human.

Phase 2: Iteration and Improvement

Armed with this data, the team holds a retrospective. They decide to make a few improvements:

1. **Expand Training & Knowledge**: They add more Q&A pairs to Lumi's knowledge base, especially for warranty and product troubleshooting, which were problematic. They retrain the intent model to improve its accuracy from 85% to ~92% on a new validation set (they added actual user queries from week1 to training – reducing that gap).

2. **Fine-tune Automation Threshold**: They noticed that Lumi might have been a bit conservative (70% recall in taking queries). Users didn't mind when Lumi tried and failed as long as it quickly handed off. So they slightly lower the confidence threshold required for Lumi to attempt answering. This might reduce precision a bit but increase recall – covering more queries.

3. **Improve Responses**: Some CSAT issues were due to tone or insufficient detail in answers. They tweak Lumi's response templates to be more friendly and add a bit of personalization ("I see you're asking about your Model X Gadget. Let me help with that."). Soft metrics but important for CSAT.

4. **Monitoring Additions**: They add an alert: if Lumi's resolution rate drops below 60% in any day or if automation coverage goes above 50% but CSAT falls below 3.5, they want to be notified. This is to catch any regressions or if they over-extended Lumi's reach.

They roll out "Lumi 2.0" with these changes in week 3 (week 2 was used for retraining and testing internally).

Metrics after rollout:

- **Operational**: Throughput of Lumi increased (now handling 50% of chats, so more load, but they added instances – CPU usage per instance is moderate). Latency remained ~1.5s p95. No spike in errors. Good.

- **AI Performance**: The user feedback now shows **85% resolution rate** (out of all chats Lumi takes, 85% are resolved without human). Precision of automation is 85% (a slight drop from 90%, meaning a few more false alarms where Lumi tried but user wasn't satisfied), but recall of automation jumped to 85% (meaning Lumi is now attempting a much larger portion of issues). The F1 of automation went up significantly with those numbers. The team also ran an offline test on chat logs with human labeling and found Lumi's intent classification accuracy is 90+% for the covered topics.

Business: **Automation Coverage** is now ~50% (Lumi handles half of all chats end-toend). This is a big improvement in efficiency. The human team is indeed noticing the difference – their queue is smaller and they can spend more time on complex cases. CSAT for Lumi-handled chats has risen to 4.2/5, thanks to better answers and maybe the fact it's handling queries it can answer well. Humans remain at ~4.5, but the gap closed. The overall support CSAT (mixing both) actually improved a tad, because Lumi took care of many simple queries quickly (customers appreciate quick service for simple issues, even if it's a bot). Another metric: **First Response Time** – with Lumi, 50% of customers now get an instant first response (within 2s), whereas previously everyone waited on average 2 minutes for a human to pick up. That likely contributed to satisfaction.

The team sees a notable **productivity gain**: previously, 10 support agents handled 100% of chats. Now, those same 10 agents handle 50% while Lumi handles the rest. Effectively, the team's capacity doubled – or one could say we could handle the same load with half the agents (though in practice they reallocate humans to more complex tasks and improving support quality elsewhere). They estimate a cost saving of, say, $X per month in support labor thanks to Lumi, which wows the operations manager.

The **conversion uplift** isn't directly applicable here, but they do measure something akin: the retention of customers who used support. They find customers who got their issue resolved by Lumi are similarly likely to purchase again as those who were helped by a human – a sign that the bot didn't hurt retention, which was a concern initially ("will users hate the bot?"). Some even specifically commented "Love that I got an answer immediately from the bot" in feedback – intangible but encouraging.

Phase 3: Ongoing Monitoring and Surprises

Lumi is now an integral part of support. The team keeps an eye on metrics over the next months. They notice a few things:

- **Concept Drift**: After a few months, a new product launch brings in questions that Lumi wasn't trained on. They see the **entropy** of Lumi's predictions spike and the resolution rate drops to ~70% for that week. Alerts fire that low-confidence predictions are happening frequently. The team reacts by updating Lumi's knowledge base for the new product, and retrains the model incorporating those queries. Metrics return to normal.

This shows how continuous monitoring caught a drift in query topics.

- **Trade-off Decisions**: At one point, they consider letting Lumi handle returns and refunds (a more sensitive topic). Initial tests show Lumi can answer correctly 80% of the time, but when wrong, customers get *really* upset (refund issues are touchy!). The metrics predicted this: likely CSAT would drop if Lumi handles refunds with only 80% success. The balanced scorecard (resolution rate vs CSAT) suggests *not* automating that yet. They decide to hold off and let humans handle refunds until they can ensure, say, 95% success by the bot. This was a business decision guided by metrics – they chose to sacrifice some potential efficiency (automation coverage could have been 70% if including refunds) in order to keep satisfaction high.

- **Improvements Plateau**: Over time, Lumi's metrics stabilized: ~60% automation, ~90% resolution on those, CSAT ~4.3. The team reached a nice equilibrium. Further improvements were incremental. They added a small feature where if Lumi's confidence is low, it asks a clarifying question ("Can you please confirm if you're asking about billing?") which bumped resolution a bit more. They also introduced a **balanced scorecard** in quarterly reviews: one that included system uptime (Lumi had a couple of outages that taught them to add redundancy), resolution rate, and customer satisfaction as three pillars to always check.

In the end, Lumi's success was measured not by one metric, but by a **suite of metrics** that told a story:

- It maintained strong operational performance (no scaling issues or slowdowns, even as load increased – thanks to metrics-driven scaling).

- It achieved good AI performance (the team continuously improved its NLP accuracy and kept an eye on confidence, preventing it from going rogue).

- It delivered business value (handling the majority of simple queries, cutting response times, saving support costs, and keeping customers happy enough).

As a postscript, WebkinsTech's leadership was convinced by these metrics to invest in more intelligent services. They saw clear ROI from Lumi: faster support, happier customers, and lower cost. This only happened because the team could **demonstrate the impact with hard numbers**. They literally had slides like "AI Support Agent Outcome: 50% tickets automated,

CSAT maintained 95% of human level, $Y saved per month, 2x faster responses." That's powerful.

This vignette underlines how metrics guide the journey of deploying and refining an intelligent microservice. From detecting issues (like new topics Lumi couldn't handle) to validating improvements (CSAT went up after changes) to making go/no-go calls on features (not automating refunds yet), the team leaned on data. It also shows the interplay: focusing only on automation rate would have been bad if CSAT tanked, so they balanced both.

Hopefully, this story makes the abstract concepts more concrete. You can likely imagine a similar narrative for whatever domain you're in – whether it's an AI-driven recommendation engine or a smart IoT service optimizing energy usage. The specifics change, but the pattern holds: **measure, learn, adjust, and improve**.

Practical Tips for Metrics in Intelligent Services

Before we wrap up, let's distill some **practical tips and best practices** when working with evaluation metrics for intelligent microservices. These are gleaned from the experiences of many engineering teams (and the kind of wisdom our WebkinsTech team earned along the way):

Define Metrics Early: When designing your intelligent microservice, think about metrics from day one. Ask "How will we know this is working?" for each objective. If it's an AI feature, decide what success means (accuracy? user engagement?). If it's system performance, decide SLOs. Early definition ensures you build in the instrumentation and data collection needed. It's much harder to bolt on metrics tracking after deployment if you forgot something.

Instrument Everything (but keep it organized): It's better to over-collect data than under-collect. Instrument your code to log important events (predictions, decisions) and expose metrics (latencies, counts). However, tag or namespace them clearly. For example, prefix model metrics with model *and business metrics with biz* in your monitoring system to differentiate. This avoids confusion and makes dashboards easier to build. Don't worry if you have more metrics than you initially use – as the system evolves, historical data might become useful (e.g., how did accuracy trend over past 6 months?).

Beware of Vanity Metrics: A vanity metric is one that looks impressive but doesn't drive action. For instance, "number of records processed" might grow just because user base grows – it doesn't tell you if the service is better.

Focus on actionable metrics – ones that, if they change, you'd do something. If you find you're tracking something that never informs any decision, consider dropping it or replacing it with a more meaningful metric.

Use Alerts and Automation: Set up alerts on key metrics thresholds – especially system health (you want to know if latency or errors are spiking ASAP) and any critical AI metric if it goes out of bounds (e.g., a sudden drop in prediction confidence or accuracy might indicate an issue with input data or a bug). Additionally, consider automated responses: e.g., auto-scale instances if CPU is high for sustained period (many orchestration systems do this), or trigger an automated retraining if model accuracy on recent data falls below a certain point (more advanced, but some teams do implement periodic retraining pipelines that kick off based on performance metrics).

Example – Metrics-Driven Retraining: One fintech company set up a system where if their model's 7-day average AUC dropped more than 2 points, it automatically triggered a model retraining job with the latest data. This ensured the model stayed fresh with minimal human intervention. They still investigated the cause, but the metric threshold prevented prolonged degraded performance. Consider if such feedback loops make sense for your use case (they aren't trivial to build, but can pay off for rapidly changing environments).

Correlate Metrics with Each Other: When analyzing, don't look at one metric in isolation. Use correlation: e.g., if conversion went up, did you also see precision or some UX metric change? If latency spiked, did error rate or model

throughput change correspondingly? Often the interesting insights are in relationships (maybe a slight increase in latency had no impact on user engagement up to a point – good to know your tolerance). Using tools or just manual analysis to overlay graphs can reveal these dynamics.

Data Quality for Metrics: Ensure the data feeding your metrics is correct. Sounds obvious, but many a time teams chased a "drop in accuracy" only to find the way they were calculating it in production was flawed (perhaps using incomplete feedback data). Validate your logging and calculations. For example, if computing precision from production outcomes, make sure you correctly label what was a true vs false positive. Garbage in, garbage out applies – metrics are only as useful as the data pipeline behind them. Treat that pipeline with the same rigor as your production code (tests, monitoring on the monitoring system!).

Share Metrics with the Team: Metrics shouldn't live in a silo. Dashboarding tools and analytics should be accessible to the whole team (and even crossfunctional teams where appropriate). Encourage a culture where product managers peek at the dashboards too, and where SREs are aware of model metrics, etc. This fosters shared ownership. It's great when a biz person can say "I noticed our automation rate increased last week – awesome!" or an engineer can say "CSAT dipped for that segment, let's investigate." Having a common view via metrics aligns everyone on the goals.

Insight from Industry: *"In God we trust; all others bring data."* – W. Edwards Deming. In practice, whenever debates arose at WebkinsTech about whether the intelligent service was helping or harming, the team pulled up the metrics to ground the discussion. Opinions are valuable, but metrics provide the evidence to make decisions rationally.

Iterate and Refine Metrics: Over the life of a project, you might realize some metrics need tweaking or new metrics are needed. Maybe you add a new feature – define new metrics for it. Or you learn that "time to first response" is actually a big driver of satisfaction, so you start measuring that. It's okay to evolve your metric set. Just be careful with continuity (if you redefine something, note it to not confuse historical comparisons). Regularly ask, "Are we measuring what matters?" and adjust if not.

Avoid Metric Overload: While we advocate collecting a lot, when it comes to focus, don't try to improve 20 things at once. Humans can effectively focus on a handful of KPIs at a time. It might help to identify a small set of primary metrics (perhaps one per category: e.g., latency for system, F1 for model, conversion for business) that are the main ones, and others are secondary or diagnostic. Otherwise, you risk analysis paralysis or contradictory optimization directions.

Document Your Metrics: Keep a reference of what each metric means, how it's calculated, and where to find it. This is immensely helpful as team members come and go or if someone new needs to understand the dashboard.

It can be a simple Markdown in your repo or a Wiki page. Document any changes too ("we switched from version 1 to version 2 of the model on this date, hence accuracy jumped"). This context is gold when reviewing metrics later.

Example – Balanced Scorecard Use: One company created an internal "AI service scorecard" that they updated each sprint. It listed: Uptime, 99th latency, Model precision, and Business KPI (specific to their domain). They graded each with Red/Yellow/Green for that sprint. If anything was Yellow/Red, that was a focus in the next sprint to bring up. This simple practice kept the team naturally balancing the different aspects. Consider adopting something similar if you find one perspective being neglected.

In summary, treat metrics as a critical part of your product, not an afterthought. Good metrics and observability can be the difference between an AI service that flourishes and one that flops without anyone knowing why. They will guide you in debugging, tell you what users are experiencing, and justify the investments in intelligence you've made.

Finally, let's conclude with a checklist to ensure you're set up for success with metrics in your intelligent microservices.

Readiness Checklist

Use the following checklist to assess whether you have comprehensive metric coverage and monitoring in place for your intelligent microservice. This can serve as a "readiness" or prelaunch checklist, and also a periodic review list to make sure nothing's fallen through the cracks:

- **Key System SLIs Identified:** Have you identified the core service health metrics (latency, error rate, throughput, etc.) relevant to your microservice? Do you have target SLOs for them (e.g., p99 latency goal, uptime requirement)?

- **Instrumentation in Code:** Is your service code emitting the necessary metrics (via counters, timers, logs)? For example, request latency timing, error logging with proper classification, etc. If using frameworks, are the out-of-box metrics enabled?

- **Monitoring Pipeline Set Up:** Do you have a monitoring system (Prometheus, DataDog, CloudWatch, etc.) configured to collect these metrics? Are all necessary metrics being scraped/collected at appropriate intervals?

- **Dashboards Created:** Are there clear dashboards visualizing the critical metrics? (At least one for system health, and if applicable one for model performance, and possibly one for business outcomes if real-time enough). Can the team access them easily?

- **AI Metrics Tracking:** If your service includes an ML model or decision logic, do you have a way to evaluate its performance in production? This could be through online feedback loops, periodic batch evaluation, or proxy metrics. E.g., is accuracy/precision being measured on real data (even if with a delay)?

- **Business Metrics Linked:** Have you defined which business KPI(s) this service is expected to influence (conversion, cost, satisfaction, etc.) and set up a way to measure that? This might be through analytics or experiments. Importantly, have you arranged to get that data and integrate it into your evaluation (even if it's in a separate system)?

- **Alerts Configured:** Are alerts in place for critical conditions? Think of both kinds:

system (e.g., high error rate, downtime) and AI-specific (e.g., sudden drop in predictions confidence or unusual output distribution). Alerts should route to the proper on-call or team communication channel.

- **Drift Detection:** (For ML) Do you have any mechanism to detect data drift or model decay? This could be an alert on a decline in accuracy or a change in input data stats.

If not automated, is someone assigned to review model performance periodically?

- **Fallbacks for AI Failures:** Have you defined thresholds at which the system will fail gracefully or fall back? And are those thresholds monitored? E.g., "if model confidence < X, route to human" – ensure that's implemented and maybe track how often that happens as a metric.

- **Security/Privacy of Metrics:** (Often overlooked) If metrics include any sensitive info, have you sanitized them? Ensure no personal data is being inadvertently logged in clear text in metrics. Also, ensure dashboards don't expose something they shouldn't to broad audiences.

- **Load Testing and Capacity Planning Done:** Have you used metrics to verify the service can handle expected load? For instance, did you run a load test to see throughput and latency and collect those metrics? This doubles as a check that metrics collection works under stress and that you know the service's limits.

- **Baseline Recorded:** Do you have a baseline snapshot of metrics before any major changes? For example, before deploying the AI feature, did you record baseline business metrics? Or with a simpler model, got baseline accuracy/latency? This is important for comparison to prove improvements. Always establish "what is normal" so you can spot anomalies.

- **Balanced Scorecard Ready:** Have you prepared a simple summary of the few toplevel metrics across categories that define success for the service? This could be a slide or document that you can show to stakeholders: e.g., "We aim for X latency, Y accuracy, Z conversion." This ensures alignment on what matters.

- **Team Awareness:** Has the team (engineering, ML, ops, even product) been briefed on these metrics and how to interpret them? Everyone should know where to look and what the key metrics are. If something goes wrong at 3 AM, on-call should know which dashboard to open first.

- **Review Process:** Is there a process to review these metrics regularly? (e.g., weekly team review of metrics trends, or a section in sprint retro for any metric changes). This keeps metrics from becoming "background noise" and ensures proactive improvements.

- **Iteration Plan:** Do you have an iterative plan if metrics are off target? For instance, if latency is too high, do you have tasks in backlog to address it? If accuracy is below expectation, plans to collect more data or improve the model? Basically, for each important metric, have at least a high-level idea of next steps if it's not where it needs to be.

Go through this list before launching a new intelligent microservice, or as a sanity check for an existing one. If you can tick all the boxes, you're in great shape – you have a thorough metrics and monitoring setup. If not, the checklist highlights areas to work on. Metrics are your safety net and your performance gauge; investing time here pays off in smoother operations and faster improvements down the line.

Conclusion

In this appendix, we journeyed through the landscape of evaluation metrics for intelligent services – from the nuts-and-bolts of throughput and latency to the nuanced terrain of model precision and user satisfaction. By now, it should be clear that effectively running an AIpowered microservice isn't just about coding the model or the service logic; it's equally about **observing, measuring, and learning** from how it behaves in the real world.

Metrics provide that all-important feedback loop. They turn the abstract concept of

"intelligence" into tangible numbers that we can track and trust. With the right metrics, you'll know when your service is drifting off course and needs retraining. You'll see the trade-offs of a design decision and can course-correct. You'll be able to demonstrate to others (and yourself) that all the fancy AI stuff is actually delivering value – or catch it when it's not, and figure out why.

As you build intelligent microservices, embrace a mindset of **continuous evaluation and iteration**. The job isn't done when the feature ships; in a sense, it's just beginning. Use metrics to iterate: maybe your first version has 80% accuracy and yields a small conversion uplift – with monitoring and experimentation, you might get it to 90% and a huge uplift. Or maybe you realize the uplift isn't there, and metrics guide you to pivot the approach. In all cases, metrics keep you grounded in reality and results.

One final thought: **observability empowers autonomy**. In the spirit of agentic AI, monitoring your services is like giving them self-awareness (through your eyes). An autonomous agent or microservice can only be trusted if we, the humans, can observe and understand its behavior. High observability (metrics, logs, traces) builds trust, which in turn allows you to give the AI more autonomy safely. WebkinsTech's team gradually gave Lumi more responsibilities as the metrics proved it could handle them. That trust was earned through data, not blind faith.

So, treat your metrics as first-class citizens. Celebrate improvements visible in metrics, and don't be afraid of metric-driven post-mortems when things go wrong. Over time, you'll likely find that a culture of metric-based decision making leads to better engineering and better products.

In closing, remember that building intelligent microservices is as much an art as a science – and metrics are the palette of colors that paint the full picture of your system's performance. Measure what matters, keep an eye on the whole painting (not just one corner of it), and you'll navigate the journey of intelligent systems with confidence.

Keep measuring, keep improving, and let the data guide you to success. **What gets measured gets improved**, and with the right metrics in hand, there's no limit to how intelligent and effective your microservices can become. Happy monitoring and may your dashboards always glow green!

Appendix C: Migration Strategies from Traditional Microservices

Most readers of this book aren't starting from scratch – you already have a microservices architecture running critical systems in production. This appendix acknowledges that reality and provides a practical roadmap for **incrementally augmenting an existing microservices ecosystem with intelligent, agentic AI capabilities**. The goal is to evolve your architecture step by step, **without a risky "big bang" rewrite**, so that your services become smarter and more autonomous over time.

Microservices have served us well by breaking large applications into *small, independently deployable services with well-defined APIs*, yielding benefits like scalability, flexibility, and team autonomy . **However, intelligent agents represent the next evolution of these concepts** – moving from rigid, predefined logic to services that can perceive context, make autonomous decisions, and act towards goals . In other words, instead of a microservice that only follows static rules, imagine giving it a "brain" that learns and adapts. This transformation is exciting, but it also raises a big question: **How can you introduce AI agents into a mature microservice system without disrupting everything that already works?**

The answer is through **incremental migration strategies**. You don't rip out your well-tested services overnight; you **start small, build confidence, and expand gradually** . In the upcoming sections, we'll walk through a fictional case study of *WebkinsTech* – a company evolving one of its microservices using an AI agent – to illustrate a step-by-step migration. We'll then generalize those lessons into architectural patterns (like the sidecar pattern for agents and agent-assisted control loops), compare *before vs. after* service flows with simple diagrams, and discuss real-world tools that can support such a transition (feature flags, shadow deployments, AI validation, observability stacks, CI/CD pipelines, etc.). We'll also present a migration readiness checklist and sprinkle in **tips (💡)** on minimizing risk for a stable rollout. Finally, we'll cover common pitfalls to avoid, so you can approach this journey with eyes wide open.

By the end of this appendix, you should feel both **motivated** and **empowered** to begin infusing intelligence into your microservices – incrementally, pragmatically, and confidently. The road to "intelligent microservices" is an evolutionary one, and every small step will bring your architecture closer to a future where services aren't just **connected** or **automated**, but also **adaptive and intelligent**.

Case Study: Webkinstech's Journey to an Intelligent Microservice

To ground our discussion, let's follow the story of **WebkinsTech**, a (fictional) mid-size tech enterprise that already embraced microservices and is now taking the next step: augmenting them with AI. WebkinsTech has dozens of microservices running everything from e-commerce and order fulfillment to customer support and analytics. Rather than boiling the ocean, they choose **one service as a pilot** for adding intelligent agent capabilities. This pilot will serve as a template for future migrations across the company.

Phase 1: Assessment and Planning

WebkinsTech starts by assessing where an AI agent could deliver the most value with manageable risk. After some deliberation, they select the **Order Routing Service** as the pilot case. This service is responsible for taking a customer's order and deciding which warehouse or fulfillment center should process it. In the current traditional microservice implementation, Order Routing uses a set of static business rules – e.g. *assign the order to the nearest warehouse that has the items in stock; if none, fall back to a primary warehouse; if that fails, flag for human intervention*. These rules work, but the operations team knows there's room for improvement. For instance, they've noticed situations where the "nearest" warehouse isn't always fastest (due to capacity issues or regional shipping delays), and some orders get stuck waiting for a human when a simple alternative (like splitting the order across two warehouses) could have solved it. **In short, the service is functional but not very smart.**

During the assessment, the WebkinsTech team outlines what *"smart" Order Routing* would entail:

- **Dynamic decision-making**: Instead of fixed rules, use real-time data (inventory levels, warehouse workloads, shipping estimates, even external factors like weather or traffic) to route orders optimally.

- **Learning from outcomes**: For example, if the agent routed an order to Warehouse A and it ended up delayed, learn to factor that into future decisions (perhaps Warehouse B would be better next time for that region).

- **Handling exceptions autonomously**: If an item is out of stock or a warehouse goes down unexpectedly, have the service automatically find the next best option (split the order, re-route to a backup center, etc.) without immediately requiring human help.

The team conducts a **feasibility check** to make sure this vision is attainable. They confirm that the data needed for smarter decisions is available: warehouses already report stock and processing times, the shipping provider has an API for transit time estimates, etc. They also identify an existing internal ML model that predicts **demand surges** (originally developed by the data science team for inventory planning) which could be repurposed to help the routing agent anticipate busy periods. These are good signs.

Next, they consider **risks and success criteria**. They define metrics to judge the AI augmentation: e.g. *order fulfillment time*, *percentage of orders requiring human intervention*, *shipping cost per order*, and *customer satisfaction on delivery speed*. Baseline values for these metrics under the old system are recorded (for instance, on average 20% of orders needed manual intervention and the rest followed the static rules). The team's hypothesis is that an intelligent agent could cut that intervention rate way down and improve speed/cost metrics without negative side effects.

Finally, WebkinsTech scopes the project. They decide on a **minimal viable intelligent feature** for the pilot: the agent will focus on one improvement first – **automatic rerouting of orders when the initially chosen warehouse can't fulfill** (due to stock-outs or overload). This is a common pain point that currently leads to delays until a support staff member manually reassigns the order. If the AI agent can handle these reroutes on its own, it would already save time. More advanced capabilities (like proactively choosing the optimal warehouse based on predicted transit times) are noted as future enhancements, but not in the first iteration. By narrowing the scope, the team keeps the project achievable.

With a clear target in mind, WebkinsTech's leadership and engineering stakeholders give a green light. They assemble a small cross-functional team: microservice developers, a data scientist, and a DevOps engineer, all responsible for different aspects of integrating the AI. **The motto is "start small, learn fast",** aligning with expert advice to *"start with a single service… let the AI gather data and make minor optimizations. As confidence builds, gradually expand its scope – the goal isn't to automate everything overnight"* .

Phase 2: Prototype and Pilot Implementation

In this phase, the team builds an initial **AI-powered version** of the Order Routing Service, but confines it to a controlled pilot environment. They choose a codename for the intelligent component – calling it the **"Smart Router Agent."** This agent is essentially an AI-enhanced decision module that will work alongside the existing Order Routing logic.

Architecture of the Pilot: Rather than fully rewriting the Order Routing Service,

WebkinsTech opts for a **side-by-side approach**. The existing service code remains largely intact (to ensure current functionality doesn't break), and the new intelligent agent is developed as an **add-on component**. Concretely, they implement the Smart Router Agent as a separate module (in fact, a separate process) that the Order Routing Service can call when it needs a decision. This is analogous to a *sidecar pattern* – the agent runs alongside the service, not inside it, so it can be developed, deployed, and scaled somewhat independently. We'll discuss the sidecar approach in detail later, but the gist is that the Order Routing Service is now augmented, not replaced: it can either use its traditional rule engine or ask the Smart Router Agent for a recommendation.

For the pilot, the team doesn't hook the agent into live production traffic yet. Instead, they use **replay and simulation** to validate its behavior. They feed the Smart Router Agent a series of real (historical) order scenarios that had tricky outcomes in the past. For example, they simulate an order where the primary warehouse was out of stock – the agent correctly suggests an alternate warehouse (or splitting the order). In another test, they simulate all warehouses in a region being slow due to overload – the agent flags the issue and chooses a cross-region fulfillment with express shipping, which though slightly costlier might meet the delivery promise. These tests are encouraging:

the AI agent's decisions align with what an experienced human operator might do, and sometimes it even finds an optimization the team hadn't considered.

To build the agent's logic, the data scientist on the team trained a model using past order data. It's essentially a **recommendation model**: given an order and current context (stocks, loads, etc.), it predicts the "best" assignment (where "best" is defined by a combination of fastest delivery and least cost, subject to constraints). The model isn't perfect – it's a first iteration trained on last year's data – but it encapsulates more factors than the old static rules. Around this model, the developers wrap an **agent decision function** that also applies some guardrails (for safety). For instance, if the model produces a recommendation that violates a hard business rule (like using a warehouse that's offline for maintenance), the agent will fall back to a safe default or defer to the old logic. These guardrails are important to ensure the AI doesn't go rogue.

Pilot in Shadow Environment: After the offline tests, WebkinsTech deploys the new Order Routing Service *with the agent included* in a staging environment that mirrors production. They run it in **shadow mode** against live data streams. In shadow mode, the service is not actually affecting real orders; instead, it passively receives a copy of incoming order requests (or replays them from logs) and goes through the motions: the traditional logic assigns a warehouse, the new Smart Router Agent also computes its recommendation, and the results are compared. All of this happens behind the scenes – the actual production orders are still being routed by the unchanged production service.

This shadow phase is incredibly valuable. It reveals a few mismatches: in 5% of cases, the

Smart Router Agent recommended a different fulfillment center than the rule-based system. The team investigates each of these differences carefully. Sometimes the agent was *correct*, finding a better option (e.g. it chose a warehouse with slightly longer distance but that had the item ready to ship immediately, saving a day in handling time – a net win). Other times, the agent made a questionable call (e.g. it picked two warehouses to split an order when it wasn't really necessary, which would increase shipping cost – the rules didn't do that). These insights allow the team to refine the agent's decision criteria and retrain the model with a bit more penalization for unnecessary split shipments.

More importantly, the shadow mode builds confidence. After running for a couple of weeks, the data shows that **had the agent been in charge, order latency would generally improve and manual interventions would drop**, with only a handful of oddball decisions avoided by the safety rules. This aligns with industry wisdom that running an AI system in parallel is a great way to catch discrepancies and issues early . In fact, companies like Amazon and Netflix have famously used the practice of routing a portion of traffic to a new system in shadow mode to build confidence, gradually shifting more load as the new system proves itself . WebkinsTech is effectively following this playbook.

Internally, the success of the pilot in the test environment begins to win over skeptics. The ops team, initially nervous about an "AI black box" making decisions, can see in the logs exactly what the agent would do and why. One ops engineer remarks that the agent is behaving a lot like their best human operators would – a reassuring sign. This transparency and the *observability* of the agent (logging its inputs, decisions, and rationale scores) is key to trust.

By the end of Phase 2, WebkinsTech has an **agent-augmented Order Routing Service** that appears to work well in principle. But it's still not serving real customers. The next step is to gently introduce it into production.

Phase 3: Gradual Rollout in Production

With positive shadow testing results, WebkinsTech moves to a **gradual production rollout**. The strategy is to start with a very limited scope, watch metrics like a hawk, and progressively expand if all goes well – a textbook example of a canary release. To do this, they employ a **feature flag** in the Order Routing Service's configuration: a simple toggle (let's call it USE_SMART_ROUTING) that controls whether an order will use the new AI agent or stick to the old rules. By default, it's "off" for all orders.

They deploy the updated Order Routing Service (now containing the integrated Smart Router Agent) into production, but with the feature flag disabled – so initially, nothing has changed from the customers' perspective or system behavior. This is another safeguard: the code is live, but dormant in terms of AI functionality. Now the team can turn on the smart routing **for a tiny subset of traffic**. They decide the first trial will be **5% of new orders** to start. Using their feature flag management tool, they enable USE_SMART_ROUTING for 5% of incoming requests (either chosen randomly or, in their case, they chose a specific small region's traffic to enable entirely, to simplify analysis).

For those 5% of orders, the service now actually calls the Smart Router Agent and uses its decision to route the order. The other 95% continue through the familiar rule-based path. This side-by-side operation in production is incredibly useful. It's essentially an A/B test: the team can compare outcomes between AI-routed orders and rule-routed orders in real time. They set up **dashboards** to monitor key metrics split by these two cohorts (e.g., average time to ship, percent of orders rerouted, any increase in errors or customer complaints).

At first, the difference isn't dramatic – which is actually a good thing. Nothing catastrophic happens in the 5% AI-driven orders. In fact, things look normal. Over the first couple of days, as orders flow, they see a slight improvement in the AI group's fulfillment times and a slightly lower rate of manual escalations. It's early and the sample is small, but the trend is in the hoped-for direction. Confidence grows and they bump the feature flag up to 20%, then 50% over the course of a week, continuously monitoring. The **incremental cutover** limits the blast radius of any potential issue and allows easy rollback – if a severe problem had been observed at 5%, they would have toggled it off immediately (thus containing the impact).

Controlled Rollout: Using feature flags and canary releases, WebkinsTech limited the blast radius of the new AI agent. By starting with just 5% of traffic, they could catch issues on a small scale. This incremental rollout strategy *"builds confidence"* and *"allows rollback if needed,"* as practiced by Amazon and Netflix when introducing new systems . Always have the ability to quickly disable the AI path and fall back to the stable path until you're sure of its performance.

During the gradual ramp-up, the team pays special attention to any anomalies. One evening, they notice a spike in split shipments on AI-handled orders – higher than expected. Investigating the logs, they find the agent started splitting orders more aggressively after encountering a string of out-of-stock events (basically, it learned from a few bad scenarios in a way that made it over-cautious). It wasn't a critical bug, but it could eat into shipping budget if left unchecked. Thanks to their monitoring, they caught this early. The fix was straightforward: adjust a parameter to make the agent a bit less trigger-happy on splitting, and roll that model update out the next day (again starting at 0% -

> 5% -> … to be safe). This incident reinforces an important lesson: *intelligent systems will have different failure modes than traditional ones.* You might not hit a null-pointer exception, but you could get an odd decision – so robust monitoring and the ability to push quick adjustments are essential.

After a couple more weeks, WebkinsTech reaches a point where **100% of orders** are now processed with assistance from the Smart Router Agent. The feature flag is fully enabled. They keep the old rule-based logic in the codebase as a fallback (just in case they need to disable the agent quickly, the rules can take over seamlessly), but over time they gain confidence that they won't need to fall back. In fact, after a month of smooth operations at 100% AI rollout, the team cleans up some of the now-redundant rule code and streamlines the service to rely on the agent by default. (They still keep a simple rule skeleton for backup, but it's no longer the primary path.)

Phase 4: Outcomes and Continuous Improvement

With the intelligent Order Routing Service fully in production, WebkinsTech starts reaping tangible benefits. The metrics they defined back in Phase 1 show clear improvement:

- **Order fulfillment speed** has improved noticeably. For example, the average time from order placement to confirmation of warehouse pick went down from about 4 hours (with the old nightly batch processing and manual touches) to under 1 hour in most cases. Urgent orders are being handled in minutes, since the agent immediately routes them optimally instead of waiting in a queue.

- The **rate of human intervention** on orders dropped from ~20% to around 5%. The majority of exceptions (like stock-outs or minor payment issues) are now resolved by the agent's logic (e.g. selecting an alternate warehouse or retrying the payment after a short delay or alternative method). The 5% of orders that still need humans are the truly tricky ones (e.g. suspected fraud cases or major system outages), which is a huge relief to the support team. They can focus on the hard problems while the AI handles the mundane ones.

- **Customer satisfaction** around delivery has inched upward – there are fewer complaints of "why did my order ship from across the country causing a delay?" because the AI is better at choosing the best option for the customer. On-time delivery rates improved (WebkinsTech even quantified a ~12%

increase in on-time deliveries in the first quarter after rollout, thanks to smarter routing and faster processing).

Interestingly, the AI agent provided some *unexpected insights*. By logging its decision rationales, it highlighted patterns that the old rules never exposed. For instance, it frequently chose a slightly farther warehouse for certain high-value items because it learned that particular location had a specialist team that processed those items faster. This insight was fed back to business — who realized they could update their distribution strategy (they decided to pre-stock more high-value items at that fast location, even for customers slightly farther away, because speed mattered more than distance for those products).

These outcomes validate the team's effort. The success of the Order Routing pilot has an infectious effect internally. Other teams at WebkinsTech are now eager to apply AI agents to their services. But before rushing ahead, the architecture group wants to codify the **general migration approach** from this experience.

They document the patterns used: running the AI in shadow mode, using feature flags for canary rollout, keeping a fallback path, and monitoring everything closely. They also note organizational learnings, like the importance of involving the ops team early and providing visibility into the agent's behavior (which helped overcome the "lack of control" fear). This case becomes a reference for how WebkinsTech will approach adding intelligence elsewhere.

By starting with a manageable slice of the system and **iterating in phases – assessment → pilot → shadow → gradual rollout → full deployment –** WebkinsTech demonstrated that it's possible to teach an "old" microservice new tricks. The Order Routing Service, once entirely static, is now a *living, learning part of their architecture*. And because they did it in a disciplined way, there was never a point where the risk ran away from them; at each stage they had checkpoints and the ability to roll back or adjust.

In the next sections, we'll take the lessons from WebkinsTech's story and generalize them.

We'll examine **migration patterns** that can be applied broadly, review **tools and**

techniques that ease such transitions, and provide checklists and tips so you can replicate these kinds of successes in your own systems.

General Strategies and Patterns for Intelligent Microservice Migration

The WebkinsTech case illustrated several key patterns in action. In this section, we distill those and other generally useful **architectural strategies** for migrating a traditional microservice to an intelligent, agent-augmented service. These patterns serve as blueprints that you can adapt to your situation. We'll cover sidecar deployments, the "strangler" approach applied to AI logic, agent-assisted control loops, and more. Along the way, we include simplified diagrams and example snippets to clarify how service flows change *before vs. after* introducing an agent.

Pattern 1: Sidecar AI Agent Deployment

One of the easiest ways to add an intelligent capability to an existing microservice is via the **Sidecar Pattern**. In microservices, the sidecar pattern refers to deploying a helper component alongside a primary service to extend its capabilities (commonly used for logging, monitoring, etc.). Here, we apply it to AI: the idea is to run the AI agent as a **companion process** next to your service, rather than embedding complex ML code directly into the service.

How it works: You package the AI agent (which could be a model server, a Python process running an ML model, a small Node/Java service with AI logic, etc.) to run in the same environment as the microservice. In Kubernetes, for example, your Pod would contain two containers: the main service container and the AI sidecar container. The service and the sidecar communicate locally (e.g. via HTTP on localhost or an IPC mechanism). The sidecar can be thought of as an "AI co-pilot" for the service – it can be consulted to provide recommendations, predictions, or to take certain actions on behalf of the service.

Why use a sidecar? This approach has several advantages for migration:

- **Isolation of concerns:** The agent's code (which might involve heavy ML libraries or different runtime requirements) is kept separate from the main service code. This means you don't have to, say, load Python ML libraries into your Java microservice process. Each can run with the tools best suited for it. It also means if the agent crashes or has issues, it doesn't necessarily take down the main service – they are isolated.

- **Independent deployability:** Initially, you might choose to not deploy the sidecar to all instances. You can run the sidecar only in some instances or have it turned off, and enable it gradually. The sidecar can even be updated or scaled differently (for instance, if the AI is CPU/GPU intensive, you might scale it horizontally or give it more resources than the main service).

- **Minimal changes to the service:** The main service just needs to know how to call the agent (e.g. make a REST call to localhost:5000/routeRecommendation). If the agent is unavailable, it falls back to existing logic. This means the core business logic of the service remains intact as a safety net.

Let's visualize a simple before-and-after of using an AI sidecar:

Before (traditional flow):

```
[ Client Request ]
         |
         v
[ Order Routing Service ] -- (applies static rules) --> [ Warehouse API /
DB ]
         |
         v
   (response sent back)
```

In the traditional flow, the Order Routing Service calls the Warehouse API (or database) directly after applying its static routing rules.

After (with AI sidecar agent):

```
[ Client Request ]
         |
         v
[ Order Routing Service ] --+--> (calls) [ AI Agent Sidecar ]
         |                  |         (computes recommendation using
ML/model)
         |                  v
         |           (recommended warehouse)
         |                  |
         \--- If agent provides valid suggestion, use it ---\
              Else, fall back to rules                       |
                                                             v
                   [ Warehouse API / DB ] <-------------/
                              |
                              v
                      (order processed)
```

In the augmented flow, the Order Routing Service, upon receiving a request, reaches out to the co-located **AI Agent** for guidance. The agent, possibly loading a machine learning model or using some cognitive logic, returns a recommendation (e.g. "route to Warehouse East with expedited shipping"). The service then either accepts this recommendation or falls back to a default behavior if the agent gave no answer or one that violates a rule. Finally, the service calls the Warehouse API as before, but now with a smarter choice.

Snippet Example: Here's a pseudo-code inside the Order Routing Service after adding the sidecar agent:

```
Warehouse chosenWarehouse;
if (featureFlagUseSmartRouting && agentClient.isAvailable()) {
    Recommendation rec = agentClient.getRecommendation(order);
    if (rec != null && rec.isValid()) {
        chosenWarehouse = rec.warehouse;
        log.info("Using AI recommendation for order {} -> warehouse {}
(confidence={})",
                 order.id, chosenWarehouse.id, rec.confidence);
    } else {
        chosenWarehouse = ruleEngine.chooseWarehouse(order);
        log.info("AI had no valid suggestion, fell back to rules for order
{}", order.id);
    }
} else {
    chosenWarehouse = ruleEngine.chooseWarehouse(order);
}
executeOrder(chosenWarehouse, order);
```

In this pseudo-code, agentClient.getRecommendation(order) is a call over the network (but local) to the sidecar agent. The code checks if the agent is enabled and available, then uses it, otherwise defaults to the ruleEngine. This shows how minimal the change can be to integrate an agent – essentially an if block around existing logic.

This pattern was effectively used by WebkinsTech and is broadly applicable. Many organizations taking initial steps in AI integration favor the sidecar or **"AI-as-a-service"** approach: for example, instead of embedding a recommendation algorithm in the ecommerce website code, they deploy a separate recommendation service that the web service calls. It's the same principle at a smaller scale.

Considerations: While sidecars are great, they introduce a new network hop and potential latency. Since the agent is local (on the same host/pod), the latency is usually small (millisecond-level). You should budget a bit of extra response time for the agent call. Also, ensure the sidecar failure modes are safe – e.g., if the agent process is OOM-killed or not responding, the main service should time out and carry on with the normal path (perhaps emitting an alert that the agent is down).

Pattern 2: Strangler Fig (Incremental Replacement) for AI Logic

The **Strangler Fig pattern** is a well-known strategy from migrating monoliths to microservices: you gradually replace parts of a system by routing some requests to a new implementation, slowly "strangling" the old code as the new grows in place . We can borrow this concept for introducing AI into a microservice. Instead of replacing a *monolith* with a *microservice*, here we're replacing a *deterministic logic path* with an *intelligent agent path* – gradually.

In practice, this looks like what WebkinsTech did: run the old and new logic in parallel (initially the old actually serving the results and the new just observing), then flip who is primary over time.

How to implement:

1. **Dual-run (Shadow) Phase:** For each request, invoke both the old logic and the new AI logic, but only use the old logic's result for the real output. The AI's result is captured for evaluation. This is essentially "shadow mode" as described earlier. It allows you to verify the AI's outputs against the known-good outputs without impacting users . At this stage, you are *not* trusting the AI to serve traffic, only to show you what it *would* do.

2. **Verification and Trust Building:** Analyze where the AI's decisions differ from the old logic, and if those differences are acceptable or even improvements. Refine the AI as needed. Over time, the AI hopefully proves to be correct (or at least not worse) in most cases. This builds trust and evidence for step 3.

3. **Phase-wise Cutover via Feature Toggle:** Introduce a conditional, as in the code snippet above, to let some fraction of requests actually use the AI's result. This could be ramped up by percentage, by user segment, or by scenario. For example, you might let the AI handle **only read-only scenarios** first or non-critical transactions, as a way to limit risk. Increase the share of traffic to the AI path gradually (canary deployment style).

4. **Complete Cutover and Removal of Old Path:** Once the AI path has been handling 100% of requests for a while with good results, you can consider removing the old code path. Often you won't remove it

immediately – it might live behind a "kill switch" for a while (so you can quickly turn it back on if needed). But eventually, if all goes well, that legacy logic can be deleted, and the AI agent becomes the new normal for that service's functionality.

This pattern is essentially applying the **scientific method** to migrating logic: for a while, treat the AI as an experiment running alongside the control (the existing system). Only when you're confident in its performance do you flip it to be the primary. Industry best practices for migrations echo this: *"migrate in phases (shadow/dual-run, verify, then cutover)"* rather than all at once. The shadow phase catches discrepancies early and avoids a faulty migration from ever affecting customers.

One more twist on the strangler approach: sometimes you might keep both old and new logic running indefinitely and use them in a complementary way. For instance, some companies run an AI model to get a prediction, but *also* run a rules engine and then combine the results (like accept the AI's answer only if it passes certain rule checks). This can be seen as a partial strangler – the new logic doesn't 100% replace the old, they form a hybrid. This is perfectly fine if it meets your goals; the end state doesn't always have to be "AI does everything." In some cases, maintaining a few key rules as a safety net is wise (think of it like an automatic override or circuit breaker if the AI goes off course).

Analogy: The strangler fig (the pattern's namesake) is a vine that grows around a tree and eventually replaces the structure of the tree. Here, the "tree" is the old algorithm and the "vine" is the new AI agent-based logic. Over time, we let the vine take more and more weight until the old tree is no longer needed.

Pattern 3: Agent-Assisted Control Loop

Not every intelligent microservice use-case is a direct request/response like order routing. In many scenarios, an AI agent works as part of a **control loop** that optimizes something continuously. This is common in systems management (think auto-scaling, scheduling, or fault recovery decisions) and in personalized experiences (like an agent tweaking UI or offers over time based on user behavior). The architectural pattern here is to insert the agent into the feedback loop of the system.

A classic control loop in computing is the **Monitor-Analyze-Plan-Execute (MAPE)** loop used in autonomic systems. With an AI agent, the loop might become: **Monitor (environment) → Analyze (with AI to detect patterns or anomalies) → Plan (AI proposes an action) → Execute (system applies the action) → Repeat**. The agent basically continuously observes the system's state and takes or recommends actions to optimize some outcome.

Example: Intelligent Autoscaler (Agent in the loop): Suppose you have a microservice that scales out/in based on CPU usage. The traditional way is a static threshold: e.g. if CPU > 80% for 5 minutes, add a new instance. An agentic approach could be: an AI agent monitors CPU, network, and request trends, and *predicts* upcoming load surges, scaling out preemptively or throttling certain work. It could also learn the patterns of daily traffic to schedule scaling events optimally. Here, the agent isn't serving an external request; it's managing the system's internal state. It acts like an automated SRE (site reliability engineer) that constantly tweaks knobs for optimal performance and reliability.

To migrate to this, you'd often **start the agent in advisory mode**. For a period, let it only make suggestions (e.g., log "I would add an instance now" but don't actually do it automatically). A human or a simple rule could then execute that suggestion. Once the agent's advice has proven sound (maybe it consistently would have saved the day), you can allow it to execute actions directly (closing the loop). This **decision support to full automation** progression is another common migration approach : first the AI watches a human, then the human watches the AI, then the AI is allowed to operate on its own.

Agent-assisted control loops also show up in business workflows. For example, consider a customer support ticket triage system: traditionally, new tickets are put in a queue and maybe some rules auto-tag their priority. An AI agent could monitor incoming tickets (Monitor), use NLP to analyze sentiment and urgency (Analyze), decide on a priority or route (Plan), and auto-assign or escalate the ticket (Execute). Early on, the agent might just suggest a priority ("I think this ticket is high priority because the customer is very upset") and a human support lead makes the final call. Over time, if those suggestions are good, the agent can automatically set the priority and only unusual cases get human review.

Key insight: In agent-assisted loops, **maintaining human oversight during the transition** is often crucial. It's not binary human vs. AI; it can be a partnership. In fact, many successful intelligent system rollouts keep a "human-in-the-loop" for quite a while. This might mean alerts that require human confirmation, or having the AI double-check with a rule, etc. It's a gradient of autonomy.

From an architecture perspective, to implement this you often use **event-driven integration**. The microservices emit events or metrics (for monitor step), the agent subscribes to these events and processes them, then either triggers another event or calls a service to enact changes. Many modern platforms (like using Kafka or similar pub/sub) make this pattern easier – the AI agent can just be another consumer on the event bus and an authorized actor in the system.

Pattern 4: External Brain (Centralized Intelligence) vs. Local Agents

Thus far, we often talked about "an agent per service" (like a sidecar per service). Another strategic decision is whether to have **multiple small agents or a centralized intelligence service**. Both approaches have pros and cons, and sometimes you'll use a mix.

- **Local Agent (per service or component):** This is akin to the sidecar idea – each service gets its own AI helper that is specialized. The advantage is specialization and encapsulation. The order service's agent is focused on orders, the support service's agent is focused on support. They can be developed independently by different teams and don't become one giant AI that does everything. This aligns well with microservice philosophy (each service + its agent tackle a bounded context). The downside might be duplication of effort or models, and each agent might only have a narrow view unless you design them to share knowledge.

- **Central Brain (AI Platform Service):** In this model, you create a new microservice which is an "AI Decision Service" or some sort of central brain that many other services call into for intelligence. For example, instead of each service having its own ML model, they all call a central AI service that has various models or agents within it. This is somewhat analogous to how many companies have a single "Recommendations

service" that serves many parts of a website, or a single "Fraud detection service" that all transactions go through. The benefit is you concentrate AI expertise and data in one place – easier to manage perhaps, and you can get holistic insights. The risk is that you introduce a **dependency** that can become critical – if that central AI service is down, many parts of your system might be impacted. It can also become a development bottleneck if not structured carefully (the AI team might become overloaded serving all others).

A hybrid approach is to have a central **AI platform** (providing model hosting, features, etc.) but still deploy small agents alongside services for the real-time decision. For example, you might have a central *feature store* or *model repository* (common in MLOps) where all models are stored and trained, but at runtime each service pulls in the model it needs and runs it locally.

When migrating, a practical approach is often: **start with one or two local agents** (to prove value in isolation). If you later find overlap or want consistency, you might refactor common parts into a central service. For instance, after WebkinsTech added agents to Order Routing and later to Inventory Management, they realized both sometimes needed to access a *"global demand prediction"*. Initially, each had its own logic; later they factored that out into a small shared service that both agents call. That became a mini "AI platform" for demand forecasting used across the company.

The pattern decision here will depend on your organization. If you have a strong central AI team, you might build shared infrastructure. If teams are independent, you might see more sidecar/local agents sprout up. Both can coexist – just keep an eye on not reinventing the wheel unnecessarily.

Pattern 5: Text-Based and Diagrammatic Thinking

This is less an architecture and more a practice: when introducing complex AI behavior into a system, it's extremely helpful to produce **clear documentation and diagrams of the before/after flows**. We mention it here as a "pattern" because time and again it proves invaluable during migrations.

Before flipping any switches, draw the sequence of how a request flows through your service *now*, and how it *will* when the agent is integrated. Identify points of decision, data needed, and failure modes. A simple flowchart can uncover issues (e.g., "What if the agent doesn't respond in time at this point? We need a timeout and fallback here.").

WebkinsTech's team, for instance, drew an architecture diagram of the Order Routing process with the agent involved and used it in meetings with ops and business stakeholders to ensure everyone understood the changes. It also helped in threat modeling (identifying new security considerations) and in planning monitoring (figuring out where to put probes to measure the agent's performance).

Some common diagrams/pictures that help in these migrations:

- **Before vs After call flow** (like the ones we sketched for sidecar): shows components and interactions.
- **Sequence diagrams** for key scenarios (normal operation, agent failure, agent disagreement with rules, etc.).
- **Data pipeline diagrams** if you are adding training data flows or periodic model update flows in the architecture (to ensure the new ML components get the data they need).
- **Dependency graphs** highlighting any new external services (e.g. "our service will now call out to an external AI API – show that on a diagram and note latency, security aspects").

By visualizing the new system, you create a shared understanding and can reason about it more clearly, minimizing surprises during rollout.

Summary of Patterns

To summarize, migrating to intelligent microservices often involves combining these patterns:

- **Sidecar deployment** of AI capabilities for minimal disruption and easy rollback.

Strangler-fig incremental enablement using dual-run and feature toggles, so the old logic is only slowly phased out as the new proves itself.

- **Agent-assisted control loops** where AI gradually goes from advisory to autonomous in operational decisions.
- **Deciding on local vs. central agents** depending on context (with a bias to start local for simplicity).
- **Using diagrams and explicit orchestration** to ensure everyone knows how things will change.

In practice, you might use *multiple* patterns together. For example, a sidecar agent is introduced and initially runs in shadow (strangler style), then goes autonomous (control loop closed) and eventually that pattern is applied across many services (some of which might leverage a central AI service).

Now that we've covered the *how* from a design standpoint, let's look at the *enablers* – the tools and processes that help make these migrations successful.

Tooling and Practices to Support the Transition

Migrating to an AI-augmented microservice isn't just about code and architecture. It also heavily involves tooling in areas like deployment, testing, and monitoring. In fact, having the right tools can mean the difference between a smooth, confident rollout and a nervewracking, blind flight. Let's explore some categories of tools (both open-source and commercial) that WebkinsTech and other organizations have used to support intelligent microservices migration.

Feature Flags and Progressive Delivery Tools

Feature Flags (Feature Toggles) are essentially conditional switches in your code that can be turned on or off (often at runtime or without redeploying) to enable/disable features. We saw how WebkinsTech used a feature flag for USE_SMART_ROUTING. This ability to control the AI feature rollout gradually is crucial. There are several tools and services for managing feature flags:

- **LaunchDarkly:** A popular commercial feature flagging platform. It allows targeting flags to percentages of users, specific segments, etc., and provides a dashboard to turn features on/off in real time. For an AI rollout, one can use LaunchDarkly to slowly dial up a feature's percentage and even do targeted rollouts (e.g., internal users only, or particular geographic region).

- **Unleash:** An open-source feature flag platform . Companies use Unleash self-hosted to manage feature toggles without a vendor. It supports gradual rollouts and custom strategies.

- **Flagsmith, FeatBit, Flipper,** and others are also open-source or low-cost alternatives. The key is not which one, but *using one*. Even a simple config file toggle can work in a pinch, but dedicated tools add ability to change flags dynamically, schedule changes, and have audit logs of who changed what (important for coordinating a risky launch).

- **Config management systems** (like **Spring Cloud Config**, **Consul**, or even environment variables and Kubernetes ConfigMaps) can also be used to toggle features if you don't have a fancy feature flag service. The concept remains the same.

Alongside feature flags, **Progressive Delivery** tools help automate and manage the rollout of new versions of services (a concept from DevOps that includes canary deployments, bluegreen deployments, etc.). For example:

- **Argo Rollouts:** An open-source Kubernetes controller for progressive delivery. You can define a strategy for a new version of a service (like v2 of Order Service with AI) to receive, say, 10% of traffic, then 30%, then 100%, with checks at each step. Argo Rollouts can even integrate with metric checks (like, if error rate goes up by X, abort the rollout and revert to v1). This is very handy if your AI changes are deployed as a new version of the service.

- **Flagger (Weaveworks):** Another open source tool that works with service meshes (like Istio/Linkerd) to do automated canary analysis. It can shift traffic between old and new gradually and watch metrics.

- **Service Meshes (Istio, Linkerd, etc.):** These can be configured for canary routing or mirroring traffic to a shadow service, etc. For instance, Istio can mirror 5% of traffic to the new AI-powered service while the main service still handles 100% – that's another way to do shadowing at the network level.

- **Commercial deployment tools** like Harness, Spinnaker, or Cloud vendor solutions (AWS CodeDeploy, Google's Anthos Service Mesh, etc.) also provide support for phased rollouts.

In essence, feature flags control behavior *within* a service, while progressive delivery controls the rollout of *service versions*. Often, you use them together: deploy the new code, then use a flag to turn on the AI behavior gradually.

For an intelligent microservice migration, these are lifesavers. They allow you to implement the **"gradual ramp-up"** pattern without constantly redeploying or maintaining two separate codebases. You just ship one version capable of both behaviors and then dial up the new behavior.

Testing and Validation Tools for AI Changes

When adding AI, traditional testing (unit, integration tests) is still necessary but not sufficient. The nondeterministic nature of AI means you also need to validate *behavior and outcomes* more extensively:

- **Unit Testing the Integration:** You should still write unit tests for the glue code (like the snippet we showed). E.g., test that if the agent client returns null, the service does fall back to rules. These tests ensure the integration logic (feature flags, fallbacks) work as expected. Standard testing frameworks (JUnit, pytest, etc.) apply here.

- **Model Validation and Offline Evaluation:** Before deploying an ML model (the core of your agent's intelligence), use tools to evaluate its performance on historical data. This might involve calculating accuracy, precision/recall, or business-specific metrics. There are open-source libraries to help with this:

- **Great Expectations:** Not for model accuracy, but for data validation. It can ensure the data feeding your model in production stays within expected ranges (to catch data drift or anomalies that could confuse the model).

- **sklearn, TensorFlow, PyTorch evaluation utilities:** Most ML frameworks provide ways to test a model on a test dataset and compute metrics. Do that and attach the results to your deployment artifact (so you know which version of the model is going out and with what expected performance).

- **Custom scripts for simulation:** As WebkinsTech did, you might replay historical events through the new logic. This can be scripted in Python or using your system's existing simulation capabilities. The key is to compare outcomes: e.g., take 1000 past orders, run old vs new and see how many differences and whether those differences would have been good or bad. This can even be automated as a test.

- **A/B Testing and Statistical Analysis:** If you have enough traffic, you can run an A/B test (like a randomized control trial) where some portion of traffic uses the AI and some doesn't, and then compare metrics. This requires careful design to be fair. Tools like **AB testing platforms** (Optimizely, Google Optimize (though deprecated), or even using feature flag cohorts as proxies) can support the assignment and analysis. You'll want to ensure you have logging to tie outcomes to which variant (AI or non-AI) was used.

- **Shadow Testing Tools:** If your system is complex, you might invest in tooling to automate shadow mode comparisons. For example, Netflix created the concept of *"Comparative canaries"* for microservices – one could imagine extending that to AI by deploying a "shadow" instance of the service and comparing responses. This often ends up being custom, but it's worth mentioning that some service mesh / load testing setups can facilitate automated shadow runs.

- **AI Specific Testing:** There are emerging tools to test AI systems specifically:

Microsoft's Checklist or Amazon's Deequ: These are more for NLP or data quality but can be adapted to ensure your model handles edge cases.

- **Adversarial testing frameworks:** If, say, your agent does NLP, you'd want to test it against variations in input. There are libraries to generate adversarial examples or perturbations.

- **Bias/Fairness checks:** If your AI decisions involve customers, ensure you test for bias. Tools like IBM AI Fairness 360 or Microsoft Fairlearn can be used offline on your model to detect unfair biases.

- **Model Explainability:** Libraries like SHAP or LIME can help you understand why the model is making decisions. This isn't exactly a test, but it's a validation aid – if the explanations don't make sense, you might catch a problem (e.g., the model latched onto an irrelevant feature).

- **Performance Testing:** Don't forget to test the performance of the new system under load. The AI sidecar might introduce latency or CPU overhead. Use your standard load testing tools (JMeter, Locust, k6, etc.) to simulate high load both with and without the AI enabled, to see if you meet your SLAs. Sometimes an ML model might slow things down – better to know before users do. If you find issues, you might need to optimize the model (quantize it, cache results, do partial updates asynchronously, etc.).

One useful practice is to use a **checklist or quality gate for the AI** before it goes live: e.g., *"Has the model been evaluated on test data? Have we done a fairness check? Have we run it in shadow for at least X days? Do we have unit tests for integration? Do we have monitoring in place?"* We will provide a general readiness checklist later in this appendix, but creating a specific one for your AI component is wise.

Finally, consider a **"gamble test"**: if you're really cautious, one approach is to run the AI in a mode where it makes a prediction, and occasionally (at random or for a certain slice) use the old logic instead – then see which performed better. This is like the AI is betting it can beat the old logic, and you verify it. If the AI consistently wins those bets (outcomes better), you gain trust. If not, you either adjust or decide maybe the AI isn't worth fully enabling yet.

Observability and Monitoring for Intelligent Services

"**You can't improve what you don't measure**" – this is doubly true when you add AI to microservices. Observability (logging, metrics, tracing) must evolve to cover the AI's behavior, not just traditional metrics.

Here's what to put in place:

Basic Service Metrics: Continue to monitor all the usual suspects – throughput (requests per second), error rates, latency, CPU/memory usage. Adding the AI should not wildly increase error rates or latency beyond acceptable bounds. If you see a regression when the feature flag is on, that's a red flag. Tools like

Prometheus/Grafana (open source) or cloud monitoring (AWS CloudWatch, etc.) can be used to track these metrics. Set up dashboards that break down metrics by "AI on vs off" if possible (e.g., using a feature flag dimension or route tag).

- **AI-Specific Metrics:** Introduce new metrics that the agent can emit. For example: **Decision confidence** – if your model provides a score or confidence, track the average confidence. Are there many low-confidence decisions? That could predict errors.

- **Override count** – how often did the system have to fall back to old logic because the AI was not confident or not available.

- **Outcome metrics** – e.g., in WebkinsTech's case, they tracked "orders autorerouted by AI" and how many of those still ended up needing human fix (that would indicate the AI maybe didn't fully solve it).

- **Drift indicators** – if you have a model, you can log the distribution of inputs (e.g., average value of key features) and monitor if they drift from the training data distribution. If your agent suddenly is seeing very different data, its decisions might become less reliable (a phenomenon known as data drift or concept drift).

- **Business KPIs** – tie into your analytics to see end-to-end impact. E.g., track if the conversion rate or task success rate for AI-handled interactions is as good or better than the old way.

- **Logging and Tracing:** Log the agent's decisions in detail (without flooding, of course). If the agent chooses X over Y, log that, including why if possible (e.g., "Chose warehouse East – stock higher and predicted delivery 1 day faster than West"). These logs are gold for post-mortem analysis if something goes wrong. They also help in explaining the AI's behavior to stakeholders. Enable distributed tracing if the agent is a separate service – so a trace ID can tie a user request through the main service to the agent and back. If using something like **OpenTelemetry**, ensure the agent's spans are instrumented, so you can see e.g. that "this request spent 20ms in the AI agent call".

- **AI Observability Platforms:** There's a growing category of tools specifically for monitoring models in production (often called **ML Observability** or **Model Monitoring** platforms). Examples: **Arize AI**, **WhyLabs**, **Evidently**, **Aporia**, etc. These tools can automatically track things like data drift, model prediction distributions, and even alert if the model's accuracy seems to be degrading (which they infer if you can provide ground truth later, or via proxy metrics). For instance, Arize can log every prediction and later you can tag which were good vs bad outcomes to see if your precision is dropping . If your use-case is high-stakes or you have many models, such a platform might be worthwhile. In WebkinsTech's case, for one pilot they did not use a dedicated platform; they built custom dashboards. But as they scale up, they might adopt one of these to centralize AI performance tracking.

Alerts and Automated Responses: Set up alerts on the key metrics. For example, if the agent's error rate (say it throws exceptions or returns "I don't know" too often) exceeds X, alert. If order fulfillment times increase instead of decrease, alert (maybe the AI is causing a slow-down somewhere). Also consider automated rollback triggers: some progressive delivery setups allow you to automatically disable a feature or roll back a deploy if metrics go out of bounds. You might wire in something like: if AIenabled requests have 2x latency of others for >10 minutes, auto-disable the feature flag. This kind of safeguard can make the rollout safer by not relying on a human to catch every issue.

In summary, treat the AI part as an integral component to be monitored, not a mystical black box. **Measure inputs, outputs, and outcomes.** As one guide suggests, *"monitor input metrics (data characteristics), output metrics (results of AI decisions), and set thresholds for alarms"*. This includes standard system health metrics as well as domain-specific ones.

One more aspect: **Explainability and Transparency**. If possible, surface the reasoning of the agent to those who operate the system. This might be through logs as mentioned, or even dashboards showing, for example, "Top factors influencing the agent's decisions this week". Some teams even build internal tools for support engineers to see why the AI did something on a particular case, which can dramatically help in troubleshooting and in communicating with affected users. While not every project can invest in a full explainability module, even a simple log of key features or a score can help.

CI/CD and MLOps Integration

Continuous Integration/Continuous Deployment (CI/CD) practices will need to account for the AI components as well as the code. Many organizations treat the "model" as just another artifact that flows through a pipeline. This is often referred to as **MLOps** (Machine Learning Operations) – extending DevOps practices to machine learning.

Some points to consider:

- **Version Control for Models and Data:** Your code is in Git, but what about the ML model file (e.g., a .pkl or .pt PyTorch file)? You should version your models, either by storing them in a model registry or at least naming them with version numbers or hashes. Tools like **MLflow** or **DVC (Data Version Control)** can help track models and even the data used to train them. In a migration, you might iterate on the model quickly; being able to reproduce and rollback to a prior model is as important as rolling back code.

- **CI for ML:** Include tests for the model in your CI pipeline. For example, after training a new model, automatically run evaluation on a test dataset and check that accuracy or other metrics meet a threshold. Only then mark that model as "candidate for prod". This prevents a bad model (maybe due to a bug or bad data during training) from even being considered for deployment.

Infrastructure as Code: If you add new components like a sidecar or a new service, manage them with your usual infrastructure tooling (Terraform, Helm charts, etc.). For instance, define the agent's deployment YAML alongside the service's YAML so it's clear they go together.

- **Continuous Deployment with Gates:** As discussed, progressive delivery can be part of CD. You might automate that after deploying to staging and running shadow tests, it deploys to prod at 0% traffic, then waits for a manual or automated approval to ramp up.

- **Model Deployment Tools:** If your agent is essentially an ML model server, there are specialized tools that can manage that. For example, **Seldon Core** or **KFServing (KServe)** can deploy models to Kubernetes with logging and scaling. Cloud providers have their own (AWS Sagemaker, GCP AI Platform, Azure ML). These might handle aspects like loading the model, providing an API, scaling it, etc. WebkinsTech's pilot was small enough that they just embedded the model in a custom service, but if you have many models or want to separate concerns, using a dedicated model serving platform can help. Just ensure it integrates well with your microservice (network calls etc.).

- **Continuous Training / Updates:** In some cases, you might not just deploy one model and be done; you might want to update it regularly as new data comes in (to prevent model drift). This leads to the idea of a **continuous training pipeline**. Tools like Kubeflow Pipelines or Apache Airflow can schedule retraining jobs. When a new model is ready, your CI/CD should take that artifact and deploy it (after tests). For instance, retrain weekly, run eval, if better than last model, push it to production (maybe via that feature flag turning on new model ID).

- **Canary for Models:** Just as we can canary code, you can canary models. Some systems support serving two models and splitting traffic between them to compare (shadow or A/B). If you roll out a new model version, you might initially route x% of requests to it (the rest to the old model) and compare outcomes. This is advanced but doable with proper routing logic in the agent or using a platform like Seldon which has an "experiment" feature.

- **Continuous Delivery of Rules/Policies:** Interestingly, if you keep some rule-based fallback or hybrid logic, treat those rules as code too. We've seen cases where business rules are in a config file; when AI comes, you might still update those rules occasionally (like threshold for confidence to accept). Those should also be under version control and go through pipeline (even if it's just a config map update).

- **Security and Governance in CI/CD:** Adding AI might mean handling new types of artifacts and data. Ensure your pipeline secures sensitive data (e.g., training data may contain PII – don't leave that lying around in logs or artifacts). Also, access control: perhaps only certain people can approve moving the AI from shadow to live.

Incorporate those approval steps in your process.

In summary, extend your DevOps pipeline to handle "data + model + code" together. Many organizations create cross-disciplinary teams (the fabled "DevOps + Data Science" collaboration) to ensure model updates flow just as smoothly as code updates. If you lack such, start with small manual steps (like an engineer manually updating the model file in the container) and then automate those steps gradually.

One more thing worth mentioning is **Rollback strategy for models**: If you push a new model and it underperforms, have a quick path to revert to the previous model. This could be as easy as flipping a feature flag to use the older model (if you kept it loaded), or redeploying the previous version. In any case, plan for it. This is analogous to code rollbacks, but sometimes data scientists unfamiliar with production might not think of it – so instill that mindset.

MLOps Maturity: Treat your AI model like any other piece of the microservice. Check it into source control (or store it versioned), test it, and deploy it with the same rigor as application code. For example, use a CI pipeline that automatically evaluates a new model and rejects it if it underperforms a baseline. This prevents regressions and ensures your AI enhancement actually *improves* over time, rather than drifting unnoticed.

Tools Summary Table

For quick reference, here's a summary of example tools in each category:

- **Feature Flags / Rollout**: LaunchDarkly (commercial), Unleash (open-source), Argo Rollouts (open-source, K8s), Istio/Envoy mirroring (open-source), Harness (commercial).

- **Testing/Validation**: Great Expectations (data validation, open), PyTorch/TensorFlow (model eval, open), custom scripts for A/B, Fairlearn or AIF360 (bias, open), unit test frameworks (JUnit, etc. for integration tests), fuzz testing tools.

- **Observability**: Prometheus & Grafana (open), Elastic Stack (ELK for logs, open),

OpenTelemetry (tracing, open), Arize AI or WhyLabs (ML monitoring, commercial), Custom dashboards on Splunk/Datadog (commercial monitoring that can ingest custom metrics).

- **CI/CD & MLOps**: Jenkins/CircleCI/GitHub Actions (CI, open/commercial), MLflow

(open-source model management), DVC (open-source data/model versioning), Seldon Core / KFServing (open model serving on K8s), Kubeflow Pipelines or Airflow (open workflow orchestration for retraining), Terraform/Helm (infra as code, open), Feature stores (like Feast, open, if needed for consistent data for models).

Using these tools isn't mandatory, but they each solve certain problems that will crop up during migration. The takeaway is: **invest time in automation and monitoring upfront** – it pays huge dividends when you start deploying and scaling intelligent microservices.

Migration Readiness Checklist

Embarking on a migration from a traditional microservice to an AI-augmented one requires preparation. Below is a **checklist** WebkinsTech used (and which you can adapt) to ensure all bases are covered before and during the transition. Treat this as a pre-flight checklist before you switch on that intelligent feature for real users:

- **Problem Definition & Success Metrics:** Have we clearly defined what improvement we expect from adding AI to this service (e.g., reduce manual work by 50%, improve latency by 100ms, etc.)? Are the metrics to measure this in place and baselined?

- **Data Availability:** Do we have access to the data needed to power the AI agent? (E.g., training data for models, real-time feeds of inputs the agent needs at runtime like sensor data or user context.) If external data is needed (third-party API, etc.), is access secured and reliable?

- **Model/Logic Validated:** If using an ML model, has it been trained on representative data and tested offline for accuracy/performance? Do domain experts agree the agent's logic (ML or rule-based AI) makes sense?

- **Fallback Ready:** Is there a mechanism to bypass or turn off the AI and revert to the old logic instantly if things go wrong (feature flag, config toggle, etc.)? And is the old path still healthy and tested in the new deployment?

- **Shadow Plan:** Have we run (or planned to run) the new AI in shadow mode or pilot phase to collect comparisons without impacting production? (This might include collecting logs of AI decisions versus actual outcomes.)

- **Incremental Release Plan:** Do we have a rollout strategy (canary percentages, specific user segment, time window, etc.) for enabling the AI gradually? And do we have the tooling in place to execute that (feature flag service, routing rules, etc.)?

- **Monitoring & Alerting:** Are all relevant metrics (system and business) being tracked with the ability to break down by AI vs non-AI traffic? Are alerts set on any anomaly (e.g., error rate spike, throughput drop, unusual decision patterns)?

- **Team Training & Alignment:** Are the ops/support teams aware of the changes and trained on the new system (e.g., do they know how to interpret AI decisions, or troubleshoot the new components)? Does everyone know who to page if the AI behaves badly at 2 AM?

- **Security & Compliance Review:** Does the AI introduction raise any new security concerns? (e.g., sending data to an external AI API – check data privacy; an agent making autonomous calls – ensure it respects auth/z). If a model is involved, has it been checked for bias or undesirable outcomes that could pose ethical or legal issues?

- **Governance & Audit:** Is there logging sufficient to audit decisions later? (Important for sensitive domains – e.g., can we explain why the AI rejected a loan application if needed? Or why it routed an order a certain way.)

- **Resource Assessment:** Do we have enough infrastructure for this (CPU for the model, maybe a GPU if needed, memory overhead)? Have we load-tested to ensure we can handle peak traffic with the new overhead?

- **User Experience Considerations:** If the AI directly or indirectly impacts user experience (like response time or content shown to user), have we considered the UX? For example, if the AI might delay a response to think a bit longer, is that okay? If the AI changes some UI element (say an agent that adjusts prices on the fly), is it doing it in a user-acceptable way?

- **Post-Launch Review Plan:** Set a date (or criteria) to review the migration's success. For example, "After 2 weeks at 100% AI, we will evaluate metrics vs baseline and decide to fully adopt or roll back." Having this checkpoint ensures you don't just push the AI and forget – you validate it delivered value.

This checklist isn't static – add items specific to your context. The key is to avoid charging ahead without doing the homework. In WebkinsTech's case, going through such a checklist surfaced a few things: for example, they realized they needed to brief the customer support team that order routing was changing, so if customers called asking "Why did my order ship from X instead of my local warehouse?", the support folks had an explanation (in this case, "Our system determined that would get it to you faster because your local facility was busy – an AI is helping optimize your delivery"). Little communication steps like that can save headaches.

Think of the checklist as the convergence of **technical readiness, data readiness, and organizational readiness**. All three need to be "green" for a low-drama rollout.

Common Pitfalls and How to Avoid Them

Even with best-laid plans, migrations can stumble. Let's discuss some common pitfalls that teams encounter when introducing intelligent agents to microservices, and strategies to avoid or mitigate them.

Pitfall 1:

Big-Bang Release of AI Features

What can go wrong: Deploying the new AI-powered version to 100% of users all at once (or otherwise doing a wholesale switch) is tempting fate. If something's wrong – and with complex AI logic, something *will* be at first – the impact is broad and fast. The team might get overwhelmed triaging issues, and worst-case, users suffer significantly or you have to do an emergency rollback under pressure.

Avoidance: Use the *incremental rollout* and *strangler patterns* we covered. This pitfall is basically not following those. Always have a phase where the AI runs in parallel or for a small slice before full deployment. It's fine to be bold with experimental features, but when they affect core business processes, caution wins. As one expert succinctly put it, *"incremental cut-over limits the blast radius of any issues and allows rollback if needed"* . Embrace that mindset.

If you find yourself in a situation (maybe due to external pressure) where a big-bang release happened, immediately consider turning it into a pseudo-canary: e.g., maybe region A can be temporarily routed back to old version to reduce impact and gather yourself. Feature flags, even if not used to gradually ramp originally, can be used post-release to turn off aspects causing trouble.

Pitfall 2:

Lack of Fallbacks and Safeguards

What can go wrong: Over-relying on the AI such that if it fails or produces weird output, the whole service fails or behaves badly. Examples: the agent process crashes and your service hangs waiting for it; or the agent gives an outlandish recommendation and there's no rule to catch it (e.g., it sets a price to $0 by mistake and you blindly take it).

Avoidance: Always implement fallback logic. That could mean timeouts (if agent doesn't respond in X ms, use default), sanity checks on outputs (if agent's suggested price < min_price, clamp it or ignore it), and circuit breakers (if agent has failed 5 times in a row, stop calling it for a while and just use defaults). Essentially, the AI should augment, but if it's unavailable or untrustworthy at a moment, the system should gracefully degrade to something reasonable.

Testing for these scenarios is critical. Simulate the agent being down or returning nonsense. Does the service still run? Also monitor the frequency of fallbacks – if you see a lot of them, it means the AI is effectively off a lot of the time (either due to errors or lack of confidence), which is itself an issue to investigate.

Pitfall 3:

Inadequate Monitoring and No Instant Feedback

What can go wrong: You deploy the AI, and it seems fine. But you aren't really measuring the detailed outcomes, and a subtle issue goes unnoticed for too long, causing slow harm (e.g., gradually the AI is drifting and making worse decisions, but no one realizes until a customer or a biz metric screams, by which time maybe damage is done).

This is like flying an airplane with your instrument panel off. Maybe it feels okay until you crash into a mountain you didn't see coming.

Avoidance: As stressed in the Observability section, instrument everything. Set up alerts for things like "no agent decisions in last 10 minutes" (could indicate it's dead) or "throughput dropped" or "customer complaints volume increased". And crucially, have someone on the team (or rotate) actively watch the metrics during the rollout period. It's common to have a "launch war room" or chat channel for a major release where engineers are live-monitoring and reacting. Do that for your AI launch – treat it as seriously as a new product launch or a major version release.

Additionally, seek *fast feedback loops*. For instance, if the AI's goal is to reduce something, try to measure that something as early as possible. In WebkinsTech's case, they could see within hours if orders were getting stuck or not. If your AI's impact is long-term (say it predicts inventory purchase which affects out-of-stock rates a month later), then try to simulate or find leading indicators, because waiting a month is a long time to realize it's not working. Maybe in that case, you run an offline simulation of a month using historical data to see what would've happened. The faster you can detect a misbehaving agent, the less damage it will do.

Pitfall 4:

Team and Culture Mismatch

What can go wrong: Sometimes the biggest challenges aren't technical but human. The software team might deploy a brilliant AI service, but the ops team doesn't know how to handle it and when an incident occurs at 3 AM, they panic or take the wrong action due to unfamiliarity. Or maybe the domain experts don't trust the AI and constantly override it, negating its benefits. In another angle, data scientists might throw a model over the wall that works in a notebook but not in real system conditions.

Avoidance: Involve all stakeholders early and often. This means cross-training – teach the ops folks about the new agent, what logs to check, what toggles exist. Maybe prepare a runbook: "If X happens, do Y (e.g., flip off feature flag, call on-call ML engineer, etc.)." Also, get domain experts (like experienced operations managers, customer support leads, etc.) to review the AI's decisions in pilot phase and give feedback. This will both improve the AI (because they'll spot dumb decisions) and increase their trust (because they see their input was used).

A specific cultural pitfall is the "not invented here" or turf war: ML team vs software team dynamic. Bridge that by forming blended teams for the project. Everyone's goal is the same – better business outcomes – make sure all parties see how the AI helps that, and that their expertise is still valued. For example, highlight that the AI frees them from drudge work so they can focus on more strategic tasks, rather than messaging it as "AI will replace you" which causes understandable resistance.

Also, prepare management and business side: ensure they know that an AI rollout is not a guaranteed win on day one. It's an iterative learning process. Set expectations that there might be some bumps and that gradual rollout is the plan. This avoids pressure that could cause premature full releases (tying back to pitfall 1).

Pitfall 5:

Security, Privacy, or Compliance Oversights

What can go wrong: In excitement to deploy the AI, someone might feed real customer data to an external AI API without proper privacy review, or a model might inadvertently expose sensitive info (like if it was trained on internal data and it ends up regurgitating a customer's phone number in a prediction – it has happened with large language models, for instance). Or an agent might have more privileges than it needs and could be exploited.

Avoidance: Threat-model the new components. Ask questions like:

- Does the agent need access to sensitive data? If yes, secure it. Maybe anonymize or tokenize data before the agent sees it if possible.

- Are we sending data to a third-party AI service or cloud? If so, does that comply with our data policies? If not sure, engage legal/compliance. For example, using a SaaS sentiment analysis API on support tickets might violate privacy if those tickets contain personal info – maybe you need an anonymization layer or use a self-hosted model instead.

- Could the agent be a security hole? E.g., if it calls external URLs or runs code (imagine an AI that executes scripts), ensure it's sandboxed. An agent making autonomous decisions should be subject to the same security controls as any automated process. If it can trigger transactions, ensure limits are in place (just like you wouldn't let a human operator issue million-dollar refunds without oversight, don't let the AI do it either).

- Authentication between the service and agent: use secure channels/keys, not some open port that anything could send requests to pretending to be the service.

- If using a pre-trained model from an open source or vendor, consider the supply chain – was it downloaded from a trustworthy source? Model tampering is a thing (AI trojans). This is a bit advanced concern, but in high-security contexts, you'd checksum and validate models similar to how you do for software packages.

One more compliance angle: if you operate in a regulated industry (finance, healthcare), deploying AI might require auditability. Some regulations (like EU GDPR's "right to explanation" or upcoming AI regulations) could require you to explain an automated decision. If your microservice's decisions become AI-driven, you may need to log reasoning or provide a way to override decisions in certain cases (for example, a human can review and overturn an AI decision about a loan or medical diagnosis). Bake that in from the start if applicable.

Pitfall 6:

Model or Data Staleness (Set and Forget Syndrome)

What can go wrong: The team does all this work, the AI feature goes live and performs great... initially. But over months, the real-world changes:

customer behavior shifts, competitors do something new, seasonality hits, etc. The model that was once accurate is now making less optimal choices. If no one is actively maintaining it, the performance could quietly degrade. This is known as model drift. A once intelligent microservice can become dumb or even harmful if its "brain" is not kept up-to-date.

Avoidance: Plan for **continuous improvement**. Treat the AI part as an iterative product, not a one-off project. This means scheduling periodic reviews of its performance, retraining the model with new data periodically, or at least monitoring for drift indicators (as mentioned in observability). Many teams set up alerts on key metrics that would catch drift – e.g., if the agent's suggestions are getting overridden by humans more often than before, something's off.

Also, keep an eye on data pipeline health. Maybe the model is fine but the data fed to it changed (e.g., an upstream service started sending empty values for a field after a schema change – and now the agent's decisions are based on incomplete info). Data quality issues can silently kill AI performance. Tools like Great Expectations in production can detect if, say, suddenly 30% of inputs to the model are null where it used to be <1%.

Have an owner for the AI component even after launch. This could be a role on the team or an assigned "model custodian". That person (or rotation) is responsible for checking in on it, retraining if needed, and responding to issues that specifically relate to the AI logic.

Pitfall 7:

Over-engineering the Solution

What can go wrong: Sometimes in enthusiasm, teams might incorporate an AI where a simpler solution could suffice, adding unnecessary complexity and maintenance cost. For example, imagine using a complex reinforcement learning agent to tune a parameter that you could have just set via a config and occasional manual adjustment. The pitfall here is doing AI for AI's sake and making the system harder to understand and debug for arguably marginal benefit.

Avoidance: Always align AI use with clear value. If a simpler heuristic or rule works 95% as well, maybe that's enough – or at least consider whether the extra complexity is justified. This isn't to dissuade using AI (after all, we're advocating it where it brings adaptability), but be wary of novelty chasing. A good practice is to start with the simplest approach and only introduce more complex AI if needed. Also, if you do introduce it, ensure the team is committed to maintaining it – otherwise that fancy agent might become technical debt.

One symptom of over-engineering is when the original people leave and the remaining team treats the AI component as a mysterious box they fear to touch. If you sense that dynamic, invest in documentation and training, or simplify the solution. Sometimes scaling back is okay. Perhaps an ML model with 100 features is hard to maintain; maybe a simpler model with 10 top features would be almost as good and way easier to retrain and reason about.

By being mindful of these pitfalls, you can increase the chances of a smooth migration. It's often said that "failures are the best teachers" – hopefully by discussing them here, you can learn vicariously and preempt them.

Lessons Learned: One of the best things you can do is conduct a *pre-mortem* and, later, a *post-mortem*. Before launch, imagine it's 6 months later and the project failed – list possible reasons (many will resemble the pitfalls above) and mitigate them now. After launch, do a retrospective: what went well, what didn't, and document those tribal lessons for the next AI project. Continuous learning applies to organizations as much as to AI models!

Conclusion: Evolving One Step at a Time

Migrating from traditional microservices to intelligent, agent-driven microservices is a journey of **incremental evolution**. As we've seen with WebkinsTech's story and the surrounding guidance, the key is to break the journey into manageable steps, just as microservices themselves broke monoliths into manageable components.

By starting with a clear value goal, introducing AI gradually (first as a sidekick, then allowing it more autonomy), and leveraging the right patterns and tools, you can **add intelligence without losing stability**. It's about augmentation, not wholesale replacement – at least not until the new approach has earned its stripes. This approach mirrors how humans often adopt AI in real life: initially with skepticism and verification (think of a driver trusting a new car's autopilot feature – hands on the wheel at first), and eventually with confidence as trust is built.

One overarching theme is **trust** – trust in the AI by the team and stakeholders, and trust by the AI team in the existing systems. Trust isn't granted, it's earned through transparency, testing, and performance. By shadow testing, monitoring, and involving humans-in-the-loop, you create a safety net that builds trust step by step. As that trust grows, the benefits start to materialize: reduced toil, faster decisions, more adaptability to change – your system starts to behave **intelligently**, not just mechanically.

It's also worth highlighting the *conversational dance* between deterministic and intelligent components. There's no rule that says everything must be AI-driven. The *art* of these migrations is finding the right mix – where to use crisp rules, where to use learning agents, and how they complement each other. Over time, you may find some services become almost fully autonomous agents, while others use AI in a lighter touch way. That's fine – let the needs drive the solution.

As you embark on your own migrations, keep the following motivational points in mind:

- **You're not alone** – many organizations are in this transition, and we have collective knowledge (and war stories) to draw on. Use communities, share lessons, and don't hesitate to leverage prior art (frameworks, libraries, case studies).

- **Small wins compound** – every little improvement the AI brings can compound over time. 5% efficiency gain here, 10% fewer errors there – they add up. Celebrate those incremental wins; they will justify the next round of investment.

- **Failure is recoverable** – by planning rollbacks and fallbacks, you've essentially removed the fear of failure. If the AI doesn't work as expected, you can turn it off. So don't be afraid to experiment. The beauty of microservices and feature flags is that you can try things with low risk. Use that power to iterate.

- **Think long-term** – today it's one AI agent in one service. Tomorrow it might be dozens of agents and perhaps multi-agent ecosystems as hinted by forward-looking architects . By building the muscle and infrastructure now, you're preparing your architecture (and your team) for a future where adaptive, learning software is the norm. Each incremental migration is not just a one-off improvement, it's laying a brick in the road to a more **autonomous, resilient system**.

In conclusion, transforming your microservices to be more intelligent is absolutely feasible and immensely rewarding, but it must be done judiciously. **Incremental migration is possible, valuable, and arguably inevitable** as software gravitates towards more adaptive behavior. By applying the strategies detailed in this appendix – from sidecars and shadow modes to feature flags and rigorous testing – you can modernize your architecture with AI in a controlled, confidence-building manner.

The future of adaptive software systems is being written one service at a time. With careful planning and execution, your existing microservices can become part of that future, evolving **without revolt**. So take that first step: pick a pilot, set up a safety net, and let your microservice grow an intelligent side. Over time, those small steps will amount to a giant leap for your organization's capabilities. Here's to your journey toward intelligent microservices, one agent at a time!

Reviewers

Anukriti Sharma – Student Engineering, University of Cincinnati

Dhivya Nagasubramanian — Vice President AI - U.S. Bank

Vaishnavi Gudur — Senior Software Engineer — Microsoft

Alok Tibrewala — IEEE Senior Member, AI Researcher

Sahil Sachdeva —Software Engineer, LinkedIn

Udaykiran Madineni — Lead product manager, Cisco

Sweta Mulji — Co-founder and Chief Customer Officer — SalesPlay X

Ajay Krishna Borra — Lead engineer, Salesforce

Surya Teja Meesala — Industry thought leader — Rymond James, IEEE

Jyothsna Devi Dontha — Controls Engineer — E Tech Group

Ramesh Sinha — currently Sr Staff Engineer — Zscaler

Avitesh kesharwani — architecture — UNCC charlotte and Genpact

Siddhi Shreekar Gowaikar — product manager — IBM

Sweta Mulji — Co-founder and Chief Customer Officer — SalesPlay X

Ramprasad Reddy Mittana — NSF International

Avitesh Kesharwani — UnCc charlotte and sr principal consultant in genpact

Saurabh Kulkarni — Software engineering professional healthcare open source project

Viswanathan Ranganathan —Senior Distributed Systems Engineer, Netflix

Vamsidhar Reddy Kamanuru —Sambanova AI, University of California San Diego

Aditya Vilas Deshpande — Senior Software Engineer — PayPal

Saurabh Kulkarni — Full Stack Engineer - ETQ

Abanish Mishra – Director, Data Science & Analytics – Match Group

Author Biography

Anil Kumar Sharma

Distinguished Fellow & Engineering Leader

Distinguished Fellow & Engineering Leader with 24 years of experience

Leads AI-enabled modernization and large-scale distributed systems with a focus on reliability and explainability. Previously held leadership roles at VMware and earlier roles at Yahoo! and Novell/NetIQ. Research interests span agentic AI, optimization, and enterprise automation, have 5 granted patents of machine learning and distributed cloud.

Shafi Khan

CEO & Founder, AutonomOps AI

CEO & Founder, AutonomOps AI.

Engineering leader with ~20 years delivering high-availability platforms and ML-powered automation. Previously at Yahoo, VMware, and Broadcom; builds agentic AI systems for SRE/DevOps. Product-driven founder focused on measurable reliability gains and customer impact.

Made in the USA
Monee, IL
16 June 2026

54141441R00448